IUTAM SYMPOSIUM ON
ADVANCES IN NONLINEAR STOCHASTIC MECHANICS

SOLID MECHANICS AND ITS APPLICATIONS
Volume 47

Series Editor: **G.M.L. GLADWELL**
Solid Mechanics Division, Faculty of Engineering
University of Waterloo
Waterloo, Ontario, Canada N2L 3G1

Aims and Scope of the Series

The fundamental questions arising in mechanics are: *Why?*, *How?*, and *How much?* The aim of this series is to provide lucid accounts written by authoritative researchers giving vision and insight in answering these questions on the subject of mechanics as it relates to solids.

The scope of the series covers the entire spectrum of solid mechanics. Thus it includes the foundation of mechanics; variational formulations; computational mechanics; statics, kinematics and dynamics of rigid and elastic bodies; vibrations of solids and structures; dynamical systems and chaos; the theories of elasticity, plasticity and viscoelasticity; composite materials; rods, beams, shells and membranes; structural control and stability; soils, rocks and geomechanics; fracture; tribology; experimental mechanics; biomechanics and machine design.

The median level of presentation is the first year graduate student. Some texts are monographs defining the current state of the field; others are accessible to final year undergraduates; but essentially the emphasis is on readability and clarity.

For a list of related mechanics titles, see final pages.

IUTAM Symposium on

Advances in Nonlinear Stochastic Mechanics

Proceedings of the IUTAM Symposium
held in Trondheim, Norway,
3–7 July 1995

Edited by

A. NAESS

Department of Structural Engineering,
Norwegian University of Science and Technology,
Trondheim, Norway

and

S. KRENK

Division of Mechanics,
Lund Institute of Technology,
Lund University,
Sweden

KLUWER ACADEMIC PUBLISHERS
DORDRECHT / BOSTON / LONDON

A C.I.P. Catalogue record for this book is available from the Library of Congress.

ISBN-13: 978-94-010-6630-3   e-ISBN-13: 978-94-009-0321-0
DOI : 10.1007/978-94-009-0321-0

Published by Kluwer Academic Publishers,
P.O. Box 17, 3300 AA Dordrecht, The Netherlands.

Kluwer Academic Publishers incorporates
the publishing programmes of
D. Reidel, Martinus Nijhoff, Dr W. Junk and MTP Press.

Sold and distributed in the U.S.A. and Canada
by Kluwer Academic Publishers,
101 Philip Drive, Norwell, MA 02061, U.S.A.

In all other countries, sold and distributed
by Kluwer Academic Publishers Group,
P.O. Box 322, 3300 AH Dordrecht, The Netherlands.

PREFACE

The IUTAM Symposium on Advances in Nonlinear Stochastic Mechanics, held in Trondheim July 3-7, 1995, was the eighth of a series of IUTAM sponsored symposia which focus on the application of stochastic methods in mechanics. The previous meetings took place in Coventry, UK (1972), Southampton, UK (1976), Frankfurt/Oder, Germany (1982), Stockholm, Sweden (1984), Innsbruck/Igls, Austria (1987), Turin, Italy (1991) and San Antonio, Texas (1993).

The symposium provided an extraordinary opportunity for scholars to meet and discuss recent advances in stochastic mechanics. The participants represented a wide range of expertise, from pure theoreticians to people primarily oriented toward applications. A significant achievement of the symposium was the very extensive discussions taking place over the whole range from highly theoretical questions to practical engineering applications. Several presentations also clearly demonstrated the substantial progress that has been achieved in recent years in terms of developing and implementing stochastic analysis techniques for mechanical engineering systems. This aspect was further underpinned by specially invited extended lectures on computational stochastic mechanics, engineering applications of stochastic mechanics, and nonlinear active control. The symposium also reflected the very active and high-quality research taking place in the field of stochastic stability. Ten presentations were given on this topic of a total of 47 papers. A main conclusion that can be drawn from the proceedings of this symposium is that stochastic mechanics as a subject has reached great depth and width in both methodology and applicability.

The Opening Address of the symposium was given by The Secretary General of the IUTAM, Dr. F. Ziegler. On the first day of the symposium, a reception in the Archbishop's Palace was hosted by the City of Trondheim. This was preceded by an organ concert hosted by Golar Nor Offshore AS in the Nidaros Cathedral. The symposium banquet was hosted by Dr. Steen Krenk and Dr. Arvid Naess in the Banquet Hall of the Ringve Museum of Musical History, with Dr. Y.K. Lin serving as the banquet speaker. The Closing Address was delivered by Dr. K. Sobzcyk, Member of the IUTAM Symposium Committee.

On behalf of the Scientific and Organizing Committees, we would like to thank the IUTAM Bureau for promoting and sponsoring this symposium. We are also greatly in debt to the other sponsors, whose contributions made this symposium possible. The contributions of the authors and participants in the symposium are, of course, instrumental in producing a successful meeting and its proceedings, and we deeply appreciate their efforts. The assistance in processing the manuscripts provided by Mrs. A.M. Soeberg of the Norwegian Institute of Technology and Ms. M. Jönsson of Lund University is gratefully acknowledged.

A. Naess　　　　　　　　　　　　　　　　　　　　　S. Krenk

Opening Address of the Secretary General of IUTAM

Chairmen, Ladies and Gentlemen,

On behalf of the International Union of Theoretical and Applied Mechanics it is my duty and my pleasure to welcome you at the IUTAM Symposium on Advances in Nonlinear Stochastic Mechanics. The proposal to arrange this Symposium was accepted in August 1992 by the General Assembly of IUTAM. Since that time both chairmen, Professor S Krenk and Professor A Naess have done an excellent job in preparing the scientific program. The chairmen have been advised by the Scientific Committee in selecting and inviting participants from all over the world.

There are sometimes some misunderstandings about the appointment of the Scientific Committee by IUTAM: If you look back to the related IUTAM Symposia held in Turin 1991 and San Antonio 1993, to mention just the two recent ones, you see different members of the Scientific Committees with only a small overlap to the current one: Professor S T Ariaratnam, Professor P Bernard, Professor V Bolotin, Professor R N Iyengar, Professor Y K Lin, Professor J B Roberts, Professor W Schiehlen (ex officio), Professor K Sobczyk and Professor P Spanos. Since the value and quality of IUTAM is reflected by its Symposia, we are very happy to have the series of proceedings and meetings in the field of Stochastic Mechanics. There are many more meetings referring to that important subject, some people even speak of an inflation, however, the quality of the IUTAM Symposia is hardly reached elsewhere.

Norway is a member of IUTAM since 1949. Its National Committee is represented in the Union by the Chairman, Professor B Gjevik, University of Oslo. There have been many important conferences organized in Norway, I just refer to the recent 15th International Congress of Acoustics (Norwegian Institute of Technology, Trondheim, Norway, 26-30 June 1995) which is related to our Union. However, we are opening the first IUTAM Symposium to be held in this beautiful country.

We look forward to a dense, but exciting program, divided into 16 Sessions. I am especially happy to see sessions devoted to evolving fields like Stochastic Finite Elements and System Identification and Control. Stochastic Mechanics will be present at the 19th ICTAM, Kyoto, 25-31 August 1996, to celebrate 20 years of IUTAM engagement in that field that have passed since the first IUTAM Symposium held in 1976 in Southampton, England. The General Assembly held during the Congress will decide on future Symposia for the years 1998/99, depending on the quality of the proposals received. It would be a good idea to consider the future of Stochastic Mechanics in the tenure of our Symposium. We shall start with an exciting session with the specially invited lecture given by Professor Masanobu Shinozuka on Computational Stochastic Mechanics.

Thank you for your kind attention.

Franz Ziegler

Final Address at the Closing Ceremony on behalf of IUTAM

Ladies and Gentlemen,

The Secretary General of IUTAM - Professor F Ziegler- and the former Secretary General - Professor W Schiehlen - who were here with us during the first part of the Symposium had to leave earlier because of other important commitments. So, being recently associated with IUTAM as a member of the IUTAM Symposium Panel, I have the honor and pleasure to say a few words on behalf of IUTAM, and in the name of Professor F Ziegler, particularly. Above all, I wish to express thanks and gratitude to the chairmen of this Symposium - Professor Arvid Naess and Professor Steen Krenk - for all the arrangements which they have made in connection with our Symposium. I also wish to express the appreciation of IUTAM to the Local Organizing Committee for all the technical and administrative help in bringing this Symposium to its successful end.

We have spent five exciting days together, presenting papers, participating in discussions and deliberations on stochastic mechanics - our favourite field of science. But we also had the occasion - which is extremely important as well - to familarize ourselves a little with this unique and beautiful country. The visit to Ringve Museum of Musical History, followed by the Symposium Dinner, will remain in our memories as an impressive sign of the Norwegian culture and hospitality.

Let me now say a few words on the future IUTAM Symposium on Stochastic Mechanics. IUTAM encourages all of you to continue your engagement in this series of Symposia. What is especially needed now is preparation of good proposals for the next symposium of this type. According to the Secretary General, various competitive proposals are welcome. Proposals for the Symposium in 1998/1999 should be sent to the Secretary General by the end of April of next year. The decision will be made by the General Assembly of IUTAM, after careful evaluation of the proposals by the IUTAM Symposia Panel.

In closing, I wish to thank you - Arvid and Steen - once again for this very successful Symposium. I wish you all a good journey back home and pleasant summer vacations.

K Sobczyk

CONTENTS

EXPERIMENTAL RANDOM EXCITATION OF NONLINEAR SYSTEMS WITH MULTIPLE INTERNAL RESONANCES

A. A. Afaneh and R. A. Ibrahim
Wayne State University,
Department of Mechanical Engineering, Detroit, MI 48202

1. Introduction

Small nonlinear modal interactions may have considerable effect on the pattern of forced vibration response of certain types of dynamic systems possessing low damping characteristics. The design of these systems may create two or more linear algebraic relationships between the system's normal mode frequencies. These relationships are referred to as internal (or autoparametric) resonances, and usually take the form $\sum K_j \omega_j = 0$. The constant $K = \sum |k_j|$ is referred to as the order of internal resonance, where k_j are integers, and ω_j are the system normal mode frequencies. Third-order internal resonance results from quadratic nonlinear coupling of normal modes, while fourth-order is due to cubic nonlinearities. Cubic nonlinearities can also cause second-order internal resonance (i.e., one-to-one). Simultaneous internal resonances are classified as independent and interacting. Independent internal resonances usually take the form

$$\sum_j^m K_j \omega_j = 0, \qquad \sum_\ell^n K_\ell \omega_\ell = 0 \tag{1}$$

such that j=1,2,.., m, and ℓ=m+1, m+2, ..., n
while interacting internal resonances usually have common frequencies, i.e.,

$$K_{11}\omega_1 + K_2\omega_2 + ... + K_m\omega_m = 0 \tag{2a}$$
$$K_{21}\omega_1 + K_{m+1}\omega_{m+1} + ... + K_n\omega_n = 0 \tag{2b}$$

The problem of multiple internal resonances can exist in many engineering systems, such as coupled beams, structure-liquid systems, cables, plates, and road vehicles. The early deterministic work of Ibrahim [1] on a liquid-structure system showed that in some cases of four-mode interaction governed by two simultaneous internal resonance conditions of the type $\omega_3 = \omega_1 + \omega_2$, $\omega_3 = \omega_4$, the system may exhibit a continuous increase in its response amplitudes. This growth could lead to model failure if the shaker were not stopped. The response amplitudes were very large and exhibited no regular trend. These complex responses were verified by using numerical simulations of the system equations of motion. All amplitudes were

1

A. Naess and S. Krenk (eds.), IUTAM Symposium on Advances in Nonlinear Stochastic Mechanics, 1–10.
© 1996 *Kluwer Academic Publishers.*

increasing with time, indicating the system's response instability. When the same model was internally tuned with different types of internal resonance conditions among three modes instead of four, both experimental and analytical results yielded steady-state responses, except those in the vicinity of exact external resonance. Later, Bux and Roberts [2] and Cartmell and Roberts [3] considered simultaneous internal resonances in a structure consisting of two orthogonal beams modeled by four normal modes and reported sililar results.

The primary objective of this paper is to examine the random response characteristics of coupled nonlinear oscillators in the presence of simultaneous internal resonances. A model of two coupled beams with nonlinear inertia interaction is considered. The primary beam is directly excited by a random support motion, while the coupled beam is indirectly excited through autoparametric coupling and parametric excitation. It is found that the two beams exchange energy in the neighborhood of internal resonances. Furthermore, there is an upper limit for the excitation level above which the system experiences unbounded response in the neighborhood of simultaneous internal resonances.

2. Analytical Modeling

The system under investigation is shown schematically in Figure (1). It consists of a primary beam OB of length L and a secondary beam of length ℓ coupled perpendicularly with the primary beam at point B. The two beams are arranged such that the bending deflection of each beam is orthogonal to the other. The other end of the primary beam is rigidly supported at point O. The primary beam is subjected to a random support motion $\ddot{Z}(t)$ in the OZ direction. For simplicity, the transverse motion of the primary beam is considered to be in the plane OYZ. This planar motion is described by transverse and rotational motions q_1 and q_2, respectively. The coupled beam carries a block of mass m_0 at position $z = \ell$. The displacements of the coupled beam elastic axis in the x, y, and z directions, at $z=\ell$, are described by $U_0(\ell,t)$, $V_0(\ell,t)$, and $W_0(\ell,t)$, respectively. The torsional oscillation of the beam cross-section is given by $\Phi_0(z,t)$.

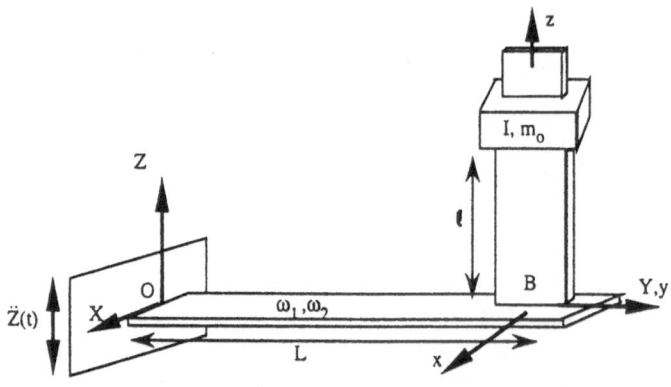

Figure 1 Schematic diagram of two coupled beam showing coordinate frames

Neglecting the extension of the beam elastic axis, and considering the beam cross-section to be thin and slender one can express the vertical drop $W_0(\ell,t)$ and the lateral displacement $V_0(\ell,t)$ in terms of $U_0(\ell,t)$ and $\Phi_0(\ell,t)$. Considering only the first mode in bending and torsion, and applying Lagrange's equation with respect to the four generalized coordinates associated with the coordinates U_0, Φ_0, q_1, and q_2 gives the following set of equations of motion:

secondary beam equations of motion

$$\ddot{U}_0 + \omega_b^2 U_0 = -B_3\{\ddot{q}_1 - \ddot{Z}(t)\}U_0 + \ell B_4 \ddot{q}_2 \Phi_0 \tag{1}$$

$$\ddot{\Phi}_0 + \omega_t^2 \Phi_0 = \ell B_4 \ddot{q}_2 U_0 \tag{2}$$

primary beam equations of motion

$$(m_{11} + m_0)\ddot{q}_1 - \frac{1}{2}m_{12}\ddot{q}_2 + k_{11}q_1 - k_{12}q_2 =$$

$$-m_0\{B_3(\dot{U}_0^2 + U_0\ddot{U}_0) + \ell(\dot{q}_2^2 + q_2\ddot{q}_2)\} - (m_0 + \frac{1}{2}M_b)\ddot{Z}(t) \tag{3}$$

$$(m_{22} + \ell^2 m_0)\ddot{q}_2 - \frac{1}{2}m_{12}\ddot{q}_1 + k_{22}q_2 - k_{12}q_1 =$$

$$m_0\ell\{B_4(\ddot{U}_0\Phi_0 + 2\dot{U}_0\dot{\Phi}_0 + U_0\ddot{\Phi}_0) + \ddot{q}_1 q_2 + \dot{q}_1\dot{q}_2\} + \frac{1}{12}M_b L\ddot{Z}(t) \tag{4}$$

where $B_1 = \dfrac{3\,E\,I_{yy}}{\ell^3}$, $B_2 = \dfrac{GJ}{\ell}$, $B_3 = \dfrac{6}{5\ell}$, $B_4 = \dfrac{1}{4}$, $\omega_b^2 = \dfrac{B_1}{m_0}$, $\omega_t^2 = \dfrac{GJ}{I\ell}$, $m_{11} = \dfrac{6}{5L}\rho I_{xx} + \dfrac{39M_b}{105}$,

$k_{11} = \dfrac{12\,E\,I_{xx}}{L^3}$, $m_{12} = \dfrac{11}{105}LM_b + \dfrac{\rho}{5}I_{xx}$, $m_{22} = \dfrac{L}{15}\rho I_{xx} + \dfrac{M_b}{105}L^2$,

$k_{22} = \dfrac{L_{11}}{3}L^2$, $k_{12} = k_{21} = \dfrac{1}{2}k_{11}L$

I_{xx} is the area moment of inertia of the primary beam cross-section about x-axis, I is the mass polar moment of inertia of the tip mass m_0, and M_b is the mass of the horizontal beam. EI_{yy} and GJ are the flexural and torsional rigidities of the vertical beam, respectively.

The terms $q_2\Phi_0$ in equation (1) and $q_2 U_0$ in equation (2) represent coupled autoparametric terms of the non-planar motion by purely rotational motion q_2 at point B. The term $q_1 U_0$ in equation (1) represents autoparametric interaction between the vertical beam bending U_0 and the linear acceleration q_1 of the coupling point. The first expression in the braces on the right-hand side of equation (3) represents the feedback inertia force of the coupled beam on the primary beam. The terms in the first parenthesis on the right-hand side of equation (4) represent the feedback moment of the coupled beam on the primary beam due to the coupled beam's bending-torsional motions. The other nonlinear inertia terms in equations (3) and (4) are due to the axial shortening of the primary beam.

In view of this linear coupling, it is convenient to transform the generalized coordinates q_1 and q_2 into principal coordinates ζ_1 and ζ_2 through the linear transformation

$$\left\{ \begin{matrix} q_1 \\ q_2 \end{matrix} \right\} = [R] \left\{ \begin{matrix} \zeta_1 \\ \zeta_2 \end{matrix} \right\} \tag{5}$$

where $[R]$ is the modal matrix $[R] = \begin{bmatrix} 1 & 1 \\ n_1 & n_2 \end{bmatrix}$, and $n_i = \dfrac{2(k_{11} - (m_{11} + m_0)\omega_i^2)}{k_{11}L - m_{12}\omega_i^2}$, i=1,2.

The normal mode frequencies of the horizontal beam are given by the following expression

$$\omega_i^2 = \frac{1}{M_i}[k_{11} - k_{12}n_i + k_{22}n_i^2], \; i = 1, 2 \tag{6}$$

where $M_i = (m_0 + m_{11}) - \frac{1}{2}n_i m_{12} + n_i[(m_{22} + \ell^2 m_0)n_i - \frac{1}{2}m_{12})]$

The dependence of the natural frequencies of the system on the tip mass position is shown in Figure (2), where solid curves are the theoretical values and symbols represent the values measured experimentally. Applying the coordinate transformation (5) to equations (3) and (4) gives the following set of nonlinear coupled differential equations:

$$\ddot{U}_0 + \omega_0^2 U_0 = [B_4 \ell n_1 \Phi_0 - B_3 U_0]\ddot{\zeta}_1 + [B_4 \ell n_2 \Phi_0 - B_3 U_0]\ddot{\zeta}_2 + B_3 U_0 \ddot{Z}(t) \tag{7a}$$

$$\ddot{\Phi}_0 + \omega_\Gamma^2 \Phi_0 = B_4 \ell \frac{m_0}{I}[n_1 \ddot{\zeta}_1 + n_2 \ddot{\zeta}_2]U_0 \tag{7b}$$

$$\ddot{\zeta}_1 + \omega_1^2 \zeta_1 = \frac{1}{M_1} \Big\{ [B_{31}\ddot{Z}(t) + B_4 \ell n_1 m_0(U_0 \ddot{\Phi}_0 + 2\dot{U}_0 \dot{\Phi}_0 + \ddot{U}_0 \Phi_0)$$

$$- B_3 m_0(\dot{U}_0^2 + U_0 \ddot{U}_0) + 2\ell m_0 n_1^2(\dot{\zeta}_1^2 + \zeta_1 \ddot{\zeta}_1)$$

$$+ \ell m_0 2 n_1 n_2 \zeta_2 \ddot{\zeta}_1 + (2n_1^2 + 3n_1 n_2)\dot{\zeta}_1 \dot{\zeta}_2 + (n_1^2 + n_1 n_2)\zeta_1 \ddot{\zeta}_2 + \ell m_0(n_1 n_2 + n_2^2)(\dot{\zeta}_2^2 + \zeta_2 \ddot{\zeta}_2) \Big\} \tag{7c}$$

$$\ddot{\zeta}_2 + \omega_2^2 \zeta_2 = \frac{1}{M_2}\Big\{ -B_{41}\ddot{Z}(t) + B_4 \ell n_2 m_0(U_0 \ddot{\Phi}_0 + 2\dot{U}_0 \dot{\Phi}_0 + \ddot{U}_0 \Phi_0) - B_3 m_0(\dot{U}_0^2 + U_0 \ddot{U}_0)$$

$$+ 2\ell m_0 n_2^2(\dot{\zeta}_2^2 + \zeta_2 \ddot{\zeta}_2)$$

$$+ \ell m_0(n_1^2 + n_1 n_2)(\dot{\zeta}_1^2 + \zeta_1 \ddot{\zeta}_1) + \ell m_0(2n_1 n_2 \zeta_1 \ddot{\zeta}_2 + (n_1 n_2 + n_2^2)\ddot{\zeta}_1 \zeta_2 + (n_2^2 + 3n_1 n_2)\dot{\zeta}_1 \dot{\zeta}_2) \Big\} \tag{7d}$$

In order to solve these equations, the acceleration terms on the right-hand sides of equations (7), involved with nonlinear expressions, must be eliminated using successive elimination. Introducing viscous linear damping to account for energy dissipation and the nondimensional parameters

$$X_1=\frac{L}{b}\frac{U_o}{\ell}, \ X_2=\frac{L}{b}\Phi_o, \ Y_i=\frac{L}{b}\frac{\zeta_i}{\ell} \ (i=1,2), \ \tau=\omega_b t, \ r_t=\frac{\omega_t}{\omega_b}, \ r_1=\frac{\omega_1}{\omega_b}, \ r_2=\frac{\omega_2}{\omega_b} \ \text{and} \ \varepsilon=\frac{b}{L}$$

equations (7) take the form,

$$X_1''+2\eta_b X_1'+X_1=\varepsilon\Big\{r_1^2(b_{11}X_1-b_{12}X_2)Y_1+r_2^2(b_{11}X_1-b_{13}X_2)Y_2+b_{14}X_1\xi(\tau)\Big\} \tag{8a}$$

$$X_2''+2\eta_t r_t X_2'+r_t^2 X_2=\varepsilon\Big\{(b_{21}r_1^2 Y_1+b_{22}r_2^2 Y_2)X_1\Big\} \tag{8b}$$

$$Y_1''+2\eta_1 r_1 Y_1'+r_1^2 Y_1=\varepsilon\Big\{b_{31}\xi(\tau)-b_{36}(X_1'^2+X_1^2)+b_{37}(2X_1'X_2'-(r_t^2+1)X_1X_2)$$

$$+b_{38}Y_1'Y_2'+b_{32}(Y_1'^2-r_1^2 Y_1^2)+b_{35}(Y_2'^2-r_2^2 Y_2^2)-b_{39}Y_1Y_2\Big\} \tag{8c}$$

$$Y_2''+2\eta_2 r_2 Y_2'+r_2^2 Y_2=\varepsilon\Big\{b_{41}\xi(\tau)-b_{45}(X_1'^2-X_1^2)+b_{46}(2X_1'X_2'-(r_t^2+1)X_1X_2)$$

$$+b_{48}Y_1'Y_2'+b_{42}(Y_1'^2-r_1^2 Y_1^2)+b_{44}(Y_2'^2-r_2^2 Y_2^2)-b_{49}Y_1Y_2\Big\} \tag{8d}$$

where a prime denotes differentiation with respect to τ, $\xi(\tau)=\ddot{Z}/l\omega_b^2$, and η_i are the damping factors associated with each mode.

The excitation $\xi(\tau)$ possesses a smooth spectral density S_o up to some frequency greater than the highest natural frequency of the system. In order to decide the value of the excitation bandwidth, one should know the frequencies of the interacting modes. From Figure (3) and the study of Bux and Roberts (1986), system (8) can possess the two simultaneous internal resonances $\omega_1=2\omega_b$ and $\omega_2=\omega_t+\omega_b$. One may introduce internal detuning parameters σ_1 and σ_2 defined by the relations

$$r_1=2+\sigma_1 \ \text{and} \ r_2=r_t+1+\sigma_2 \tag{9}$$

In view of the high dimensionality of this case, analytical methods are not promising to predict response statistics or stochastic stability of fixed solutions. Both Gaussian and non-Gaussian closures have failed to yield any useful results. Alternatively, Monte Carlo simulation and experimental tests are considered. Numerical simulation is carried out to estimate the system response statistics. Because of space limitations experimental results are presented in this paper. The reader may refer to Afaneh and Ibrahim [4] for Monte Carlo simulation results.

3. Experimental Investigation

The experimental model used in actual random tests consists of two vanadium tool steel beams. The primary beam is 0.39 m long, with a cross section 0.0254 x 0.003572 m. The secondary beam is $\ell=0.22$-m long, with a cross section 0.01905 x

0.000803 m. It is attached vertically to the free end of the primary beam by means of two-right angle support brackets, arranged such that its bending displacement is perpendicular to the plane of the primary beam flexural oscillation, as indicated in Figure (1). The tip mass ($m_o = 0.106\,kg$) is a steel block attached to the secondary beam at a pre-selected position determined by the desired internal detuning.

The system's natural frequencies and damping ratios are measured by using free and forced sine vibration tests. The natural frequencies are measured at different tip mass positions while the measured damping ratios are averaged over the detuning length due to the irregular nature of the damping forces. The average damping ratios are $\zeta_1 = 0.009$, $\zeta_2 = 0.01$, and $\zeta_b = 0.002$, $\zeta_t = 0.015$. The measured and calculated system natural frequencies as a function of tip mass position ℓ are shown in Figure (2). At $\ell = 0.16$ m, the four measured natural frequencies have values of $\omega_b \approx 5\,Hz$ (for the secondary beam bending mode), $\omega_t \approx 37\,Hz$, (for the secondary beam torsion mode), $\omega_1 \approx 9.7\,Hz$, and $\omega_2 \approx 42\,Hz$ (for the first and second natural frequencies of the primary beam, respectively). These measured values are slightly different from those required for exact internal resonances. The excitation upper frequency limit is set to exceed the highest natural frequency of the four modes.

ILS software is used to estimate a number of statistical parameters for the response and excitation. ILS is also installed on an in-house SparcStation 10 Sun computer system, which receives the acquired data via a network line for fast processing. In addition to the subroutines available in the ILS software, other programs have been developed to estimate mean squares versus time and probability density functions.

Based on preliminary tests, it is decided to set the time duration of each test at 375 seconds, which has been found to be enough to exhibit all possible dynamic characteristics of the system and to establish acceptable stationarity for each signal. The analysis of each test signal is usually preceded by data qualification in terms of three basic characteristics. These are the stationarity of the data, the presence of periodicities, and the Gaussianity of data.

Figure (3) shows typical time history response records for both the shaker excitation and response signals. It shows that the main beam's first two modes (Y_1 and Y_2) and the coupled beam bending mode X_1 are essentially narrow-band random processes, while the coupled beam torsion mode exhibits a multi-frequency signal. It is believed that the lower frequency components in the torsion signal are detected from other modes. The corresponding mean squares, power spectra, and probability density functions of Figure (3) are estimated and plotted in Figures (4) through (6), respectively. Figure (4) shows that the excitation reaches a stationary state very quickly, while the primary beam's first two modes and the torsion response of the coupled beam attain stationarity after a transient period. It is the bending mode of the coupled beam which exhibits nonstationarity. This may be because this mode is excited parametrically and autoparametrically. The power spectra given in Figure (5) confirm our remarks concerning the time history records of Figure (4). For example, the first bending mode of the main beam, about 9.7-Hz, appears in the spectrum of the coupled beam torsion mode. The other frequency components form a sideband spectrum surrounding the torsion frequency. The response probability density curves deviate from the Gaussian curves (plotted by dotted curves in Figure (6)).

A series of tests was conducted at different excitation levels to examine the dependence of mean squares responses on the excitation level and the bifurcation of the non-excited modes. The results are plotted in Figures (7), which show two possible response regimes. The first is the response of the primary beam with zero motion for the coupled beam. This regime takes place over an excitation level range going from zero up to critical level. The first regime can be divided into linear and nonlinear responses of the main beam's first two modes. The linear response is evident in the lower portion of the excitation level, while the nonlinear response takes place at relatively higher excitation levels which are still below the critical level. Above this critical level the second regime begins and is characterized by non-zero motions of the coupled beam as a result of energy transfer from the primary beam due to the inertia nonlinear coupling. The bifurcations of the bending and torsional modes of the coupled beam do not take place at the same critical excitation level, because the bending mode is excited parametrically and autoparametrically while the torsion mode is only excited autoparametrically. The coexistence of more than one internal resonance may lead to an irregular pattern of energy transformed between the four modes. The excitation is kept under a certain level above which the system response becomes very large. Running a test under this can destroy the model.

In order to determine the effectiveness of the nonlinear modal coupling in the presence of simultaneous internal resonances, another series of random excitation tests was conducted for different tip mass positions. The corresponding mean square responses for each position were estimated and plotted versus the tip mass position (internal detuning parameter). The primary beam mean square response was first measured at a specified excitation level when the tip mass is mounted at the primary beam (far away from exact detuning). This excitation level was kept the same for all tests. The measured mean squares of the primary beam modes were taken as normalizing values for all other measurements at different tip mass positions. This means that the effectiveness of the autoparametric interaction was measured by the departure of the normalized main beam mean squares from unity. Figure (8) shows the dependence of normalized mean squares of the responses on the tip mass position. It can be seen that there is a nonlinear absorbing effect in the primary beam due to energy transfer to the coupled main in the neighborhood of exact detuning.

4. Conclusions

The random response statistics of two coupled beams described by four nonlinear ordinary differential equations have been investigated experimentally for the case of simultaneous internal resonance conditions. The primary beam is directly exited by a random support motion while the bending mode of the coupled beam is excited parametrically and autoparametrically. Measured results reveal that the coupled beam bifurcates in the-mean-square and modal interaction in the neighborhood of internal resonances takes place in a form of energy exchange. Note that the nonlinear response of the system can be more dependent on one of the internal resonance conditions than on the other. This dependence is governed by the level of excitation. The coexistence of multiple internal resonances and nonlinear coupling may cause complicated behavior in both experimental and numerical analysis, especially at relatively high excitation levels. The experimental tests revealed that above a critical excitation level the response of the indirectly excited modes becomes very large and grows without bound if the shaker were not stopped.

Acknowledgment. This research is supported by a grant from the National Science Foundation under grant No. MSS-9203733 and by additional funds from the Institute for Manufacturing Research at Wayne State University.

5. References

1. Ibrahim, R. A.: Multiple internal resonance in a structure-liquid system, *ASME J. Engineering for Industry* **98**, 1092-1098, 1976.
2. Bux, S. L. and Roberts, J. W.: Nonlinear vibrators interactions in systems of coupled beams, *J. Sound and Vibration* **104**(3), 497-520, 1986
3. Cartmell, M. P. and Roberts, J. W.: Simultaneous combination resonances in and autoparametrically resonant system, *J. Sound and Vibration* **123**, 1988, 81-101, 1988.
4. Afaneh, A. A. and Ibrahim, R. A.: Random excitation of coupled oscillators with single and simultaneous internal resonances, submitted for publication.

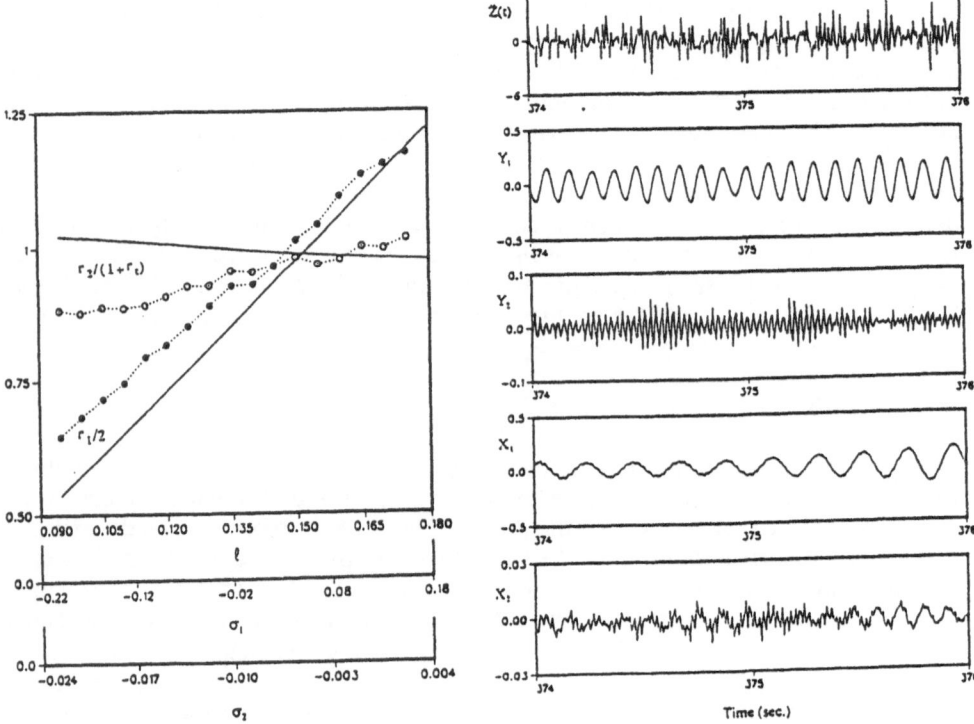

Figure 2 Dependence of internal detuning ratios on the tip mass positio

Figure 3 Section of time history records

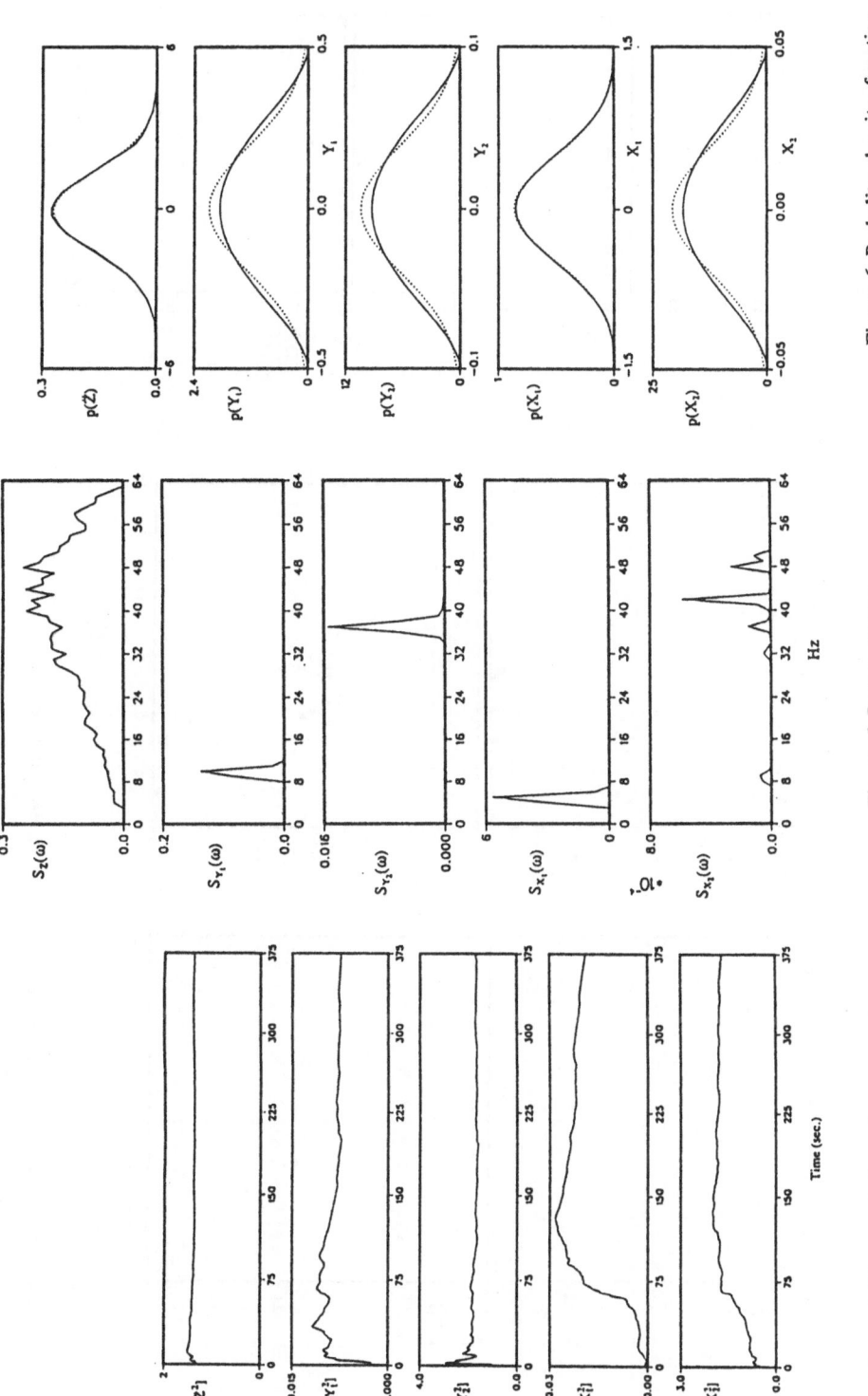

Figure 6 Probality density functios

Figure 5 Power spectra

Figure 4 Time evolution of mean squares

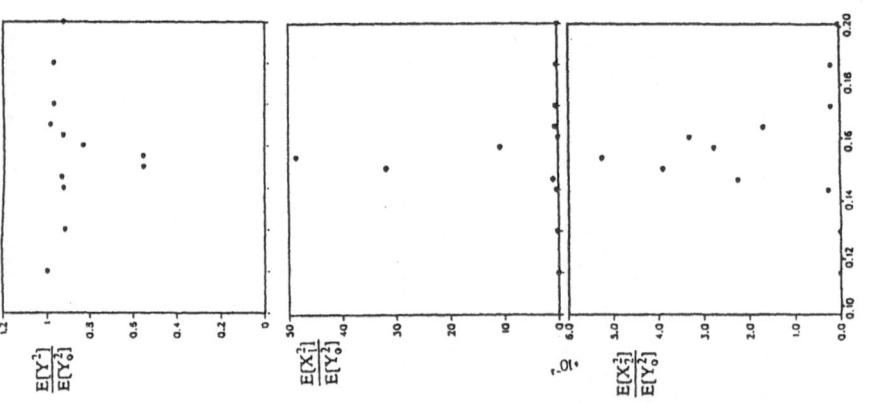

Figure 8 Dependence of mean square responses on tip mass position

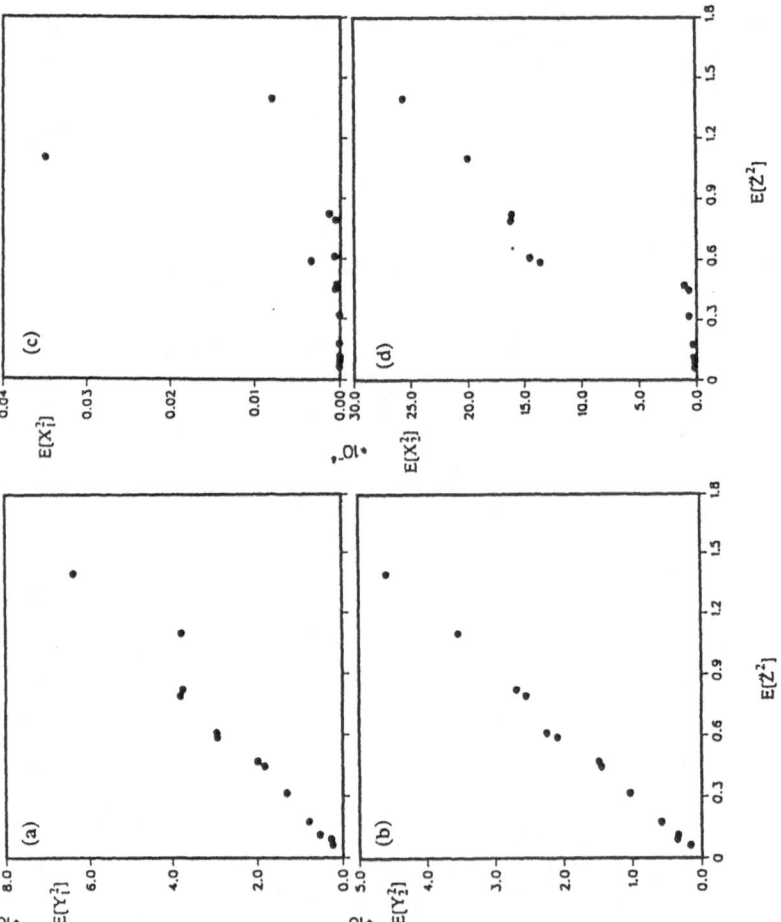

Figure 7 Dependence of mean square response on excitation level
(a) primary beam first mode
(b) primary beam second mode
(c) coupled beam first bending mode
(d) coupled beam first torsion beam

STOCHASTIC STABILITY OF VISCOELASTIC SYSTEMS UNDER BOUNDED NOISE EXCITATION

S. T. ARIARATNAM
Solid Mechanics Division
Faculty of Engineering
University of Waterloo
Waterloo, Ontario
Canada, N2L 3G1

Abstract.

The almost-sure stochastic stability of linear viscoelastic systems, parametrically forced by a bounded noise excitation, is investigated. By the use of the averaging method for integro-differential equations, the top Lyapunov exponent is evaluated asymptotically when the intensity of the excitation process is small. The stability region, which corresponds to negative values of the top Lyapunov exponent, is sketched in the parameter plane in the form of a "Strutt diagram". It is found that noise can have a stabilizing effect under certain conditions.

1. Introduction

Investigations of the dynamic stability of viscoelastic systems such as plates, columns and cylindrical shells subjected to time-dependent axial loads often lead to the study of an integro-differential equation of motion of the form

$$\ddot{x}(t) + \beta \int_0^t h(t-s)\dot{x}(s)ds + \omega_0^2[1 + \xi(t)]x(t) = 0 \qquad (1)$$

In the above equation $\xi(t)$ represents the fluctuation in the axial load and $h(t)$ the relaxation function of the viscoelastic material; the remaining parameters are constants. Eq. (1) also arises as the variational equation in the study of the stability of steady state solutions of nonlinear systems.

11

A. Naess and S. Krenk (eds.), IUTAM Symposium on Advances in Nonlinear Stochastic Mechanics, 11–18.
© 1996 *Kluwer Academic Publishers*.

In the particular case when $\xi(t)$ is a deterministic sinusoidal function of time, i.e. $\xi(t) = \mu \cos \nu t$, and when the relaxation function has the form $h(t_1 - t_2) = h_1(t_1)h_2(t_2)$, the stability of the solution of Eq. (1) was studied by Cederbaum *et al* (1991) by numerical calculation of the top Lyapunov exponent. If, however, the constants β and μ are assumed to be small, i.e. $\beta = O(\epsilon)$, $\mu = O(\epsilon)$, $0 < \epsilon << 1$, approximate analytical conditions for stability may be obtained by the use of perturbation or asymptotic methods. Moskvin et al (1984) obtained approximate stability boundaries using the Floquet theory and the method of Hill's determinants for arbitrary forms of the relaxation function. The present author (Ariaratnam (1995)) obtained the same stability conditions in the vicinity of the fundamental parametric resonance frequency $\nu = 2\omega_0$ by the use of the averaging theorem for integro-differential equations due to Larionov (1969) and evaluation of the top Lyapunov exponent.

When the excitation $\xi(t)$ in Eq. (1) is a wide-banded stationary stochastic process of small intensity, Ariaratnam (1993) obtained the condition for almost-sure stability, valid in the first approximation, by the method of stochastic averaging. The excitation process $\xi(t)$ in the above investigation is not a bounded function and may take arbitrarily large values with small probability, and hence may not be realistic in many physical applications. In the present paper, $\xi(t)$ is taken to be a bounded stochastic process of the form $\xi(t) = \mu \sin[\nu t + \sigma B(t) + \gamma]$, where μ, ν, σ are constants, $B(t)$ is the standard Wiener process and γ is a random variable uniformly distributed in the interval $[0, 2\pi)$. Such a process was first considered by Stratonovich (1967) and has since been used in certain applications by Dimentberg (1988) and Lin and Cai (1995). The mean-square value of the process is fixed at $\mu^2/2$, while its spectral density function can be made to approximate the well-known Dryden and von Karman spectra of wind turbulence by suitable choice of the constants μ, ν, and σ. The method of averaging for integro-differential equations is employed to derive an expression for the top Lyapunov exponent of the response of Eq. (1) when the parameters β and μ are small. The stability boundaries in the (μ, ν)-parameter plane are found by setting the top Lyapunov exponent to zero. The case of deterministic periodic excitation follows from the results obtained by letting the parameter σ tend to zero. It is found that, in the vicinity of fundamental parametric resonance, i.e. when $\nu \simeq 2\omega_0$, the one-sided Fourier cosine and sine transforms of the relaxation function $h(t)$ evaluated at the mean frequency ω_0 play the roles of equivalent viscous damping and additional stiffness, respectively, in the first approximation. It is also found that, under certain circumstances, parametric random excitation may have a stabilizing effect on the system response.

2. Averaged Equations

The excitation process is taken in the form

$$\xi(t) = \mu \sin\left[\nu t + \sigma B(t) + \gamma\right] \tag{2}$$

where $B(t)$ is the standard Wiener process and γ is a random variable. If γ is uniformly distributed in the interval $[0, 2\pi)$, then $\xi(t)$ is a stationary process having auto-correlation function

$$R(\tau) = \frac{1}{2}\mu^2 \exp\left(-\frac{\sigma^2 \tau}{2}\right) \cos \nu \tau \tag{3}$$

and spectral density function

$$S(\omega) = \int\limits_{-\infty}^{\infty} R(\tau)\, e^{i\omega\tau} d\tau = \frac{1}{2}\mu\sigma^2 \frac{\omega^2 + \nu^2 + \sigma^2/4}{[(\omega^2 - \nu^2 - \sigma^2/4)^2 + \sigma^4 \omega^2]} \tag{4}$$

Furthurmore $|\xi(t)| \le \mu$ for all values of the time t and hence is a bounded stochatic process.

In order to apply the averaging method to the integro-differential Eq. (1), we make the transformation to the amplitude and phase variables a, ϕ by means of the relations

$$x(t) = a \cos \Phi, \quad \dot{x}(t) = -a\omega_0 \sin \Phi, \quad \Phi = \frac{1}{2}\nu t + \phi(t) \tag{5}$$

Substituting these relations in Eq. (1) leads to the pair of first-order stochastic integro-differential equations

$$\dot{a} = \frac{1}{2}a\omega_0 \xi(t) \sin 2\Phi - \beta \sin \Phi \int\limits_0^t h(t-s)a(s) \sin \Phi(s)ds \tag{6}$$

$$\dot{\phi} = \Delta + \frac{1}{2}\omega_0 \xi(t)(1 + \cos 2\Phi) - \frac{\beta}{a} \cos \Phi \int\limits_0^t h(t-s)a(s) \sin \Phi(s)ds \tag{7}$$

where $\Delta = \omega_0 - \nu/2$.

We re-write Eq. (2) for $\xi(t)$ in the form

$$\xi(t) = \mu \sin(\nu t + \psi), \quad \psi = \sigma B(t) + \gamma \tag{8}$$

and substitute in Eqs. (6) and (7) to get

$$\dot{a} = \frac{1}{2}a\mu\omega_0 \sin(\nu t + \psi)\sin(\nu t + 2\phi) - \beta\sin\Phi\int_0^t h(t-s)a(s)\sin\Phi(s)ds$$

$$(9)$$

$$\dot{\phi} = \Delta + \frac{1}{2}\mu\omega_0 \sin(\nu t + \psi)[1 + \cos(\nu t + 2\phi)] - \frac{\beta}{a}\cos\Phi \times$$

$$\int_0^t h(t-s)a(s)\sin\Phi(s)ds \qquad (10)$$

$$\dot{\psi} = \sigma\dot{B}(t) \qquad (11)$$

where $\dot{B}(t)$ is the formal derivative of the Wiener process, i.e. $\dot{B}(t)$ is a white noise process of unity intensity.

Eqs. (9) - (11) are exactly equivalent to Eqs. (1) and (2) and cannot be solved in analytical form. However, solutions valid asymptotically in the first approximation may be obtained if certain coefficients appearing on the right-hand side of Eqs. (9) and (10) are assumed to be small. The averaging method for integro-differential equations developed by Larionov (1969) can then be applied to obtain the so-called averaged equations. Thus, we assume that $\beta = O(\epsilon)$, $\mu = O(\epsilon)$, $\Delta = O(\epsilon)$, $0 < \epsilon \leq 1$. The assumption on the detuning parameter Δ effectively restricts the analysis to those excitation frequencies ν that are in the vicinity of the frequency $2\omega_0$ of fundamental parametric resonance.

After averaging the right-hand sides of Eqs. (9) and (10), we obtain the averaged equations

$$\dot{a} = \left[\frac{1}{4}\mu\omega_0 \cos(\psi - 2\phi) - \frac{1}{2}\beta H_c(\nu/2)\right]a \qquad (12)$$

$$\dot{\phi} = \Delta + \frac{1}{4}\mu\omega_o \sin(\psi - 2\phi) + \frac{1}{2}\beta H_s(\nu/2) \qquad (13)$$

where

$$H_c(\omega) + iH_s(\omega) = \int_0^\infty h(s)\,e^{i\omega s}\,ds \qquad (14)$$

and no distinction has been made between the averaged and the original non-averaged variables a, ϕ.

In averaging the convolution integrals it is necessary to make the physically acceptable assumption that the functions $h(t)$ and $th(t)$ are integrable over the range $(0, \infty)$; the detailed evaluation of these averages has been carried out in Ariaratnam (1995).

The change of variables $\rho = \log a$, $\quad \theta = \phi - \psi/2$, transforms Eqs. (11) - (13) into the following pair of averaged stochastic differential equations:

$$d\rho = \left[\frac{1}{4}\mu\omega_0 \cos 2\theta - \frac{1}{2}\beta H_c(\nu/2)\right] dt \tag{15}$$

$$d\theta = \left[\Delta + \frac{1}{2}\beta H_s(\nu/2) - \frac{1}{4}\mu\omega_0 \sin 2\theta\right] dt - \frac{1}{2}\sigma dB(t) \tag{16}$$

3. Lyapunov Exponent and Stochastic Stability

The Lyapunov exponent of the system given by Eq. (1) may be defined as

$$\lambda = \lim_{t\to\infty} \frac{1}{2t} \log\left[x^2(t) + \frac{1}{\omega_0^2}\dot{x}^2(t)\right] \tag{17}$$

which, on making use of Eq. (5), becomes

$$\lambda = \lim_{t\to\infty} \frac{1}{t} \log a(t) = \lim_{t\to\infty} \frac{1}{t}\rho(t) \tag{18}$$

The Lyapunov exponent is a measure of the average exponential growth rate of the amplitude process $a(t)$ for large values of t and is a deterministic number with probability one (w.p.1) for the system given by Eqs. (15) and (16). Depending on the initial values of the vector $\{x(t), \dot{x}(t)\}$, there will in general be two such values for λ. If both Lyapunov exponents are negative, the trivial solution of Eq. (1) will be stable w.p.1 while if the larger exponent is positive, the response will be unstable w.p.1.

In order to calculate λ, we integrate both sides of Eq. (15) to obtain

$$\rho(t) - \rho(0) = \frac{1}{4}\mu\omega_0 \int_0^t \cos 2\theta(t)\, dt - \frac{1}{2}t\beta H_c(\nu/2) \tag{19}$$

so that, from Eq. (18),

$$\lambda = \frac{1}{4}\mu\omega_0 \lim_{t\to\infty} \frac{1}{t} \int_0^t \cos 2\theta(t)\, dt - \frac{1}{2}\beta H_c(\nu/2) \tag{20}$$

The random process $\theta(t)$ given by Eq. (16) can be shown to be ergodic, in which case we can write

$$\lim_{t\to\infty} \frac{1}{t} \int_0^t \cos 2\theta(t)\, dt = E[cos2\theta] \quad w.p.1. \tag{21}$$

where $E[\ ]$ denotes the expectation operator. Thus, $w.p.1$

$$\lambda = \frac{1}{4}\mu\omega_0 E[\cos 2\theta] - \frac{1}{2}\beta H_c(\nu/2) \tag{22}$$

It now remains to evaluate $E[\cos 2\theta]$ in order to obtain λ. To this end, we set up the following Fokker-Planck equation governing the invariant (or stationary) probability density function $p(\theta)$ of the process $\theta(t)$ defined by Eq. (16):

$$\sigma^2 \frac{d^2 p}{d\theta^2} - \frac{d}{d\theta}\left[(4\Delta_0 - \mu\omega_0 \sin 2\theta)p\right] = 0 \tag{23}$$

where $\Delta_0 = \Delta + \frac{1}{2}\beta H_s(\nu/2)$. The solution of Eq. (23) satisfying the periodicity condition $p(\theta) = p(\theta + 2\pi)$, as obtained by Stratonovich (1967), is

$$p(\theta) = C^{-1} \exp\left[(8\Delta_0\theta + \mu\omega_0 \cos 2\theta)/\sigma^2\right] \times$$
$$\int_{\theta}^{\pi+\theta} \exp\left[-(8\Delta_0\psi + \mu\omega_0 \cos 2\psi)/\sigma^2\right] d\psi \tag{24}$$

where the normalizing constant C is given by

$$C = 2\pi^2 \exp(-\pi q) \mid I_{iq}(\mu\omega_0/\sigma^2) \mid^2 \tag{25}$$

$I_{iq}(z)$ being the Bessel function of imaginary argument and imaginary order, and $q = 4\Delta_0/\sigma^2$. Using Eqs. (24) and (25), the value of $E[\cos 2\theta]$ is found to be

$$E[\cos 2\theta] = F\left(\mu\omega_0/\sigma^2, 4\Delta_0/\sigma^2)\right) \tag{26}$$

where

$$F(z, q) = \frac{1}{2}\frac{d}{dz} \log\left[I_{iq}(z) I_{-iq}(z)\right] \tag{27}$$

which can be written in the form

$$F(z, q) = \frac{1}{2}\left[\frac{I_{1+iq}(z)}{I_{iq}(z)} + \frac{I_{1-iq}(z)}{I_{-iq}(z)}\right] \tag{28}$$

Hence, using Eq. (22), the expression for λ becomes

$$\lambda = \frac{1}{4}\mu\omega_0 F(\mu\omega_0/\sigma^2, 4\Delta_0/\sigma^2) - \frac{1}{2}\beta H_c(\nu/2) \tag{29}$$

The Lyapunov exponent λ obtained here using the invariant probability density $p(\theta)$ of Eq. (24) is the larger of the two Lyapunov exponents. When the modified detuning parameter $\Delta_0 = 0$, we obtain from Eqs. (28) and (29),

$$\lambda = \frac{1}{4}\mu\omega_0 \frac{I_1(\mu\omega_0/\sigma^2)}{I_0(\mu\omega_0/\sigma^2)} - \frac{1}{2}\beta H_c(\nu/2) \tag{30}$$

The stability boundary, which corresponds to $\lambda = 0$, is given by

$$\mu\omega_0 \, F(\mu\omega_0/\sigma^2, \, 4\Delta_0/\sigma^2) = 2\beta H_c(\nu/2) \qquad (31)$$

Using a table of Bessel functions, the stability boundary in the $(\mu, \, \nu/2\omega_0)$-parameter space can be plotted numerically for different values of β and σ. Depending on the relations amongst the parameters $\mu\omega_0/\sigma^2$, Δ_0/σ^2 and unity, various asymptotic expansions of the Bessel functions involved can be employed to simplify Eqs. (29) and (30). For example, when the noise intensity σ is so small that $\mu\omega_0/\sigma^2 \gg 1$ and $\Delta_0/\sigma^2 \gg 1$, we obtain the asymptotic result

$$\lambda = \frac{1}{4}\mu\omega_0 \, (1 - \alpha^2)^{1/2} - \frac{1}{8}\sigma^2 \, (1 - \alpha^2)^{-1} - \frac{1}{2}\beta H_c(\nu/2) \qquad (32)$$

where $\alpha = 4\Delta_0/(\mu\omega_0)$, which when the noise parameter σ tends to zero leads to the result

$$\lambda = \frac{1}{4}\mu\omega_0 \, (1 - \alpha^2)^{1/2} - \frac{1}{2}\beta H_c(\nu/2) \qquad (33)$$

obtained previously by Ariaratnam (1995) for the case of purely periodic excitation. The asymptotic result in Eq. (32) can be shown to be valid also when $\alpha = 0$, i.e. when $\Delta_0 = 0$, by direct expansion of the terms in Eq. (30) to yield

$$\lambda = \frac{1}{4}\mu\omega_0 - \frac{1}{8}\sigma^2 - \frac{1}{2}\beta H_c(\nu/2) \qquad (34)$$

This shows that stability may be improved in the vicinity of $\Delta_0 = 0$, i.e. near $\nu = 2\omega_0 + \beta H_s(\omega_0)$, by the addition of noise. For other relations amongst the various parameters, asymptotic expressions for the Lyapunov exponent λ, similar to that of Eq. (32), can be obtained in the manner shown by Stratonovich (1967). A sketch of the region of instability is presented in the figure below.

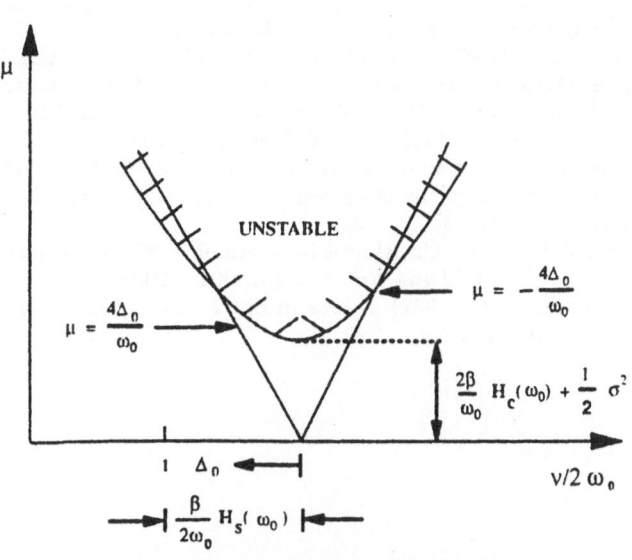

4. Conclusions

The stochastic stability of a linear viscoelastic system parametrically excited by a bounded stationary stochastic process in the form of a sinusoidal function with noise modulated phase has been studied.

When the mean frequency of the excitation is in the vicinity of the frequency of fundamental parametric resonance, the condition for almost-sure stability of the trivial equilibrium state has been obtained, which is asymptotically valid for small values of the excitation amplitude and the viscoelastic damping constant. It was found that the one-sided Fourier cosine and sine transforms of the viscoelastic relaxation function, evaluated at the mean natural frequency of the system, played the roles of effective viscous damping and additional elastic stiffness, respectively. It is also shown that near the frequency of fundamental parametric resonance, modulation of the phase of the excitation by random noise may improve the stability of the system.

The stability condition for a purely viscously damped system can be obtained from Eq. (29) by taking $h(t)$ to be the Dirac delta function, in which case $H_c = 1$ and $H_s = 0$. The method of analysis presented here can be readily adapted to a viscoelastic system modelled differently such as that in Potapov (1994).

References

Ariaratnam, S. T. (1993) Stochastic Stability of Linear Viscoelastic Systems, *Probabilistic Engineering Mechanics* **8**, 153–155.

Ariaratnam, S. T. (1995) Investigation of Parametric Instability of Viscoelastic Systems using Lyapunov Exponents, in B. Tabarrok and S. Dost (eds.), *Proceedings of the Canadian Congress of Applied Mechanics*, Victoria, B.C., May 22 - June 1, 920 - 921.

Cederbaum, G., Aboudi, J. and Elishakoff, I. (1991) Dynamic Instability of Shear-Deformable Viscoelastic Laminated Plates by Lyapunov Exponents, *Int. J. Solids and Struct.* **28** (3), 317 – 327.

Dimentberg, M. F. (1988) *Statistical Dynamics of Nonlinear and Time-Varying Systems*, Research Studies Press, Taunton, England Wiley.

Larionov, G. S. (1969) Investigation of the Vibration of Relaxing Systems by the Averaging Method, *Mechanics of Polymers* (English Translation) **5**, 714 – 720.

Lin, Y. K. and Cai, G. Q. (1995) *Probabilistic Structural Dynamics – Advanced Theory and Applications*, McGraw-Hill, New York, 252.

Moskvin, V. G., Semenov, V. A. and Smirnov, A. I. (1984) On the Investigation of Parametric Vibrations of Systems with Hereditary Friction, *Mech. of Solids* (English Translation) **19** (3), 31 – 36.

Potapov, V. D. (1994) On Almost Sure Stability of a Viscoelastic Column under Random Loading, *J. Sound and Vib.* **173** (3), 301 – 308.

Stratoinovich, R. L. (1967) *Topics in the Theory of Random Noise*, Volume II, Gordon and Breach, New York, 289, 294 – 302.

FIXED POINTS AND ATTRACTORS FOR RANDOM DYNAMICAL SYSTEMS

Application to stochastic bifurcation theory

L. ARNOLD AND B. SCHMALFUSS
Institut für dynamische Systeme
Universität Bremen
Postfach 330440
28334 Bremen
Germany

1. Introduction

We have been pursuing for quite some time the study of systems under randomness within the framework of *dynamical systems* (i.e. flows of mappings on some state space). For a non–technical introduction and a survey of results see Arnold [2].

One of the outcomes of these efforts is the following formal definition of a random dynamical system which covers most systems under random perturbations presently of interest, in particular random and stochastic difference and differential equations.

Definition 1.1 *A (continuous) random dynamical system (henceforth abbreviated as RDS) and state space $\mathbb{R}^d$ consists of the following two ingredients:*

1. A metric dynamical system $(\Omega, \mathcal{F}, \mathbb{P}, (\theta_t)_{t \in \mathbb{R}})$ (for short: θ), i.e. a probabilistic space $(\Omega, \mathcal{F}, \mathbb{P})$ with a measurable flow of measure preserving transformations $\theta_t : \Omega \to \Omega$, i.e. $\theta_0 = \mathrm{id}$, $\theta_{t+s} = \theta_t \circ \theta_s$ for all t, $s \in \mathbb{R}$, $(t, \omega) \mapsto \theta_t \omega$ measurable and $\theta_t \mathbb{P} = \mathbb{P}$. We assume that θ_t is ergodic. [This is the model of a stationary stochastic process, like real or white noise, as a dynamical system.]

2. A cocycle ϕ over θ of continuous mappings of $\mathbb{R}^d$, i.e. a measurable mapping

$$\phi : \mathbb{R}^+ \times \Omega \times \mathbb{R}^d \to \mathbb{R}^d, \quad (t, \omega, x) \mapsto \phi(t, \omega, x),$$

for which $\phi(t, \omega) := \phi(t, \omega, \cdot)$ is continuous and satisfies the cocycle property

$$(i) \qquad \phi(0, \omega) = \mathrm{id}_{\mathbb{R}^d},$$

19

A. Naess and S. Krenk (eds.), IUTAM Symposium on Advances in Nonlinear Stochastic Mechanics, 19–28.
© *1996 Kluwer Academic Publishers.*

$$(ii) \qquad \phi(t+s,\omega) = \phi(t,\theta_s\omega) \circ \phi(s,\omega).$$

[This is a model of a dynamical system perturbed by noise; in case ω is not present ϕ is just a classical continuous dynamical system.]

Note that for the purposes of this note we only allow ϕ to have one sided time $T = \mathbb{R}^+$ as we want to treat non–invertible mappings. However, this noise θ is always assumed to have two–sided time $T = \mathbb{R}$ which is basically without loss of generality due to the existence of a natural extension from $\mathbb{R}^+$ to $\mathbb{R}$ (see Arnold [1], Appendix A.1). The set–up permits us to study ϕ *forward* in time on the *whole* two–sided time axis. For example, $\phi(t,\theta_{-t}\omega)$ is considered mapping a point x at time $-t$ forward to a point $\phi(t,\theta_{-t}\omega)x$ at time $t - t = 0$. It makes generally a big difference for a non–autonomous system like ours between moving from 0 to t, and moving from $-t$ to 0.

The purpose of this note is to present the random analogue of two concepts which are fundamental for deterministic dynamical systems, namely fixed points, and attractors, and present some recent results and examples.

Definition 1.2 *Given an RDS ϕ. We call a random variable $u : \Omega \to \mathbb{R}^d$ (random)* fixed point *(or stationary solution) of the RDS ϕ if it is invariant under ϕ, i.e.*

$$\phi(t,\omega)u(\omega) = u(\theta_t\omega) \quad \text{for all } t, \omega.$$

The definition of attractors, a concept which belongs to topological dynamics, is more involved and will be given in the next section.

2. Attractors for random dynamical systems

Attractors capture information on long term evolution of dynamical systems. Their structure can tell us whether a system has e.g. laminar or turbulent behavior, so that the study of attractors and their change if parameters are varied is part of bifurcation theory, cf. Temam [14]. As an attractor of an RDS will be a random set $\omega \mapsto A(\omega)$ (i.e. a set set–valued random variable) we need the following preparations (see also Arnold [1], section 1.6.).

Definition 2.1 (i) *We call a mapping $\omega \mapsto B(\omega)$ (for short: B) into the compact non–void subsets of $\mathbb{R}^d$ a* random compact set *if for any $x \in \mathbb{R}^d$ the mapping*

$$\omega \mapsto d(x, B(\omega)), \qquad d(x,B) := \inf_{y \in B} |x - y|,$$

is measurable.

(ii) *A random closed set B is called* forward invariant *for the RDS ϕ if*

$$\phi(t,\omega)B(\omega) \subset B(\theta_t\omega) \quad \text{for all } t \in \mathbb{R}^+, \, \omega \in \Omega$$

(equivalently, $\phi(t, \theta_{-t}\omega)B(\theta_{-t}\omega) \subset B(\omega)$). It is called invariant *for ϕ if*

$$\phi(t, \omega)A(\omega) = A(\theta_t\omega) \quad \text{for all } t \in \mathbb{R}^+, \omega \in \Omega$$

(equivalently, $\phi(t, \theta_{-t}\omega)A(\theta_{-t}\omega) = A(\omega)$).

Now an attractor is only defined relative to a universe of sets which are attracted (its domain of attraction), and the construction of an attractor is usually done via an intermediate set called absorbing. Here are the random analogue of these notations.

Definition 2.2 *(i) Let $\mathcal{D}$ be a family of random compact sets which is closed with respect to inclusions (i.e. if $D_1 \in \mathcal{D}$ and $D_2(\omega) \subset D_1(\omega)$ then $D_2 \in \mathcal{D}$). A random compact set $B \in \mathcal{D}$ is called* absorbing *in $\mathcal{D}$ for the RDS ϕ if*

(a)   *B is forward invariant,*

(b)   *B is absorbing sets in $\mathcal{D}$:*

For all $D \in \mathcal{D}$ there exists a time $t_0(\omega, D)$ such that

$$\phi(t, \theta_{-t}\omega)D(\theta_{-t}\omega) \subset B(\omega) \text{ for all } t \geq t_0.$$

(ii) A (global, random) attractor *of ϕ in $\mathcal{D}$ is a random compact set $A \in \mathcal{D}$ withe the following properties:*

(a)   *A is invariant,*

(b)   *A is attracting set in $\mathcal{D}$:*

For all $D \in \mathcal{D}$

$$\lim_{t \to \infty} d(\phi(t, \theta_{-t}\omega)D(\theta_{-t}\omega) \mid A(\omega)) = 0,$$

where $d(A \mid B) := \sup_{x \in A} d(x, B)$ is the semi–Hausdorff metric. $\mathcal{D}$ is then called a domain of attraction *of A.*

We would like to stress that both the definition of absorbing and attracting are based on moving points from time $-t$ to time 0 (and *not* from 0 to t !), and letting $t \to \infty$. This *pullback* convergence allows to consider Omega-limit sets that lie in fixed ω–*fibers*. Random attractors have been studied by Crauel and Flandoli [5], [6], Debussche [7] and Schmalfuß [9], [10], [12]. The following theorem is a particular application of Schmalfuß [11].

Theorem 2.3 *Let ϕ be an RDS. If B is an absorbing set for ϕ in the family $\mathcal{D}$ then ϕ possesses a unique attractor $A \in \mathcal{D}$ with domain of attraction $\mathcal{D}$ which is given by*

$$A(\omega) = \bigcap_{t \geq 0} \overline{\phi(t, \theta_{-t}\omega)B(\theta_{-t}\omega)}.$$

The set A is connected.

Remark 2.4 (i) The forward invariance of B implies that $\phi(t, \theta_{-t})B(\theta_{-t}\omega)$ is decreasing, hence

$$A(\omega) = \bigcap_{t \geq 0} \overline{\phi(t, \theta_{-t}\omega)B(\theta_{-t}\omega)} \neq \emptyset.$$

(ii) If C is a forward invariant random compact set then

$$A(\omega) = \bigcap_{t \geq 0} \overline{\phi(t, \theta_{-t}\omega)C(\theta_{-t}\omega)}$$

is a random set which is invariant.

3. Examples

In this section we consider RDS which are solutions of stochastic differential equations. For this we model white noise (Brownian motion) as a metric dynamical system as follows (we only treat the scalar case): Let $(W_t)_{t \in \mathbb{R}}$ be standard Brownian motion in $\mathbb{R}$, i.e. a stochastic process which has stationary increments, $W_{t+h} - W_t \sim \mathcal{N}(0, |h|)$, $W_0 = 0$, and continuous trajectories. Put $\Omega = \{\omega \in C(\mathbb{R}, \mathbb{R}) : \omega(0) = 0\}$, $\mathcal{F}$ the Bore-l σ–algebra of Ω, $\mathbb{P}$ the measure on $\mathcal{F}$ generated by W (so called *Wiener measure*). The shift

$$\theta_t\omega(\cdot) := \omega(t + \cdot) - \omega(t)$$

is measure preserving and ergodic, and $W_t(\omega) = \omega(t)$ is Brownian motion (with which we work). The generalized derivative of W_t is called *white noise*.

3.1. EXAMPLE 1 A NONLINEAR STRATONOVICH EQUATION

Consider the following Stratonovich stochastic differential equation in $\mathbb{R}^d$ with scalar multiplicative noise

$$dx_t = -Ax_t\, dt + f(x_t)\, dt + \sigma x_t \circ dW_t, \qquad (1)$$

where the $d \times d$ matrix is assumed to be positive definite,

$$(Ax, x) \geq c_1\|x\|^2 \quad \text{for all } x \in \mathbb{R}^d \text{ and some } c_1 > 0,$$

the nonlinearity $f \in C^1$ satisfies the growth condition

$$(f(x), x) \leq c_2\|x\|^2 + c_3, \; c_2 < c_1, \; c_3 > 0, \qquad (2)$$

and σ is a strength parameter. Under these assumptions, the stochastic differential equation (1) is known to generate through its solution $\phi(t, \omega)x$ (satisfying $\phi(0, \omega)x = x \in \mathbb{R}^d$) an RDS over white noise θ, cf. Gichman and Skorochod [8] chapter I.6, remark 3.

Theorem 3.1 *The RDS ϕ generated by (1) has a unique attractor with domain of attraction*

$$D(\omega) \subset K(0, r(\omega)), \quad r(\theta_{-t}\omega)e^{-\delta t} \to 0$$

for $t \to \infty$ and any $\delta > 0$ where $K(0, r) = \{x \in \mathbb{R}^d : \|x\| \le r\}$.

Proof: The only task is to find an absorbing set. By the assumptions on A and f and the chain rule,

$$d\|x_t\|^2 = (-2(c_1 - c_2)\|x_t\|^2 + 2c_3 + \xi_t)dt + 2\sigma\|x_t\|^2 \circ dW_t, \qquad (3)$$

where $\xi_t \le 0$. Comparing (3) with the affine equation

$$dy_t = (-2(c_1 - c_2)y_t + 2c_3)dt + 2\sigma y_t \circ dW_t \qquad (4)$$

starting at $y_0 = \|x\|^2$ yields

$$\|\phi(t, \omega)x\|^2 \le \psi(t, \omega)\|x\|^2, \quad t \ge 0,$$

where ψ is the RDS generated by (4). Moreover, ψ has a unique globally stable fixed point

$$\eta(\omega) = 2c_3 \int_{-\infty}^0 \exp(2(c_1 - c_2)t + 2\sigma W_t)dt$$

(see Arnold and Crauel [4] or Schmalfuß [13] section 3. It is not difficult to show that

$$B(\omega) = \{x \in \mathbb{R}^d : \|x\|^2 \le 2\eta(\omega)\}$$

is an absorbing set of ϕ in $\mathcal{D}$ $\square$

3.2. EXAMPLE 2: A RANDOM PITCHFORK BIFURCATION

Consider the one dimensional Stratonovich stochastic differential equation with parameter $\alpha \in \mathbb{R}$

$$dx_t = (\alpha x_t - x_t^3 + g(x_t))dt + \sigma x_t \circ dW_t. \qquad (5)$$

Assume that g is C^1 and satisfies

$$g(0) = 0, \quad \frac{g(x)}{x} > 0 \quad (x \neq 0) \qquad (6)$$

$$|g'(x)| < 2x^2 \quad (x \neq 0). \qquad (7)$$

It follows from (6), (7) that (5) generates a unique RDS ϕ for any $\alpha \in \mathbb{R}$. Clearly $u \equiv 0$ is a fixed point, so that $(0, \infty)$ and $(-\infty, 0)$ are invariant sets due to uniqueness of solution.

Arnold and Boxler [3] studied (5) for $g \equiv 0$ and proved that it undergoes a stochastic pitchfork bifurcation at $\alpha = 0$: While $u \equiv 0$ is the only globally attracting fixed point for $\alpha \leq 0$, there are unique (globally attracting) fixed points $u^+ \in (0, \infty)$ and $u^- \in (0, \infty)$ for $\alpha > 0$ with $\mathbb{E}(u^\pm)^2 = \alpha$. They only studied the case that $\mathcal{D}$ contains nonrandom single points sets.

We will now investigate the case $g \not\equiv 0$ and obtain the following result.

Theorem 3.2 *Let ϕ be the RDS generated by (5). Then*
(i) ϕ has a unique attractor $A_\alpha = [a_\alpha^-, a_\alpha^+]$ on the state space $\mathbb{R}$ for any $\alpha \in \mathbb{R}$, and $\mathcal{D}$ is the family of all random compact sets fulfilling the conditions of theorem 3.1.
(ii) In particular, for $\alpha < 0$, $A_\alpha(\omega) = \{0\}$.
(iii) For $\alpha > 0$, ϕ restricted to $(0, \infty)$ has an invariant compact interval $A(\omega) = [a_1(\omega), a_2(\omega)]$ with $0 < a_1(\omega)$ (whose boundary points are thus random fixed points). Similarly for ϕ restricted to $(-\infty, 0)$.

Proof: (i) Equation (5) has a unique solution. In particular,

$$f(x) = 2\alpha x - x^3 + g(x)$$

satisfies conditions (2) for any $\alpha > 0$ where A is given by $-\alpha$. The case $\alpha < 0$ follows similarly.

(ii) This statement follows directly from writing (5) as

$$dx_t = \left(\alpha - x_t^2 + \frac{g(x_t)}{x_t} \right) x_t dt + \sigma x_t \circ dW_t$$

and noting that $-x^2 + g(x)/x \leq 0$ which yields

$$|\phi(t, \omega)x| \leq \exp(\alpha t + \sigma W_t)|x|.$$

In particular, 0 is globally attracting any random compact set $D \in \mathcal{D}$.

(iii) Since by our assumptions for $\alpha > 0$

$$\alpha x - x^3 + g(x) \leq -\alpha x + c(\alpha), \quad x > 0 \tag{8}$$

where $0 < c(\alpha) = O(\alpha^{\frac{3}{2}})$, the random interval

$$\left(0, c_\alpha \int_{-\infty}^0 \exp(\alpha t + \sigma W_t) dt \right]$$

is forward invariant by (8).

On the other hand, the map $y = 1/x^2$ transforms (5) on $(0, \infty)$ into

$$dy_t = (-2\alpha y_y + 2 + h(y_t))dt - 2\sigma y_t \circ dW_t, \tag{9}$$

where $h(y) = -2g(x)/x^3 = -2y^{3/2}g(y^{-1/2}) < 0$. Comparing with a corresponding affine equation as in example 1 we find that the interval

$$\left(0, 2\int_{-\infty}^{0} \exp(2\alpha t - 2\sigma W_t)dt\right] \tag{10}$$

is forward invariant for the RDS generated by (9), hence

$$\left[\left(2\int_{-\infty}^{0} \exp(2\alpha t - 2\sigma W_t)dt\right)^{-\frac{1}{2}}, \infty\right)$$

is forward invariant for ϕ. Finally

$$\begin{aligned} B(\omega) &= [b_1(\omega), b_2(\omega)] \\ &= \left[\left(2\int_{-\infty}^{0} \exp(2\alpha t - 2\sigma W_t)dt\right)^{-\frac{1}{2}}, c_\alpha \int_{-\infty}^{0} \exp(\alpha t + \sigma W_t)dt\right] \end{aligned}$$

is forward invariant for ϕ, hence by remark 2.4 (ii) the interval

$$A(\omega) = \bigcap_{n\in\mathbb{Z}^+} \phi(t, \theta_{-t}\omega)B(\theta_{-t}\omega) = [a_1(\omega), a_2(\omega)]$$

is invariant for ϕ. Since $\phi(t, \omega)$ is a homeomorphism on $(0, \infty)$, $\phi(t, \omega)a_i(\omega) = a_i(\theta_t\omega)$, hence the a_i are fixed points of ϕ □
We will prove in the next section that $A(\omega)$ is in fact one–point, completing the proof that (5) undergoes a pitchfork bifurcation of fixed points.

4. A random fixed point theorem

We now formulate a random version of the Banach fixed point theorem for differentiable RDS. This theorem has the advantage of *not* imposing a contraction constant which is uniformly smaller than 1 for all ω, an assumption very unrealistic for the RDS.

Definition 4.1 *Let B be a forward invariant set for an RDS ϕ, and $u \in B$ a fixed point of ϕ. u is called* exponentially stable uniformly in B *(with rate γ) if there exists a $\gamma > 0$ such that*

$$\lim_{t\to\infty} \sup_{y\in B(\omega)} \|\phi(t, \omega)y - u(\theta_t\omega)\|e^{\gamma t} = 0$$

Schmalfuß [13] proved the following theorem.

Theorem 4.2 *Let ϕ be an RDS such that $x \mapsto \phi(t,w)x$ is C^1. Let B be a forward invariant convex set which is contained in the open ball with radius $r(w)$. Assume*

$$(i) \qquad \mathbb{E}\log^+ r(\theta_1 w) < \infty,$$

$$(ii) \qquad \mathbb{E}\sup_{0 \le t \le 1} \log \|D\phi(t,w,x)\| < \infty,$$

$$(iii) \qquad \mathbb{E}\sup_{x \in B(w)} \log \|D\phi(1,w,x)\| =: \Lambda.$$

Then ϕ has a unique exponentially stable fixed point in B with rate γ.

Assumption (i) is needed to control the size of $B(\theta_t w)$. Assumption (ii) is used for reducing the proof to discrete time $\mathbb{Z}^+$, and

$$\lambda(w) = \lim_{t \to \infty} \frac{1}{t} \sup_{x \in B(w)} \log \|D\phi(t,w,x)\|$$

is the uniform Lyapunov exponent of $D\phi$ on the random set B bounded by Λ

We will apply this theorem to show that the interval in theorem 3.2 (iii) is actually a point. For B we choose the interval constructed in the proof of theorem 3.2 (iii) which satisfies assumption (i). We now check that assumptions (iii) holds for equation (5) and this choice of B. Differentiating (5) gives

$$D\phi(t,w,x) = \exp\left(\alpha t + \int_0^t (-3(\phi(s,w)x)^2 + g'(\phi(s,w)x))ds + W_t\right).$$

It remains to show that

$$\Lambda := \alpha + \mathbb{E}\sup_{x \in B(w)} \int_0^1 (-3(\phi(t,w)x)^2 + g'(\phi(t,w)x))dt < 0.$$

By $A(w) \subset B(w)$ it suffices to prove that $\mathbb{E}b_1^2 \ge \alpha$. Let c_2 be the right endpoint of the interval (10) which is the unique fixed point of a certain affine equation. We are able to verify that $c_2^{-1} \in L^1$. The chain rule gives

$$\log c_2(\theta_t w) + 2\alpha t = \log c_2(w) + \int_0^t \frac{2}{c_2(\theta_\tau w)} d\tau + w(t).$$

Multiplying this equation with t^{-1} and letting $t \to \infty$ gives

$$\alpha = \mathbb{E}\frac{1}{c_2} = \mathbb{E}b_1^2. \qquad (11)$$

because $(\log c_2(\theta_t w))/t$ has a limit which has to be 0. We have proved

Corollary 4.3 *For any $\alpha > 0$ the RDS ϕ generated by (5) possesses a unique positive globally attracting fixed point u^+ where $D \in \mathcal{D}$ are compact measurable sets contained in random intervals with measurable boundaries $[r_1, r_2]$, $r_1 > 0$ such that*

$$r_1^{-2}(\theta_{-t}\omega)e^{-\delta t} \to 0 \text{ and } r_2(\theta_{-t}\omega)e^{-\delta t} \to 0$$

for $t \to \infty$. Similar, we can find a unique negative globally attracting fixed point u^-.

Let us mention that $A(\omega) = [u^-(\omega), u^+(\omega)]$ is an attractor of ϕ in the sense of section 2, and that

$$(\mathbb{E}(u^\pm)^2)^{\frac{1}{2}} \geq O(\alpha^{\frac{1}{2}}) \quad \text{for } \alpha \to 0.$$

This property follows from (11). Note further that in order to establish a stochastic pitchfork bifurcation for (5) we need to impose *global* conditions on g, in contrast to the local conditions (on the derivatives of g at 0) one needs the deterministic case. The reason is that random fixed points can move around the whole state space in the course of time.

References

1. Arnold, L. *Random Dynamical Systems*. Preliminary version 2, November 1994.
2. Arnold, L. Six lectures on random dynamical systems (CIME Summer School). Report Nr. 314, Institut für Dynamische Systeme, Universität Bremen, August 1994. Submitted.
3. Arnold, L. and Boxler, P. Stochastic bifurcation: Instructive examples in dimension one. In Mark Pinsky and Volker Wihstutz, editors, *Diffusion processes amd related problems in analysis, Vol. II: Stochastic flows. Progress in Probability*, Vol. 27, pages 241–256, Boston Basel Stuttgart, 1992. Birkhäuser.
4. Arnold, L. and Crauel, H. Iterated function systems and multiplicative ergodic theory. In Mark Pinsky and Volker Wihstutz, editors, *Diffusion processes and related problems in analysis, Vol. II: Stochastic flows. Progress in Probability*, Vol. 27, pages 283–305, Boston Basel Stuttgart, 1992. Birkhäuser.
5. Crauel, H. and Flandoli, F. Attractors for random dynamical systems. *Prob. Th. Related Fields*, 100(2):365–393, 1994.
6. Crauel, H. and Flandoli, F. Hausdorff dimension of invariant sets for random dynamical systems. submitted, 1994.
7. Debussche, A. On the finite dimensionality of random attractors. Technical Report 10. 95, Scuola Normale Superiore di Pisa, 1995. to appear in Stochastic Analysis and Applications.
8. Gichman, I.I. and Skorochod, A.W. *Stochastische Differentialgleichungen*. Akademie Verlag, Berlin, 1971.
9. Schmalfuß, B. Backward cocycles and attractors of stochastic differential equations. In V.Reitmann, T. Riedrich, and N. Koksch, editors, *International Seminar on Applied Mathematics–Nonlinear Dynamics: Attractor Approximation and Global Behaviour*, pages 185–192, 1992.
10. Schmalfuß, B. Stochastische Attraktoren des stochastischen Lorenz–Systems. *ZAMM*, 74(6):T626–T627, 1994.

11. Schmalfuß, B. Attractors for the non autonomous Navier–Stokes equation. in preparation, 1995.
12. Schmalfuß, B. Measure attractors and stochastic attractors. Technical Report 332, Universität Bremen, Institut für Dynamische Systeme, 1995.
13. Schmalfuß, B. A random fixed point theorem based on Lyapunov exponents. Technical Report 349, Universität Bremen, Institut für Dynamische Systeme, 1995.
14. Temam, R. *Infinite-Dimensional Dynamical Systems in Mechanics and Physics.* Springer–Verlag, Berlin–Heidelberg–New York, 1988.

NONLINEAR WAVE PROPAGATION IN COMPLEX STRUCTURES MODELLED BY RANDOM MEDIA WITH SELF-STRESSES

A. K. BELYAEV[1]*, H. IRSCHIK[1] and F. ZIEGLER[2]

1. Institut für Technische Mechanik und Grundlagen der Maschinenlehre,
Johannes Kepler University of Linz, A-4040, Linz, Austria
2. Institut für Allgemeine Mechanik, Technical University of Vienna,
A-1040, Vienna, Austria
* On leave from the State Technical University of St. Petersburg, Russia

Abstract - General modelling for complex engineering structures is proposed. In order to describe such intrinsic properties of complex structures as vibration localisation and parameter uncertainty, complex structures are modelled by random media with self-stresses, the role of the latter being played by the internal degrees of freedom of structural members. The rheological model is composed of an infinite number of elastic-plastic elements in parallel. The Dyson integral equation is applied to solve the problem of wave propagation in essentially heterogeneous random media. It is shown that the considerable spatial decay of the propagating wave is caused by (i) dispersion and scattering, (ii) resonant absorption in secondary systems attached to the primary structure and (iii) nonlinear material damping and dry friction between the structural members. The latter causes various nonlinear effects, e.g. the vibration saturation in complex structures. Structures are also shown to exhibit another nonlinear effect, namely, maximum distance which the wave propagates down the structure.

1. Introduction

Wave propagation in complex structures is addressed. A deterministic modelling is known to result in inadequate description of actual complex structures because of the existence of many inherently uncontrolled factors, i.e. uncertainties [1,2]. In structural dynamics, uncertainties about precise dynamic properties of structural components arise from stiffness, mass and damping fluctuations caused by variations in material properties as well as variations resulting from manufacturing and assemblage. In addition to these uncertainties, the essential heterogeneity and the complex interior have to be taken into account. The concept of random media [1] seems to be suitable to embrace essential heterogeneity, complexity and uncertainties of complex mechanical structures.

Any complex structure is actually an assemblage of structural members. The structural members at high frequencies are weakly coupled which causes the localisation of vibration within structural members [3-5]. On the other hand despite the localisation the disturbance propagates through the structures. The idea of the paper is to apply the modal analysis to describe the localised vibrations and wave approach to model the wave propagation through the structure. To this aim the mechanics of continuum with self-stresses is used. In this regard the paper is to be considered as combination of the modal

29

A. Naess and S. Krenk (eds.), IUTAM Symposium on Advances in Nonlinear Stochastic Mechanics, 29–38.
© *1996 Kluwer Academic Publishers.*

[6] and wave [7] approaches to Statistical Energy Analysis (SEA).

The paper is organised as follows. In Section 2 the boundary value problem for complex structures is derived in time and frequency domains. Non-linear state equation for elastic-plastic material is linearized which results in an amplitude-dependent complex Young's modulus. Since the conventional asymptotic methods are not suitable for analysis of essentially heterogeneous media the Dyson method is applied. The Dyson integral equation is solved in Section 3 under the assumption that the field of heterogeneity is Gaussian and statistically homogeneous. An asymptotic solution of the Dyson equation exhibits some interesting non-linear properties. In Section 4 the vibration saturation in complex structures is observed provided that the yield strength distribution function is taken in the form of the microplasticity distribution function. Local nonlinear vibration is analysed in Section 5.

2. Boundary value problem

Consider a typical substructure L_n of a chain-like structure. Since the vibration localises within the substructure its absolute displacement $u_n(x,t)$ is sought in the form of expansion in terms of the substructure's normal modes $v_{nk}(x)$

$$x \in L_n, \quad u_n(x,t) = U(x,t) + \sum_{k=1}^{\infty} v_{nk}(x) \, q_{nk}(t) \tag{1}$$

where $q_{nk}(t)$ are the generalised coordinates and $U(x,t)$ describes the travelling wave. Equation (1) may be understood in such a manner that the modal sum in the right hand side of eq. (1) acts as internal degrees of freedom with respect to the travelling wave. The normal modes are specified as the vibratory modes of the traction-free substructure.

Using eq. (1) the expressions for kinetic T and potential P energy are given by

$$T = \frac{1}{2} \sum_{n=1}^{N} \int_{L_n} \mu \dot{u}_n^2 \, dx = \frac{1}{2} \int_L \mu \dot{U}^2 dx + \sum_{n=1}^{N} \sum_{k=1}^{\infty} \dot{q}_{nk}^2 \tag{2}$$

$$P = \frac{1}{2} \sum_{n=1}^{N} \int_{L_n} C\left(u_n'\right)^2 dx = \frac{1}{2} \int_L C(U')^2 dx + \sum_{n=1}^{N} \sum_{k=1}^{\infty} \left[q_{nk} \int_{L_n} CU'v_{nk}' dx + \frac{1}{2} \Omega_{nk}^2 q_{nk}^2 \right] \tag{3}$$

Here $\mu(x)$ is the mass per unit of length and Ω_{nk} are the natural frequencies. $C(x)=EA$ is the axial rigidity where E is Young's modulus and A is the cross-sectional area. Distributed forces are assumed to be absent and the normal modes of each substructure are normalised. Equations (2) and (3) are approximate in the regard that the projection of the travelling wave velocity on the velocities of the localised modes is neglected. In other words, the travelling wave and the vibrations localised in structural members are assumed to exchange the energy via potential energy alone. In this sense the vibrations localised in each substructure are to be interpreted as the self-stresses with respect to the wave propagating in the primary structure.

The boundary value problem is derived by means of Hamilton's variational principle

$$x \in L_n, \quad \left(EAU'\right)' - \mu \sum_{k=1}^{\infty} \Omega_{nk}^2 v_{nk} q_{nk} - \mu \ddot{U} = 0 \; ; \quad n = 1,2,..N \tag{4}$$

$$\tilde{q}_{nk} + \Omega_{nk}^2 q_{nk} = -\int_{L_n} EA v'_{nk} U' \, dx \quad , \quad k = 1,2..\infty \qquad x=0, \quad EAU' = F_0 \qquad (5)$$

The end $x=0$ is assumed to be under the corresponding external forces F_0. Because of the vibration localisation in secondary systems the variables q_{nk} and U are considered as independent ones while deriving eqs (4) and (5).

The following modelling of the complex structures is therefore proposed. The complex structure is supposed to be a random medium with self-stresses. The rheological model for the complex structure is composed of an infinite number of Prandtl elements (the generalised Prandtl model), Fig. 1a. Each Prandtl element consists of an elastic element Edh in series with a dry friction damper which has a maximum allowable load $Ehdh$, where h is the dimensionless yield strength ($0<h<\infty$). The model is known to be a suitable one to model the intrinsic nonlinear properties of material damping caused by nonlinear stress-strain behaviour as well as to account for the mechanical energy dissipation due to dry friction between the structural members during their relative motions.

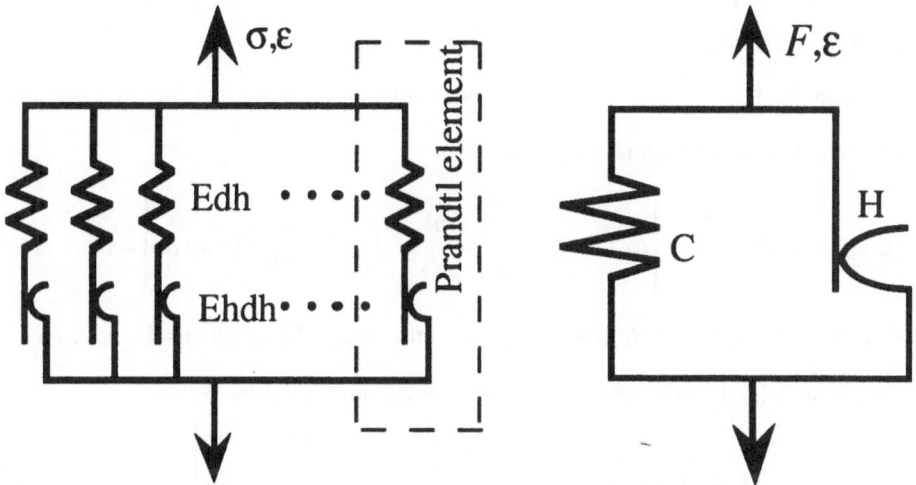

Fig. 1a Generalised Prandtl's rheological model

Fig. 1b Rheological model for non-anchored coupling

The summation over all Prandtl's elements gives the following state equation

$$\sigma = E\varepsilon - E\int_0^\infty \varepsilon_h R(h) dh \quad ; \quad \varepsilon = h \, \text{sign} \dot{\varepsilon}_h + \varepsilon_h \qquad (6a, b)$$

where σ and ε are the stress and the deformation in the material, $d\sigma_h$ and ε_h are the stress and the deformation of the element having yield strength h, respectively, $R(h)$ is the density of the yield strength distribution, $\int_0^\infty R(h) dh = 1$.

To apply the spectral decomposition

$$U(x,t) = \int_{-\infty}^{+\infty} \tilde{U}(x,\omega) \, e^{i\omega t} \, d\omega \; ; \; \varepsilon(x,t) = \int_{-\infty}^{+\infty} \tilde{\varepsilon}(x,\omega) \, e^{i\omega t} \, d\omega \; ; \; \varepsilon_h(x,t) = \int_{-\infty}^{+\infty} \tilde{\varepsilon}_h(x,\omega) \, e^{i\omega t} \, d\omega \qquad (7)$$

the nonlinearity sign $\dot{\tilde{\varepsilon}}_h$ in eq. (6b) is linearized by means of equivalent linearization,

i.e. sign $\dot{\tilde{\varepsilon}}_h = 4[\pi\omega a_h]^{-1}\dot{\tilde{\varepsilon}}_h$, where $a_h = |\tilde{\varepsilon}_h|$. Equation (6b) is then rewritten as

$$\tilde{\varepsilon}_h = \tilde{\varepsilon}\left[1 + i\, 4h/(\pi a_h)\right]^{-1} \; ; \quad a_h = \sqrt{a^2 - (4h/\pi)^2} \quad . \tag{8a, b}$$

where $a = |\tilde{\varepsilon}|$ is the amplitude of the total deformation. As follows from eq. (8b) $0 < h < \pi a/4$, otherwise $a_h = 0$. Substitution of eq. (8b) into eq. (8a) yields

$$\tilde{\varepsilon}_h = \left[1 - (4h/(\pi a))^2 - i\, 4h\,(\pi a)^{-1}\sqrt{1 - (4h/(\pi a))^2}\right]\tilde{\varepsilon} , \quad 0 < h < \pi a/4 \tag{9}$$

which finally results in the complex Young's modulus [8]

$$\tilde{\sigma} = \hat{E}\,\tilde{\varepsilon}; \quad \hat{E} = E\left[1 - \int_0^1 \left(1 - \eta^2 - i\eta\sqrt{1 - \eta^2}\right)R(\pi a\eta/4)\,\pi a\,d\eta/4\right] = E\,\psi\,(a) \tag{10}$$

where $\psi\,(a)$ is a frequency-independent complex function.

Expressing $\tilde{q}_{nk}(\omega)$ by means of eq. (5) and substituting it into eq. (4) one obtains the following integro-differential equation for $\tilde{U}$

$$x \in L_n, \quad \left(\hat{E}A\tilde{U}'\right)' + \mu\left[\omega^2\tilde{U} + \sum_{k=1}^{\infty}\frac{\hat{\Omega}_{nk}^2 v_{nk}}{-\omega^2 + \hat{\Omega}_{nk}^2}\int_{L_n}\hat{E}Av_{nk}'\tilde{U}'(x,\omega)\,dx\right] = 0 \tag{11}$$

wherein E and Ω_{nk} are replaced by $\hat{E}$ and $\hat{\Omega}_{nk} = \Omega_{nk}\sqrt{\psi\,(a)} = \Omega_{nk}(1 + i\chi)$, respectively, and $\chi(a)$ is the material damping. Integration by parts yields

$$x \in L_n, \quad \left(\hat{E}A\tilde{U}'\right)' + \mu\left[\omega^2\tilde{U} + \sum_{k=1}^{\infty}\frac{\hat{\Omega}_{nk}^4 v_{nk}}{-\omega^2 + \hat{\Omega}_{nk}^2}\int_{L_n}\mu v_{nk}\tilde{U}\,dx\right] = 0 \quad . \tag{12}$$

The system of the normal modes $v_{nk}(x)$ is complete in L_n, i.e. the function $\tilde{U}$ may be represented in the series expansion

$$\tilde{U}(x,\omega) = \sum_{k=1}^{\infty} v_{nk}(x)\beta_{nk}(\omega) , \quad \beta_{nk}(\omega) = \int_{L_n}\mu v_{nk}\tilde{U}\,dx \tag{13}$$

which permits the following form of eq. (12)

$$x \in L_n, \quad \left(\hat{E}A\tilde{U}'\right)' + \mu\,\omega^2 s(\omega)\,\tilde{U} = 0, \quad s(\omega) = 1 + \sum_{k=1}^{\infty}\frac{\hat{\Omega}_{nk}^2}{\omega^2}\frac{\hat{\Omega}_{nk}^2}{-\omega^2 + \hat{\Omega}_{nk}^2} \tag{14}$$

Parameter s reflects the inertial and spectral properties of the structure and is a superposition of single-degree-of-freedom resonance curves. The width of each resonance curve is $2\chi\Omega_{nk}$ at the half-power level. Provided that the resonant width is large compared to the natural frequency separation $\Delta\Omega_{nk}$ (the case of high modal overlap)

$$\Delta\Omega_{nk} = \Omega_{nk+1} - \Omega_{nk} \le \chi \left(\Omega_{nk+1} + \Omega_{nk}\right) \quad \text{or} \quad \Delta\Omega_{nk} / \Omega_{nk} \le 2\chi \tag{15}$$

the resonance curves in eq (14) merge to form a smooth frequency function. In this case the sum in eq (14) can be replaced by an integral over the high-frequency domain

$$s(\omega) = 1 + \omega^{-2} \int_{\Theta}^{\infty} \frac{\Phi(\Omega)d\Omega}{-\omega^2 + (1+i\chi)^2\Omega^2} = (1 - i\kappa(\omega))^2 \tag{16}$$

where $\Phi(\Omega)$ is a smooth function of the eigenfrequency distribution. The modal density of any mechanical system is known to increase with the growth of the ordinal number of the eigenfrequency [9]. Hence, there exists a critical frequency Θ, and for frequencies $\omega > \Theta$ the condition (15) and the replacement of the sum by the integral is valid. Obviously Θ is specific for each mechanical structure and it depends primarily on the relative density of eigenfrequencies and the damping value (cf. eq (15)). Equation (16) introduces a non-dimensional frequency-dependent absorption κ of vibration by the structure. In the case of light damping $\chi \ll 1$ and local smooth function $\Phi(\Omega)$ one can evaluate the integral in eq (16), [9]

$$\kappa(\omega) = \frac{1}{\omega^2} \int_{\Theta}^{\infty} \frac{\chi\Omega^2 \Phi(\Omega)\, d\Omega}{\left(\Omega^2 - \omega^2\right)^2 + 4\chi^2\Omega^4} \approx \frac{\pi}{2} \frac{\Phi(\omega)}{\omega^3} \tag{17}$$

Hence, the absorption of high-frequency vibration depends primarily on the distribution function $\Phi(\omega)$. It means that the vibration localised in the structural members acts as an absorber for the travelling wave. Since the resonance curves in eq. (14) merge, a considerable absorption of the energy of the travelling wave is observed.

3. Mean amplitude

To solve eq. (14) we make use of the Liouville substitution [10]. To this end we introduce a new independent variable y and a new dependent variable $\tilde{V}$

$$y = \langle c \rangle \int_0^x d\xi / c\,(\xi) \; ; \quad \tilde{V}(y) = \sqrt{\rho c}\; \tilde{U} \; ; \quad c = \sqrt{E/\rho} \tag{18}$$

where c is the velocity of sound. The medium is assumed to be statistically homogeneous, hence, the mean value $\langle c \rangle$ does not depend on x. Substituting (18) into (14) yields

$$\frac{d^2\tilde{V}}{dy^2} + \left[\lambda^2 + \varepsilon(y)\right]\tilde{V} = 0; \quad \lambda^2 = (\omega/\langle c \rangle)^2 s(\omega) + \langle e \rangle; \quad e\,(y) = -\frac{1}{\sqrt{\rho c}}\frac{d^2}{dy^2}\sqrt{\rho c} \tag{19}$$

where λ is a wave number, $e(y)$ is the heterogeneity and $\varepsilon(y) = e(y) - \langle e \rangle$ is a new centred random function. We model the complex structure by a random-heterogeneous medium with the following independent random parameters: the velocity of sound c and the impedance ρc. The following single differential equation

$$\frac{d^2\tilde{V}}{dy^2} + \left[\lambda^2 + \varepsilon(y)\right]\tilde{V} = \delta(y - y_0) \tag{20}$$

is equivalent to eq. (19) completed by the Neumann boundary condition at $y=y_0$.

In the case of slightly heterogeneous media ε is supposed to be much smaller than the first term in square brackets in eq. (20), thus, the WKB-method [10] or other conventional asymptotic methods can be applied. However in the case of essentially heterogeneous media the heterogeneity may be compared with or even may exceed the first component in the square brackets. This prevents the application of the formalism of the aforementioned asymptotic methods. The solution of eq. (20) will be found by use of the Dyson integral equation, that was originally applied in quantum mechanics' scattering theory and later was applied to the wave propagation in [11-13].

The intent of the study is to find the mean field of the amplitude of the propagating wave $\langle\tilde{V}(y)\rangle$. $\tilde{V}(y)$ is the solution of eq. (20) having a Dirac delta-function in the right hand side. Hence, $\tilde{V}(y) = G(y,y_0)$ where $G(y,y_0)$ is the Green function and $\langle\tilde{V}(y)\rangle = <G(y,y_0)>$ where $<G(y,y_0)>$ denotes the averaged Green's function. The function $<G(y,y_0)>$ is known to be the solution of the Dyson integral equation [11, 12]

$$\langle G(y,y_0)\rangle = G_0(y,y_0) + \iint G_0(y,r_1)M(r_1,r_2)\langle G(r_2,y_0)\rangle dr_1 dr_2 \ . \tag{21}$$

Here G_0 is Green's function or homogeneous medium and M is the kernel of an integral operator, which is called the quantum-mechanical mass operator in quantum field theory. This kernel contains an infinite number of terms and it takes on an especially simple form for a centred normal function ε, [11, 12]

$$M = \langle\varepsilon K\varepsilon\rangle - \left[\langle\varepsilon K\varepsilon K\varepsilon K\varepsilon\rangle - \langle\varepsilon K\varepsilon\rangle K\langle\varepsilon K\varepsilon\rangle\right] + \dots \tag{22}$$

where K is an integral operator with the kernel G_0.

Consider the case of the field of heterogeneity to be statistically homogeneous. The correlation function $B_\varepsilon(y_1,y_2)$ of the random function $\varepsilon(y)$ as well as G_0 and M depend only on the difference of arguments, so the integral equation with a difference kernel (21) can be solved by means of the spatial Fourier transform

$$G_0(r) = \int_{-\infty}^{+\infty} G_0(k)e^{ikr}dk; \quad \langle G(r)\rangle = \int_{-\infty}^{+\infty} \langle\tilde{G}(k)\rangle e^{ikr}dk; \quad M(r) = \int_{-\infty}^{+\infty} \tilde{M}(k)e^{ikr}dk \tag{23}$$

that results in the following formulae

$$\langle\tilde{G}(k)\rangle = \tilde{G}_0(k)\left[1 - 4\pi^2\tilde{G}_0(k)M(k)\right]^{-1}; \quad \langle G(r)\rangle = \int_{-\infty}^{+\infty} \tilde{G}_0(k)\left[1 - 4\pi^2\tilde{G}_0(k)M(k)\right]^{-1}e^{ikr}dk \tag{24}$$

Substituting $\tilde{G}_0(k) = 1/\left[2\pi(\lambda^2 - k^2)\right]$ into eq. (24) yields

$$\langle G(r)\rangle = \frac{1}{2\pi}\int_{-\infty}^{+\infty} \frac{e^{ikr}dk}{\lambda^2 - \int_{-\infty}^{+\infty} M(r)\,e^{-ikr}dr - k^2} \tag{25}$$

Further analysis is performed for a Gaussian function ε. Essentially heterogeneous media are conventionally modelled by random media with small-scale heterogeneity of considerable magnitude. We take the exponential correlation function of the heterogeneity ε, i.e. $B_\varepsilon(r) = \sigma^2 \exp(-\tau|r|)$, where $\sigma = \sqrt{\langle \varepsilon^2 \rangle}$ is the standard deviation of the heterogeneity (or its intensity), τ^{-1} is an integral radius of correlation (or the scale of heterogeneity). The latter is assumed to be small compared to the wave length, i.e. $|\lambda|/\tau \ll 1$, which permits to keep only the first term in the expansion for M, eq. (22), i.e. $M(r) = B_\varepsilon(r)G(r)$. Evaluating of the definite integral (25) by means of residue calculus renders

$$\langle G(r) \rangle = -\frac{i}{2} \left[\frac{k_1^{-1} e^{ik_1|r|}}{1 + \dfrac{\tau + i\lambda}{i\lambda} \dfrac{\sigma^2}{\left[(\tau + i\lambda)^2 + k_1^2\right]^2}} + \frac{k_2^{-1} e^{ik_2|r|}}{1 + \dfrac{\tau + i\lambda}{i\lambda} \dfrac{\sigma^2}{\left[(\tau + i\lambda)^2 + k_2^2\right]^2}} \right] \qquad (26)$$

where k_1 and k_2 are the poles of the integrand in (25) in the upper half-plane, namely

$$k_{1,2} = \sqrt{\lambda^2 - (\tau + i\lambda)^2 \pm \sqrt{\left(\lambda^2 + (\tau + i\lambda)^2\right)^2 + 4\frac{\tau + i\lambda}{i\lambda}\sigma^2}}$$

Since $|\lambda| \ll \tau$, asymptotically $k_1 = -\lambda + i\tau$, $k_2 = -\lambda\left[1 - i\sigma^2/(2\lambda^3\tau)\right]$ so the first term in eq. (26) is getting small even within the correlation radius. It represents the near field which can be neglected. Therefore the influence of heterogeneity is taken into account by k_2 in eq. (26). The imaginary part of k_2 determines the travelling wave's decay that depends upon the heterogeneity intensity σ and can be considerable for an essentially heterogeneous medium ($\sigma \gg 1$). The following equation for the mean field in the complex structure holds

$$\langle U(y) \rangle = -(2i\lambda_e)^{-1} \exp(-i\lambda_e|y|) ; \quad \lambda_e = \lambda(1 - i\xi) ; \quad \xi = \sigma^2/\left[2\lambda^3\tau\right] \qquad (27)$$

where λ_e is an effective wave number.

4. Deformation amplitude

Equation (27) does not yield the mean field since λ_e depends via the complex Young's modulus upon the deformation amplitude a and the latter, in its turn, is to be found through the displacement $\tilde{U}$. Therefore, at the most, eq. (27) may be considered as the equation which determines a. This equation is an integral one since the complex Young's modulus is a functional of a, cf. eq. (10). Obtaining $\langle a \rangle^2$ from eqs. (18) and (27) and taking the logarithmic derivative we obtain the integro-differential equation

$$\frac{1}{\langle a\rangle}\frac{d\langle a\rangle}{dx}=-\frac{\omega}{\langle c\rangle}\left[\xi+\kappa+\frac{1}{2}\int_0^1\eta\sqrt{1-\eta^2}\,R\!\left(\frac{\pi\langle a\rangle\eta}{4}\right)\frac{\pi\langle a\rangle}{4}\,d\eta\right] \qquad (28)$$

The distribution function $R(h)=\beta\tilde{R}h^{\beta-1}$ ($\tilde{R}>0$, $\beta>0$) is a standard one in the theory of internal material damping, cf. [8]. Evaluation of the integral in eq. (28) renders

$$\frac{1}{\langle a\rangle}\frac{d\langle a\rangle}{dx}=-\frac{\omega}{\langle c\rangle}[\xi+\kappa+\gamma\langle a\rangle^\beta]\;;\quad \gamma=\frac{\beta\tilde{R}}{2}\left(\frac{\pi}{4}\right)^\beta B\!\left(\frac{\beta+1}{2};\frac{3}{2}\right) \qquad (29)$$

where $B(\ ;\ )$ is Euler's beta-function. The solution of eq. (29) is

$$\langle a(x)\rangle=\left[\left(\langle a(0)\rangle^{-\beta}+\frac{\gamma}{\xi+\kappa}\right)\exp\left(\frac{\beta\alpha(\xi+\kappa)}{\langle c\rangle}x\right)-\frac{\gamma}{\xi+\kappa}\right]^{-1/\beta} \qquad (30)$$

The structure demonstrates the nonlinear properties, e.g. the following inequality holds

$$\langle a(x)\rangle<A(x)=\left\{\frac{\gamma}{\xi+\kappa}\left[\exp\left(\frac{\beta\alpha(\xi+\kappa)x}{\langle c\rangle}\right)-1\right]\right\}^{-1/\beta} \qquad (31)$$

i.e. the vibration saturation is observed. The latter formula indicates an upper limit of the deformation amplitude at any point of the structure, even if the power of the external excitation is unbounded. The upper limit or majorant $A(x)$ does not depend on the power of the source, but it is a function of the spatial distance from the source and the mechanical characteristics of the structure.

The mean deformation amplitude $\langle a(x)\rangle$, eq. (30), and the majorant $A(x)$ are presented in Fig. 2. The following parameters are taken: frequency 500 Hz, $<c>=2\cdot10^3$ m/s, $\xi+\kappa=0.2$. The yield strength h is assumed to be uniformly distributed $\beta=1$, hence the material damping factor is $\gamma\langle a(x)\rangle$. We take $\gamma=5$ that corresponds to the material damping factor equal to 0.05 while the deformation amplitude is $\langle a(x)\rangle=0.01$.

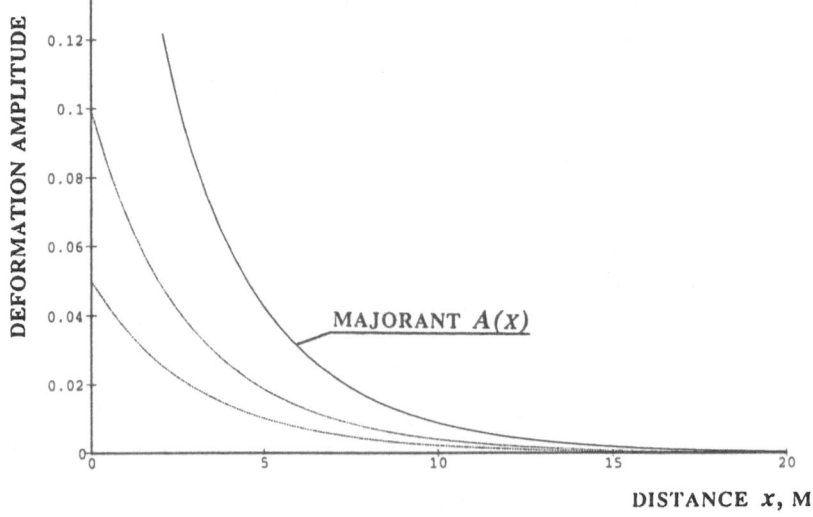

Fig. 2 Deformation field and its majorant

5. Local vibrations in complex structures

The formalism proposed allows us to analyse the local vibrations as well. As an example we consider wave propagation through non-anchored coupling. Local principle in structural dynamics, e.g. [14], allows one to describe various parts of structures by means of different approaches. Wave propagation along the primary structures may be described as above while the local effects of the wave transition through structural joints may be described by conventional methods of structural dynamics. Non-anchored coupling of structural members [15] is known to be an effective measure against the transition of intensive external loads (earthquake, aircraft-crash-induced waves) into the building interior. This type of coupling has the rheological model depicted in Fig. 1b and can be described by the following state equation

$$F = C\varepsilon + H \operatorname{sign} \dot{\varepsilon} \, , \tag{32}$$

where F is the axial force. Equation (32) may be linearized which results in the following dynamic equation for the deformation

$$\frac{\partial^2}{\partial x^2}\left(C\varepsilon + \frac{H_*}{\omega}\frac{\partial\varepsilon}{\partial t}\right) - \mu\frac{\partial^2\varepsilon}{\partial t^2} = 0; \quad H_* = \frac{4}{\pi}H \tag{33}$$

The solution of this equation is given by, e.g. [8]

$$\varepsilon(x,t) = \begin{cases} (\varepsilon_0 - \varepsilon_1 x)\cos(\omega t - \lambda x - \varphi), & 0 < x < x_{max} \\ 0 & , \quad x \geq x_{max} \end{cases} \tag{34}$$

where

$$\varepsilon_0 = \frac{1}{C}\sqrt{Z^2 - H_*^2} \; ; \; \varepsilon_1 = \frac{\lambda H_*}{2C} \; ; \; \lambda = \frac{\omega}{c} \; ; \; x_{max} = \frac{2}{\lambda}\sqrt{(Z/H_*)^2 - 1} \; ; \; \tan\varphi = \frac{Z}{C\varepsilon_0} \tag{35}$$

Here Z is the magnitude of the force applied at the support in question, its value may be obtained using the previous formalism. As follows from eq. (35) there exists a maximum distance x_{max} which the wave propagates. Wave motion is arrested when the maximum distance is exceeded. Such a result becomes useful for the estimation of the size of non-anchored supports in power plants to prevent the spread of waves to the interior sensitive components. Deformation magnitude is plotted in Fig. 3 for the following dimensionless parameters: $\varepsilon_0 = 0.03, \varepsilon_1 \lambda = 0.004$.

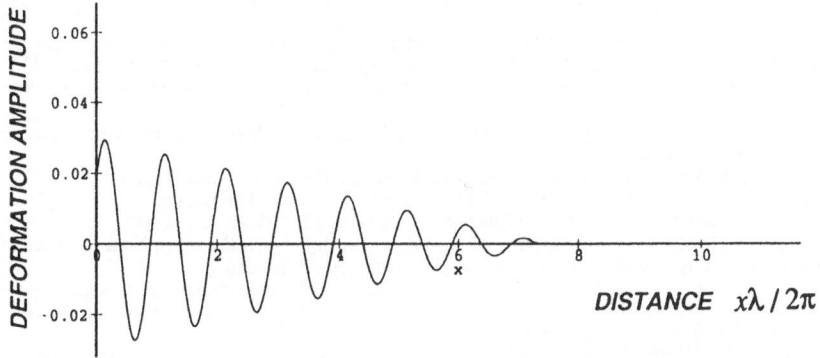

Fig 3. Wave propagation through non-anchored coupling of structural members

6. Conclusions

General modelling of complex engineering structures at high frequencies is proposed and propagation of uniaxial waves in chain-like complex structures is considered in some detail. The complex structure is modelled by a random medium with self-stresses. Modal analysis is used to describe the vibration that localises in structural members while the wave approach is applied to model the wave propagation through the structure. The rheological model of the medium is composed of an infinite number of elastic-plastic elements. The model allows one to describe such intrinsic properties of complex structures as parameter uncertainty, essential heterogeneity, nonlinear energy dissipation and resonant absorption at high frequencies.

Considerable spatial decay of the propagating wave was observed and the main causes for the decay are indicated. Firstly, essential heterogeneity causes dispersion and scattering. Secondly, the vibrations localised in structural members act as dynamic absorbers with respect to the travelling wave which causes resonant absorption of mechanical energy. Finally, internal material damping and dry friction between the structural members cause various nonlinear effects that depend on the function of the yield strength distribution. For example, the vibration saturation is observed for a distribution function of dynamic microplasticity. One of the merits of the present approach is the possibility to combine it with conventional methods of the vibration theory. It is demonstrated for a particular problem of the vibration transition through a non-anchored coupling of structural members. Another nonlinear effect, namely, the maximum distance which the wave propagates was revealed.

Although not addressed here, a study of the random local vibrations in structural members of complex structures can also be developed based on this concept.

Acknowledgement - The study was financially supported by the Austrian Scientific Research Promotion Fund (FWF) under the contract P09533-TEC.

REFERENCES

1. Ibrahim, R.A. (1987) Structural dynamics with parameter uncertainties, *Applied Mechanics Review* **40**, 309-328 .
2. Fahy, F.J. (1994) Statistical energy analysis: a critical overview, *Philosophical Transactions of the Royal Society London A* **346**, 431-447.
3. Pierre, C. (1990) Weak and strong vibration localization in disordered structures: a statistical investigation, *Journal of Sound and Vibration* **139**, 111-132.
4. Li, D. and Benaroya, H. (1992) Dynamics of periodic and near-periodic structures, *Applied Mechanics Review* **45**, 447-459.
5. Xie, W.-C. and Ariaratnam, S.T. (1994) Numerical computation of wave localization in large disordered beam-like lattice trusses, *American Institute of Aeronautics and Astronautics Journal* **32**, 1724-1732.
6. Lyon, R. H. (1975) *Statistical Energy Analysis of Dynamical Systems: Theory and Applications*, MIT Press, Cambridge, MA.
7. Langley, R.S. (1992) A wave intensity technique for the analysis of high frequency vibrations, *Journal of Sound and Vibration* **159**, 483-502.
8. Palmov, V.A. (1976) *Vibrations of Elastic-Plastic Bodies* (in Russian), Nauka, Moscow .
9. Bolotin, V.V. (1984) *Random Vibrations in Elastic Systems*, Nijhoff, The Hague.
10. Heading, J. (1962) *An Introduction to Phase-Integral Methods*, Wiley, New-York.
11. Tatarsky, V.I. (1969) *Wave Propagation in Turbulent Atmosphere* (in Russian), Nauka, Moscow.
12. Sobczyk, K. (1985) *Stochastic Wave Propagation*, Elsevier, Amsterdam.
13. Belyaev, A.K. and Ziegler, F. (1995) Effective loss-factor of heterogeneous elastic solids and fluids, in J.F. Rodrigues and M.M. Marques (eds.) *Proceedings of 9th Symposium on Trends in Application of Mathematics to Mechanics*, Longman Scientific and Technical.
14. Belyaev, A.K. (1990) On the application of the locality principle in structural dynamics, *Acta Mechanica* **83**, 213-222.
15. Krutzik, N.J. (1988) Reduction of the dynamic response by aircraft crash on building structures, *Nuclear Engineering and Design* **110**, 191-200.

STOCHASTIC LINEARIZATION AND LARGE DEVIATIONS

P. BERNARD
*Laboratoire de Mathématiques Appliquées. URA CNRS 1501,
Université Blaise Pascal - Clermont Ferrand.
63177,Aubière Cedex. France*

Abstract. Stochastic equivalent linearization is the most popular approximation method for the dynamic of a non-linear system under random excitation. A complete presentation of this method can be found in [4]. Despite the fact it was introduced 40 years ago, the first justification was proposed by F.Kozin [3] in 1987. Another approach was recently introduced by the author in collaboration with L. Wu [2], based on the use of a large deviation principle. The goal of this contribution is to present this approach to the stochastic dynamic engineering public.

1. Introduction

Modelling numerous problems in random vibrations leads to the dynamic of a nonlinear oscillator driven by a White Noise, which is described by the second order stochastic differential equation:

$$\ddot{x}(t) + f(x(t), \dot{x}(t)) = \sigma \dot{w}, \tag{1}$$

where $\dot{w}$ denotes a normalized Gaussian White Noise.
In the phase space, equation (1) takes the form of the Itô stochastic differential system:

$$\begin{cases} dx_t = y_t dt, \\ dy_t = \sigma dw_t - f(x_t, y_t)dt. \end{cases} \tag{2}$$

We are concerned by the weak solutions of this system, and more precisely by the stationary solution, for which we assume existence and uniqueness. To state more precisely what we mean, let us introduce some notations. Let $\Omega = C(\mathbb{R}, \mathbb{R}^2)$ be the space of continuous functions on $\mathbb{R}$ with

39

A. Naess and S. Krenk (eds.), IUTAM Symposium on Advances in Nonlinear Stochastic Mechanics, 39–46.
© *1996 Kluwer Academic Publishers.*

values in $\mathbb{R}^2$, endowed with the topology of uniform convergence on compact subsets, for which Ω is a Polish Space (metrizable, separable and complete). By a process, we mean a probability measure P on the Borel σ-field of Ω. If P is a stationary process, that is P is invariant by translations, m_P denotes the marginal of order one of P, that is, m_P is the probability distribution of $(x(t), y(t))$ for fixed t; it is also the invariant probability measure of the Markov process defined by (2).

As an illustration, we will consider the example of the generalized Duffing oscillator, characterized by $f(x, y) = g(x) + cy$, where g is positive, skewsymmetric, with polynomial growth. The primitive G of g such that $G(0) = 0$ is the elastic potential of the vibrating structure. Existence and uniqueness of a stationary solution of (2) can be proved, and an explicit formulation of density ϕ of the invariant probability measure is:

$$\phi(x, y) = K exp(-2\frac{c}{\sigma^2}H(x, y)), \tag{3}$$

where $H(x, y) = G(x) + \frac{1}{2}y^2$ is the Hamiltonian of the associated conservative system.

The explicit expression of the transition probability is unknown for this example; in fact, it is only known in the linear case.

In spite of its apparent simplicity, and excepted for the linear case, only very little is known concerning the solution of (2). That is why the method of Statistical Linearization was introduced . The idea is to look for a linear oscillator whose displacement, when driven by a White Noise isonomic to the one driving (1), could be used as an approximation of the stationary displacement of (1). Let us recall briefly for convenience the process (for a complete account, see the book [4]):

Equation (1) is rewritten as:

$$\ddot{x}(t) + kx(t) + c\dot{x}(t) + [f(x(t), \dot{x}(t)) - (kx(t) + c\dot{x}(t))] = \sigma\dot{w}. \tag{4}$$

One then computes the positive constants $\bar{k}$ and $\bar{c}$ minimizing the expectation with respect to the probability measure m_P of the square of the term between brackets. Hence:

$$(\bar{c}, \bar{k}) = argmin\{\int(f(x, y) - (kx + cy))^2 m_P(dx, dy); c > 0, k > 0\}. \tag{5}$$

This problem is convex in the variables (c, k), and, as $\int xym_P(dx, dy) = 0$, the necessary conditions of optimality are:

$$\begin{cases} \bar{c} = \int yf(x, y)m_P(dx, dy)/ \int y^2 m_P(dx, dy), \\ \bar{k} = \int xf(x, y)m_P(dx, dy)/ \int x^2 m_P(dx, dy). \end{cases} \tag{6}$$

The solution obtained this way is called "true linearization"[3]. As the probability measure m_P is unknown in most cases, the calculation is made using the marginal m_Q of the stationary solution of the linear system with coefficients $\bar{c}$ and $\bar{k}$. This is a fixed point problem, which solution (whenever there is existence and unicity) will define the equivalent linear oscilltor. In this case, we speak of "Gaussian Linearization". The resulting SDE is:

$$\begin{cases} dx_t = y_t dt, \\ dy_t = \sigma dw_t - (\bar{k}x_t + \bar{c}y_t)dt. \end{cases} \quad (7)$$

In this formulation, there is a lack of justification: in particular, the link existing between the solution P of the initial problem and the solution $Q^{\bar{c},\bar{k}}$ of the linearized problem is unclear. Let us remark that the quantity to minimize is an expectation with respect to m_P, which is unknown. If we minimize the expectation of the same quantity with respect to the invariant probability $m_{Q(c,k)}$ of the linear approximation, this probability depends on the coefficients c and k, and hence the necessary conditions of optimality will not be the same: one should also consider the derivatives of the measure $m_{Q(c,k)}$ with respect to c and k in the calculation of first order conditions.

$$(\bar{c},\bar{k}) = argmin\{\int (f(x,y) - (kx + cy))^2 m_Q^{(c,k)}(dx, dy); c > 0, k > 0\}. \quad (8)$$

This problem is different from both the true and the Gaussian linearization. We will come back to this question later.

The goal of this contribution is to introduce linearization methods based on the concept of Entropy, which are very near to the practical methods described just before, and mathematically justified.

2. Hypotheses and notations

In the continuation, we assume the following:
$H1$) There is existence and unicity of a strong solution of (2), non-explosion of the solution, and existence of a stationary solution .
$H2$) $f(x, y)$ is skewsymmetric on $\mathbb{R}^2$.
$H3$) $xf(x, y) \geq 0$ and $yf(x, y) \geq 0$ for all $(x, y) \in \mathbb{R}^2$.
$H4$) The following integrals exist with respect to every centered Gaussian probability measure ν on $\mathbb{R}^2$ with diagonal covariance, and exist with respect to the invariant probability measure of (2):

$$\int f^k(x, y)\nu(dx, dy), \quad k = 1, 2; \quad \int x^2 \nu(dx, dy), \quad \int y^2 \nu(dx, dy). \quad (9)$$

2.1. RELATIVE ENTROPY

Let Ω be a Polish space, Σ its Borel σ-field. $B(\Sigma)$ denotes the space of real valued Σ-measurable bounded functions on Ω endowed with the topology of uniform convergence, $C_b(\Omega)$ the subspace of bounded continuous functions on Ω. If λ and μ are two probability measures on (Ω, Σ), the relative entropy of λ with respect to μ is defined as follows:

$$h(\mu; \lambda) = sup\{\int \phi(x)\lambda(dx) - Log(\int exp(\phi(x))\mu(dx)); \ \phi \in B(\Sigma)\}. \quad (10)$$

Lemma 1 $h(\mu; \lambda) < \infty$ if and only if $\lambda \lll \mu$ and $f = \frac{d\lambda}{d\mu}$ is such that $f Log f$ is μ-integrable. Under these circumstances, $h(\mu; \lambda) = \int f(x) Log f(x)\mu(dx)$.

Let λ, μ be two probability measures on (Ω, Σ) and Σ_0 be a subfield of Σ. $h_{\Sigma_0}(\mu; \lambda)$ will be defined using the restrictions of λ, μ to Σ_0. If $\Sigma_1 \subset \Sigma_2$, then $h_{\Sigma_1}(\mu; \lambda) \leq h_{\Sigma_2}(\mu; \lambda)$. Moreover, the difference can be interpreted as an entropy; if λ_ω and μ_ω are versions of the regular conditional probability distribution of λ and μ given Σ_1, then:

$$h_{\Sigma_2}(\mu; \lambda) = h_{\Sigma_1}(\mu; \lambda) + E_\lambda[h_{\Sigma_2}(\mu_\omega; \lambda_\omega)]. \quad (11)$$

2.2. DONSKER-VARADHAN ENTROPY OF A STATIONARY PROCESS WITH RESPECT TO A MARKOV PROCESS

Let E be a Polish space. Ω now denotes the space of all continuous functions on $\mathbb{R}$ with values in E, endowed with the topology of uniform convergence on compact subsets. Let Ω_t^+ be the quotient of Ω which elements are functions on $[t, \infty)$ with values in E, and $\mathcal{F}_t^s$ the σ-field on Ω generated by $\{\omega(u) : s \leq u \leq t\}$. Ω_t^+ can be identified with Ω endowed with the σ-field $\mathcal{F}_\infty^t$. (θ_t) is the group of shifts on Ω: $\theta_t\omega(s) = \omega(t+s)$.

Let $(P_x)_{x \in E}$ be the family of probability distributions on Ω_0^+ associated to some Markov process. Let Q be a stationary probability distribution on Ω, $Q_{0,\omega}$ being the family of regular conditional distributions of Q given $\mathcal{F}_0^{-\infty}$. The following result is due to Varadhan ([5]):

Theorem 1 Define $H(t, Q) = E_Q\{h_{\mathcal{F}_t^0}(P_{0,\omega(0)}; Q_{0,\omega})\}$, for all $t > 0$. Then, either $H(t, Q) \equiv \infty$ for all $t > 0$, or there exists a constant $H(Q) < \infty$ such that $H(t, Q) = t H(Q)$ for all $t > 0$. $H(Q)$ (also denoted $H(Q; P)$) is the Donsker-Varadhan entropy of Q relatively to $(P_x)_{x \in E}$.

2.3. EQUIVALENCE OF DIFFUSIONS

The following result is a straightforward application of Girsanov theorem.

Lemma 2 *Let Z_t be a diffusion process with values in $\mathbb{R}^2$, solution of*

$$dZ_t = b(Z_t)dt + \Sigma dW_t, \tag{12}$$

where $\Sigma = \begin{bmatrix} 0 \\ \sigma \end{bmatrix}$, with initial probability distribution ν. Let us assume existence, unicity and non explosion of a strong solution of (12). If X_t is the solution of (2) with initial probability distribution μ, then the probability distributions of X and Z are equivalent if and only if μ and ν are equivalent and $b_1(Z_1, Z_2) = Z_2$; the density $\frac{dP_Z}{dP_X}(t, X)$ and the relative entropy $h_{\mathcal{F}_1^0}(P_X; P_Z)$ are then given by:

$$\frac{dP_Z}{dP_X}(t, X) = \frac{d\nu}{d\mu}(X_0)exp\{\frac{1}{\sigma}\int_0^t (b_2 + f)(X_s)dW_s - \frac{1}{2\sigma^2}\int_0^t (b_2 + f)^2(X_t)dt\},$$

$$h_{\mathcal{F}_1^0}(P_X; P_Z) = \frac{1}{2\sigma^2}E^{P_Z}(\int_0^1 (b_2 + f)^2(X_t)dt) + h(\mu; \nu).$$

3. Large Deviations and Linearization

The idea is to construct a family of empirical processes with samples of the Markov process $(P_{0,x})$. These processes will be obtained by time averaging trajectories of $(P_{0,x})$, and asymptotically converge to a stationary process. (Other solutions can be obtained using sample averaging [2]). We then study the large deviations for this convergence. The knowledge of the rate function, which is related to entropy, allows one to determine the stationary Gaussian process having the "highest probability" to be realized by the empirical process. It can be proved that this Gaussian process is a Markov process, and is solution of a SDE (7).

3.1. FIRST METHOD

Let us define a family of empirical processes on Ω obtained from the Markov process $(P_{0,x})$ by averaging in time, on the interval $[0, t]$, trajectories issued from x. That is to say, for every $t > 0$ and $\omega \in \Omega$, a process $R_{t,\omega}$ is defined on Ω by $R_{t,\omega}(A) = \frac{1}{t}\int_0^t 1_A(\theta_s\omega)ds$ for all $A \in \mathcal{F}_\infty^0$. For every $t > 0$, $\omega \mapsto R_{t,\omega}$ is a mapping on Ω with values in $\mathcal{M}_1(\Omega_0^+)$, the convex set of probability measures on $\mathcal{F}_\infty^0$, which is $\mathcal{F}_\infty^0$-measurable. For every $t > 0$ and $x \in E$, this mapping induces a probability measure $\Gamma_{t,x}$ on $\mathcal{M}_1(\Omega_0^+)$ by $\Gamma_{t,x}(B) = P_{0,x}\{\omega \in \Omega : R_{t,\omega} \in B\}$ for all mesurable $B \subset \mathcal{M}_1(\Omega_0^+)$. If the Markov process $(P_{0,x})$ is ergodic with invariant probability measure μ on E, and if $P_\mu \in \mathcal{M}_S(\Omega)$, space of all stationary probability measures on Ω, is the stationary Markov process with marginal μ, then, by the ergodic theorem, $\Gamma_{t,x} \Rightarrow \delta_{P_\mu}$ for all x when $t \to \infty$. Varadhan ([5]) showed that the weak principle of large deviations with rate function $H(Q)$ is available for this problem. This means that, for any stationary probability Q, if $N(Q)$ is a closed convex neighbourhood of Q in $\mathcal{M}_S(\Omega)$, such that: $\{Q'_0 = Q'(\omega(0) \in \cdot); Q' \in N(Q)\}$ is compact in $\mathcal{M}_1(\Omega)$, and $inf\{H(Q'); \ Q' \in N(Q)\} \geq H(Q) - \epsilon$ then, as $t \to \infty$,

$$exp[-tH(Q) + o(t)] \leq P_\mu(R_t \in N(Q)) \leq exp[-t(H(Q) - \epsilon) + o(t)]. \quad (13)$$

This relation can be interpreted as follows: the Gaussian stationary centered probability distribution $\bar{Q}$ which has the highest probability to be realized by the empirical measure $R_{t,\omega}$ is a solution of the variational problem

$$H(\bar{Q}) = inf\{H(Q) : Q \text{ Gaussian stationary centered}\}. \quad (14)$$

For any Gaussian probability measure ν on E, let us define:

$$S_G(\nu) = \{Q \in \mathcal{M}_S : Q \text{ Gaussian centered}, m_Q = \nu\},$$
$$J_G(\nu) = inf\{H(Q); Q \in S_G(\nu)\}; (+\infty \text{ if } S_G \text{ is empty}). \quad (15)$$

The following result is fundamental:

Theorem 2 *Assume H1) to H4). Then:*
1) $J_G(\nu) < \infty$ if and only if the Gaussian probability measure ν has a diagonal variance matrix
$\begin{bmatrix} \sigma^2/2kc & 0 \\ 0 & \sigma^2/2c \end{bmatrix}$. *We note $\nu = \nu^{k,c}$, and, concerning this probability $\nu^{k,c}$, the variational problem (15) has one and only one solution: $Q^{k,c}$, the solution of (7), and*

$$H(Q^{k,c}) = \frac{1}{2\sigma^2} \int [f(x,y) - (kx + cy)]^2 \nu^{k,c}(dx, dy) = E_1(k,c). \quad (16)$$

2) The variational problem (14) has a solution $\bar{Q}$ if and only if the optimisation problem $inf\{E_1(k,c) : k > 0, c > 0\}$ has a solution; the same equivalence is available for unicity.
3) The solutions $Q^{\bar{k},\bar{c}}$ obtained this way are also values for which $H(Q)$ reaches its minimum in the class of all stationary Gaussian probability measures on Ω, that is to say are solutions of the variational problem (14).

A proof of this result can be found in [2].

3.2. SECOND METHOD

It is the same that the first one, excepted that the roles of P and Q are inversed. This is possible because Q is also a Markov process, and hence one can consider the entropy of the stationary process P_μ with respect to the Markov process $Q^{k,c}$. The Gaussian Markov process $Q^{k,c}$ being given by (7), let us compute (k,c) such that $Q^{k,c}(R_t \in N(P_\mu))$ is asymptotically maximum as $t \to \infty$, where $N(P_\mu)$ is a small neighbourhood of P_μ in $\mathcal{M}_1(\Omega)$.

The large deviations theory asserts that:

$$Q^{k,c}(R_t \in N(P_\mu)) = exp\{-t[H(P_\mu; Q^{k,l}) \pm \epsilon] + o(t)\},$$

hence the variational problem is to minimise $H(P_\mu; Q^{k,c})$ in $k > 0, l > 0$. One obtains:

$$H(P_\mu; Q^{k,c}) = \frac{1}{2\sigma^2} \int [f(x,y) - (kx + cy)]^2 \mu(dx, dy) = E_2(k,c). \quad (17)$$

Let us remark that this result is exactly the "true" linearization. There is existence and unicity of the solution $(\bar{k}, \bar{c})$, which is given by (6).

4. The example of Duffing oscillator

As an example, let us consider the very classical Duffing oscillator, characterized by: $f(x,y) = c_0 y + k_0(x + \lambda x^3)$, with $c_0, k_0, \lambda > 0$. Hypothese H1) to H4) are satisfied. Moreover, in this case, the density of the invariant probability measure is known:

$$\phi(x,y) = K \, exp[-2\frac{c_0}{\sigma^2}(k_0(\frac{x^2}{2} + \lambda\frac{x^4}{4}) + \frac{y^2}{2})] = \phi_1(x)\phi_2(y).$$

The unique solution of the "true" linearization problem is given by:

$$\bar{c} = c_0; \quad \bar{k} = k_0(1 + \lambda\frac{E(X^4)}{E(X^2)}). \quad (18)$$

The mean is here relative to the invariant probability measure of the displacemant x, that is with respect to ϕ_1. Concerning the Gaussian linearization, the solution is:

$$\hat{c} = c_0; \quad \hat{k} = k_0(1 + 3\lambda\hat{E}(X^2)), \quad (19)$$

where the mean $\hat{E}$ is relative to the invariant probability measure of the equivalent linear system.

The results of the three linearizations are compared through the Power Spectral Densities in *Figure 1*. [1], where the values of the parameters in the model are as follows:

$$c_0 = 0.05; \ k_0 = 1; \lambda = 10; (\sigma)^2 = 0.1 \ .$$

The computed values of the parameters of the linearized oscillator are, respectively for the true linearization, for the Gaussian linearization, and for the new linearization:

$$k = 5.2937; \ k = 6; \ k = 3.1084;$$

$$c = 0.05; \ c = 0.05; \ c = 0.2317 \ .$$

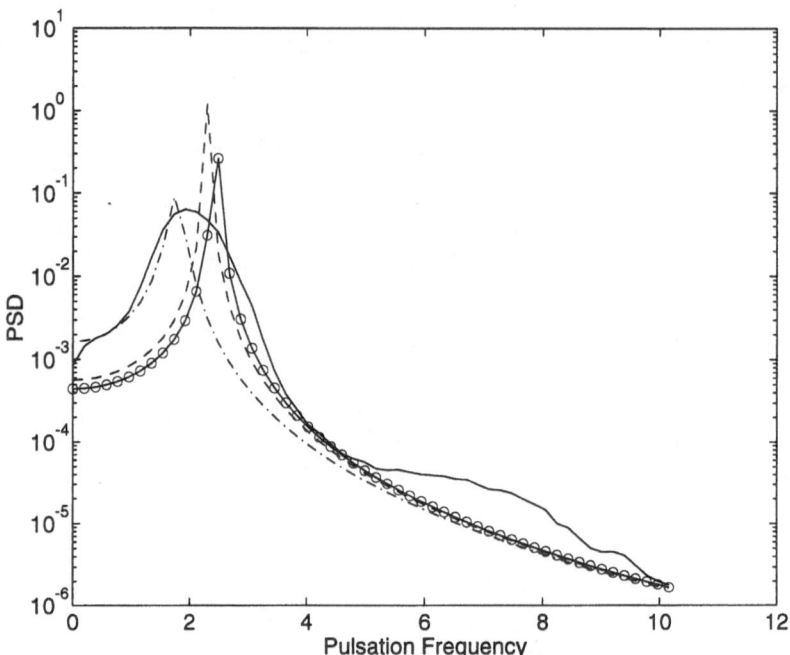

Figure 1. Duffing oscillator.
Simulation(———), true linearization(— — ——),
Gaussian linearization(— ○ — ○ —), new linearization(— · — · —)

References

1. Alaoui Ismaili M. (1995) PHD thesis, Université Blaise Pascal.
2. Bernard P., Wu L.(1995) Linéarisation d'un oscillateur excité par un bruit blanc: un point de vue entropique, Report of the Laboratoire de Mathématiques Appliquées, Université Blaise Pascal.
3. Kozin F.(1988) The Method of Statistical Linearization for Non-linear Stochastic Vibrations, in *Nonlinear Stochastic Dynamic Engineering Systems*, F.Ziegler, G.I.Schueller Editors, Springer Verlag.
4. Roberts J.B., Spanos P.D.(1990) Random Vibration and Statistical Linearization, J.Wiley & Sons.
5. S.R.S.Varadhan S.R.S.(1984) Large Deviations and Applications, SIAM Publications 46.

MEANDERING PROPAGATION OF FATIGUE CRACKS THROUGH SOLIDS

WITH RANDOMLY DISTRIBUTED PROPERTIES

V.V.BOLOTIN
Russian Academy of Sciences, Institute of Mechanical Engineering,
101830, Moscow, Centre, Russia

Abstract. A problem is discussed of fatigue crack propagation in solids with mechanical properties considered as a random field. Local kinking and branching of cracks are included into the study to estimate the influence of random meandering on the distribution of the averaged crack growth rate and the total fatigue life. Parametric analysis is performed to compare the contribution of various factors into the meandering propagation of fatigue cracks.

1. Introduction

Kinking and branching of fatigue cracks is a well-known experimental fact. Cracks change their trajectories under the influence of different factors: nonstationarities of the applied stress-state field, various material nonhomogeneities (both bulk and local), instability effects accompaning dynamic crack propagation, etc. Kinking of cracks in quasi-static monotonous loading is studied extensively in linear fracture mechanics where a number of criteria are proposed to predict crack trajectories under mixed mode loading. To estimate fatigue crack growth rate, the well-known semi-empirical equations are widely used such as the Paris-Erdogan equation with the substitution, instead of the mode I stress intensity factor range, the range of a certain effective loading parameter. The recent survey of literature can be found in the paper by Ramulu and Kobayashi (1994) and Pook (1994).

Random fatigue phenomena are discussed in literature rather widely. Most studies are based on the randomization of the known semi-empirical equations. For example, the Paris-Erdogan equation frequently is used with the replacement of its right-hand side with a properly chosen random function of the crack size, or the applied stress,

47

A. Naess and S. Krenk (eds.), IUTAM Symposium on Advances in Nonlinear Stochastic Mechanics, 47–58.
© 1996 *Kluwer Academic Publishers.*

or both. Another approach is based on the use of appropriate probabilistic models suited to describe any irreversible process, and among them the fatigue crack propagation. These models are discussed in detail by Sobczyk and Spencer (1992).

A theory of fatigue crack growth based on the synthesis of fracture mechanics and continuum damage mechanics is proposed by Bolotin (1983). More extended presentation of this proposal with the application to various fatigue phenomena can be found in the book and survey papers by Bolotin (1989, 1990, 1994). The application of the theory to fatigue cracks subjected to kinking and branching is discussed in the paper by Bolotin (1985). However, the pure deterministic case is discussed in that paper. The planar crack propagation in a body with randomly distributed mechanical properties is considered, in the framework of the discussed theory, in the paper by Bolotin (1993).

In this paper we consider the fatigue crack growth with account of randomly distributed local nonhomogeneities. Such nonhomogeneities are inherent to any polycrystalline, polymer or artificial composite material. The scale of nonhomogeneities varies in a wide scale between the orders of magnitude of the grain size and the plastic process zone. In any case, the scales of nonhomogeneities and their correlation scale may be assumed small compared with the cross section size of a specimen or a structural component. It means that mechanical properties within a body form a random field with small characteristic lengths. In principle, all material parameters vary from grain to grain including their complience characteristics such as Young modulus, yield stress, etc. However, concerning fatigue phenomenona, the resistance of material to the dispersed damage accumulation and quasi-brittle fracture are of primary significance. Hence, we assume all material properties but the mentioned above resistance parameters as deterministic. Even under this assumption, the total number of parameters treated as random fields is still large. In particular, we must discriminate fracture toughness against crack growth in various modes as well as different resistance of the material to microcracking under tensile and shear loading.

2. Kinking and branching of fatigue cracks

Most studies in crack kinking and branching concern the first small deviations from the planar shape of a crack. A number of papers are published, where small parameter and related techniques are used to evaluate stress intensity factors and strain energy release rates for cracks with small kinks. Difficulties arise when we approach to the second, third, etc. kinkings, moreover when a crack propagates in a meandering way. In principle, the techniques of computational mechanics allow to calculate stress intensity factors for cracks and crack-like defects of arbitrary shape. But fatigue problems are more complicated due to the necessity to study the stress-strain field in a wide domain ahead of the crack to evaluate the damage produced by cyclic loading. To avoid computational difficulties which could be overcome only, with the use of powerful computers we replace the actual meandering crack (Figure 1,a) with a shallow elliptical crack which tips coincide with the tips of the actual crack (Figure 1,b). In this case we can consider the crack propagation through a body with randomly distributed properties as a sequence of first small kinkings and branchings.

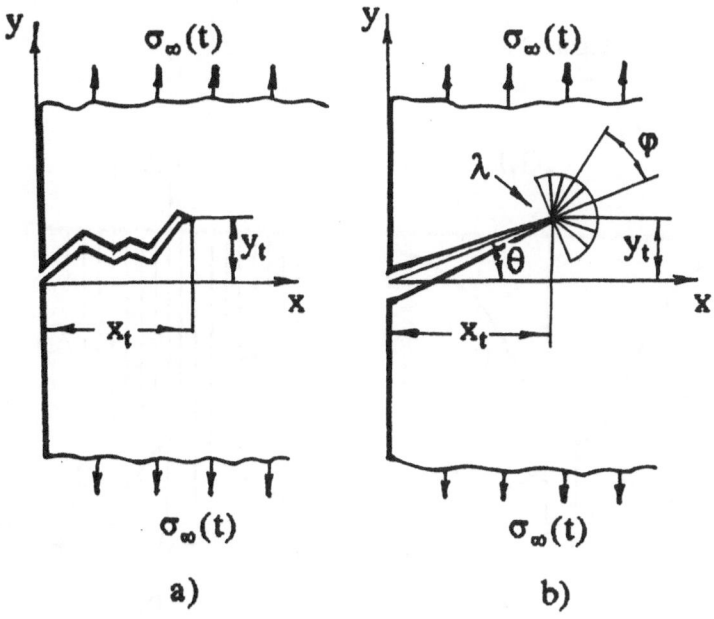

Figure 1. Meandering fatigue crack: a - real crack, b - its schematization.

Evidently, it is coarse approximation to actual fatigue cracks, especially from the viewpoint of the stress distribution in the vicinity of crack tips. To get a better approximation one ought to replace the actual meandering crack with a Z-shape, zig-zag crack, etc. But there are no reliable data on the stress distribution around such cracks. On the other hand, the plane crack with small kinks is acceptable on the stage of growth of a primarily plane crack as well as on the later stage when a crack propagates with small deviations from the planar shape.

Consider an inclined crack in the plane stress state, e.g. an edge crack shown in Figure 1,b. Let the applied stress $\sigma_\infty(t)$ vary in time cyclically. The maximal stress level is sufficiently low to provide the inequality $G < \Gamma_0$. Here G is the generalized driving force corresponding to maximal applied stress, and Γ_0 is the generalized resistance force for the nondamaged material. At a certain cycle number N_* the equality $G = \Gamma$ is primarily attained due to the damage accumulated near the crack tip. Denote the initial crack length a, the initial angle of crack inclination θ, and the angle of first kinking φ(Figure 1,b). The angle φ can be found from condition

$$\min_{\varphi,N} \{\Gamma(a,\theta,\varphi,N) - G(a,\theta,\varphi,N)\} = 0 \qquad (2.1)$$

which is illustrated in Figure 2,a.

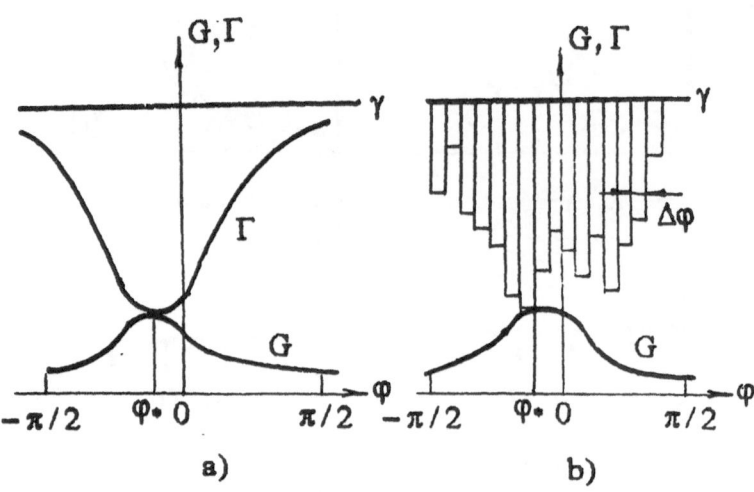

Figure 2. Relationship between driving generalized force G, resistance generalized force Γ and the angle of kinking φ : a - deterministic case, b - stochastic case.

In the case of linear elastic material, the generalized driving force G is equal to the strain energy release for a crack with a small kink. Before kinking the stress intensity factors are

$$K_I = Y_I(\theta)\sigma_\infty(\pi a)^{1/2}, \quad K_{II} = Y_{II}(\theta)\sigma_\infty(\pi a)^{1/2}. \qquad (2.2)$$

with form-factors Y_I and Y_{II} . These factors depend on the initial angle θ and, maybe, on the normalized length, such as the ratio a/b when a plate of finite width b is considered.

When the crack obtains a small kink with the length $\lambda \ll a$, the stress intensity factors may be presented in the form

$$k_I(\alpha,\theta,\varphi) = K_I(\alpha,\theta)f_{11}(\varphi) + K_{II}(\alpha,\theta)f_{12}(\varphi)$$

$$(2.3)$$

$$k_{II}(\alpha,\theta,\varphi) = K_I(\alpha,\theta)f_{21}(\varphi) + K_{II}(\alpha,\theta)f_{22}(\varphi)$$

Here $K_I(a,\theta)$ and $K_{II}(a,\theta)$ are the stress intensity factors defined according to (2.2). Functions $f_{jk}(\varphi)$ depend on the angle of kinking from the initial plane. There are now quite agreeable numerical results obtained with various approaches (although most of these results concern a central inclined cracks). It was also shown by that the Irwin equation

$$G = \frac{k_I^2(a,\theta,\varphi) + k_{II}^2(a,\theta,\varphi)}{E} \qquad (2.4)$$

presents a good approximation for the stress energy release rate. This issue was checked by a number of authors.

In any case, we assume that equation (2.4) is valid, at least as a good approximation, for fatigue cracks when the material is linear elastic and the fatigue microdamage effects on the fracture toughness only (Bolotin and Kovekh, 1993). As to the generalized resistance forces, assume that they depend on two scalar damage measures similar to the parameters used in the continuum damage mechanics (Krajcinovic and Lemaitre, 1987). The first measure notated ω_σ characterizes damage contributed by opening mode microcracks. The second one is notated ω_τ. It characterizes the shear mode microcracking. At the initiation stage when the first kinking is considered, one may suppose that the damage rates $d\omega_\sigma/dN$ and $d\omega_\tau/dN$ depend on the ranges of the corresponding circumferential stresses near the tip. As in the earlier publications (Bolotin, 1985, 1990), we use the power-threshold law of damage accumulation, i.e. equations

$$\frac{d\omega_\sigma}{dN} = \left(\frac{\Delta\sigma_n - \Delta\sigma_{th}}{\sigma_\omega}\right)^{m_\sigma}, \quad \frac{d\omega_\tau}{dN} = \left(\frac{|\Delta\tau_n| - \Delta\tau_{th}}{\tau_\omega}\right)^{m_\tau} \qquad (2.5)$$

Here $\Delta\sigma_n$ and $\Delta\tau_n$ are the ranges of the circumferential stresses σ_n and τ_n within a cycle. Parameters σ_ω and τ_ω characterize resistance of the material to microcracking in opening and shear modes, respectively. Parameters $\Delta\sigma_{th}$ and $\Delta\tau_{th}$ are of the meaning of threshold resistance to microcracking. At $\Delta\sigma_k < \Delta\sigma_{th}$ or $|\Delta\tau_n| < \Delta\tau_{th}$ the right-hand sides in the corresponding equations (2.5) are to be put to zero. The power exponents m_σ and m_τ are also material parameters. The listed parameters, generally, depend on temperature and other environmental conditions as well on the damage measures and acting stress ratios. Equations that allow to calculate the circumferential stresses $\sigma_n(\varphi)$ and $\tau_n(\varphi)$ in the vicinity of an elliptical crack are rather cumbersome and are omitted here.

To close the set of constitutive equations a formula for the generalized resistance force Γ is needed. As in the paper by Bolotin (1985), we assume that

$$\Gamma = \gamma\left[1 - (\psi_\sigma^{\alpha_\sigma} + \psi_\tau^{\alpha_\tau})\right] \qquad (2.6)$$

Here γ is the specific fracture work for nondamaged material, ψ_σ and ψ_τ are the damage measures at the tip, i.e. $\psi_\sigma = \omega_\sigma(x_t, y_t, \varphi)$, $\psi_\tau = \omega_\tau(x_t, y_t, \varphi)$. The tip coordinates are denoted here with x_t and y_t (Figure 1).

3. A Probabilistic Model of Solids

Let mechanical properties form a two-dimensional random field in the body. To choose an appropriate model for these properties, it is expedient to diminish as much as possible the number of stochastic variables to reduce to minimum the additional computational work performed without reliable statistical information. We consider such bulk parameters as Young modulus, Poisson ratio, etc. as deterministic and constant in a body. However, the specific fracture work of the nondamaged material is one of the most important parameters affecting fatigue crack growth process. Therefore, it is necessary to treat the specific fracture work as a random field. We assume this field as homogeneous and isotropic. The scales of correlation and variability of this field have the order of magnitude of the grain size. We assume this size to be constant in the body. This assumption allows to simplify the description of other random fields relevant to fatigue. Among them there are the stresses σ_ω and τ_ω characterizing material's resistance to damage accumulation. The specific fracture work γ may be related to a certain ultimate stress σ_U as $\gamma \sim \sigma_U^2 \lambda$ where λ is a scale parameter. The stresses σ_ω and τ_ω are to be in strong correlation with σ_U. For simplification, we assume that all these stresses are connected deterministically. Then we may write the following relationships: $\sigma_\omega / \sigma_\omega^0 = \sigma_U / \sigma_U^0$, $\tau_\omega / \tau_\omega^0 = \sigma_U / \sigma_U^0$. Here σ_ω^0, τ_ω^0, σ_U^0 are some representative (deterministic) magnitudes of random variables σ_ω, τ_ω, σ_U. This results into equations

$$\sigma_\omega = \sigma_\omega^0 (\gamma / \gamma_0)^{1/2}, \quad \tau_\omega = \tau_\omega^0 (\gamma / \gamma_0)^{1/2} \tag{3.1}$$

where $\gamma_0 = (\sigma_U^0)^2 \lambda$ is the representative specific fracture work. Similarly, we may assume the threshold resistance stresses $\Delta\sigma_{th}$ and $\Delta\tau_{th}$ to be proportional to corresponding stresses σ_ω and τ_ω. This assumption is in agreement with the common treatment of the threshold stress intensity factor ΔK_{th}. The latter enters into empirical equations of fatigue crack growth as a part of the critical stress intensity factor K_{IC}, e.g. $\Delta K_{th} = 0.05 K_{IC}$. Exponents m_σ, m_τ, α_σ and α_τ entering equations (2.5) and (2.6) we consider as deterministic.

As a result, all the relevant random variables in equations (2.1) and (2.4) - (2.6) are presented in function of a single variable, namely, the local magnitude of the specific fracture work $\gamma(x, y, \varphi)$. Here x and y are coordinates of the point in consideration, and φ is angle of the eventual crack propagation. The field $\gamma(x, y, \varphi)$ is homogeneous with respect to x, y in the body volume, and homogeneous with respect to φ in the interval $(-\pi, \pi)$.

In the further specification of the model, we assume that the magnitudes of $\gamma(x, y, \varphi)$ are independent in the points divided by the characteristic distance λ. Thus the crack trajectories are schematized as sequences of links with the length λ. Each connection point is, generally, the point of kinking. Some of these points, at least in principle, can occur to be the points of branching. But, generally, crack branchings are rather seldom phenomena especially when the quasistatic fracture and plane-stress or plane-strain states are considered. Even after branching, one of the branches does not necessary propagates.

One of the way to represent the field $\gamma(x, y, \varphi)$ is to introduce a probability density function $u(\varphi)$ at each point of kinking. This angular distribution, due to homogeneity and isotropy, is the same at each point and describes a homogeneous random function of φ given in the interval $(-\pi, \pi)$. For example, the angular distribution

$$\gamma(x, y) = \gamma_- + (\gamma_+ - \gamma_-)u(\varphi) \tag{3.2}$$

is appropriate. Here $\gamma_- > \gamma_+$ are deterministic constants, $u(\varphi)$ is a normalized random function of φ. When this function is defined on the segment $[0, 1]$, γ_- means the minimum, γ_+ the maximum specific fracture work for the nondamaged material.

Another model, suitable especially for crack meandering in polycrystalline materials is as follows. All the body consists of a number of small domains which size λ is of the order of magnitude of local material nonhomogeneities. The material parameters $\gamma, \sigma_\omega, \tau_\omega, \Delta\sigma_{th}$, and $\Delta\tau_{th}$ are constant within each domain varying independently from one domain to neighbouring ones. That makes the length λ both the non-homogeneity and correlation characteristic length. At last, we schematize each elementary domain as a isosceles triangle with the angle $\Delta\varphi$ at its apex (Figure 1,b).

For the further computation, we use the betha-distribution for the magnitudes of u within each three-angular domain . The probabiling density function for u is

$$p(u) = \frac{\Gamma(\mu+\nu)u^{\mu-1}(1-u)^{\nu-1}}{\Gamma(\mu)\Gamma(\nu)}, \quad u \in [0, 1] \tag{3.3}$$

with two shape parameters $\mu > 0$, $\nu > 0$. Equations (3.2) and (3.3) cover a wide variety of special cases, from the specific fracture work γ uniformly distributed in $[\gamma_-, \gamma_+]$ variables at $\mu = \nu = 1$ up to strongly centered γ at $\mu \gg 1, \nu \gg 1$.

4. Numerical results

The following data are used for the numerical simulation: $E = 200$ GPa, $\gamma_- = 40$ kJ$/$m^2, $\gamma_+ = 60$ kJ$/$m^2, $\mu = \nu = 4$, $\sigma_\omega^0 = 10$ GPa, $\Delta\sigma_{th} = \Delta\tau_{th} = 0$, $m_\sigma = m_\tau = 4$; $\alpha_\sigma = \alpha_\tau = 1$.

The applied stress range is $\Delta\sigma_\infty$ = 150 MPa (all cycles are in tension, and no closure effects are involved). To evaluate the effect of the material resistance to shear micro-cracking, the ratio $\tau_\omega/\sigma_\omega$ is subjected to variation. In particular, two cases are studuied in detail: $\tau_\omega/\sigma_\omega$ = 0.5 (a comparatively low resistance to shear microcracking) and $\tau_\omega/\sigma_\omega$ = 0.75 (a comparatively high resistance). As it occurs usually in the Monte-Carlo simulation of random proceeses, not only the estimates of probabilistic parameters, but just the behaviour of samples is of interest. In this problem we are interested in samples of crack trajectories as well as the crack tip coordinates and crack growth rate samples in function of the cycle number.

Some results are presented in Figures 3 and 4 for the case $\tau_\omega/\sigma_\omega$ = 0.5. Crack trajectories are shown in Figure 3,a. The initial planar crack with the length a = 1mm is inclined to the direction of the applied tension at the angle 45°. The solid line is obtai-ned by averaging of 20 sample trajectories. Two "extreme" trajectories and two "mode-rate" ones chosen from 20 samples are presented in Figure 3,a, too. Note that the scales along the vertical and horizontal axis differ in ten times. All the samples show the tenden-cy to approach to the mode I cracks. However, the initial growth is frequently oriented along the primary crack direction. The reason is, first, in a comparatively low resistance to shear microcracking, and second, due to an unfavourable distribution of tangentional stresses around the tip prior to the first kinking. All the samples as well as the mean trajectory expose a gradual transformation of cracks into the mode I cracks. As to the meandering, it proceeds also when a crack approaches rather closely the mode I.

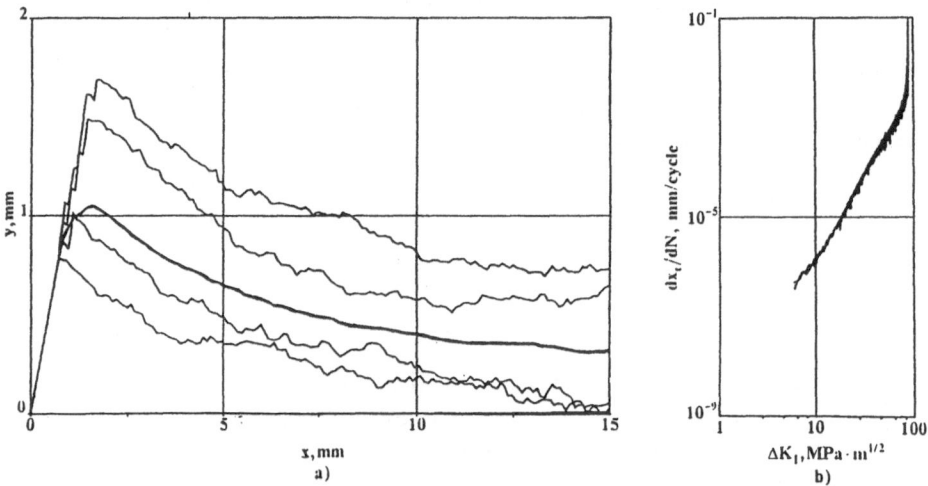

Figure 3.. Fatigue crack propagation from the initial cut with the length a = 1 mm inclined at 45° at $\tau_\omega/\sigma_\omega$ = 0.5: a - sample crack trajectories (the result of averaging upon 20 samples is plotted with the solid line); b - sample crack growth rate diagrams

The crack growth rate diagram, i.e. the relationship between the growth rate da/dN and the range ΔK of the stress intensity factor is shown in Figure 3,b. Here $\Delta K = 1.12 \Delta \sigma_\infty (\pi a)^{1/2}$, where $a = x_t$, and the correction factor corresponds to the cross-section edge crack. This relationship is similar to common diagrams, both experimental and theoretical. The scatter of crack growth is plotted in log-log scale. It is probably that a part of this scatter is originated from the numerical procedure and could be deleted with the use of a more fine mesh. However, the order of magnitude of scattering is the same as in laboratory experimentation.

An additional information is presented in Figure 4 where the crack tip coordinates x_t, y_t are plotted versus the cycle number N. As in Figure 3, the estimates of the expected values of functions $x_t(N)$ and $y_t(N)$ are drawn with solid lines. The "worst", "best" and two "moderate" samples are given here, too. The samples of x_t are rather smooth while the samples of y_t oscillate violantly. The negative part of y_t - diagram is omitted in Figure 4,b althouth some of trajectories enter in to the domain $y_t < 0$.

Figures 5 and 6 are obtained for the same numerical data as Figures 3 and 4. The only difference is that here $\tau_\omega / \sigma_\omega = 0.75$. That results into the change of the general picture: the first kinking is oriented, with minor exceptions, towards the mode I. Other conclusions are similar to those for the former case.

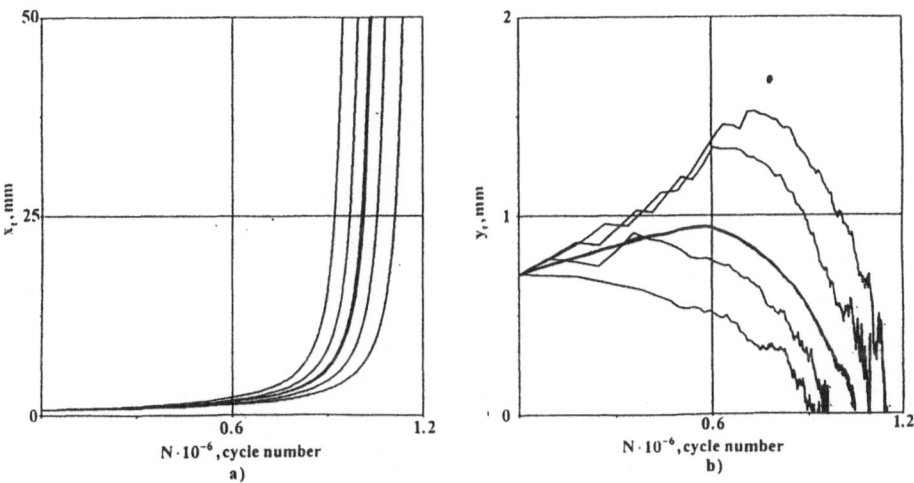

Figure 4. Samples of fatigue crack propagation as tip coordinates $x_t(N), y_t(N)$ in function of the cycle number N. Numerical data as in Figure 3.

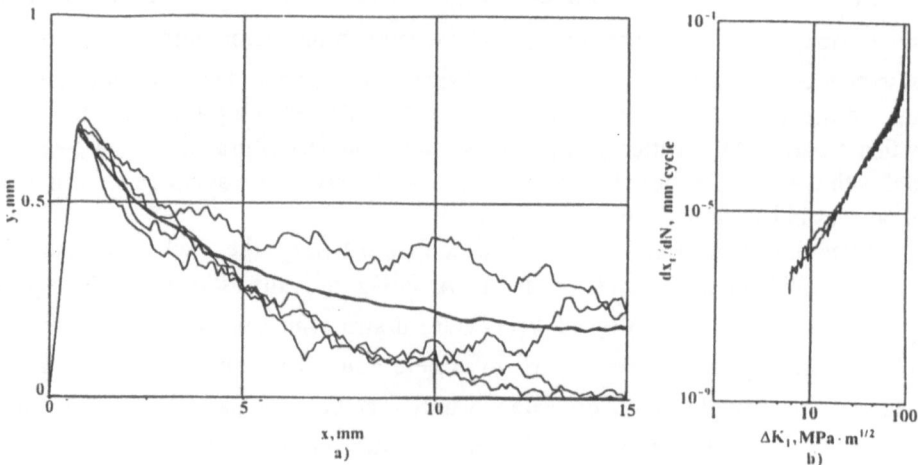

Figure 5. The same as in Figure 3 at $\tau_\omega / \sigma_\omega = 0.75$.

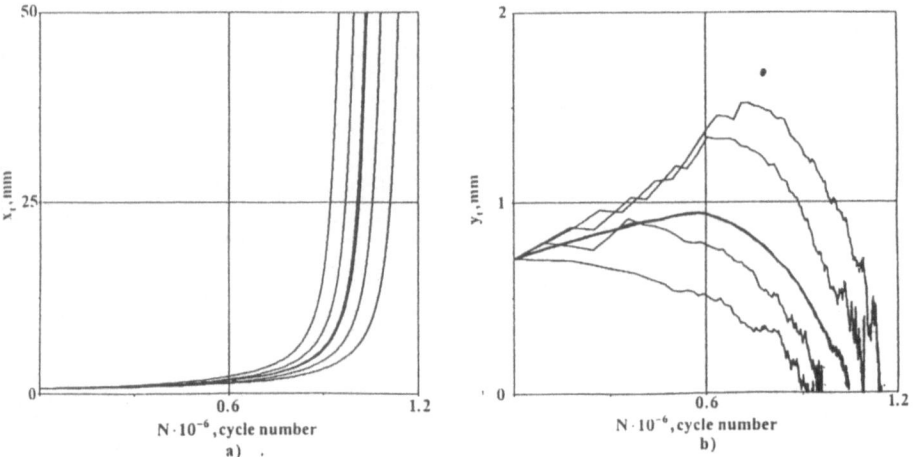

Figure 6. The same as in Figure 4 at $\tau_\omega / \sigma_\omega = 0.75$.

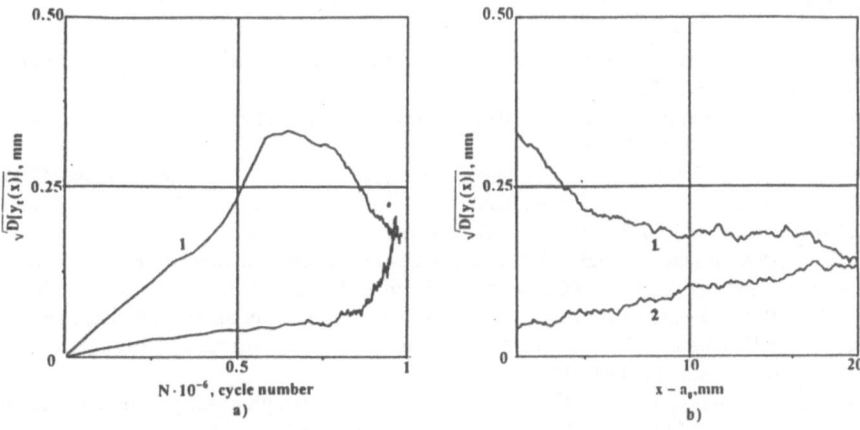

Figure 7. Estimates for strandard deviations of $y_t(x)$ and $y_t(N)$ for $\tau_\omega/\sigma_\omega$ = 0.5 and 0.75 (lines 1 and 2, respectively).

The statistical estimates of the standard deviations for random functions $y_t(x)$ and $y_t(N)$ are presented in Figure 7. The lines 1 and 2 are plotted for $\tau_\omega/\sigma_\omega$ = 0.5 and $\tau_\omega/\sigma_\omega$ = 0.75, respectively. It is to be mentioned a large difference of lines 1 and 2 at the beginning of crack growth. When the ratio $\tau_\omega/\sigma_\omega$ is comparatively low, a part of samples expose a tendency to propagate closely to the initial, inclined direction turning towards the mode I only later. As a results, the variance of $y_t(x)$ and $y_t(N)$ becomes rather large. Generally, the variances are significant on the first stage of crack growth and rather low when a crack approaches to the mode I. The right-hand parts of Figures 3-7 may be interpreted as measures of the crack meandering during the regular mode I propagation.

CONCLUSIONS

A probabilistic model was proposed to describe the fatigue crack propagation through materials with randomly distributed properties. The parametrical study of the problem is performed using the Monte Carlo simulation. The influence of some parameters on the samples of trajectories and crack growth rate diagrams is analysed. The results appear to be in a reasonable agreement with the observed fatigue crack behaviour in real polycrystalline materials.

ACKNOWLEGEMENTS

This study is partially supported by the Russian Foundation for Basic Research (grant N 93-01-16486).

References

Bolotin, V.V. (1983) Equations of fatigue crack growth, *Izv. Akad. Nauk SSSR, Mekh. Tverd. Tela (MTT)*, 4, 153-160 (in Russian).

Bolotin, V.V. (1985) A unified approach to damage accumulation and fatigue crack growth, *Engng Fracture Mech.*, 22(3), 387-398.

Bolotin, V.V. (1989) *Prediction of Service Life for Machines and Structures*, New York, ASME Press.

Bolotin, V.V. (1990) Mechanics of fatigue fracture. In *Nonlinear Fracture Mechanics*, CISM Course N 314 (M.P.Wnuk, ed.), Springer-Verlag, Wien - New York, pp.1-60.

Bolotin, V.V. (1993) Random initial defects and fatigue life prediction, in *Stochastic Approaches to Fatigue* (K.Sobczyk, ed.) CISM Course N 334 Springer-Verlag, Wien - New York, pp.126-163.

Bolotin, V.V. (1994) Mechanics of fatigue crack growth as a synthesis of micro- and macromechanics of fracture, in *Handbook of Fatigue Crack Propagation in Metallic Structures* (A.Carpinteri., ed.), Elsevier, Amsterdam, pp. 883-911.

Bolotin, V.V. and Kovekh, V.M. (1993) Numerical simulation of fatigue crack growth in a medium with microdamage, *Izv. RAN, Mekh. Tverd. Tela (MTT)*, 2, 132 (in Russian).

Krajcinovic, D. and Lemaitre, J. (1987) *Continuum Damage Mechanics - Theory and Applications*, Wien - New York, Springer-Verlag.

Pook, L.P. (1994). Mixed mode crack propagation. in *Handbook of Fatigue Crack Propagation in Metallic Structures* (A. Carpinteri, ed.), Amsterdam, Elsevier, pp. 1027-1072.

Ramulu M. and Kobayashi A.S. (1994) Numerical and experimental study of mixed mode fatigue crack propagation. in *Handbook of Fatigue Crack Propagation in Metallic Structures*. (A.Carpinteri., ed.), Amsterdam: Elsevier, pp. 1073-1123.

Sobczyk, R and Spencer B.F. (1992) *Random Fatigue: From Data to Theory*, Academic Press, Boston - New York.

THE STUDY OF BIFURCATIONS THROUGH THE SOLUTION OF THE FOKKER-PLANCK EQUATION

F.BONTEMPI
Department of Structural Engineering, Polytechnic of Milan
P.zza L.da Vinci, 32 - 20133 MILAN

AND

L.FARAVELLI
Department of Structural Mechanics, Univesity of Pavia
Via Abbiategrasso, 211 - 27100 PAVIA

Abstract

The solution of the Fokker-Planck equation of a nonlinear dynamical system is pursued by a cell-method. The purpose is to conduct a bifurcation analysis of the system. Some numerical examples are discussed.

1. Introduction

Consider a dynamical system with n degrees of freedom arranged in the vector $\mathbf{y} = [y_1, y_2, \ldots y_n]^T$. Provided the fulfilling of the usual regularity hypotheses (Birkhoff and Rota,1959), there exists a unique solution of the equation of motion

$$\dot{\mathbf{y}}(t) = \mathbf{Q}(\mathbf{y}, t, \mu) \tag{1}$$

with the initial condition $\mathbf{y}(t_0) = \mathbf{c}$. In Eq(1) t denotes the time and μ is the only parameter on which the system dynamics depends. Such a solution can be represented by a trajectory in the n-dimensional phase space S^n; it can be numerically computed by appropriate integration schemes (Griffiths and Smith, 1991). This simple picture is significantly altered if: a) the system is strongly nonlinear (Wedig, 1988), and/or b) there are random terms in Eq.(1) (Bolotin, 1969), and/or c) Eq.(1) shows sensitivity to the initial conditions, leading one to chaotic dynamics (Baker and Gollub, 1990). In these cases, the problem of determining the evolution of system (1) cannot be regarded as solved by the knowledge of a unique well defined trajectory in the space S^n. It will be satisfactorily analyzed only by the computation of the joint probability density $p(\mathbf{y}, t)$ of the dynamic variables $\mathbf{y}$: it is a function of time with known initial shape $p(\mathbf{y}, t_0)$ (Lin, 1976).

A. Naess and S. Krenk (eds.), IUTAM Symposium on Advances in Nonlinear Stochastic Mechanics, 59–68.
© *1996 Kluwer Academic Publishers.*

The concepts of stability and reliability (Bolotin, 1969) are also connected with the form of the probability density function: the exit of the system from a *safe region* in the space S^n can be regarded as the *failure* of the system, the probability of leaving this region being the probability of failure.

The simplest situation that one can meet when Eq. (1) is considered is the following. One starts with a value of the system parameter μ by which the dynamics of the system is safe; as μ increases , one meets, soon or later, a situation of *failure* in the sense previously specified. Generally speaking, the value of μ which characterizes this situation causes a *bifurcation* of the dynamics of system (1). It is worth noting that, while for simple systems (like the Duffing oscillator considered in this paper as first example) the point of bifurcation is very well marked by an abrupt change in the motion, in the case of more complex systems (like the oscillator of the second example), the same break point appear sometimes indistinct.

TABLE 1. Analysis of a nonlinear stochastic system

local equation of motion (ordinary differential equation)			global equation of motion (partial differential equation)
single system	many systems	Cell Method	Liouville Equation Fokker-Planck Equation
(Whitney, 1990; Gould and Tobochnik, 1987)	(Wedig, 1989)	(Hsu and Chiu, 1987; Kreutzer, 1985)	(Bergman and Spencer,1993; Kunert and Pfeiffer,1994; Langtangen,1991)
Monte Carlo Method	Discrete Gibbs Set	Continuum	Gibbs set
Lagrangian point of view			Eulerian point of view

Table I summarized the tools of analysis for a general nonlinear stochastic system.

(1) In order to follow the evolution of system (1), one can select a *local* description of the motion or a *global* one: the first choice is just related to the ordinary differential equation (1), while, in the second case, the problem is reformulated in terms of a partial differential equation, like the Liouville Equation or the Fokker-Planck Equation (FPE). These kinds of equations are established by continuity statements, like that governing fluid dynamics: the whole flux defined by Eq.(1) must be simultaneously considered and, thence, one is naturally conducted to introduce a probabilistic description.

In the case of a local description one has to develop several consecutive realizations, as in the classical Monte Carlo Method, or has to develop many simultaneous realizations leading to the study of a Gibbs Set (Casciati, 1993). In the latter case, the probabilistic description is achieved by considering the evolution of the collection of systems (each of them being as system (1) but characterized by the index

$k = 1, .. N_{gib}$):

$$\dot{\mathbf{y}}^k(t) = \mathbf{Q}^k(\mathbf{y}, t) \tag{2}$$

with the set of initial conditions $\mathbf{y}^k(t_0) = \mathbf{c}^k$

(2) If the motion of system (1) is studied by Monte Carlo Method or by the Gibbs Set, one can speak of description from a *Lagrangian* point of view: it means that the representative point of the system in the phase space is followed like a particle of fluid along its motion in fluid dynamics. One can also consider the study of the dynamics of system (1), from the *Eulerian* point of view. In this case, one considers in turn the evolution of a part of the phase space, i.e. how a cell of fluid, initially in a well defined portion of the phase space, moves itself. This leads to the *Cell Method* (Hsu and Chiu, 1987; Kreutzer, 1985; Tongue, 1990): it is based on the local description with ordinary differential equations, but the discrete Gibbs Set (with finite N_{gib}) is generalized to a continuum set. It can be classified as the transition between a local and a global point of view.

2. Some considerations about the Fokker-Planck Equation

Considering the variable $\mathbf{y}$ as a Markov process, the value of the probability density $p(\mathbf{y}, t)$ at time t_2 only depends on the probability distribution at the preceding instant $t_1 < t_2$. The partial differential equation that describes the evolution of the probability density $p(\mathbf{y}, t)$ is the Fokker-Planck Equation (Caughey, 1971; Casciati and Faravelli, 1991). The existence of the limits:

$$b_j = \lim_{\tau \to 0} \frac{\overline{[y_j(t + \tau) - y_j(t)]}}{\tau}' \quad a_{ij} = \lim_{\tau \to 0} \frac{\overline{[y_i(t + \tau) - y_i(t)] \cdot [y_j(t + \tau) - y_j(t)]}}{\tau} \tag{3}$$

is assumed. In Eq. (3) b_i is the intensity of the process of the first order and a_{ij} the one of second order. The intensities of higher orders are identically zero for Gaussian and diffusion processes. The bar over some terms indicates averaging on a set of realizations. The Fokker-Planck equation is then written for

$$\dot{p}(\mathbf{y}, t) = -\sum_{i=1}^{n} \frac{\partial}{\partial y_i} \left[b_i(\mathbf{y}, t) \cdot p(\mathbf{y}, t) \right] + \frac{1}{2} \sum_{i=1}^{n} \sum_{j=1}^{n} \frac{\partial^2}{\partial y_i \partial y_j} \left[a_{ij}(\mathbf{y}, t) \cdot p(\mathbf{y}, t) \right] \tag{4}$$

The functions $b_j(\mathbf{y}, t)$ characterize the average flow of the process, and then are connected with the hyperbolic character of equation (4) (i.e. the character which causes $p(\mathbf{y}, t)$ to move with inaltered shape but with the tendency to sharp gradients), while the functions $a_{ij}(\mathbf{y}, t)$ are related to the diffusion of the process, i.e. to the parabolic character of equation (4) (the one which causes a smoothing spread, with zero mean displacement in S^n, of $p(\mathbf{y}, t)$). Finally, in the case that the first member of (4) is identically zero, one has a stationary FPE.

The following boundary and normalizations conditions have to be satisfied

$$\lim_{\|\mathbf{y}\| \to +\infty} p(\mathbf{y}, t) = 0, \qquad \int_{S^n} p(\mathbf{y}, t) \cdot dy_1 \ldots dy_n = 1 \tag{5}$$

Clearly, if the dynamics of the system (1) depends on a parameter μ, $p = p(\mathbf{y}, t, \mu)$, and from the shape change of this function (from unimodal to bimodal for instance) one recognizes the bifurcation of the system behaviour.

3. The solution by cell-technique.

The technique of solution of the FPE presented in this Section is strictly related to its physical nature. Consider for sake of simplicity the two dimensional case, i.e. the equation associated with a dynamical system with two $(n = 2)$ degrees of freedom (y_1, y_2). One writes explicitly

$$\begin{cases} \dot{y}_1 = f_1(y_1, y_2, t) \\ \dot{y}_2 = f_2(y_1, y_2, t) \end{cases} \tag{6}$$

and the motion develops in the phase plane $S^n = S^2 = (y_1, y_2)$. The extension to multivariate problems is straightforward and will be illustrated in the example.

To obtain the numerical solution of Eq. (4), one must define, in the phase plane, the region $\Omega = (y_{1,min} - y_{1,max}; y_{2,min} - y_{2,max})$ where the motion of the dynamical system develops. Ω is partitioned by a grid of N_{box} rectangular boxes, each of them of size $\Delta y_1 \times \Delta y_2$, with $N_{box} = \prod_{i=1}^{n=2} \left[\frac{y_{i,max} - y_{i,min}}{\Delta y_i} \right]$.

Due to this discretization, the new unknowns are the probabilities $p_i(t) = \int_i p(y_1, y_2, t) \cdot dy_1 \cdot dy_2$ that system (1) finds itself, at instant t, in the i-box ($i = 1, ..N_{box}$). The solution of the FPE requires then to find the set of values $(p_i(t_k))$, for each instant of time $t_k = t_0 + k \cdot \Delta t$, Δt being a suitable increment of time.

Let the initial probability mass be like a δ-Dirac function:

$$p_j(t_0) = 1; \; p_i(t_0) = 0, \; for \; j \neq i, j = 1, ..N_{box} \tag{7}$$

This distribution of probability can be visualized as a prism, of base $\Delta y_1 \cdot \Delta y_2$ and height $h = p_j(t_0)/\Delta y_1 \cdot \Delta y_2$.

After an increment of time Δt, i.e. at time $t_1 = t_0 + \Delta t$, this prism due to the convective and the diffusive characters of the FPE, moves in a different part of the phase space and its shape is altered, provided that the volume is $p_j(t_0)$. To compute this evolution, one considers the motion of the base of this prism (i.e. the rectangular grid $\Delta y_1 \times \Delta y_2$), located by the four vertices: with reference to Fig.1 one sees that the rectangular base deforms itself in a quadrilateral shape, the position of the four vertices $k = 1, 2, 3, 4$ being moved according the equation of motion of the dynamical system

$$\begin{aligned} \dot{y}_1^k &= f_1(y_1^k, y_2^k, t) \\ \dot{y}_2^k &= f_2(y_1^k, y_2^k, t), \; k = 1, ..4 \end{aligned} \tag{8}$$

with initial conditions the coordinates of the base vertices. The motion of the points inside this base can be found with any interpolating technique, as usual in the Finite Element Method. One considers a normalized quadrilateral element with axis (ξ_1, ξ_2) initially superimposed over the i-th box, where the interpolating shape functions are defined:

$$N_1(\xi_1, \xi_2) = \frac{1}{4} \cdot (1 - \xi_1) \cdot (1 - \xi_2), \quad N_2(\xi_1, \xi_2) = \frac{1}{4} \cdot (1 + \xi_1) \cdot (1 - \xi_2)$$
$$N_3(\xi_1, \xi_2) = \frac{1}{4} \cdot (1 + \xi_1) \cdot (1 + \xi_2), \quad N_4(\xi_1, \xi_2) = \frac{1}{4} \cdot (1 - \xi_1) \cdot (1 + \xi_2) \tag{9}$$

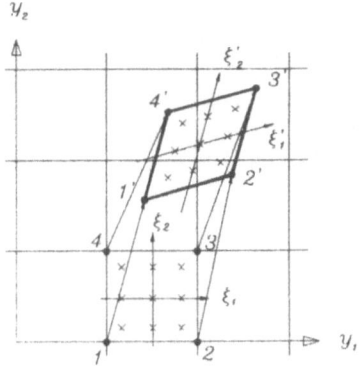

Figure 1. The motion of a single cell.

From Eq.(9) one obtains the generic point $(y_1(\xi_1, \xi_2), y_2(\xi_1, \xi_2))$ as

$$y_1(t) = \sum_{i=1}^{4} N_i(\xi_1, \xi_2) \cdot y_1^k(t), \qquad y_2(t) = \sum_{i=1}^{4} N_i(\xi_1, \xi_2) \cdot y_2^k(t) \qquad (10)$$

Now the problem is to know where the probability $p_i(t_0)$ is spreading. Use is made of a Gauss-Legendre integration technique, with a grid of 3×3 points of sampling ($[\xi_{1,1}\xi_{1,2}\xi_{1,3}] \times [\xi_{2,1}\xi_{2,2}\xi_{2,3}]$, with weights $[W_1 W_2 W_3] \times [W_1 W_2 W_3]$; their sum must be equal to 1. One assigns, to each sampling point (ξ_{1i}, ξ_{2j}), a probability $W_i \cdot W_j \cdot p_i(t_0)$. It is straightforward to know in which box the sampling point moves, and to put here the associated probability. In this way, one obtains the set of probability $(p_i(t_1))$ at instant $t_1 = t_0 + \Delta t$ from the known set $(p_i(t_0))$ at time t_0. One has implicitly computed the transition probability matrix $\mathbf{T} = [t_{i,j}]$ of the Markov process $\mathbf{y}$, which governs the evolution of the set of probability as

$$\begin{pmatrix} p_1(t_1) \\ p_2(t_1) \\ \dots \\ p_n(t_1) \end{pmatrix} = \begin{pmatrix} t_{1,1}t_{1,2}\dots t_{1,n} \\ t_{2,1}t_{2,2}\dots t_{2,n} \\ \dots \\ t_{n,1}t_{n,2}\dots t_{n,n} \end{pmatrix} \cdot \begin{pmatrix} p_1(t_0) \\ p_2(t_0) \\ \dots \\ p_n(t_0) \end{pmatrix} \qquad (11)$$

without allocate it computationally. This is an important computational effort saving, because the matrix $\mathbf{T}$ can be very large, of the order of 100 or 200 to power n. Futhermore, one only needs to follow the evolution of the non-empty cell, i.e. the cells for which $p_i > 0$: this is of computational importance, because, for very low noise intensity excitation, the dynamics of the system occurs in a very restricted portion of Ω. In particular no special attention must be devoted to the limit case of a deterministic system, usually a break-up situation for finite element or finite difference technique.

It is worth noting that if the integration in time of the single trajectory correspond to the Lagrangian description of the motion, the integration of the four vertices of each box of the grid can be regarded as an Eulerian aspect of the solution procedure.

Clearly, the computational effort increases with the number of boxes N_{box} and with the size of the domain Ω. From one side, Ω must contain all the possible

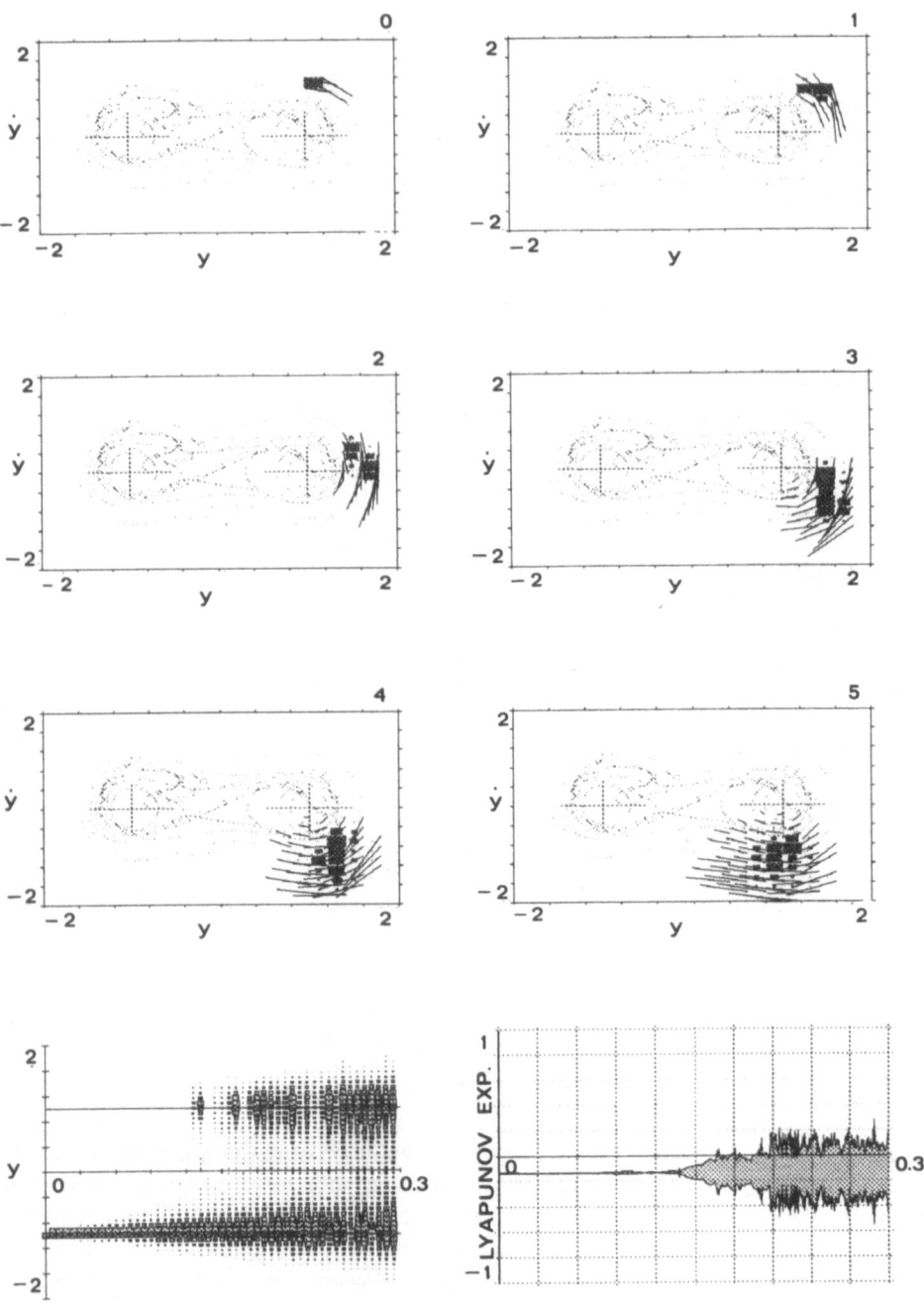

Figure 2. a) Evolution of the probability for the Duffing oscillator: the size of the square is proportional to the value of the probalibity. The abscissa shows the displacement in the range (-2, 2) and the ordinate the velocity in the range (-2, 2). b) The probability density obtained by the Cell Method as a function of the noise intensity $S_0 \in (0,0.3)$. c) Lyapunov exponents vs. noise intensity.

points visitable to the dynamical system during the motion evolution. Theoretically, in the presence of a white noise excitation, this region should be infinite. In effect, one can truncate the admissible region through some simulations showing the range of the phase space most frequently visited. Furthermore, in the further presence of a deterministic periodic forcing, the region can be easily determined with considerations on the unnoised motion.

From the other side, one can wish to have $\Delta_i \rightarrow 0$, $i = 1, 2$, to obtain a perfect knowledge of the probability density function of the status of the system in time, because

$$\lim_{\Delta_1 \rightarrow 0, \Delta_2 \rightarrow 0} \frac{p_i(t)}{\Delta_1 \Delta_2} \rightarrow p(\mathbf{y}, t) \tag{12}$$

Again, one must consider some compromise. It appears suitable from an engineering point of view to have a 1/20 or 1/10 uncertainty (i.e. an error of 5% or 10%) on the status of the system for each degree of freedom y_i.

The important characteristics of this method of solution is that, by the Cell Method, the number of unknown is only related to accuracy considerations, the stability being always achieved; this is the contrary of what must consider with discretization methods like finite difference (Kunert and Pfeiffer, 1991) or finite element (Langtangen, 1991). These techniques need anyway a very fine grid of points or of nodes for the interpolating functions in the phase space: the resulting number of unknown is of the order of $n_{un} = \prod_{i=1}^{n}[100 \text{or} 200] = [100 \text{or} 200]^n$, without which one is unable to achieve any significant solution.

4. Two numerical examples.

The solution algorithm is first discussed for the classical Duffing oscillator (Casciati,1993). The equation of motion is

$$\ddot{y} + d \cdot \dot{y} + (y - y^3) = A \cos(t) + w(t) \tag{13}$$

in which y is the position, $\dot{y}$ the velocity, and $\ddot{y}$ the acceleration of the oscillator of mass 1 and damping d ($d = 0.185$). The term $(y - y^3)$ represents the elastic non linear restoring force (connected with two stable points $y = -1$ and $y = +1$ where the deformation energy presents its local minima), while the forcing term consists of two parts, one ($A \cos(t)$) harmonic, with period $T_F = 2 \cdot \pi$, and the other ($w(t)$) random, a Gaussian white-noise with intensity $2\pi S_0$. With the positions $y_1 = y$ and $y_2 = \dot{y}$, one can rewrite the second order differential equation (Eq. (13)) as the following nonautonomous system of first order differential equations

$$\begin{cases} \dot{y}_1 = y_2 \\ \dot{y}_2 = -d \cdot y_2 - (y_1 - y_1^3) + A \cos(t) + w(t) \end{cases} \tag{14}$$

In Fig.2.a, one shows the first 8 step ($t_k = \Delta t \cdot k, k = 0, ..7, \Delta t = 2 \cdot \pi/20$) of the evolution of the probability density function, from an single initial cell ($A = 0.3$ and $S_0 = 0$: a deterministic chaotic example).

Only a stochastic excititation ($A = 0.0$ and $\mu = S_0 \in (0, 0.3)$) is considered in Fig.2.b. Here the bifurcation is characterized by the spread of the probability density function from the energy minimum around $y = -1$ to the other. In this diagram one represents, as a function of the noise intensity S_0, the density obtained by the Cell Method, The bifurcation happens for $S_0 = 0.21$: the same result

(Fig.2.c) is obtained by computing the Lyapunov exponents (Baker and Gollub 1990; Casciati, 1993).

The second application regards the determination of the probability of failure of the simplified tank model of Fig.3.a, with the only mechanical degree of freedom $\phi(t)$, subject to a non stationary random ground motion. The example is very similar to the one considered by Spencer and Bergman (1985) and Naess and Johnsen (1991). The equation of motion, including the second order effects, is

$$m \cdot (\ddot{x} + l \cdot \ddot{\phi}) \cdot l + G(\phi, \dot{\phi}, t) = mg \cdot l\phi \qquad (15)$$

where $m = 1250kg$ is the mass concentrated at the tank, $g = 1000cm/s^2$ is the gravity acceleration, $l = 3000cm$ is the height of the tank, and $G(\phi, \dot{\phi}, t)$ is the hysteretic, degrading restoring force, modelled by the Bouc-Wen model (Wen, 1986):

$$G(\phi, \dot{\phi}, t) = d \cdot \dot{\phi} + \alpha \cdot k \cdot \phi + (1 - \alpha) \cdot k \cdot z \qquad (16)$$

$$z = \frac{1}{\eta} \cdot \left[A \cdot \frac{\dot{\phi}}{\phi_{max}} - v \cdot \left(\beta \cdot \frac{|\dot{\phi}|}{\phi_{max}} \cdot |z|^{n-1} \cdot z + \gamma \cdot \frac{\dot{\phi}}{\phi_{max}} \cdot |z|^n \right) \right] \qquad (17)$$

In Eqs. (16) and (17) the elastic stiffness is $k = 1.2268 \cdot 10^{10} N/cm^2$, the damping $d = 2.349 \cdot 10^8$ (being the critical damping = 0.01), $\phi_{max} = 0.2rad = 12^0$ and $\alpha = 1/20$. The degradation is governed by the following expressions: $A(t) = A_0 - \delta_a \cdot E_D(t)$, $v(t) = v_0 + \delta_v \cdot E_D(t)$, $\eta(t) = \eta_0 + \delta_\eta \cdot E_D(t)$, $E_D(t)$ being the work dissipated by hysteresis. The following numerical values were assumed $A_0 = 1$, $\delta_A = 0.01$, $v_0 = 1$, $\delta_v = 0.01$, $\eta_0 = 1$, $\delta_\eta = 0.01$. The fundamental period of the structure in the elastic range is $T_0 = 6.02s$. One defines the safe region by the following inequalities: $-0.25 \leq y_1 \leq +0.25$, $-0.4 \leq y_2 \leq +0.4$, $-1 \leq y_3 \leq +1$ and $0 \leq y_4 \leq 0.4 * 10^{10}$. Of course, $y_1 = \phi$, $y_2 = \dot{\phi}$, $y_3 = z$ and $y_4 = E_D$.

The structure is initially at rest; the ground acceleration is modelled by a white noise process of duration $T = 30s$. One analyzes the situation of the structure at time $t = 60s$. In Fig.3.b, one has the bifurcation diagram of the rotation $y_1 = \phi$ of the tank against the ground acceleration process intensity in the range $(0, 2\pi \tilde{S}_0) = (0, \Delta tg^2/25)$. Below that, the probability of failure is sketched.

The same results are showed in Fig.3.c but they are now obtained by the Cell Method: here the bifurcation, characterized by a non zero failure probability, happen around $g/10$.

The handling of the FPE in a $n = 4$-dimension phase space appears as a present limit from a computational point of view. For a larger number of degrees of freedom, solutions can only be pursued within stationary situations.

5. Conclusion.

The dynamics of a mechanical system subject to stochastic excitations can be described by the Fokker-Planck Eqaation (FPE): solving this equation, one obtains the evolution from an initial known density of the probability density function of finding the system around a point in the phase space. The difficulties regard the mixed parabolic-hyperbolic character of the equation itself and the large number of dimensions of the phase space where one works (for a N degree-of-freedom system $n = 2 \cdot N$ are the actual dimensions of the phase space). In this paper, some aspects

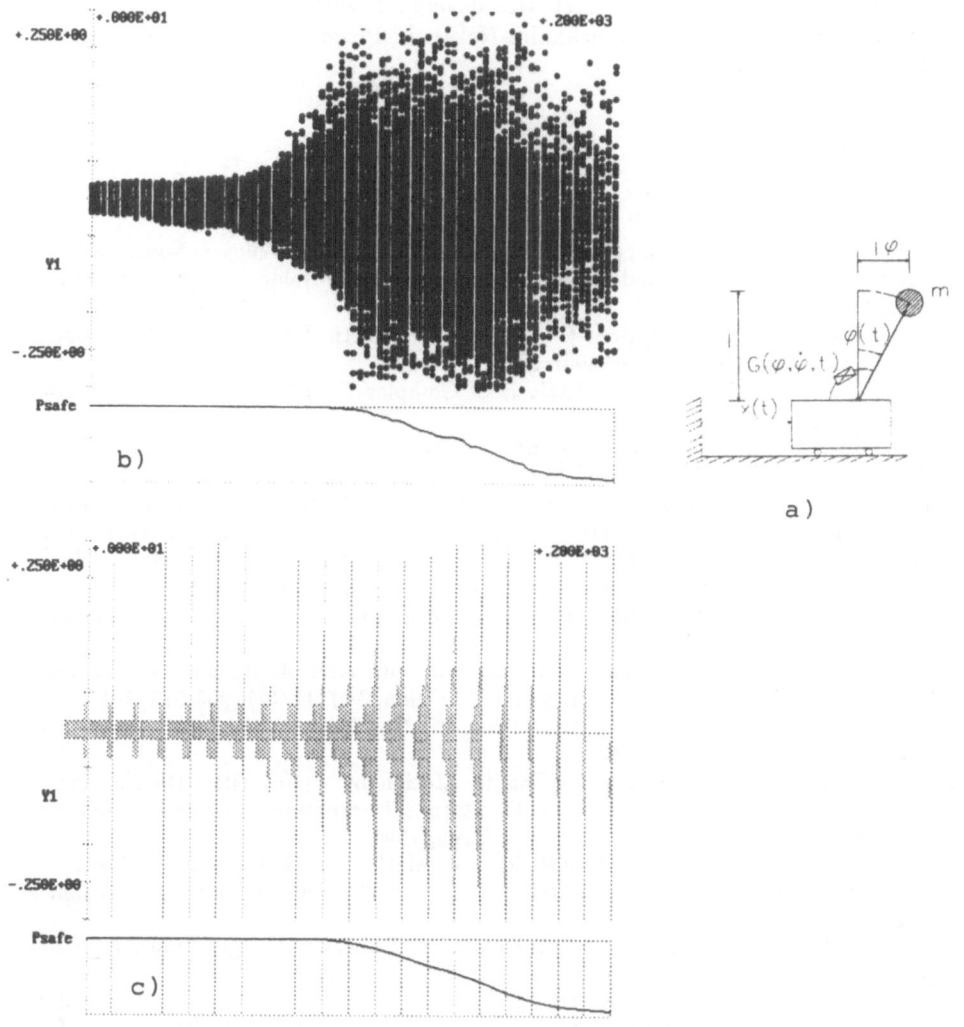

Figure 3. The simplified tank model (a) and the probability of failure, versus the noise intensity, resulting from a bifurcation diagram (b) and the Cell Method (c). The abscissa gives, for the ground acceleration process, $(2\pi S_0/\Delta t)^{\frac{1}{2}}$ in the range $(0, g/5)$.

of the FPE are reviewed and some improvement to the Cell Method are illustrated.

Acknowledgement - This research was developed within the EC Human Capital and Mobility - Stochastic Mechanics Network. Program contract number ER-BCHRXCT 9404565.

References

Baker G.L., Gollub J.P., Chaotic Dynamics. An Introduction., Cambridge University Press, 1990.
Birkhoff G., Rota G.C., Ordinary Differential Equations., John Wiley & Sons, 1959.
Bolotin V.V., Statistical Methods in Structural Mechanics., Holden-Day Inc., San Francisco, 1969.
Casciati F. (ed.), Dynamic Motion: Chaotic and Stochastic Behaviour., CISM Courses and Lectures No.340, Springer-Verlag, Berlin, 1993.
Casciati F., Faravelli L., Fragility Analysis of Complex Structural Systems., Research Studies Press, Tauton, 1990.
Caughey T.K., Nonlinear Theory of Random Vibrations., in Advances in Applied Mechanics Vol.11, Chia-Shu Yih (ed.), Academic Press, 1971.
Crandall S.H., Chandiramani K.L. and Cook R.G., Some First-Passage Problems in Random Vibration, Journal of Applied Mechanics, Transaction of ASME, 532-538, 1966.
Gould H., Tobochnik J., An Introduction to Computer Simulation Methods., Addison-Wesley Publishing Company, Inc., Reading Massachusetts, USA, 1987.
Griffiths D.V., Smith I.M., Numerical Methods for Engineers., Blackwell Scientific Publications, 1991.
Hsu C.S., Chiu H.M., Global Analysis of a System with multiple Responses including a strange attractor., Journal of Sound and Vibration, Vol. 114(2), 203-218, 1987.
Kreutzer E., Analysis of Chaotic Systems Using the Cell Mapping Approach., Ingenieur-Archiv, Vol. 55, 285-294, 1985.
Kunert A., Pfeiffer F., Description of Chaotic Motion by an Invariant Distribution at the Example of the Driven Duffing Oscillator., International Series of Numerical Mathematics, Vol.97, Birkhauser Verlag Basel, 1991.
Langtangen H.P., A General Numerical Solution Method for Fokker-Planck Equations with Applications to Structural Reliability., Probabilistic Engineering Mechanics, Vol.6, No.1, 33-48, 1991.
Lin K., Probabilistic Theory of Structural Dynamics., Robert E. Krieger Publishing Company, 1976.
Naess A., Johnsen J.N., 1991, The Path Integral Solution Technique applied to the Random Vibration of Hysteretic System., in Computational Stochastic Mechanics, P.D. Spanos and C.A. Brebbia eds., Computational Mechanics, Southampton.
Spencer B.F., Bergman L.A., On the Reliability of a simple Hysteretic System., Journal of Engineering Mechanics, ASCE, Vol.111, No.12, December, 1985.
Spencer B.F., Bergman L.A., On the Numerical Solution of the Fokker-Planck Equation for Nonlinear Stochastic Systems, Nonlinear Dynamics, Vol. 4, No. 5, 357-372, 1993.
Tongue B.H., A Multiple-Map Strategy for Interpolated Mapping., Int. J. Non-Linear Mechanics, Vol.25, No.2/3, pp.177-186, 1990.
Wedig W., Vom Chaos zur Ordnung., GAMM - Mitteilungen, Heft 2, 3-31, 1988.
Wedig W., Analysis and Simulation of Nonlinear Stochastic Systems., Proceedings of the IUTAM Symposium on Nonlinear Dynamics in Engineering Systems, Springer Verlag, Berlin, 1989.
Wen Y.K., Stochastic Response and Damage Analysis of Inelastic Structures. Probabilistic Engineering Mechanics, Vol.1, No.1, 49-57, 1986.
Whitney C.A., Random Processes in Physical Systems., John Wiley & Sons, 1990.

SPECTRAL RESPONSE OF A BEAM-STOP SYSTEM UNDER RANDOM EXCITATION

R. BOUC AND M. DEFILIPPI
Laboratoire de Mécanique et d'Acoustique CNRS
31 chemin J.AIGUIER 13402 Marseille Cedex. France

Abstract. A linearization procedure to estimate the spectral response at various points of a beam-stop system is proposed. The elastic stop is replaced by a spring with stiffness depending on the amplitude of the deflection at the impact location. Performing the expectation of the spectral density function of the linear system with respect to the probability density of the response amplitude (assumed to be a random variable), an estimate of the nonlinear response spectrum is derived. The efficiency of the method is checked by comparing the results with those of numerical simulations.

1. Introduction

The classical stochastic linearization method does not accurately predict the Power Spectral Density(PSD) of the stationary response of *lightly* damped nonlinear random systems, unless the nonlinearity is weak. If the spectral energy and resonance frequencies are correctly found, the linearization procedure generally leads to underestimated spectral-bandwidths and overestimated resonance levels.

To improve the linearization, the concept of equivalent linear system with random coefficients has been introduced for one degree of freedom lightly damped non-linear systems. A family of linear oscillators depending on a random variable related to the amplitude of motion, is considered. By taking the expectation of the spectral density function of the linear system with respect to the probability density of the random variable, an estimate of the nonlinear response spectrum is derived [1] to [7]. Explicit dependence between the frequency and the amplitude is needed and the use of the stochastic averaging method leads to an explicit formula for

A. Naess and S. Krenk (eds.), IUTAM Symposium on Advances in Nonlinear Stochastic Mechanics, 69–78.
© *1996 Kluwer Academic Publishers.*

the probability density [6], [7]. Recently the method has been extended to asymmetrical oscillators [8] and also to multidimensional nonlinear systems [9]. The numerical simulation results convincingly show how efficient this new method is with large non-linearities.

In the present paper a continuous beam-stop system is considered. Such model can be used in the study of heat exchangers and nuclear piping systems where flow-induced vibrations or seismic excitations cause impact at the vulnerable pipe-baffle interfaces. The procedure below will certainly be of use to solve the inverse problem [10]: Find the impact force from measured spectral displacement responses at some accessible points.

2. Beam-stop system. Spectral responses

In this section the effect of a pair of motion limiting stops on the bending vibration of an uniform beam of length L with moving support (excitation) is studied (Figure 1).

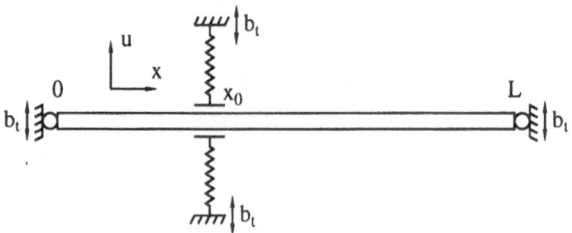

Figure 1. beam-stop system

The differential equation of motion can be written as

$$\rho S \frac{\partial^2}{\partial t^2} u(x,t) + \frac{\partial^2}{\partial x^2}\left(EI \frac{\partial^2}{\partial x^2} u(x,t)\right) + f(u_0(t))\delta(x - x_0) = -\rho S \ddot{b}_t, \qquad (1)$$

$u(x,t)$ denotes the relative deflection of the beam, E, I, ρ and S are the elasticity modulus, moment of inertia, mass density and area of the section, respectively. The impact force is a function of the deflection of the beam at x_0, $u_0(t) = u(x_0,t)$, and is defined by

$$f(u) \;=\; k(u+e) \text{ if } -\infty < u \le -e \qquad (2)$$
$$f(u) \;=\; 0 \text{ if } -e < u < +e \qquad (3)$$
$$f(u) \;=\; k(u-e) \text{ if } +e \le u \le +\infty \qquad (4)$$

k is the spring stiffness and e the clearance, $\delta(x - x_0)$ the Dirac delta function. The acceleration process of the moving support, $\ddot{b}_t$, is a wide-band (not necessarily the Gaussian white noise) stationary stochastic process with mean value zero. The Green's function of the beam without stops, which satisfies the (assumed) homogeneous boundary conditions at the ends [11], will be denoted by $G_0(x, x'; \omega)$ where ω is the frequency. Small amount of modal damping will be introduced, obtaining the "damped" Green's function which can be expressed as

$$G_\xi(x, x'; \omega) = \frac{EI}{\rho S L} \sum_m \frac{\phi_m(x)\phi_m(x')}{\omega_m^2 - \omega^2 + 2i\xi_m\omega_m\omega}, \quad i = \sqrt{-1} \qquad (5)$$

where ω_m, ϕ_m are the natural frequencies and normal modes of the beam without stops, respectively.

Our problem is the following: Let $S_{\ddot{b}}(\omega)$ be the PSD of the acceleration input, find $S_u(\omega; x)$, the corresponding PSD of $u(x, t)$ or $S_u(\omega; x, y)$, the cross-PSD between $u(x, t)$ and $u(y, t)$ where x an y are two arbitrary points of the beam.

2.1. THE CLASSICAL STOCHASTIC LINEARIZATION PROCEDURE

Following the well-known stochastic linearization procedure, equation (1) is replaced by the linear equation

$$\rho S \frac{\partial^2}{\partial t^2} u(x, t) + \frac{\partial^2}{\partial x^2}\left(EI \frac{\partial^2}{\partial x^2} u(x, t)\right) + K_{eq} u_0(t)\delta(x - x_0) = -\rho S \ddot{b}_t \qquad (6)$$

where the equivalent spring stiffness is given by

$$K_{eq} = Arg \min_K E[f(u_0) - K u_0]^2 \qquad (7)$$

so that

$$K_{eq} = \frac{E[f(u_0)u_0]}{E[u_0^2]}. \qquad (8)$$

Expectation in (7) is calculated using Gaussian law with mean value zero and variance $E[u_0^2]$. It follows that K_{eq} is a function of $E[u_0^2]$ which can be calculated from the spectrum $S_{u_0}(\omega)$.

Let

$$H_l(\omega; x) = \frac{\rho S}{EI} \int_0^L G_\xi(x, x'; \omega)dx' \qquad (9)$$

be the transfert function $-\ddot{b}_t \rightarrow u(x, t)$ of the beam without stops. The corresponding (equivalent) one from (6) solves

$$H_{eq}(\omega; x, K_{eq}) = -\frac{K_{eq}}{EI} G_\xi(x, x_0; \omega) H_{eq}(\omega; x_0, K_{eq}) + H_l(\omega; x) \qquad (10)$$

from which we deduce

$$H_{eq}(\omega; x_0, K_{eq}) = \frac{H_l(\omega; x_0)}{1 + \frac{K_{eq}}{EI} G_\xi(x_0, x_0; \omega)} \qquad (11)$$

$$H_{eq}(\omega; x, K_{eq}) = H_l(\omega; x)[1 - \frac{K_{eq}}{EI} \frac{G_\xi(x, x_0; \omega)}{1 + \frac{K_{eq}}{EI} G_\xi(x_0, x_0; \omega)}]. \qquad (12)$$

It follows that

$$S_u(\omega; x, y) = H_{eq}(\omega; x, K_{eq}) S_{\ddot{b}}(\omega) H_{eq}^*(\omega; y, K_{eq}) \qquad (13)$$

$$E[u_0^2(t)] = \frac{1}{2\pi} \int_{-\infty}^{\infty} | H_{eq}(\omega; x_0, K_{eq}) |^2 S_{\ddot{b}}(\omega) d\omega \qquad (14)$$

K_{eq} is then determinated by solving iteratively equations (14) and (8) using Gaussian law.

2.1.1. *Modal analysis associated to the equivalent system (6)*
The eigenvalue problem: Find $(\Omega, \Psi(x))$ such that all the boundary conditions are satisfied and

$$\Psi^{(iv)}(x) - \Omega^2 \frac{\rho S}{EI} \Psi(x) + \frac{K_{eq}}{EI} \Psi(x_0) \delta(x - x_0) = 0, \qquad (15)$$

allows us obtaining other useful expressions for PSD responses. If, for some integer k, $\phi_k(x_0) = 0$ then $\Omega_k = \omega_k$ and $\Psi_k(x) = \phi_k(x)$ are solutions of (15), where ω_k, ϕ_k are the natural frequencies and normal modes of the beam without stops respectively. Using the Green's function $G_0(x, x_0; \omega)$ we look for other solutions according to the form, [11],

$$\Psi_k(x) = -\frac{K_{eq}}{EI} \Psi_k(x_0) G_0(x, x_0; \Omega_k). \qquad (16)$$

For nontrivial solutions the equivalent frequencies, Ω_k, are given solving

$$1 + \frac{K_{eq}}{EI} G_0(x_0, x_0; \Omega_k) = 0, \ k = 1, 2, \ldots \qquad (17)$$

whereas the normalization condition gives

$$\Psi_k(x_0) = \frac{EI}{K_{eq}} [\frac{1}{L} \int_0^L G_0^2(x, x_0; \Omega_k) dx]^{-\frac{1}{2}}. \qquad (18)$$

The kth normal mode is then given by (16). Orthogonality-property of normal mode is easily deduced.

From the expansion

$$u(x,t) = \sum_k \Psi_k(x) q_k(t) \tag{19}$$

follows

$$\ddot{q}_k + 2\xi_k \omega_k \dot{q}_k + \Omega_k^2 q_k = -\nu_k \ddot{b}_t, \ \ k = 1, 2, \ldots \tag{20}$$

where modal damping is added and $\nu_k = \frac{1}{L} \int_0^L \Psi_k(x) dx$. Expression for PSD and cross-PSD result in

$$S_u(\omega; x) = \sum_{m,n} \Psi_m(x) H_m(\omega) S_{\ddot{b}}(\omega) H_n^*(\omega) \Psi_n(x) \tag{21}$$

$$S_u(\omega; x, y) = \sum_{m,n} \Psi_m(x) H_m(\omega) S_{\ddot{b}}(\omega) H_n^*(\omega) \Psi_n(y) \tag{22}$$

where

$$H_k(\omega) = \nu_k [\Omega_k^2 - \omega^2 + i2\xi_k \omega_k \omega]^{-1}. \tag{23}$$

The above expansions can be truncated considering only a finite number of modes, to calculate $E[u_0^2(t)]$ and K_{eq}.

We shall see in section 3 that the stochastic Gaussian linearization method leads to poor approximation of the nonlinear PSD response of the beam with stops.

2.2. LINEAR SYSTEM WITH RANDOM SPRING STIFFNESS

The main idea of this section is briefly pointed out in the introduction of the present paper, following the procedure detailed in [6] for 1-DOF system and in [9] for n-DOF.

We want to express K_{eq} in term of the amplitude of the deflection at x_0. To do this, it is assumed that the first mode q_1 is dominant in the expansion (19). The deflection at x_0 is momentarily written $u_0(t) = \Psi_1(x_0, a) q_1(t)$ where $q_1(t) = a \cos \Phi_1(t)$, $\Phi_1(t) = \Omega_1(a)t + \varphi_1$. Later, a will be considered as a random variable with probability density $p(a)$. Now, the expectations in (7) and (8) are calculated, *for fixed a*, with respect to the phase Φ_1 assumed to be uniformly distributed over $(0, 2\pi)$ (see [6]), yielding

$$K_{eq}(\Psi_1(x_0, a)a) = \frac{1}{2\pi \mid \Psi_1(x_0, a)a \mid} \int_0^{2\pi} f(\Psi_1(x_0, a)a \cos \Phi) \cos \Phi d\Phi. \tag{24}$$

The mode $\Psi_1(x, a)$ and the frequency $\Omega_1(a)$ (and also $\Psi_k(x, a), \Omega_k(a)$ for $k = 2, 3, \ldots$) are deduced, *for each value of a*, solving the nonlinear coupled equations (24), (16), (17) and (18) where the constant K_{eq} in (16), (17) and (18) is replaced by the new expression (24). Damped modes q_k are given by (20) with Ω_k replaced by $\Omega_k(a)$ and ν_k by $\nu_k(a)$. Conditional PSD responses, $S_u(\omega; x, y \mid a)$, are given by (13) or (21), (22), (23) with $K_{eq}(\Psi_1(x_0, a)a)$, $\Psi_k(x, a)$, $\Omega_k(a)$ in place of K_{eq}, $\Psi_k(x)$, Ω_k.

Assuming the results in [6] hold for the present case, the probability density of the random variable a is given by

$$p(a) = Ca\Omega_1(a)D_1(a)\exp[-\int_0^a \frac{4\xi_1\omega_1 a\Omega_1^2(\alpha)D_1(\alpha)}{\nu_1^2(\alpha)S_{\ddot{b}}(\Omega_1(\alpha))}d\alpha] \qquad (25)$$

where $D_1(a) = 1 + a\frac{\Omega_1'(a)}{2\Omega_1(a)}$, $\Omega_1'(a)$ denotes the derivative of $\Omega_1(a)$ with respect to a, and C the normalization constant. Approximate PSD responses for the beam-stop system are then obtained by averaging the conditional spectrum with respect to the probability density (25). As an example $S_u(\omega; x)$ is given by

$$S_u(\omega; x) = \int_0^\infty S_u(\omega; x \mid a)B^2(a)p(a)da, \qquad (26)$$

where

$$B^2(a) = \frac{2\xi_1\omega_1 a^2\Omega_1^2(a)}{\nu_1^2(a)S_{\ddot{b}}(\Omega_1(a))} \qquad (27)$$

in order to preserve energy for fixed a: From $u(x, t) \simeq \Psi_1(x, a)a\cos\Phi_t$ we have approximately

$$E(u^2 \mid a) \simeq \frac{\Psi_1^2(x, a)a^2}{2}. \qquad (28)$$

On the other hand, by also considering only the first mode in the expansion (21), we get

$$\frac{1}{2\pi}\int_{-\infty}^\infty S_u(\omega; x \mid a)B^2(a)d\omega \simeq \frac{\Psi_1^2(x, a)\nu_1^2(a)B^2(a)S_{\ddot{b}}(\Omega_1(a))}{4\xi_1\omega_1\Omega_1^2(a)}, \qquad (29)$$

since $\ddot{b}_t$ is "wide-band" and the damping coefficient ξ_1 is "small". Equating (28) and (29) yields equation (27).

3. Example

The above procedure is applied to the beam-stop system with hinged ends already studied in [2]. Geometrical and mechanical parameters: Elasticity modulus $E = 1.96 \times 10^{11}(Nm^{-2})$, mass density $\rho = 7.8 \times 10^3(kgm^{-3})$, moment of inertia $I = 1.19 \times 10^{-6}(m^4)$, length $L = 6(m)$, section area $S = 1.37 \times 10^{-3}(m^2)$, spring-stop stiffness $k = 2 \times 10^6(Nm^{-1})$ and excitation $\ddot{b}_t$: a Gaussian white noise with unit strength, will be fixed in the following. Various location, x_0, and clearance, e, of the stop, will be considered. Boundary conditions imply: $\omega_k = (\frac{k\pi}{L})^2(\frac{EI}{\rho S})^{\frac{1}{2}}$, $\phi_k(x) = 2^{\frac{1}{2}} \sin \frac{k\pi x}{L}$.

Assuming a solution of the form $u(x,t) = \sum_{k=1}^{3} \phi_k(x) q_k(t)$, equation (1) is replaced by a 3-dimensional, second-order nonlinear differential system. After adding modal damping with coefficients $c_k = 2\xi_k \omega_k, k = 1, 2, 3$, numerical simulations are performed, using a classical step by step integration scheme and FFT procedure for PSD estimates, as detailed in [2].

Setting $U_1^0(a) = | \Psi_1(x_0, a)a |$, equation (24) leads to

$$K_{eq} = 0 \text{ if } U_1^0(a) \leq e \tag{30}$$

$$K_{eq} = \frac{2k}{\pi}(\arccos\frac{e}{U_1^0(a)} - \frac{e}{U_1^0(a)}(1 - (\frac{e}{U_1^0(a)})^2)^{\frac{1}{2}}) \text{ if } U_1^0(a) > e \tag{31}$$

so that $\Psi_k(x, a) = \phi_k(x)$, $\Omega_k(a) = \omega_k$ for $U_1^0(a) \leq e$. *For each value of a,* the nonlinear coupled equations (30),(31), (16),(17),(18) are solved using a Newton's method. To be consistent with the order of the simulated system, only three terms in the expansion of the Green's function

$$G_0(x, x_0; \Omega_k(a)) = \frac{EI}{\rho SL} \sum_{k=1}^{3} \frac{\phi_k(x)\phi_k(x_0)}{\omega_k^2 - \Omega_k^2(a)} \tag{32}$$

are kept in equations (16),(17).

Figure 2 describes the behaviour of the first two frequencies $\frac{\Omega_1(a)}{2\pi}$, $\frac{\Omega_2(a)}{2\pi}$ and modes $\Psi_1(x_0, a)$, $\Psi_2(x_0, a)$, for $x_0 = 2(m)$ and $e = 1.9 \times 10^{-3}(m)$, versus the amplitude a of the first mode. The overall range of a corresponds to the support of the probability $p(a)$ (25).

Figure 3 has been obtained for constant modal damping $\xi_k = 0.05$, k=1,2,3 with, for Figure 3(a), $e_0 = 1.9 \times 10^{-3}(m)$, $x_0 = 2(m)$ and for Figure 3(b) $e_0 = 0.39 \times 10^{-3}(m)$, $x_0 = 1.5(m)$. The first resonance of the mean PSD over $(0, L)$, $S_u(\omega) = \frac{1}{L} \int_0^L S_u(x, \omega)dx$, is shown in Figure 3(a). Comparing with the numerical simulation result, it is seen that the SGL method (dashed line) yields correct resonance frequency but overestimated associated level and underestimated spectral bandwidth. The proposed method (dotted line) gives correct results. Note that the effect of the stops moves the frequency of the first mode from $6.5Hz$ to about $15Hz$ for the considered system parameters, in particular for the considered large value of the spring-stop stiffness k. The spectral broadening and level of the first three modes for PSD at x_0 are well described by the proposed method as shown in Figure 3(b).

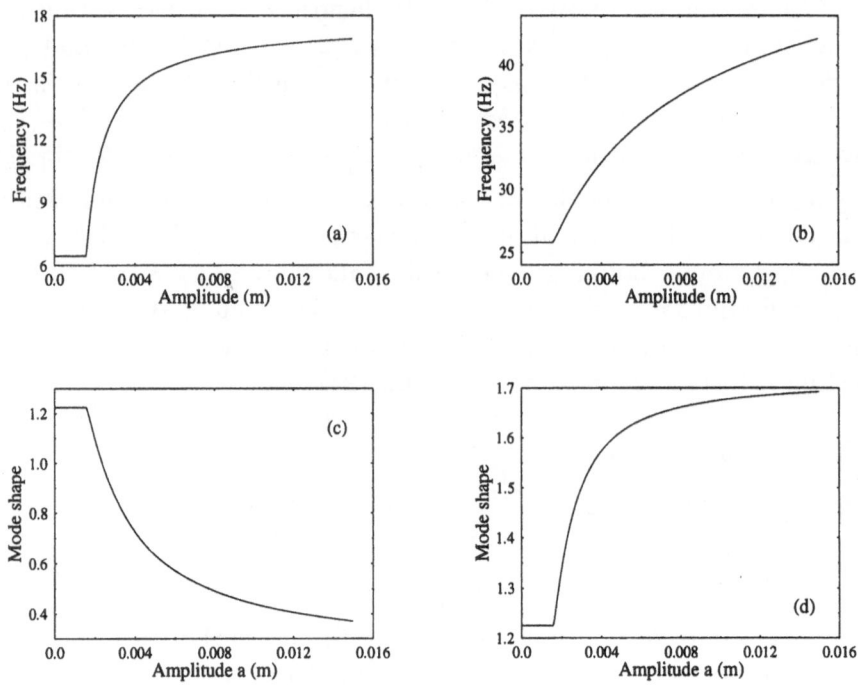

Figure 2. Frequencies and mode shape at x_0. Mode 1 (a), (c). Mode 2 (b), (d), for $x_0 = 2(m)$, $e = 1.9 \times 10^{-3} m$ and spring-stop stiffness $k = 2 \times 10^6 (Nm^{-1})$.

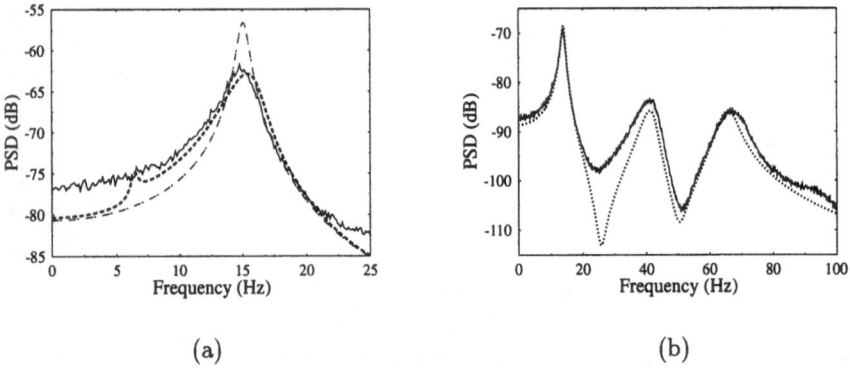

Figure 3. (a) $S_u(\omega)$ for $x_0 = 2(m)$, $e = 1.9 \times 10^{-3}(m)$ and $k = 2 \times 10^6 (Nm^{-1})$. (b) $S_u(\omega; x_0)$ for $x_0 = 1.5(m)$, $e = 0.39 \times 10^{-3}(m)$ and $k = 2 \times 10^6 (Nm^{-1})$. (Solid line: simulation; (a) dashed line: SGL method; (a), (b) dotted line: the proposed method).

Figure 4 has been obtained with $\xi_1 = 0.05$, $\xi_2 = \frac{0.05}{4}$, $\xi_3 = \frac{0.05}{9}$. The mean spectrum $S_u(\omega)$ is shown in Figure 4(a). The spectral broadening and the resonance levels are well described despite the fact that higher harmonics (which are present in the simulated response (arrow)) are ignored by the method. The level of these harmonics is greater in the PSD of the reaction force at x_0, as indicated by the arrows in Figure 4(b). For fixed a, the PSD of the linearized reaction force at x_0 is given by $S_f(\omega; x_0 \mid a) = K_{eq}^2 \mid H_{eq}(\omega; x_0, K_{eq}) \mid^2 B^2(a) S_{\ddot{b}}(\omega)$, where H_{eq} is given by (11), K_{eq} by (30), (31) and $B^2(a)$ by (27). After averaging with respect to $p(a)$ the PSD is shown in Figure 4(b) where only the third harmonic is taken into account, as detailed in [8].

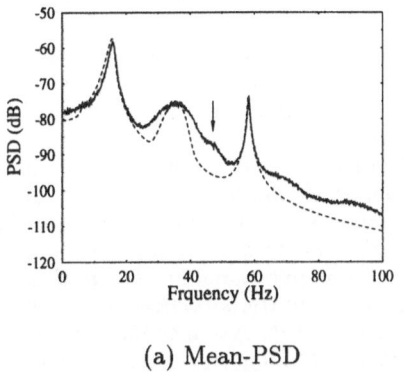

(a) Mean-PSD

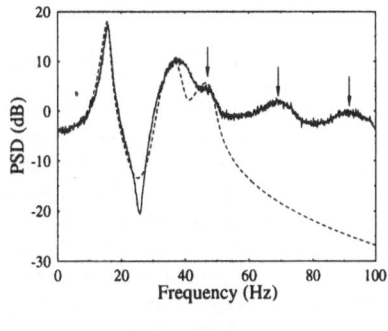

(b) PSD of the reaction force

Figure 4.   PSD for $x_0 = 2(m)$, $e = 1.9 \times 10^{-3}(m)$, $k = 2 \times 10^6 (Nm^{-1})$. (Solid line: simulation; dotted line: the proposed method)

4. Concluding remark

The broadening, level and shift of the three resonant peaks are described satisfactorily by the proposed method to estimate the PSD responses (deflection and reaction force) of the beam-stop system.

It seems possible to improve the above procedure by considering more than one mode in the expansion (19). Setting $u_0^n = \sum_{k=1}^n \Psi_k(x_0, a) q_k$, where $q_k = a_k \cos \Phi_k$, $K_{eq}(a, \Psi(x_0, a))$ is defined by

$$K_{eq}(a, \Psi(x_0, a)) = \frac{\int_0^{2\pi} \ldots \int_0^{2\pi} f(u_0^n) u_0^n d\Phi_1 d\Phi_2 \ldots d\Phi_n}{(2\pi)^n \frac{1}{2} \sum_1^n [\Psi_k^2(x_0, a) a_k^2]}$$

where now, $a = (a_1, \ldots, a_n)$ is a vector-valued random variable and $\Psi(x_0, a)$ denotes the vector $(\Psi_1(x_0, a), \ldots, \Psi_n(x_0, a))$. The same modal analysis allows to calculate the frequencies $\Omega_k(a)$ and the modes $\Psi_k(x, a)$ explicitly.

As in [9], the object of present research is to find a suitable expression for the joint probability density function, $p(a) = p(a_1, a_2, \ldots, a_n)$.

References

1. Miles, R.N. (1989) An approximation solution for spectral response of Duffing's oscillator with random input, *Journal of sound and vibration* 132 **1**, 43-49.
2. Guihot, P. (1990) Analyse de la réponse de structures non-linéaires sollicitées par des sources d'excitation aléatoires, application au comportement des lignes de tuyauteries sous l'effet d'un seisme, *Thèse de Doctorat de l'Université Paris VI, INSTN*
3. Bouc, R. (1991) The power spectral density of a weakly damped strongly nonlinear random oscillation and stochastic averaging *Publication du LMA: Colloque Contrôle actif Vibro-acoustique et dynamique stochastique, Marseille, ISSN 0750-7356*, 127
4. Soize, C. (1991) Sur le calcul des densités spectrales des réponses stationnaires pour des systèmes dynamiques stochastiques non-linéaires *Publication du LMA: Colloque Contrôle actif Vibro-acoustique et dynamique stochastique, Marseille, ISSN 0750-7356,* 127
5. Miles, R.N. (1993) Spectral response of a bilinear oscillator, *Journal of sound and vibration* 163 **2**, 319-326.
6. Bouc, R. (1994) The power spectral density of response for a strongly nonlinear random oscillator, *Journal of sound and vibration* 175 **3**, 317-331.
7. Fogli, M., Bressolette, P. and Bernard, P. (1994) Dynamics of the stochastic oscillator with impact. To appear in the *European Journal of Mechanics.*
8. Bellizzi, S. and Bouc, R. (1994) Spectral response of asymmetrical random oscillators. To appear in *Probabilistic Engineering Mechanics.* This paper has been also presented in the *Second International Conference on Computational Stochastics Mechanics/Athens/Greece/ 12-15 June, 1994.* A.A. Balkema/Rotterdam/Brookfield/1995.
9. Bellizzi, S. and Bouc, R. (1995) Spectre de puissance de systèmes non-linéaires vibrants sous sollicitations aléatoires, *Second Colloque National en Calcul des Structures Giens (France), 16-19 mai, 1995.* Editions Hermès, Paris.
10. Lin, S.Q. and Bapat, C.N. (1993) Extension of clearance and impact force estimation approaches to a beam-stop system, *Journal of sound and vibration* 163 **3**, 423-428.
11. Nicholson, J. W. and Bergman, L. A. (1986) Free-vibration of combined dynamical system, *Journal of Engineering Mechanics* 112 **1**, 1-13.

HIGHER ORDER APPROXIMATIONS FOR MAXIMA
OF RANDOM FIELDS

KARL BREITUNG

Department of Civil Engineering
University of Calgary
2500 University Drive N.W.
Calgary, Alberta, T2N 1N4

1. Introduction

In many applications random influences are modelled by random fields. Examples can be found in [3] and [11].

The distribution of the maxima of differentiable Gaussian random processes and fields can in general not computed exactly. Therefore approximations for these distributions were developed.

For univariate stationary processes the book of Leadbetter et al. [8] gives an overview of the results. Homogeneous random fields are studied in Adler [1]. For stationary vector processes some approximations are derived in Breitung [6], chapter 8.

In this paper a higher order approximation for the mean number of maxima over a high level for homogeneous fields is given. Then a higher order approximation for the maximum of a non-homgeneous zero mean process with constant variance is derived.

2. The maxima of a homogeneous field

We will use in the following the notation of the book of Adler [1]. Given is a twice continuously differentiable Gaussian random field on a set $T \subset \mathbb{R}^n$. The covariance function of the random field is defined by

$$R(s,t) = \mathbb{E}(X(s)X(t)).$$ (1)

A. Naess and S. Krenk (eds.), IUTAM Symposium on Advances in Nonlinear Stochastic Mechanics, 79–88.
© *1996 Kluwer Academic Publishers.*

We assume that the field is homogeneous, i.e. that $R(s,t)$ depends only on the distance $t - s$ and that for all $t \in S$

$$\mathbb{E}(X(t)) = 0 \qquad (2)$$
$$\mathrm{var}(X(t)) = 1. \qquad (3)$$

We get then for the first partial derivatives

$$\mathbb{E}(X_i(t)) = 0 \qquad (4)$$
$$\mathbb{E}(X_i^2(t)) = -R_{ii}(t) = \lambda_i. \qquad (5)$$

By a suitable rotation we can achieve always that the matrix of the covariances $\mathrm{cov}(X_i(t), X_j(t))$ is a diagonal matrix Λ with diagonal elements $\lambda_1, \ldots, \lambda_n$. For the covariances between the process and the first and second derivatives we obtain then

$$\mathrm{cov}(X(t), X_i(t)) = 0 \qquad (6)$$
$$\mathrm{cov}(X_i(t), X_{kl}(t)) = 0 \qquad (7)$$
$$\mathrm{cov}(X(t), X_{kl}(t)) = \delta_{kl}\lambda_k \qquad (8)$$
$$\mathrm{cov}(X_i(t), X_j(t)) = \delta_{ij}\lambda_i \qquad (9)$$

Further we have for the covariances between the second derivatives (see [1], p. 31)

$$\mathrm{cov}(X_{ij}(t), X_{kl}(t)) = \left. \frac{\partial^4 R(t,t)}{\partial t_i \partial t_j \partial t_k \partial t_l} \right|_{t=0}. \qquad (10)$$

As in [1] the matrix $X''(t)$ is also considered as an $n(n+1)/2$ dimensional vector in the computations where the means and covariances of $X''(t)$ are needed. The expected number of local maxima above a level β of a homogeneous n-dimensional Gaussian random field on a compact set $T \subset \mathbb{R}^n$ is given by the following integral (see [1], p. 123)

$$\mathbb{E}(M_\beta(T)) = (-1)^n \lambda(T) \int_\beta^\infty \int_{G(x'')} \det(x'') \varphi(x, o, x'') \, dx'' \, dx. \qquad (11)$$

Here $\varphi(x, o, x'')$ is the joint normal p.d.f. of $X(t)$, $X'(t)$ and $X''(t)$ and

$$G(x'') = \{x''; x'' \text{ is a negative definite matrix}\}. \qquad (12)$$

The Lebesgue measure of the set T is denoted by $\lambda(T)$.

In [1] an asymptotic approximation for this expected value as $\beta \to \infty$ is derived. In the following we will derive a two term approximation where the first term is identical to the approximation in Adler.

Since X' and (X, X'') are independent, we can write the density in the last equation as the product of the p.d.f. of X' at $\mathbf{o}$, the p.d.f. of X and the conditional p.d.f. $\varphi(x''|x)$ of X'' given $X = x$. This gives then for the integral the form

$$\varphi(x, \mathbf{o}, x'') = (2\pi)^{-(n+1)/2} |\det(\Lambda)|^{-1/2} \exp\left(-\frac{x^2}{2}\right) \varphi(x''|x). \qquad (13)$$

Let the covariance matrix of the second derivatives be denoted by Σ. The conditional covariance matrix Σ_c of the second derivatives under the condition $X(t) = x$ is then

$$\Sigma_c = \Sigma - \Lambda\Lambda^T. \qquad (14)$$

The conditional mean values are given by

$$\mathbb{E}(X''|X = x) = x\Lambda. \qquad (15)$$

Therefore the conditional p.d.f. of the second derivatives can be written as

$$\varphi(x''|x) = |\det(\Sigma_c)|^{-1/2} \exp\left(-\frac{1}{2}(x'' - x\Lambda)\Sigma_c^{-1}(x'' - x\Lambda)\right). \qquad (16)$$

Using the explicit form of the conditional density, we can rewrite equation 11 (see [1], p. 134)

$$\mathbb{E}(M_\beta(S)) = \frac{(-1)^n \lambda(T)}{(2\pi)^{n/2} |\det(\Lambda)|^{1/2} |\det(\Sigma_c)|^{1/2}} \qquad (17)$$

$$\times \underbrace{\int_\beta^\infty \int_{G(x'')} \det(x'') \exp\left(-\frac{1}{2}(x^2 + (x'' - x\Lambda)^T \Sigma_c^{-1}(x'' - x\Lambda))\right) dx'' dx}_{=I(\beta)}. \qquad (18)$$

We evaluate the function $J(\beta) = \beta^{-(n+1)n/2-(n+1)} I(\beta)$ using the Laplace method. Making now the transformation $x \mapsto w = \beta^{-1}x$ and $x'' \mapsto w'' = \beta^{-1}(x'' - x\Lambda)$, we get

$$J(\beta) = \int_D \det(w'' + w\Lambda) \exp\left(-\frac{\beta^2}{2}(w^2 + w''^T \Sigma_c^{-1} w'')\right) dw'' dw \qquad (19)$$

The integration domain is defined by

$$D = \{w \geq 1, G(w'' + w\Lambda)\}. \tag{20}$$

The logarithm of the density has the form

$$\ln(\ldots) \propto -\beta^2(l(w, w'')) \propto -\frac{\beta^2}{2}(w^2 + w'' \Sigma_c^{-1} w''). \tag{21}$$

In the domain D the logarithm has a unique global maximum at the point $(1, \mathbf{0}_{nn})$. The determinant $\det(w'' + w\Lambda)$ at this point is $\det(\Lambda)$.

To find the Taylor expansion of the determinant at this point, we consider the determinant $\det(A)$ of a symmetric $n \times n$ matrix A. The first and second derivatives of $\det(A)$ are then the determinants of certain submatrices of A. Let A_{ij} be the submatrix of A obtained by deleting the i-th row and j-th column. We have

$$\left(\frac{\partial \det(A)}{\partial a_{ij}} \right)_{i,j=1,\ldots,n} = (2A_{ij} - \text{diag}(A_{ii}))_{i,j=1,\ldots,n} \tag{22}$$

with A_{ij} the cofactor of a_{ij}

$$A_{ij} = (-1)^{i+j} \det(A_{ij}). \tag{23}$$

For the second derivatives we obtain

$$\frac{\partial^2 \det(A)}{\partial a_{ij} \partial a_{kl}} = \begin{cases} 0 & \text{if } i = j = k = l \\ A_{ii;kk} & \text{if } i = j \text{ and } k = l \text{ but } i \neq k \\ -2A_{ij;kl} & \text{elsewhere} \end{cases} \tag{24}$$

Here

$$A_{ij;kl} = \det(A_{ij;kl}), \tag{25}$$

i.e. the determinant of the submatrix obtained by deleting the i-th and k-th row and the j-th and l-th column. In the case of a diagonal matrix $\Lambda = \text{diag}(\lambda_1, \ldots, \lambda_n)$ we have then

$$\Lambda_{ij;kl} = \begin{cases} \prod_{k=1, m\neq i,k}^{n} \lambda_m & \text{if } i = j \text{ and } k = l \text{ but } i \neq k \\ -2\prod_{m=1, m\neq i,j}^{n} \lambda_m & \text{if } i = k \text{ and } j = l \text{ but } i \neq k \\ 0 & \text{elsewhere} \end{cases} \tag{26}$$

This gives the following Taylor expansion for the determinant at $(1, \mathbf{o}_{nn})$

$$\det(\mathbf{w}'' + \mathbf{w}\Lambda) = \det(\Lambda)\left(1 + n(w-1) + \sum_{i=1}^{n}\frac{w_{ii}}{\lambda_i}\right.$$

$$\left. +2\sum_{i=1}^{n}\sum_{j=i+1}^{n}\frac{w_{ii}w_{jj}}{\lambda_i\lambda_j} - 2\sum_{i=1}^{n}\sum_{j=i+1}^{n}\frac{w_{ij}^2}{\lambda_i\lambda_j} + r(w, \mathbf{w}'')\right)$$

$$= \det(\Lambda)\left(1 + n(w-1) + \sum_{i=1}^{n}\frac{w_{ii}}{\lambda_i} - 2\sum_{i=1}^{n}\sum_{j=i+1}^{n}\frac{w_{ii}w_{jj} - w_{ij}^2}{\lambda_i\lambda_j} + r(w, \mathbf{w}'')\right) . (27)$$

The last term is asymptotically of order $o(\varphi(\beta)\beta^{-2})$ and can be neglected for an expansion up to order $\varphi(\beta)\beta^{-2}$.

$$J(\beta) = \det(\Lambda)\left[\int_D \exp(-\frac{\beta^2}{2}l(w, \mathbf{w}'')) \, d\mathbf{w}''dw\right.$$

$$+ \quad n\int_D (w-1)\exp(-\frac{\beta^2}{2}l(w, \mathbf{w}'')) \, d\mathbf{w}''dw$$

$$+ \quad \frac{1}{2}\sum_{i=1}^{n}\sum_{j=i+1}^{n}\int_D \frac{w_{ii}w_{jj} - w_{ij}^2}{\lambda_i\lambda_j} \exp(-\frac{\beta^2}{2}l(w, \mathbf{w}'')) \, d\mathbf{w}''dw$$

$$+ \quad \int_D r(\mathbf{w}'', w)\exp(-\frac{\beta^2}{2}l(w, \mathbf{w}'')) \, d\mathbf{w}''dw. \qquad (28)$$

Evaluating these integrals with the Laplace method gives

$$I(\beta) = \beta^{n-1}(2\pi)^{n(n+1)/2}|\det(\Sigma_c)|^{1/2}\det(\Lambda)$$

$$\times\left[1 + \beta^{-2}\left(n + \frac{1}{2}\sum_{i=1}^{n}\sum_{j=i+1}^{n}\frac{\mathbb{E}(X_{ii}X_{jj}|X=\beta) - \mathbb{E}(X_{ij}^2)}{\lambda_i\lambda_j} + o(\beta^{-2})\right)\right] . (29)$$

Here the covariance between X_{ii} and X_{jj} depends on X, but not the variance of X_{ij}, since the matrix Λ of the covariances between X and the second derivatives was chosen as a diagonal matrix. This conditional covariance is here

$$\mathbb{E}(X_{ii}X_{jj}|X=\beta) = \text{cov}(X_{ii}X_{jj}) - \lambda_i\lambda_j. \qquad (30)$$

But equation 10 shows that the covariance is equal to the variance $\text{var}(X_{ij})$. Therefore all terms in the summation are equal to -1. This gives then

$$I(\beta) = \det(\Lambda)\left(1 - \frac{(n-2)n}{2\beta^2} + o(\beta^{-2})\right). \qquad (31)$$

The final result is then

$$\mathbb{E}(M_\beta(T)) = \frac{\lambda(T) |\det(\Lambda)|^{1/2} \beta^{n-1}}{(2\pi)^{(n+1)/2}}$$

$$\times \exp\left(-\frac{\beta^2}{2}\right)\left[1 - \frac{(n-1)n}{2\beta^2} + o(\beta^{-2})\right]. \tag{32}$$

For the case of a two-dimensional random field we have the form

$$\mathbb{E}(M_\beta(T)) = \frac{\lambda(T) |\det(\Lambda)|^{1/2} \beta}{(2\pi)^{3/2}} \exp\left(-\frac{\beta^2}{2}\right)\left[1 - \beta^{-2} + o(\beta^{-2})\right]. \tag{33}$$

3. The Karhunen-Loève expansion for random fields

Under some conditions it is possible to make an eigenfunction expansion of the covariance function $R(s,t)$ which leads to an eigenfunction representation of the random field. The details for the univariate case can be found in [9], p. 412-416 and for the multivariate case in [1], chap. 3.3. In the following we will not assume that $X(t)$ is homogeneous, only that the mean is equal to zero and the variance is equal to unity.

An eigenfunction $\phi(t)$ of of the covariance function $R(t,s)$ with eigenvalue λ is a function satisfying the equation

$$\int_T R(t,s)\phi(s)\,ds = \lambda\phi(t). \tag{34}$$

Mercer's theorem states that there exists a series of orthonormal eigenfunctions $\phi_j(t)$ with eigenvalues λ_j such that

$$R(t,s) = \sum_{j=1}^{\infty} \lambda_j \phi_j(t)\phi_j(s). \tag{35}$$

If such a sequence $\phi_1(t), \phi_2(t), \ldots$ for the covariance function is given, the random field $X(t)$ has the following expansion

$$X(t) = \sum_{j=1}^{\infty} h_j(t) X_j \tag{36}$$

with

$$h_j(t) = \lambda_j^{1/2} \phi_j(t) \tag{37}$$

and the X_i's are a sequence of independent standard normal random variables.

4. An approximation for the global maximum

In the following we will assume that the covariance function has a finite expansion in the form

$$R(t, s) = \sum_{j=1}^{k} \lambda_j \phi_j(t) \phi_j(s). \tag{38}$$

Then the random field has a finite expansion in the form

$$X(t) = \sum_{j=1}^{k} h_j(t) X_j \tag{39}$$

with the $X_1, \ldots, X_k$ again k independent standard normal random variables. We define now the vector function $h : T \to I\!R^k, t \mapsto h(t)$ by

$$h(t) = (h_1(t), \ldots, h_k(t)). \tag{40}$$

Since we have $\text{var}(X(t)) = 1$, we get due to the definition of the $h_j(t)$ that

$$1 = \text{var}(X(t)) = \sum_{i=1}^{k} h_i^2(t)\text{var}(X_i) = \sum_{i=1}^{k} h_i(t)^2 = |h(t)|. \tag{41}$$

Therefore the length of the vector $h(t)$ is unity. Let $S^{k-1} = \{x \in I\!R^k; |x| = 1\}$ be the unit sphere in the k-dimensional space. Since $|h(t)| = 1$ for all $t \in T$ the function $h(t)$ is in fact a function $T \to S^{k-1}, t \mapsto h(t)$. The set $h(T)$ is therefore a submanifold of this sphere.

Now we have for probability that the maximum of the random field is larger than the value β the expression

$$P(\beta) = \text{P}(\max_T X(t) > \beta) = \text{P}(\max_T \langle h(t), X \rangle > \beta) \tag{42}$$

with $\langle x, y \rangle$ the scalar product of x and y. Under the assumptions made the function is injective, since elsewhere we would have two different points t and s with the same function value $h(t) = h(s)$ and this would mean $\text{corr}(X(t), X(s)) = 1$, i.e. $X(t)$ and $X(s)$ are equal with probability one.

Let $X = (X_1, \ldots, X_k)$ be a standard normal random vector. Then the random vector $U = (X_1/|X|, \ldots, X_k/|X|)$ and the random variable $|X| =$

$\sqrt{\sum_{j=1}^{k} X_i^2}$ are independent. We have $X = |X|U$. The length $|X|$ has a χ-distribution with k degrees of freedom. Its p.d.f. is

$$f_k(x) = \underbrace{\frac{2}{2^{k/2}\Gamma(k/2)}}_{=c_k} x^{k-1} e^{-x^2/2}. \qquad (43)$$

The vector U has a uniform distribution on the unit sphere S^{k-1}. Using this, we can rewrite now the probability on the right hand side of (42) in the form

$$P(\beta) = \int_{\beta}^{\infty} P(\max_{T}\langle h(t), xU\rangle > \beta) f_k(x) \, dx \qquad (44)$$

$$= \int_{\beta}^{\infty} P(\max_{T}\langle h(t), U\rangle > \frac{\beta}{x}) f_k(x) \, dx. \qquad (45)$$

Making the substitution $x \to u = x/\beta$, we get

$$P(\beta) = \beta \int_{1}^{\infty} P(\max_{T}\langle h(t), U\rangle > u^{-1}) f_k(\beta u) \, du \qquad (46)$$

$$= c_k \beta^k \int_{1}^{\infty} P(\max_{T}\langle h(t), U\rangle > u^{-1}) u^{k-1} e^{-(\beta u)^2/2} \, du. \qquad (47)$$

For this integral now an asymptotic expansion can be made as $\beta \to \infty$ using Watson's lemma (see [2], p. 102 or [6], p. 45), if we have an expansion of the function $P(\max_{T}\langle h(t), U\rangle > u^{-1})$ as $u \to 1+$. Since U has a uniform distribution over the sphere, the probability is just the surface integral over the set $\{u \in S^{k-1}; \inf_T |u - h(t)| \le \arccos(1 - u^{-2}/2)\}$ divided by the surface area of the sphere. This set is a neighborhood of the n-dimensional submanifold $H = \{(h_1(t), \ldots, h_k(t)); t \in T\}$ of the sphere S^{k-1}.

For the general case of $n > 2$, Sun [10] derives an two term asymptotic approximation for the integral, but his result expresses the second term of the expansion in way rather unsuitable to computation as the total or scalar curvature of the submanifold without deriving its form as a function of the mapping h.

As an example we will consider a two-dimensional field where $X(t_1, t_2)$ is given by

$$X(t_1, t_2) = \sum_{i=1}^{k} h_i(t_1, t_2) X_k. \qquad (48)$$

Let a coordinate vector field be given by

$$E_i(t_1, t_2) = \left(\frac{\partial h_1(t_1, t_2)}{\partial t_i}, \ldots, \frac{\partial h_k(t_1, t_2)}{\partial t_i} \right). \tag{49}$$

Integrating over the sphere S^{k-1} the following result is derived by Knowles and Siegmund [7]

$$P(\max_T X(t_1, t_2) > \beta) = \beta \varphi_1(\beta) \frac{\lambda(M)}{2\pi} + \varphi_1(\beta) \frac{\lambda(\partial M)}{\sqrt{2\pi}} \tag{50}$$

$$+ \quad \frac{\varphi_1(\beta)}{\beta} \frac{\chi(M)}{2\pi} + o(\varphi_1(\beta)/\beta) \tag{51}$$

with $\varphi_1(x) = (2\pi)^{-1/2} \exp(-x^2/2)$ the standard normal density. Here $\lambda(M)$ is the two-dimensional surface area given by

$$\lambda(M) = \int_0^1 \int_0^1 \det(E_i(t_1, t_2)^T \cdot E_j(t_1, t_2))^{1/2} dt_1 dt_2. \tag{52}$$

The value of $\lambda(\partial M)$, the length of the boundary of M, is given by

$$\lambda(\partial M) = \int_0^1 \sqrt{E_1^2(t_1, 0)} \, dt_1 \quad + \quad \int_0^1 \sqrt{E_1^2(1, t_2)} \, dt_2 \tag{53}$$

$$+ \int_0^1 \sqrt{E_1^2(t_1, 1)} \, dt_1 \quad + \quad \int_0^1 \sqrt{E_1^2(0, t_2)} \, dt_2. \tag{54}$$

$\chi(M)$ is the Euler -Poincaré characteristic of the set M. Its value is an even integer less equal two. Further details can be found in the article [7].

5. Summary and Conclusions

In this paper two different methods for higher order approximations for the maxima of Gaussian random fields were presented. The first result is a two term approximation for the expected number of local maxima of a homogeneous Gaussian field over a high level. The second method uses the eigenfunction expansion of the covariance function for obtaining a two term approximation for the global maximum of a non-homogeneous Gaussian field with constant variance.

It should be possible to derive similar results for non-stationary Gaussian processes and fields. Until now for them there are only upcrossing rate approximations for the maxima ([4] and [5]).

References

1. R.J. Adler. *The Geometry of Random Fields*. Wiley, New York, 1981.
2. N. Bleistein and R.A. Handelsman. *Asymptotic Expansions of Integrals*. Dover Publications Inc., New York, 1986. Reprint of the edition by Holt, Rinehart and Winston, New York, 1975.
3. V.V. Bolotin. *Wahrscheinlichkeitsmethoden zur Berechnung von Konstruktionen*. VEB-Verlag für das Bauwesen, Berlin (GDR), 1981.
4. K. Breitung. Asymptotic approximations for the extreme value distribution of non-stationary differentiable normal processes. In *Transactions of the Tenth Prague Conference on Information Theory, Statistical Decision Functions and Random Processes 1986*, volume I, pages 207–215, Prague, Czech. Rep., 1988. Academia.
5. K. Breitung. The extreme value distribution of non-stationary vector processes. In A. H.-S. Ang, M. Shinozuka, and G.I. Schuëller, editors, *Proceedings of ICOSSAR '89 5th Int'l Conf. on structural safety and reliability*, volume II, pages 1327–1332. American Society of Civil Engineers, 1990.
6. K. Breitung. *Asymptotic Approximations for Probability Integrals*. Springer, Berlin, 1994. Lecture Notes in Mathematics, Nr. 1592.
7. M. Knowles and D. Siegmund. On Hotelling's approach to testing for a nonlinear parameter in regression. *International Statistical Review*, 57(3):205–220, 1989.
8. M.R. Leadbetter, G. Lindgren, and H. Rootzén. *Extremes and Related Properties of Random Sequences and Processes*. Springer, New York, 1983.
9. A. Papoulis. *Probability, Random Variables and Stochastic Processes*. McGraw-Hill, New York, third edition, 1991.
10. J. Sun. Tail probabilities of the maxima of Gaussian random fields. *The Annals of Probability*, 21(1):34–71, 1993.
11. D. Veneziano, M. Grigoriu, and C.A. Cornell. Vector process models for system reliability. *Journal of the Engineering Mechanics Division ASCE*, 103(3):441–460, 1977.

EXTENSION OF THE STOCHASTIC DIFFERENTIAL CALCULUS TO COMPLEX PROCESSES

S.CADDEMI* and G.MUSCOLINO**

*Dipartimento di Ingegneria Strutturale & Geotecnica
Università di Palermo, Viale delle Scienze,
I-90128 Palermo, Italy

**Facoltà di Ingegneria
Università di Messina, Contrada Sperone 31
I-98166 S.Agata, Messina, Italy

Abstract

In structural engineering complex processes arise to predict the first excursion failure, fatigue failure, etc. Indeed to solve these problems the envelope function, which is the modulus of a complex process, is usually introduced. In this paper the statistics of the complex response process related to the envelope statistics of linear systems subjected to parametric stationary normal white noise input are evaluated by using extensively the properties of stochastic differential calculus.

1. Introduction

For Gaussian inputs and linear systems, the first and the second order moments completely define the statistics of the response. However, in many cases, such as prediction of the first excursion failure, fatigue failure, etc., we are concerned with the statistics of the envelope process. The envelope process [1,2] for narrow band stationary and non-stationary processes can be seen as the modulus of the pre–envelope [1] which is a complex process whose real part is the given process while its imaginary part is related to the real one in such a way that the complex process exhibits power in the positive frequency range only. It follows that the statistics of the envelope are related to the covariances of the pre–envelope.

For linear systems the pre–envelope covariances are usually evaluated by means of an integral formulation [3,4]. Alternatively for non–white input processes the pre–envelope covariances can be evaluated as solution of a set of first order differential equations [5,6] whose input is the pre-envelope of the input process. These differential equations for real input processes represent the differential equations able to provide the traditional second order moments.

For non-linear systems a quite different situation appears: (i) the pre-envelope of the response cannot be evaluated as the solution of differential

A. Naess and S. Krenk (eds.), IUTAM Symposium on Advances in Nonlinear Stochastic Mechanics, 89–98.

equations subjected to the pre-envelope of the input but directly as the pre-envelope of the response; (ii) the integral solution cannot be used, hence the formulation of differential equations governing the evolution of the pre-envelope statistical moments is required.

In this paper the approach (ii) is applied in order to evaluate the pre-envelope of the response for linear systems sujected to a parametric white noise input process, for such input the pre–envelope differential equations are determined by an extension to complex processes of the stochastic differential calculus. In particular an Itô stochastic differential equation is presented in order to make available the extension to complex field of the stochastic differential calculus.

An application to quasi-linear systems is studied in order to particularise the quantities appearing in the moment differential equations as a consequence of the extension of the Itô rule to complex processes.

Here, for clarity's sake all the formulation is presented for a first order SDOF system, nevertheless the extension to greater order and MDOF systems is straightforward.

2. Linear systems subjected to external stationary input

The equations of motion of a monodimensional linear system subjected to stationary input can be written as follows

$$\dot{X}(t) = \alpha X(t) + \beta W(t) \qquad (1)$$

where α and β are deterministic parameters which characterise the system, $W(t)$ is a zero mean stationary normal white noise stochastic process characterised by the intensity q, $X(t)$ is the response function and the dot over a variable means time derivative.

For linear systems the envelope process of the response process can be evaluated according to two alternative ways. In the first, the envelope process can be evaluated as the modulus of a complex process (the so-called pre-envelope process) whose imaginary part is the Hilbert transform of the response process. In the second, the envelope process can be evaluated as the modulus of the complex response of a linear system subjected to the pre-envelope white noise input. The latter is a complex input having its real part as a white noise process while its imaginary part is the Hilbert transform of the real part, that is

$$F(t) = \frac{1}{\sqrt{2}} \left[W(t) + i\hat{W}(t) \right] \qquad (2)$$

where $F(t)$ is the pre-envelope input process and the accent $\hat{}$ means Hilbert transform [7] (also indicated as $\mathcal{H}[\cdot]$ in the following)

$$\hat{W}(t) = \mathcal{H}[W(t)] = \frac{1}{\pi} \int_{-\infty}^{\infty} \frac{W(\rho)}{t - \rho} \, d\rho \qquad (3)$$

where ρ is real and the integral is a Cauchy principal value.

It is useful to emphasise that the imaginary part of the pre-envelope process cannot be considered as a general process because it is strictly related to the real part of the process, as a consequence the following relationships hold

$$E\left[W(t)\right] = E\left[\hat{W}(t)\right] = 0 \qquad (4a)$$

$$R_{WW}(t - t') = E\left[W(t)W(t')\right] = q\,\delta(t - t') \qquad (4b)$$

$$R_{\hat{W}\hat{W}}(t - t') = R_{WW}(t - t') \qquad (4c)$$

$$R_{\hat{W}W}(t - t') = -R_{W\hat{W}}(t - t') = \hat{R}_{WW}(t - t') \qquad (4d)$$

Equation (4c) can be proved by recalling the definition of the Dirac's delta as the following convolution integral

$$\delta(t) = \frac{1}{\pi^2} \int_{-\infty}^{\infty} \frac{d\tau}{\tau(t - \tau)} \qquad (5)$$

Moreover equation (4d) can be written in explicit form as follows [8]

$$\hat{R}_{WW}(t - t') = \frac{q}{\pi(t - t')} \qquad \text{for} \qquad t \neq t'$$
$$\hat{R}_{WW}(0) = 0 \qquad \text{for} \qquad t = t' \qquad (6)$$

It follows that $W(t)$ and $\hat{W}(t)$ are statistically dependent processes with correlation function equals to zero for $t=t'$.

On this base the following points have to be remarked: (i) the processes $W(t)$ and $\hat{W}(t)$ possess the same correlation functions (in view of equation (4c)) and both can be considered as white noise processes; (ii) however they are different processes, in fact the cross-correlation function is different from the auto-correlation function of the two processes; (iii) in view of the linearity of the system, the response of equation (1) in which $W(t)$ is replaced with $\hat{W}(t)$ coincides with the Hilbert transform of the output.

To obtain the statistics of the complex output we have to evaluate $E[X^2]$, $E[X\hat{X}]$, $E[\hat{X}^2]$ only, since $E[X]=E[\hat{X}]=0$. In order to do this we write equation (1) and its Hilbert transform in incremental form as follows

$$dX(t) = \alpha X(t)dt + \beta\,dB(t)$$
$$d\hat{X}(t) = \alpha \hat{X}(t)dt + \beta\,d\hat{B}(t) \qquad (7)$$

where $B(t)$ is the Wiener process and $\hat{B}(t)$ its Hilbert transform, related to $W(t)$ and $\hat{W}(t)$ by means of the following relationships

$$dB(t) = W(t)dt \qquad ; \qquad d\hat{B}(t) = \hat{W}(t)dt \tag{8}$$

respectively . It can be proved that the increment of $B(t)$ satisfies the following relationships

$$
\begin{aligned}
\mathrm{E}\left[dB(t)dB(t)\right] &= \mathrm{E}\left[d\hat{B}(t)d\hat{B}(t)\right] = q\,dt \\
\mathrm{E}\left[dB(t)d\hat{B}(t)\right] &= 0
\end{aligned}
\tag{9}
$$

which means that dB and $d\hat{B}$ are infinitesimals of order $dt^{\frac{1}{2}}$.

By using equations (7) and (9) we can evaluate the second order moments as solution of differential equations which can be determined by evaluating the increment of X^2, $X\hat{X}$ and $\hat{X}^2$, neglecting infinitesimals of order greater than dt, as follows

$$
\begin{aligned}
\Delta(X^2) &= X\,dX + dX\,X + dX^2 \cong 2\alpha X^2(t)dt + 2\beta^2\left(dB(t)\right)^2 \\
\Delta(\hat{X}^2) &= \hat{X}\,d\hat{X} + d\hat{X}\,\hat{X} + d\hat{X}^2 \cong 2\alpha\hat{X}^2(t)dt + 2\beta^2\left(d\hat{B}(t)\right)^2 \qquad (10) \\
\Delta(X\hat{X}) &= X\,d\hat{X} + dX\,\hat{X} + dX\,d\hat{X} \cong 2\alpha X\hat{X}\,dt
\end{aligned}
$$

By averaging equations (10) and dividing by dt we obtain the differential equations governing the evolution of the second order stochastic moments as follows

$$
\begin{aligned}
\dot{\mathrm{E}}[X^2] &= 2\alpha\mathrm{E}[X^2] + 2\beta^2 q \\
\dot{\mathrm{E}}[\hat{X}^2] &= 2\alpha\mathrm{E}[\hat{X}^2] + 2\beta^2 q \qquad\qquad (11) \\
\dot{\mathrm{E}}[X\hat{X}] &= 2\alpha\mathrm{E}[X\hat{X}]
\end{aligned}
$$

Since the strenght of stationary input is not time dependent, the stationary solution of equations (11) is

$$\mathrm{E}[X^2] = \mathrm{E}[\hat{X}^2] = -\frac{\beta^2 q}{\alpha} \quad ; \quad \mathrm{E}[X\hat{X}] = 0 \tag{12}$$

Equation (12) shows that for linear systems the second order moments of X and $\hat{X}$ are coincident.

3. Linear systems subjected to stationary parametric input

The equations of motion of a linear system subjected to stationary parametric white input can be written as follows

$$\dot{X}(t) = \alpha X(t) + g(X(t))W(t) + f(t) \tag{13}$$

where $g(X(t))$ is a function of the response process $X(t)$, and $f(t)$ is a deterministic time dependent function.

Since

$$\mathcal{H}\Big[g(X(t))W(t)\Big] \neq g(X(t))\, \hat{W}(t) \tag{14}$$

it follows that, for the parametric nature of the input, the envelope process of the response does not coincide with the response of the same system subjected to the pre-envelope white input, on the contrary of the case of external input. Hence the latter approach, based on the pre-envelope of the input, does not provide suitable results for our purpose. It follows that we evaluate the the second order moments $E[X^2]$, $E[\hat{X}^2]$, $E[X\hat{X}]$ and higher order moments by considering $\hat{X}$ as the Hilbert transform of the output process X. In particular our attention is devoted to evaluating the statistical moments which involve $\hat{X}$ by extending the traditional stochastic differential calculus to complex processes. In order to do this, first we recall the main two steps of the traditional stochastic differential calculus [9,10]: (i) conversion of the differential equation of motion into an Itô type differential equation where, on the contrary of the traditional differential calculus infinitesimal of different order appear; (ii) use of the Itô differential rule allowing the complete characterisation the response process by means of the evaluation of the stochastic moments.

Hence according to step (i) equation (13) can be written in Itô form as follows

$$dX(t) = \alpha X(t)dt + g(X(t))dB(t) + f(t)\,dt + \frac{1}{2}g(X(t))g'(X(t))\,q\,dt \tag{15}$$

where the last term is the so-called Wong-Zakai correction term.

To evaluate the statistics of the envelope the Hilbert transform of the response has to be computed. The latter can be evaluated numerically as the discrete Hilbert transform of $X(t)$ or analitically by transforming both sides of equation (15) as follows

$$\begin{aligned} d\hat{X}(t) =& \alpha \hat{X}(t)dt + \mathcal{H}\Big[g(X(t))dB(t)\Big] \\ &+ \frac{1}{2}\mathcal{H}\Big[g(X(t))g'(X(t))\Big]\,q\,dt + \hat{f}(t)\,dt \end{aligned} \tag{16}$$

In order to evaluate the moments of X and $\hat{X}$ the Itô differential rule can be applied. It consists in writing the increment of a real valued function $\phi(X(t), \hat{X}(t); t)$ continuously differentiable with respect to t and twice differentiable with respect to X as follows

$$\Delta\phi = \frac{\partial\phi}{\partial t}dt + \frac{\partial\phi}{\partial X}dX + \frac{\partial\phi}{\partial\hat{X}}d\hat{X}$$
$$+ \frac{1}{2}\left[\frac{\partial^2\phi}{\partial X^2}(dX)^2 + 2\frac{\partial^2\phi}{\partial X\partial\hat{X}}dXd\hat{X} + \frac{\partial^2\phi}{\partial\hat{X}^2}(d\hat{X})^2\right] \tag{17}$$

where the second order terms of the series are retained because both $(dB(t))^2$ and $(d\hat{B}(t))^2$ are infinitesimals of order dt. Substituting equations (15) and (16) into equation (17), and retaining infinitesimals of order dt only, we can write

$$\Delta\phi = \frac{\partial\phi}{\partial t}dt + \frac{\partial\phi}{\partial X}\left[\alpha X dt + g(X)dB(t) + \frac{1}{2}g(X)g'(X)qdt + f(t)\,dt\right]$$
$$+ \frac{\partial\phi}{\partial\hat{X}}\left[\alpha\hat{X}dt + \mathcal{H}\Big[g(X)dB(t)\Big] + \frac{1}{2}\mathcal{H}\Big[g(X)g'(X)\Big]qdt + \hat{f}(t)\,dt\right]$$
$$+ \frac{1}{2}\left[\frac{\partial^2\phi}{\partial X^2}g^2(X)qdt + 2\frac{\partial^2\phi}{\partial X\partial\hat{X}}g(X)dB\,\mathcal{H}\Big[g(X)dB(t)\Big]\right.$$
$$\left.+ \frac{\partial^2\phi}{\partial\hat{X}^2}\Big(\mathcal{H}\Big[g(X)dB(t)\Big]\Big)^2\right] \tag{18}$$

By selecting the suitable function $\phi(X(t),\hat{X}(t);t)$, making the stochastic average of both sides of equation (18) and dividing by dt the differential equations governing the evolution of the statistical moments can be obtained, in particular the case of quasi-linear systems will be treated in the next section.

4. Application to quasi-linear systems

The importance of studying quasi-linear systems has been recently shown [11] and it relies on the circumstance that any non-linear system can be optimally approximed by a quasi-linear one.

For quasi-linear systems $g(X(t))=\gamma X(t)$, it follows that equation (18) takes the form

$$\Delta\phi = \frac{\partial\phi}{\partial t}dt + \frac{\partial\phi}{\partial X}\left[(\alpha + \frac{1}{2}\gamma^2 q)Xdt + \gamma\,XdB + f(t)dt\right]$$
$$+ \frac{\partial\phi}{\partial\hat{X}}\left[(\alpha + \frac{1}{2}\gamma^2 q)\hat{X}dt + \gamma\,\mathcal{H}\Big[XdB\Big] + \hat{f}(t)dt\right]$$
$$+ \frac{1}{2}\left[\frac{\partial^2\phi}{\partial X^2}X^2\gamma^2 qdt + 2\frac{\partial^2\phi}{\partial X\partial\hat{X}}\gamma^2\,XdB\,\mathcal{H}\Big[XdB\Big]\right. \tag{19}$$
$$\left.+ \gamma^2\frac{\partial^2\phi}{\partial\hat{X}^2}\Big(\mathcal{H}\Big[XdB\Big]\Big)^2\right]$$

For sake of simplicity and to show the fundamental aspects of the proposed procedure, we focus the attention on the second order moments only. Hence by selecting $\phi(X)=X, \hat{X}, X^2, X\hat{X}, \hat{X}^2$ in equation (19) and averaging, the following increments can be obtained respectively

$$E[\Delta(X)] = \left(\alpha + \frac{1}{2}\gamma^2 q\right)E[X]\, dt + \gamma\, E[X\, dB] + f(t)\, dt$$

$$E\left[\Delta(\hat{X})\right] = \left(\alpha + \frac{1}{2}\gamma^2 q\right)E[\hat{X}]\, dt + \gamma\, E\left[\mathcal{H}[X\, dB]\right] + \hat{f}(t)\, dt$$

$$E[\Delta(X^2)] = 2\left(\alpha + \frac{1}{2}\gamma^2 q\right)E[X^2]\, dt + 2\gamma E[X^2\, dB]$$
$$+ 2E[X]f(t)\, dt + E[X^2]\gamma^2 q\, dt$$

$$E\left[\Delta\left(\hat{X}^2\right)\right] = 2\left(\alpha + \frac{1}{2}\gamma^2 q\right)E[\hat{X}^2]\, dt + 2\gamma E\left[\hat{X}\,\mathcal{H}[X\, dB]\right] \qquad (20)$$
$$+ 2E[\hat{X}]\hat{f}(t)\, dt + \gamma^2 E\left[\left(\mathcal{H}[X\, dB]\right)^2\right]$$

$$E\left[\Delta\left(X\hat{X}\right)\right] = 2\left(\alpha + \frac{1}{2}\gamma^2 q\right)E[X\hat{X}]\, dt + \gamma E[\hat{X}X\, dB]$$
$$+ E[\hat{X}]f(t)\, dt + \gamma E\left[X\,\mathcal{H}[X\, dB]\right] + E[X]\hat{f}(t)\, dt$$
$$+ \gamma^2 E\left[X\, dB\,\mathcal{H}[X\, dB]\right]$$

Notice that equations (20) can be written in a more suitable form by considering that the following relationships hold

$$E\left[X\, dB\right] = E\left[\mathcal{H}[X\, dB]\right] = E\left[X^2\, dB\right] = 0$$
$$E\left[X\,\mathcal{H}[X\, dB]\right] = -E\left[\hat{X}X\, dB\right] \qquad (21)$$
$$E\left[\hat{X}\,\mathcal{H}[X\, dB]\right] = E\left[X\, dB\,\mathcal{H}[X\, dB]\right] = 0$$

The differential equations governing the evolution of the first and second order moments can be obtained from equations (20) dividing by dt and accounting for equations (21) as follows

$$\dot{E}[X] = \alpha E[X] + \frac{1}{2}E[X]\gamma^2 q + f(t)$$

$$\dot{E}[\hat{X}] = \alpha E[\hat{X}] + \frac{1}{2}E[\hat{X}]\gamma^2 q + \hat{f}(t)$$

$$\dot{E}[X^2] = 2(\alpha + \gamma^2 q)E[X^2] + 2f(t)E[X] \qquad (22)$$

$$\dot{E}[X\hat{X}] = 2\left(\alpha + \frac{1}{2}\gamma^2 q\right)E[X\hat{X}] + \hat{f}(t)E[X] + +f(t)E[\hat{X}]$$

$$\dot{E}[\hat{X}^2] = 2\left(\alpha + \frac{1}{2}\gamma^2 q\right)E[\hat{X}^2] + \gamma^2\, G(t)\, q + 2\hat{f}(t)E[\hat{X}]$$

where

$$G(t) = \frac{1}{qdt} E\left[\left(\mathcal{H}[XdB]\right)^2\right] = \frac{1}{\pi^2 q} \int_0^\infty \frac{E[X^2(\rho)d\rho]}{(t-\rho)^2} d\rho \qquad (23)$$

The integral in equation (23), which depends on the real part of the response, can be evaluated by means of a numerical procedure. In particular, let $t_0=0, t_1, \ldots, t_n$ be an equispaced subdivision of the time axis, and $\Delta t = t_s - t_{s-1}$ the time step selected so that $t_k = k\Delta t$. Moreover suppose that the function $E[X^2(\rho)]$ is linear in the time interval (t_s, t_{s+1}). Under this hypothesis an approximation of equation (23) can be written as

$$G(t_k) \cong \frac{1}{\pi^2} \sum_{\substack{s=k \\ s \neq k-1,k}}^{k+n} \left[\left(\frac{E[X^2(t_k)]}{k-s-1} - \frac{E[X^2(t_k)]}{k-s}\right)\right.$$
$$+ \left(E[X^2(t_k)] - E[X^2(t_{k-1})]\right)\ln\left(\frac{k-s-1}{k-s}\right)\right] \qquad (24)$$
$$+ \left(E[X^2(t_{k+1})] - E[X^2(t_k)]\right)$$

However the function $G(t)$ provided by equation (23) becomes constant and takes the value $G(t)=1$ for linear systems subjected to external input.

In the following a quasi-linear system with $\alpha=-0.5$, $\gamma=0.5$ and $f(t)=1$, subjected to a zero mean white noise with $q=1$, has been considered.

In order to prove the validity of the proposed approach the results have been tested against a Monte-Carlo simulation. The Monte-Carlo simulation has been performed on the base of a thousand samples where for each of them the analysis has been conducted according to the procedure presented in [12] and the Hilbert trasforms evaluated by means of the numerical procedure proposed in [13].

First, the function $G(t)$ defined by equation (23) has been evaluated by means of the numerical procedure represented by equation (24) and compared with the result obtained by means of Monte-Carlo simulation. The results, plotted in Fig.1, are in good agreement.

Then the differential equations (22) governing the evolution of the first and second order moments have been integrated, the results are reported in Figs.2 and 3 where they are compared with the Monte-Carlo results.

Figs.2 and 3 show that first and second order moments are in good agreement with the results provided by the Monte-Carlo simulation. However, while $E[X]$ and $E[X^2]$ are subjected to zero initial conditions, the evolution of $E[\hat{X}]$, $E[X\hat{X}]$ and $E[\hat{X}^2]$ needs suitable initial conditions even though they do not affect the stationary solution reached after a transient.

5. Conclusions

In stochastic analysis complex processes have to be considered if the statistics of the envelope process of the response need to be evaluated. Indeed the latter process can be seen as the modulus of a complex process called pre-envelope process.

In this paper the stochastic moments of the pre-envelope process of the response of linear systems subjected to a parametric white noise have been evaluated. In particular, an extension of the stochastic differential calculus to complex processes has been presented in order to formulate the differential equations governing the evolution of the stocastic moments.

An application to quasi-linear systems has been presented in order to evidentiate the problems related to the integration of the pre-envelope moments differential equations.

A good accuracy of the solution, in comparison to the Monte-Carlo simulation, has been reached for the first two order statistical moments of the pre-envelope response.

References

1. Dugundji, J., (1958) Envelope and Pre–envelope of Real Waveforms, *IRE Transaction on Information Theory*, **4**, 53–57.
2. Krenk, S., Madsen, H.O. and Madsen, P.H., (1988) Stationary and Transient response envelopes, *Journ. of Engineering Mechanics Division, ASCE*, **109**, 263–276.
3. Di Paola, M. and Petrucci, G., (1990) Spectral Moments and Pre–Envelope Covariances of non–Stationary Processes, *Journ. of Applied Mechanics, ASME*, **57**, 218–224.
4. Muscolino, G., (1991) Non–Stationary Pre–Envelope Covariances of non–Classically Damped Systems , *Journ. of Sound and Vibrations*, **149**, 107–123.
5. Langley, R.S., (1986) Structural Response to Non–Stationary Non–White Stochastic Ground Motions, *Earthquake Engineering and Structural Dynamics*, **14**, 909–924.
6. Muscolino, G., (1992) Pre–Envelope Covariance Differential Equations, *Probabilistic Mechanics and Structural and Geotechnical Reliability*, 6-th Special Conference EM, ST, GT, Div/ASCE, Denver, CO/July 8–10.
7. Papoulis, A., (1965) *Probability Random Variables and Stochastic Processes*, Mc Graw–Hill, Kogokusha,Tokio.
8. Di Paola, M., (1985) Transient Spectral Moments of Linear Systems, *S.M.Archives*, **10**, 225–243.
9. Itô, K., (1951) On a Formula Concerning Stochastic Differential, *Nagoya Math. J.*, **3**, 55–65.
10. Itô, K., (1961) *Lectures of Stochastic Processes*, Tata Institute Fundamental Research, Bombay, India.
11. Di Paola, M. (in press) Quasi-linear Oscillator under Poisson Pulses, *Journ. of Applied Mechanics, ASME*.
12. Caddemi, S. and Di Paola, M. (in press) Itô and Stratonovich Integrals for Deterministic Impulsive Parametric Input, *Journ. of Applied Mechanics, ASME*.
13. Di Paola, M. and Muscolino, G. (1986) On the Convergent Parts of High Order Spectral Moments of Stationary Structural Responses, *Journ. of Sound and Vibration*, **110**, 233–245.

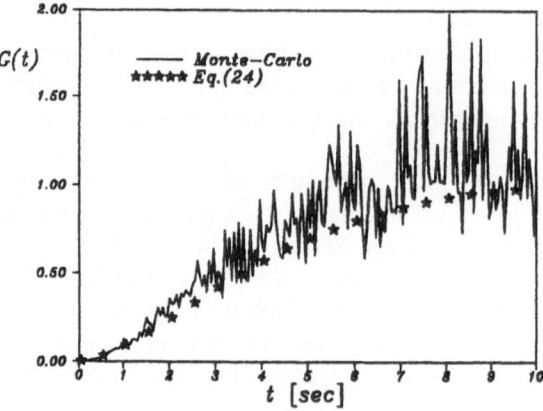

Fig.1. Evolution of the term $G(t)$

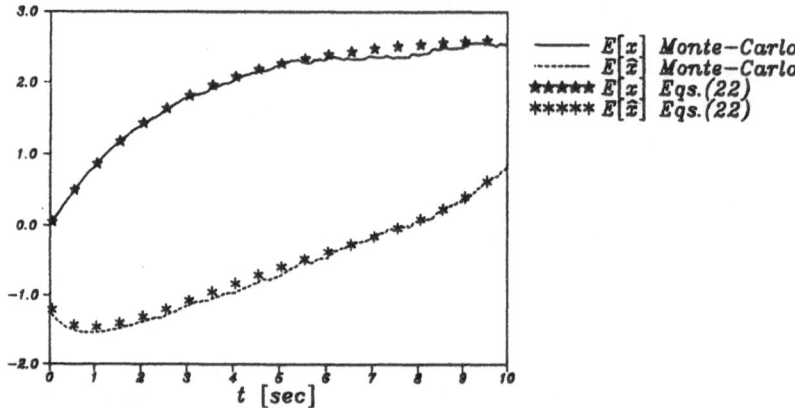

Fig.2. Evolution of the first order moments

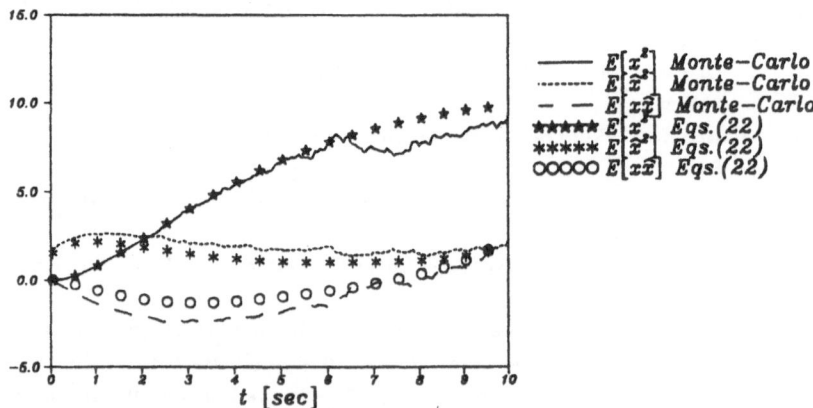

Fig.3. Evolution of the second order moments

RESPONSE OF A HYSTERETIC SYSTEM UNDER NON-STATIONARY EARTHQUAKE EXCITATIONS

G. Q. CAI AND Y. K. LIN

Center for Applied Stochastics Research
Florida Atlantic University, Boca Raton, FL 33431, U.S.A.

1. Introduction

Under strong earthquake excitations, a structure is likely to become nonlinear and inelastic. The term hysteresis is used to describe a type of inelastic behavior in which the restoring force depends not only on the instantaneous deformation, but also the past history of the deformation. Consider an engineering structure idealized as a single-degree-of-freedom system governed by

$$\ddot{X} + 2\zeta\dot{X} + \alpha X + \eta\, Z(t) = f_k(X, \dot{X})\, \xi_k(t) \tag{1}$$

where $\xi_k(t)$ are ground accelerations, and $Z(t)$ is a hysteretic restoring force, described by the Bouc-Wen model (Bouc, 1967; Wen, 1976, 1980)

$$\dot{Z} = -\gamma\, |\dot{X}|\, Z\, |\dot{Z}|^{n-1} - \beta\dot{X}\, |Z|^n + A\dot{X} \tag{2}$$

which has the following desirable properties: (i) the deformation-force relationship is smooth, (ii) the parameters in the model can be adjusted to match a real hysteresis behavior, and (iii) it is more amenable to analytical treatments.

The right hand side of equation (1) suggests that the excitations can be either additive, or multiplicative, or both. In the case of a column, for example, the vertical ground acceleration gives rise to a multiplicative excitation and the horizontal ground acceleration to an additive excitation. Since an earthquake ground acceleration is a non-stationary stochastic process, a versatile model, capable of incorporating seismic wave propagation and local geological features, is that of a random pulse train

$$\xi_k(t) = \sum_{j=1}^{N(t)} (Y_k)_j\, h(t - \tau_j) \tag{3}$$

99

A. Naess and S. Krenk (eds.), IUTAM Symposium on Advances in Nonlinear Stochastic Mechanics, 99–108.
© *1996 Kluwer Academic Publishers.*

where τ_j is the random time at which the jth pulse is initiated, $(Y_k)_j$ is the random magnitude of the jth pulse, $N(t)$ is a Poisson process, $h(t-\tau_j)$ is a deterministic pulse shape function which may be determined from the knowledge of the physical features of the ground, and the subscript k denotes a particular direction. For different j, the magnitudes $(Y_k)_j$ are assumed to be independent, but have the same probability distribution as Y_k. It is known that such a $\xi_k(t)$ process possesses an evolutionary spectral density (Lin, 1986)

$$\hat{\Phi}_{kk}(t,\omega) = \frac{1}{2\pi} E[Y_k^2] |a(t,\omega)|^2 \qquad (4)$$

where E[ ] denotes a statistical average,

$$a(t,\omega) = \int_{-\infty}^{\infty} h(u) \sqrt{v(t-u)}\, e^{-i\omega u}\, du \qquad (5)$$

and $v(t)$ is the average arrival rate of the random pulses per unit time. Equations (3) through (5) imply that both the horizontal and vertical ground accelerations have the same pulse shape and the same arrival rate, but different statistical properties of the pulse magnitudes. If this is not the case, then different sets of $h(u)$ and $v(t-u)$ functions can be selected for different k.

In the present paper, a modified version of quasi-conservative averaging procedure is applied to the hysteretic system (1) under both additive and multiplicative excitations. The total energy of the system is approximated as a Markov process, and its time-dependent probability density functions is obtained by using the numerical procedure of path-integral. A numerical example of a hysteretic column subjected to both vertical and horizontal ground excitations is given for illustration.

2. Modified Quasi-Conservative Averaging

The hysteretic force Z in equation (1) plays two different roles: it dissipates energy and, in the absence of external loads, returns the system to a local equilibrium. In what follows, Z is replaced by an equivalent damping force $h(X,\dot{X})$ plus an equivalent spring force $u(X)$, namely,

$$Z = h(X,\dot{X}) + u(X) \qquad (6)$$

One reasonable choice for the equivalent spring force $u(X)$ is the so-called "backbone", which passes the extremities of every hysteresis loop, as illustrated in Fig. 1. For a given set of parameters A, n, γ and β, this equivalent spring force can be obtained analytically or numerically. The equivalent damping $h(X,\dot{X})$ is then given by $Z - u(X)$.

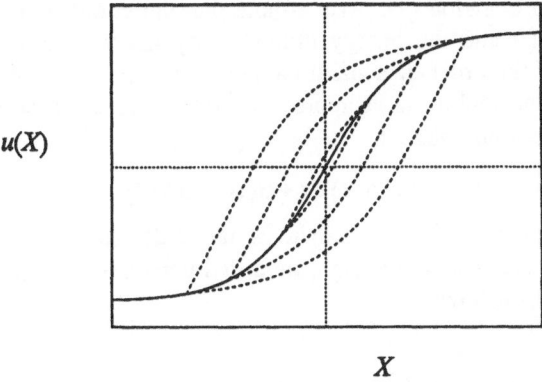

Figure 1. Backbone of a hysteresis model.

The equivalent potential energy and the total energy of the system follow as

$$U(X) = \frac{1}{2}\alpha X^2 + \eta \int_0^X u(y)\,dy, \qquad \Lambda = \frac{1}{2}\dot{X}^2 + U(X) \qquad (7)$$

Letting

$$\mathrm{sgn}X\sqrt{U(X)} = \sqrt{\Lambda}\cos\phi, \qquad 0 \le \phi < 2\pi$$
$$\dot{X} = -\sqrt{2\Lambda}\sin\phi \qquad (8)$$

where $\mathrm{sgn}X$ is the sign of X, equation (1) can be replaced by

$$\dot{\Lambda} = -4\zeta\Lambda\sin^2\phi + \eta\sqrt{2\Lambda}\sin\phi\, h(\Lambda,\phi) - \sqrt{2\Lambda}\sin\phi\, f_k(\Lambda,\phi)\,\xi_k(t) \qquad (9)$$

$$\dot{\phi} = -2\zeta\sin\phi\cos\phi + \frac{\eta}{\sqrt{2\Lambda}}\left[h(\Lambda,\phi)\cos\phi + \frac{u(\Lambda,\phi)}{\cos\phi}\right] - \frac{\cos\phi}{\sqrt{2\Lambda}}f_k(\Lambda,\phi)\,\xi_k(t) \qquad (10)$$

in which $h(\Lambda,\phi)$, $u(\Lambda,\phi)$ and $f_k(\Lambda,\phi)$ are obtained from $h(X,\dot{X})$, $u(X)$ and $f_k(X,\dot{X})$ upon replacing X and $\dot{X}$, respectively, by functions of Λ and ϕ according to (8). In passing, we note that if the excitations were additive and periodic, and the system were performing a steady-state periodic motion, then the total energy Λ would be conserved, and the energy dissipated by the hysteretic force would correspond to the area of a hysteresis loop.

We now return to equation (1). We assume that the correlation functions of the random excitations $\xi_k(t)$ and the energy dissipated by the viscous damping and the hysteresis force are of the order of a small parameter ε. Then the total energy $\Lambda(t)$ is slowly varying with time, and it can be approximated as a Markov process, governed by an Itô stochastic differential equation

$$d\Lambda = m(\Lambda,t)\,dt + \sigma(\Lambda,t)\,dB(t) \tag{11}$$

in which $m(\Lambda,t)$ and $\sigma(\Lambda,t)$ are known as the drift and diffusion coefficients, respectively, and they are obtained using a modified version of the quasi-conservative averaging procedure as follows

$$m(\Lambda,t) = -4\zeta\Lambda\left\langle \sin^2\phi \right\rangle_t + \eta\frac{A_r(\Lambda)}{T} + \int_{-\infty}^{0}\left\langle \sqrt{2\Lambda}\,\sin\phi(t+\tau)f_j(t+\tau)\sin\phi(t)\frac{\partial[\sqrt{2\Lambda}\,f_k(t)]}{\partial\Lambda}\right.$$

$$\left. + \cos\phi(t+\tau)f_j(t+\tau)\frac{\partial[\sin\phi(t)f_k(t)]}{\partial\phi}\right\rangle_t R_{jk}(t,\tau)\,d\tau \tag{12}$$

$$\sigma^2(\Lambda,t) = 2\Lambda\int_{-\infty}^{\infty}\left\langle \sin\phi(t+\tau)f_j(t+\tau)\sin\phi(t)f_k(t)\right\rangle_t R_{jk}(t,\tau)\,d\tau \tag{13}$$

where A_r is the area of the hysteretic loop corresponding to a total energy level Λ, $f_k(t)$ is an abbreviation for $f_k[\Lambda(t),\phi(t)]$, T is called a quasi-period, which is equal to the period of the corresponding undamped free motion, and $< \cdot >_t$ denotes time-averaging over a quasi-period T. The first two terms on the right-hand-side of (12) represent the dissipated energies by the linear damping and the hysteresis mechanism per unit time, respectively. The original quasi-conservative averaging procedure due to Landa and Stratonovich (1962) and Khasminskii (1964) is applicable to white-noise excitations; it was extended by Roberts (1982) to non-white additive excitations, and was further extended by Cai (1995) to additive and multiplicative non-white excitations. Since the total energy Λ in the system is slowly varying, X and $\dot{X}$ are approximately periodic functions with a quasi-period T. Thus, functions $\sin\phi$, $\cos\phi$ and f_j in (12) and (13) are also approximately periodic with period T. We then expand these functions into Fourier series, and obtain expressions of (12) and (13) in terms of Fourier coefficients. For details, see Cai (1995).

3. Probability Density of the Energy Process

The probabilistic evolution of the energy process $\Lambda(t)$ is described by the transition probability density $q(\lambda, t | \lambda', t')$, which is governed by the Fokker-Planck equation

$$\frac{\partial}{\partial t} q = -\frac{\partial}{\partial \lambda}[m(\lambda, t) q] + \frac{1}{2}\frac{\partial^2}{\partial^2 \lambda}[\sigma^2(\lambda, t) q] \qquad (14)$$

where λ is the state variable for $\Lambda(t)$, and the symbols λ', t' behind a vertical bar indicate a given condition $\Lambda(t') = \lambda'$, $t' \le t$. Equation (14) is solved subject to the condition

$$[q(\lambda, t | \lambda', t')]_{t=t'} = \delta(\lambda - \lambda') \qquad (15)$$

Since the evolutionary spectral densities of the random excitations are assumed to be slowly varying with time, the drift and diffusion coefficients will not change appreciably in a short time interval. In a sufficiently short time step $\Delta t = t - t'$, the solution for (14) is approximately Gaussian, namely

$$q(\lambda, t | \lambda', t') = \frac{1}{\sqrt{2\pi\,\sigma^2(\lambda', t')\Delta t}}\exp\{-\frac{[\lambda - \lambda' - m(\lambda', t')\Delta t]^2}{2\sigma^2(\lambda', t')\Delta t}\} \qquad (16)$$

Given an initial probability density $p(\lambda, t_0)$, the probability density $p(\lambda, t)$ can be calculated successively as follows, using short time solutions in the form of (16),

$$p(\lambda, t) = \int_0^\infty q(\lambda, t | \lambda', t') p(\lambda', t')\, d\lambda' \qquad (17)$$

and starting with $t' = t_0$. In the step-by-step calculation, values of the probability density $p(\lambda, t)$ are obtained at discrete points, and its values between two neighboring points can be obtained with a suitable interpolation scheme. The above procedure is known as path integration (see, e.g., Wehner and Wolfer, 1983; Naess and Johnsen, 1993).

4. A Hysteretic Column under Earthquake Excitations

As en example, consider a massless column, supporting a concentrated mass at the top, and is clamped rigidly below in the ground. If the column deformation is dominated by a single mode, then the non-dimensional governing equation for the horizontal displacement of the concentrated mass may be expressed as follows (Lin and Cai, 1995)

$$\ddot{X} + 2\zeta\dot{X} + [1 - \eta - \kappa - \xi_1(t)]X + \eta\,Z(t) = \xi_2(t) \qquad (18)$$

where ζ, η and κ are constants characterizing the dynamical properties of the system, Z represents the hysteretic component of the restoring force, and $\xi_1(t)$ and $\xi_2(t)$ are vertical and horizontal ground accelerations. The parameter η represents the relative contribution of the hysteresis component. In particular, $\eta = 0$ corresponds to a purely elastic column, in which case κ represents the weight of the concentrated mass expressed as a fraction of the static buckling load of the column. The ground accelerations $\xi_1(t)$ and $\xi_2(t)$ are assumed to be evolutionary Kanai-Tajimi processes (Lin and Yong, 1987), which can be modeled in the form of equation (3) with a pulse shape function

$$h(t-\tau) = \omega_g \exp[-\zeta_g \omega_g (t-\tau)]\{\frac{1-2\zeta_g^2}{(1-\zeta_g^2)^{1/2}} \sin[\omega_{gd}(t-\tau)] + 2\zeta_g \cos[\omega_{gd}(t-\tau)]\}, \quad t > \tau \quad (19)$$

where $\omega_{gd} = \omega_g(1-\zeta_g^2)^{1/2}$. In an attempt to simulate a record of the 1985 Mexico City earthquake, Lin and Yong (1987) selected $\omega_g = \pi$ rad/s, $\zeta_g = 0.3$, and a varying average pulse arrival rate

$$\nu(t) = t(1-\cos\frac{\pi t}{30}), \quad 0 \le t \le 60 \quad (20)$$

Equation (20) implies that the earthquake ground motion has a duration of 60 seconds. Fig. 2 depicts the evolution of the evolutionary spectral density $\hat{\Phi}_{\xi\xi}(t,\omega)$ with respect to time, calculated for $\omega_g = \pi$ rad/s and $\zeta_g = 0.3$. In the present example, the mean-square ground acceleration reaches its maximum at $t = 35$s.

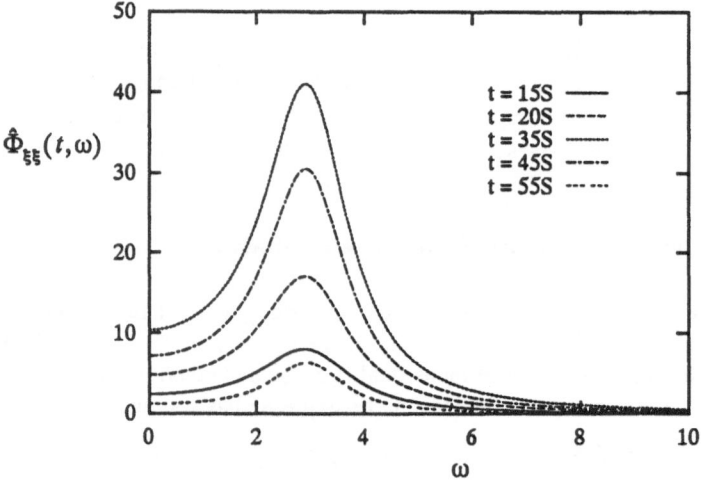

Figure 2.   Time-evolution of spectral density of an earthquake model.

Numerical results have been obtained for a hysteretic column characterized by structural parameters $\kappa = 0.04$, $\zeta = 0.025$, $n = 1$, $A = 1$ and $\beta = \gamma = 0.5$. Fig. 3 depicts the probability density function $p(\lambda, t)$ of the energy process $\Lambda(t)$ at two time instants $t = 30$s and $t = 40$s for the case of a moderate hysteresis $\eta = 0.5$. The mean square values of the random pulse magnitudes are assumed to be $E[Y_1^2] = 0.0005$ in the vertical direction and $E[Y_2^2] = 0.002$ in the horizontal direction. Results obtained from Monte Carlo simulations are also included in the figure for comparison, and they are in good agreement with the analytical results. When carrying out the Monte Carlo simulations, samples for excitations $\xi_1(t)$ and $\xi_2(t)$ are obtained according to (3), (19) and (20). An algorithm for generating a Poisson process with a time-varying average arrival rate is employed.

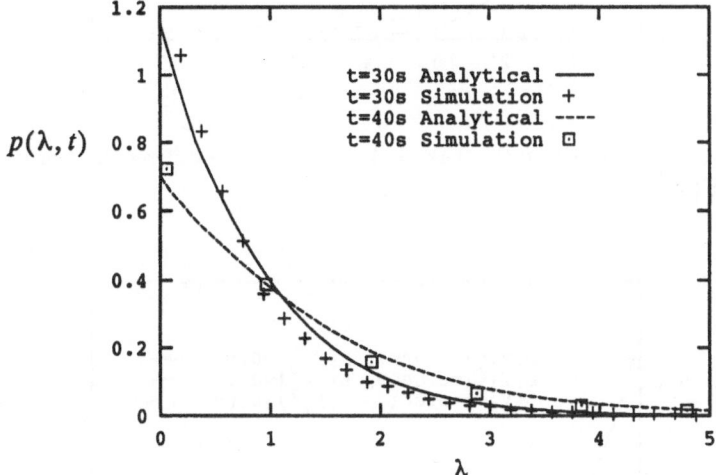

Figure 3.   Probability density $p(\lambda, t)$ at $t = 30$s and 40s
$(E[Y_1^2] = 0.0005,\ E[Y_2^2] = 0.002,\ \eta = 0.5)$.

Fig. 4 shows the probability densities of the energy process at $t = 40$s for $E[Y_1^2] = 0.0005$, $E[Y_2^2] = 0.002$, computed for three different levels of hysteresis: $\eta = 0.2, 0.5$ and 0.8. It is seen that the response total energy tends to decrease as the hysteresis level increases. Fig. 5a and Fig. 5b show the probability densities at $t = 40$s within two different ranges of energy level λ. Three different sets of excitations were assumed in the computation. As expected, the level of the response total energy increases with an increase of the intensity of either the vertical or horizontal ground acceleration. The

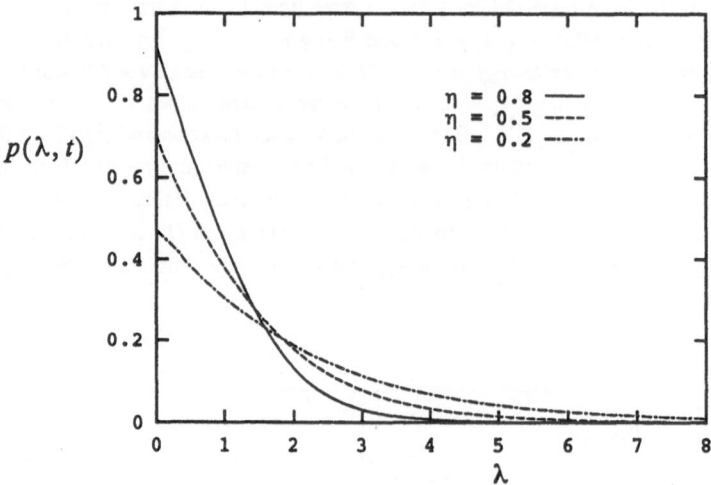

Figure 4.   Probability density $p(\lambda, t)$ at $t = 40$s computed for different hysteresis levels
($\text{E}[Y_1^2] = 0.0005$, $\text{E}[Y_2^2] = 0.002$).

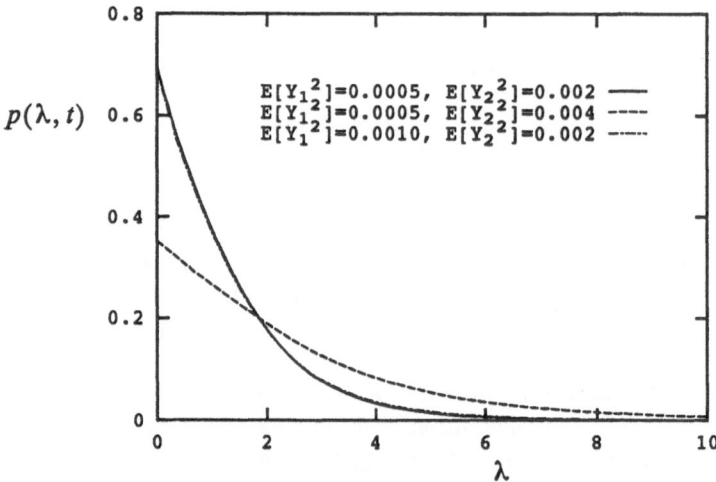

Figure 5a.   Probability density $p(\lambda, t)$ at $t = 40$s computed for different excitation levels
within a lower λ range ($\eta = 0.5$).

vertical excitation has a minor effect on the probability density in the lower response range (Fig. 5a); yet it significantly increases the probability density in the higher range (Fig. 5b). Since system failure is mainly caused by a higher response level, the vertical ground motion has a major effect on system reliability, which, unfortunately, has not been fully appreciated in the past.

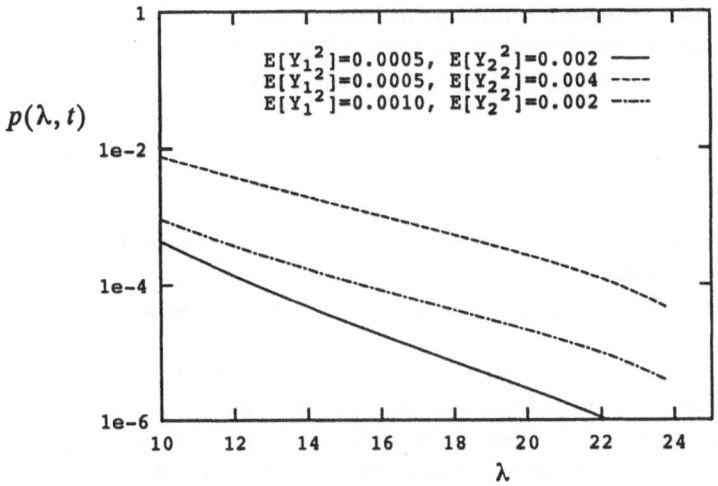

Figure 5b. Probability density $p(\lambda, t)$ at t = 40s computed for different excitation levels within a higher λ range (η = 0.5).

5. Concluding Remarks

Under the assumption that the ground accelerations are broad-band evolutionary processes and that the energy dissipation is low, the response total energy of a nonlinear single-degree-of-freedom structure may be approximated as a Markov Process. The Fokker-Planck equation for the probability density of this process can be derived using a modified version of quasi-conservative averaging, and solved using the numerical procedure of path-integration. The results so obtained agree quite well with those of Monte Carlo simulations. For the specific problem of a hysteretic column under both vertical and horizontal excitations, an increase of the level of either type of excitation increases the level of the response energy. The vertical ground acceleration is found to have a significant effect on higher range of the response, of importance to the assessment of structural reliability. The hysteresis component of the structural model also plays an important role.

6.   Acknowledgement

The work reported in this paper is supported by the National Science Foundation under Grant BCS-9312640. Opinions, findings and conclusions expressed are those of the writers, and do not necessarily reflect the views of NSF. We also wish to thank Mr. J. S. Yu for the assistance of carrying out the numerical calculation.

7.   References

Bouc, R. (1967) Forced vibration of mechanical system with hysteresis, Abstract, *Proceedings of 4th Conference on Nonlinear Oscillation*, Prague, Czechoslovakia.

Cai, G. Q. (1995) Random vibration of nonlinear systems under non-white excitations, *J. Eng. Mech. ASCE*, 121, 633-639.

Khasminskii, R. Z. (1964) On the behavior of a conservative system with small friction and small random noise, *Prikladnaya Matematika i Mechanica* (Appl. Math and Mech.), 28, 1126-1130 (in Russian),.

Landa, P. S. and Stratonovich, R. L. (1962) Theory of stochastic transitions of various systems between different states, *Vestnik MGU* (Proc. of Moscow University), Series III (1), 33-45 (in Russian).

Lin, Y. K. (1986) On random pulse train and its evolutionary spectral representation, *Prob. Eng. Mech.*, 1, 219-223.

Lin, Y. K. and Cai, G. Q. (1995) *Probabilistic Structural Dynamics: Advanced Theory and Applications*, McGraw-Hill, New-York.

Lin, Y. K. and Yong, Y. (1987) Evolutionary kanai-tajimi earthquake models, *J. Eng. Mech. ASCE*, 113, 1119-1137.

Naess, A. and Johnsen, J. M. (1993) Response statistics of nonlinear, compliant offshore structures by path integral solution method, *Prob. Eng. Mech.* 8, 91-106.

Roberts, J. B. (1982) A stochastic theory for nonlinear ship rolling in irregular seas, *J. Ship Res.* 26, 229-245.

Wehner, M. F. and Wolfer, W. G. (1983) Numerical evaluation of path-integral solution to Fokker-Planck equation, *Physical Review A*, 27, 2663-2670.

Wen, Y. K. (1976) Method for random vibration of hysteretic systems, *J. of Eng. Mech. Div. ASCE*, 103, 249-263.

Wen, Y. K. (1980) Equivalent linearization for hysteretic systems under random excitation, *J. Appl. Mech.*, 47, 150-154.

NONLINEAR ACTIVE CONTROL AND STOCHASTIC EXCITATION

F. CASCIATI
Department of Structural Mechanics, Univesity of Pavia
Via Abbiategrasso, 211 - 27100 Pavia - Italy

Abstract

Standard strategies of active control applied to civil engineering structures can offer an excellent response performance, but require a significant energy supply and a correct identification of the structure behaviour. In order to release these two constraints, the possible exploitation of suitable nonlinearities is under discussion. This contribution emphasizes the special features of the nonlinear structural control of large structural systems in the presence of environmental actions modelled as stochastic processes. It also underlines the way robustness and stability can be pursued and how the trade-off between performance and cost dominates the technical decisions.

1. Introduction

The attempts to apply standard strategies of active control to civil engineering structures excited by environmental loads, as wind and earthquake, is quite recent (Kobori, 1988; Soong, 1990; Housner et al. , 1992; Housner and Masri, 1993). If one controls the motion, it is likely that he will maintain the dispacements very small and, hence, geometry and material nonlinearities are prevented. This means to design the controller for linear structural systems. Linear control is a mature subject that offers a variety of powerful methods and a long experience of applications (Kwakernaak and Sivan, 1972; Anderson and Moore, 1990). Thence, classical stochastic formulations of control theory only covers linear stochastic control (Astrom, 1970) when one must allow for stochastic disturbances.

But there is evidence (Casciati, 1995) that civil engineering structures of standard conception seem to reject linear active control schemes. Indeed, their active control would require a significant energy supply to preserve the small range operation. Moreover, in designing linear controllers it is usually necessary to assume that the parameters of the system model are reasonably well known, but civil structural engineering problems

A. Naess and S. Krenk (eds.), IUTAM Symposium on Advances in Nonlinear Stochastic Mechanics, 109–116.

involve several uncertainties in the model parameters. This is pushing one to discuss a possible exploitation of suitable nonlinearities (Slotine and Li, 1991). In fact,

- nonlinear control may permit the use of less expensive components, as actuators and sensors, with nonlinear characteristics;
- nonlinearities can be intentionally introduced into the controller so that model uncertainty can be tolerated;
- the adoption of base isolation or passive damper devices could dissipate most of the incoming (external) energy. Their use alone does not solve all the control problems. However their coupling with active control devices (hybrid control) shows cost and performance optimality;
- the topic of nonlinear control design is attracting particular attention because the advent of powerful microprocessors has made the implementation of nonlinear controllers a relatively simple matter.

Nonlinearity was first considered at the end of the eighties in the control of chemical plants and in robot manifacturing. Still it was not related to a structural feature. Later, a sequence of events are driving toward nonlinear stochastic structural control:

- some international workshops (Housner et al., 1992; Housner and Masri, 1993) prepared the way to an intensive research activity within Civil Engineering applications;
- the American National Science Foundation is regarding control systems as basis for a rebuilding and enhancing nation's infrastructures: *"The control systems of today can learn, they can change based upon need, they can anticipate a need, and they can correct a mistake."*
- the International Association of Structural Control (IASC) organized the 1st World Conference on Structural Control in Pasadena, California, on August 1994;
- the Association for Control of Structures (ACS) organizes the 1st European Conference on Structural Control in Barcelona, Spain, on May 1995.

2. Nonlinear control

Let a physical system to be controlled and the specifications of its desired behaviour be given. One is required to construct a feedback control law to make the closed-loop system display the desired behaviour (see Figure 1).
A nonlinear dynamic system is governed by a set of first-order differential equations:

$$\dot{\mathbf{x}} = \mathbf{f}(\mathbf{x}, \mathbf{u}, t) \tag{1}$$

where the mathematical model $\mathbf{f}$ depends on the vector of the state variables, $\mathbf{x}$, on the vector of the control action, $\mathbf{u}$, and on the parameter t, the time.
One distinguishes nonlinear regulation and nonlinear tracking. In the first case a stabilizer is to be designed so that the state of the closed-loop system will be stabilized around an equilibrium point. A control law $\mathbf{u}$ is pursued such that, starting from any

Active Structural Control
(closed loop)

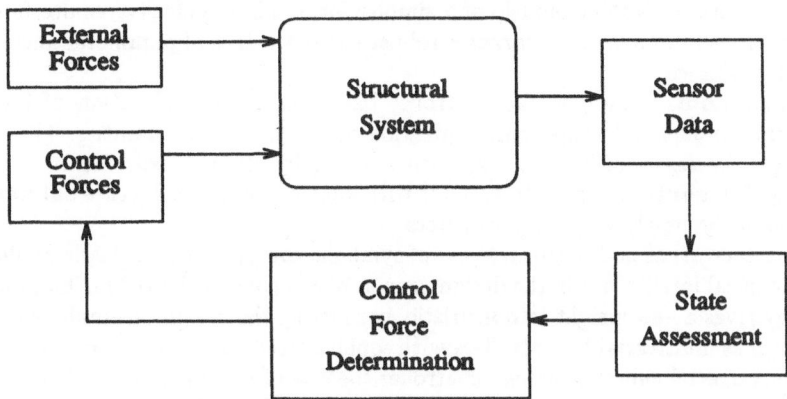

Figure 1.   Active structural control scheme.

initial state, $\mathbf{x}$ tends to $\mathbf{0}$ as $t \rightarrow \infty$ (asymptotic stabilization problem).
The definition of the asymptotic tracking problem requires a second equation be added through the model $\mathbf{h}$:

$$\mathbf{y} = \mathbf{h}(\mathbf{x}) \tag{2}$$

and a desired output trajectory $\mathbf{y}_d$. The problem is then to find a control law $\mathbf{u}$ such that, starting from any initial state, the tracking error $\mathbf{y}(t) - \mathbf{y}_d(t)$ go to zero, while the whole state $\mathbf{x}$ remains bounded.

The response of a nonlinear system to one command does not reflect its response to another command. This makes a systematic specification for nonlinear systems almost impossible. Indeed, one looks for some qualitative specifications of the desired behaviour: stability, tracking accuracy, robustness and cost. The above qualities conflict to some extent and a good control can be obtained only based on effective trade-off in terms of stability / robustness, stability / performance, cost/performance and so on.

A control design requires to go through the following standard procedure, possibly with a few iterations (Slotine and Li, 1991):

1. specify the desired behaviour and select actuators and sensors;
2. model the structural system by a set of differential equations;
3. design a control law for the system;
4. analyse and simulate the resulting control system;
5. implement the control system in hardware.

Nevertheless, in the analysis of nonlinear control systems there is no general method for designing nonlinear controllers; one has just a rich collection of alternative and complementary techniques:

- **feedback linearization** deals with techniques for transforming original system models into equivalent models of a simpler form. They typically require full state measurement and do not guarantee robustness in the face of parameter uncertainty or disturbances.
- **robust control** designs the controller based on the consideration of both the nominal model and some characterization of the model uncertainties. They apply to specific classes of nonlinear systems and require state measurements.
- **adaptive control** applies to systems with known dynamic structure but unknown constant or slowly-varying parameters.
- **expert control** makes effective use of symbolic computation as classic application of artificial intelligence in the design process of any control algorithm. This methodology gives a new insight into heuristic, permitting the design of simple regulators, as well as multivariable controllers with sophisticated control laws (Pedrycz, 1989). A very special subset of expert controller consists of the fuzzy controllers based on fuzzy logic reasoning (Faravelli and Yao, 1995a and 1995b).

3. Control stability

Stability is a basic property of a closed-loop system; it guarantees that the system will reach an equilibrium state after external or internal disturbances. Stability tests require that a mathematical model of the system be available, but for complex structural systems such a mathematical model is unknown and/or is too expensive to detemine: these are the systems for which fuzzy controllers are more appropriate.

3.1. EXPERT CONTROL STABILITY

Following Pedrycz (1989), given a fuzzy controller establishing a dynamical incremental fuzzy relation G_k on the phase variables, the motion fuzzy model is stable in the limit if a relation G_∞ exists such that:

$$\lim_{k \to \infty} G_k = G_\infty \tag{3}$$

This notion concern aymptotic stability. In practice one requires that stability is achieved in a long enough period T. The transient response depends on the fuzzy relation G_k; if after several iteration its value does not change, the controller is designed appropriately. Of course, the length of the period T depends on the structural problem under investigation.

In contrast to several classical control engineering techniques, only a few methods are available that guarantee or check stability for fuzzy controllers. In practice validation is performed with simulations and massive tests. The state-space approach is the simplest approach to the problem. One introduces a finite number of cells in the state space covered by the membership functions. Cell by cell, then, the trajectory resulting from the action of the fuzzy controller is obtained, the equilibrium points identified and the stability area is eventually detected. Other, more sophisticated, but less general, methods are listed and briefly illustrated in (Mikut and Bretthauer, 1994).

3.2. LYAPUNOV STABILITY

When the mathematical model is available, one makes reference to classical stability theory based on the existence of a Lyapunov function V. It must be positive definite and its time derivative must be negative semi-definite. Then, the equilibrium point around which this function exists is stable in the sense of Lyapunov. Moreover, if V is decrescent, the equilibrium point is uniformly stable (and if the dervative is negative definite the equilibrium point is uniformly asymptotically stable. For instance, in the linear case, the Lyapunov function can be selected as $\mathbf{x}^T \mathbf{S} \mathbf{x}$, $\mathbf{S}$ being the solution of the Lyapunov matrix equation

$$\mathbf{A}^T \mathbf{S} + \mathbf{S} \mathbf{A} = -\mathbf{Q}$$

A being the open-loop system matrix (which is stable) and $\mathbf{Q}$ is a symmetric and positive semi-definite matrix.

The constrained receding horizon strategy for the nonlinear system (1) and (2) evaluates the control forces at time t by minimizing:

$$J(t) = \sum_{i=0}^{N-1} \{L(\mathbf{x}(t+i)) + M(\mathbf{u}(t+i))\} \tag{4}$$

with respect to $\mathbf{u}(t)$, $\mathbf{u}(t+1), \ldots, \mathbf{u}(t+N-1)$ subject to the constraint $\mathbf{x}(t+N) = 0$, with $L > 0$ and $M > 0$ for any $\mathbf{x}$ (but $L(0) = 0$ and $M(0) = 0$). At time $t+1$, $\mathbf{u}(t+1)$ is found by minimizing $J(t+1)$ subject to $\mathbf{x}(t+N+1)$. The control law is of the type:

$$\mathbf{u}(t) = \kappa(\mathbf{x}(t)) \tag{5}$$

and the closed loop system is written:

$$\mathbf{x}(t+1) = \mathbf{f}_c(\mathbf{x}) = \mathbf{f}(\mathbf{x}, \kappa(\mathbf{x}(t)), t) \tag{6}$$

The optimal value of the cost function (4) is regarded as Lyapunov function to prove the asymptotic stability of the closed loop system (De Nicolao et al., 1994; De Nicolao and Scattolini, 1994).

4. Stochastic nonlinear control

Stochastic control theory deals with dynamical systems, described by differential equations, and subject to disturbances which are characterized as stochastic processes $\eta(t)$ (Casciati and Faravelli, 1991):

$$\dot{\mathbf{x}} = \mathbf{f}(\mathbf{x}, \mathbf{u}, \boldsymbol{\eta}(t), t) \tag{7}$$

$$\mathbf{y} = \mathbf{h}(\mathbf{x}) \tag{8}$$

Given a system and a criterion, the problem is to find the control law which minimizes the criterion. The classical control criteria try to minimize either the variance of the output or the expected value of a loss function. But all procedures derived from the classical LQR (linear quadratic regulator) theory show the inconvenience of lack of stability.

One studies here, for sake of notation simplicity, a single-input-single-output (SISO) system controlled by a constrained receding-horizon controller, with $f(0,0) = 0$, time discrete and $T = N \cdot \Delta t$. It models a base-isolated hybrid-control structural system, subjected to an earthquake ground acceleration a_g. The control criterion is summarized in the expression:

$$\min_{u_t, u_{t+1}, \ldots u_{t+N-1}} E[J(t)] = \sum_{i=0}^{N-1} \left[L(E[x^2(t+i)|x(t)]) + M(u(t+i)) \right]$$

$$E[x^2(t+1)|x(t)] = E[\{f(x, \kappa(x))\}^2] \tag{9}$$
$$E[x^2(t+N)] = 0$$
$$|u_i| \leq \bar{u}$$

with $L(x) > 0$ and $M(u) > 0$, but nul at zero, and $u(t) = \kappa(x)$ the resulting strategy. Let $V(\xi, t, t+N-1)$ denote the optimal value of the cost function in (9) when $x(t) = \xi$. Moreover, one introduces:

$$V(\xi) = V(\xi, t, t+N-1) \tag{10}$$
$$\bar{V}(f_c(\xi)) = V(f(\xi, \kappa(\xi)), t+1, t+N-1) \tag{11}$$
$$V(f_c(\xi)) = V(f(\xi, \kappa(\xi)), t+1, t+N) \tag{12}$$

The sequence of optimal values:

$$V(\xi) > \bar{V}(f_c(\xi)) \geq V(f_c(\xi))$$

is a Lyapunov function ensuring the asymptotic stability of the nominal closed-loop system. Indeed,

$$V(\xi) = \bar{V}(\xi) + L(\xi) + M(\kappa(\xi))$$

shows why $V(\xi) > \bar{V}(\xi)$. On the other side, the sequence $u(t+1), u(t+2), \ldots, u(t+N-1), 0$ is a feasible control for the minimization problem over $[t+1, t+N]$. Since its optimal solution is $V(f_c(\xi))$, one obtains that $\bar{V}(f_c(\xi)) \geq V(f_c(\xi))$.

5. Conclusions

This paper emphasizes the following items:

- Nonlinerity is becoming a need for Civil Engineering structural control;
- The nature of the external disturbance is well modelled in a stochastic context;
- Robustness can easily achieved by special control schemes (sliding control (stable), fuzzy control (stable?));

A technique which presents stability as a primary requisite is also illustrated and its potentialities are shown. The robustness of this technique is the subject of ongoing research.

Acknowledgement

This research was supported by grant 94.RS.74 from the Italian Space Agency (ASI).

References

Astrom K.J. (1970), *Introduction to Stochastic Control Theory*, Academic Press, New York.

Anderson B.D.O. and Moore J.B. (1990), *Optimal Control - Linear Quadratic Methods*, Prentice-Hall Inc., Englewood Cliffs.

Casciati F. (1995), Active Control of Nonlinear Structures, submitted for publication in *Microcomputers in Civil Engineering*.

Casciati, F. and Faravelli, L. (1991), *Fragility Analysis of Complex Structural Systems*, Research Studies Press, Taunton.

Casciati, F., Faravelli, L. and Yao, T. (1994), Fuzzy Control of Civil Structures subjected to Earthquake Loading, *Proceedings EUFIT'94 (2nd European Congress on Intelligent Techniques and Soft Computing)*, Aachen, Germany, Vol. 2, 1050-1054.

Casciati, F., Faravelli, L. and Yao, T. (1994), Application of Fuzzy Logic to Active Structural Control, *Proceedings 2nd European Conference on Smart Structures and Materials*, eds. A. McDonach, P.T. Gardiner, R.S. McEwen and B. Culshaw, Glasgow, 206-209

De Nicolao G., Magni L. and Scattolini R. (1994), On Robustness Properties of Constrained Receding-Horizon Controllers, *Proc. of the 33rd Conference on Decision and Control*, Lake Buena Vista, IEEE, 3023-3024

De Nicolao G. and Scattolini R. (1994), Stability and Output Terminal Constraints in Predictive Control, in D. Clarke (ed.), *Advances in Model-Based Predictive Control*, Oxford University Press, Oxford.

Faravelli L. and Yao T. (1995a), Application of an Adaptive-Network-Based Fuzzy Inference System (ANFIS) to Active Structural Control, *Proc. 1st World Conference on Structural Control*, Vol. 1, WP1-49-58.

Faravelli L. and Yao T. (1995b), Use of Adaptive Network in Fuzzy Control of Civil Structures, accepted for publication in *Microcomputers in Civil Engineering*.

Housner G.W., Masri S.F., Casciati F. and Kameda H. (eds.) (1992), *Proceedings of the U.S.-Italy-Japan Workshop/Symposium on Structural Control and Intelligent Systems*, University of Southern California CE-9210.

Housner G.W. and Masri S.F. (eds.)(1993) *Proceedings of International Workshop on Structural Control - 1993*, University of Southern California CE-9311.

Kobori T. (1988), State of the Art Report: Active Seismic Response Control, *Proc. 9th WCEE (World Conference on Earthquake Engineering)*, Vol. 8, 435-446.

Kwakernaak H. and Sivan R. (1972), *Linear Optimal Control Systems*, Wiley Interscience, New York.

Mikut R. and Bretthauer G. (1994), STABFUZ - A Programme Package for Testing the Stability of Fuzzy Systems, *Proc. EUFIT'94(2nd European Congress on Intelligent Techniques and Soft Computing)*, Vol. 1, 297-302.

Pedrycz W. (1989), *Fuzzy Control and Fuzzy Systems*, Research Studies Press, Taunton.

Slotine J.E. and Li W. (1991), *Applied Nonlinear Control*, Prentice-Hall Inc., Englewood Cliffs.

Soong, T.T. (1990), *Active Structural Control: Theory and Practice*, Longman Scientific and Technical, Harlow.

Yao, J.T-P. (1972), Concept of structural control, *Journal of the Structural Division, ASCE*, 98 (7), 1567-1574.

VISCOPLASTIC RESPONSES WITH STOCHASTIC Q-DAMPING FOR SOIL

G. DASGUPTA
Professor, Civil Engineering & Engineering Mechanics
Columbia University, New York, NY 10027-6699, USA

Abstract

Viscoelastic responses in mechanical systems are significantly influenced by material damping losses. In time dependent steady harmonic vibration problems, energy loss per cycle generally varies with the frequency of excitation. In a large number of geophysical computations, however, frequency <u>independent</u> *Q-models* — a higher value of Q indicating less damping — are routinely chosen for convenient representation of soil damping characteristics in the visco-plastic regime dominated by nonlinear effects.

The distribution of the viscous loss index — *Q-factor* — can be realistically depicted by a spatially correlated random field with bounded variation as per *in situ* experiments. Consequently, the numerical estimation of dynamic viscoplastic response statistics of a soil-structure system demands extensive and time consuming computation invoking methods of nonlinear stochastic mechanics. In this paper on frequency independent viscoplastic *Q-factors*, a systematic nonlinear formulation is presented in order to reduce computational costs without compromising the accuracy demanded by hazard mitigation problems arising out of seismic, wind and wave activities.

Symbolic computation is carried out to implement the analytical steps of continuum mechanics in order to construct quality nonlinear finite elements for the interior and boundary elements for the outer semi-infinite region. The numerical computation for production runs is anticipated to be carried out by C++ object-oriented program modules. It is noteworthy to observe that the computer algebra code segments provided an efficient framework for C++ development.

A. Naess and S. Krenk (eds.), IUTAM Symposium on Advances in Nonlinear Stochastic Mechanics, 117–126.
© *1996 Kluwer Academic Publishers.*

1. Introduction

Structural safety considerations can be adequately addressed in an
economic design process which accounts for the active participation
of the soil foundations in dissipating energy. In addition to the uncertainties associated with the forcing function, the realistic depiction
of the supporting soil mass as an anisotropic medium with randomly
distributed constitutive properties facilities physical modeling in conformity with the experimentally observed response statistics.

Certain parts of the structure undergo severe non-linear deformation as evidenced by very large deformation causing geometrical
distortions which may or may not cause overall catastrophic failure.
Hence, it is essential to analyze the structure for non-linear behavior.
For example, under appreciably high intensity of wind, wave or earthquake loading enormous amount of dynamic forces are transmitted
by the structure to the foundation. Such severe structure-foundation
interface stresses in many cases would be sufficient to introduce nonlinearity in the underlying soil masses. Such nonlinear constitutive
phenomena pertaining to a random medium poses a challenge to
an analyst since the numerically intensive tools of computational
stochastic mechanics invariably incurs high cost of design.

The structure and the soil near-field — the nonlinear region, are
represented by finite elements. Conventional finite elements where
the shape functions are chosen without any consideration of the underlying random field becomes unsuitable since the condition of static
equilibrium is violated for all cases where the constitutional variability is encountered. Thus for the soil region close to the structure-
foundation interface specially tailored elements are used whose shape
functions are calculated for each Monte Carlo sample. This is a significant step in designing the high quality computer code for nonlinear stress-strain implementation pertaining to randomly distributed
material properties. Considerable savings are made for the damping
loss being independent of frequencies, however, for the intrinsic nonlinearity formulation of a time domain solution scheme was successfully carried out where long algebraic expressions were symbolically
operated on using a *computer mathematics environment* where the
steps of algebraic simplification and transformations of calculus are
carried out by the software *Mathematica*. In the interest of clarity,
a generic one-dimensional case is illustrated in detail. Since *Mathematica* operates on arbitrary data-structures in a uniform manner
spatial dimensionality does not pose any conceptual difficulty.

2. Definition of Q Model

The severe nonlinearity associated with the failure of the supporting soil demands a *time domain* formulation. For a singledegree-of-freedom system the governing equation is:

$$m\ddot{x} + c(t)\dot{x} + kx = F(t) \tag{1}$$

where t indicates time and a time derivative is denoted by a *dot* operated on the response $x(t)$. The nonlinearity in damping is depicted by $c(t)$, the mass and the stiffness parameters, m and k, respectively, are held constant at this stage of introductory illustrative example.

The Q- model requires a frequency analog of Eq. 1, hence:

$$-\omega^2 m\hat{x} + k\left(1 + \frac{i\omega\hat{c}(\omega)}{k}\right)\hat{x} = \hat{F}(\omega) \tag{2}$$

needs to be studied, where a time dependent response $z(t)$ is Fourier transformed to $\hat{z}(\omega)$ in terms of the frequency index ω. The Q-factor is defined, in conjunction with the stiffness k from Eq. 2, by:

$$Q^{-1} = \frac{i\omega\hat{c}(\omega)}{k} \tag{3}$$

The advantage of the <u>constant</u> Q model is due to its natural definition related to the energy loss per cycle, refer to Day and Bernard Minster (1984). This poses an immediate challenge when nonlinear stiffness is encountered. During a time step around t_i the damping c_i becomes:

$$\hat{c}_i(\omega) = \frac{kQ^{-1}}{\omega} \tag{4}$$

In this paper the Q factor will be assumed to be described by a spatially correlated random field. Due to the intrinsic nonlinearity one resorts to the Monte Carlo simulation technique, vide Shinozuka (1986). A non-deterministic quantity will be represented by a *tilde*, thus the stochastic version of the above relation will be written as:

$$\hat{\tilde{c}}(\omega) = \frac{k\tilde{Q}^{-1}}{\omega} \tag{5}$$

Thus the first order term with a homogeneous representation is:

$$\hat{\bar{c}}(\omega) = \frac{k\bar{Q}^{-1}}{\omega} \qquad \text{where} \qquad c \text{ and } k \text{ are instantaneous values} \tag{6}$$

In the numerical calculation $\hat{\bar{c}}(\omega)$ was assumed to be constant within each finite element. This scalar quantity is random according to the prescribed spatial correlation over elements.

3. Stochastic Q Model in Finite Elements

Determination of the *equivalent* term $\bar{Q}$ must be in conformity with the balance of momentum condition. This problem is similar to obtaining a *equivalent* modulus of elasticity $\bar{E}$ for a bar within the context of a Monte Carlo sampling instance. This concept has been adequately described in Dasgupta and Yip (1989). For one-dimensional bar and beam problems, with respectively two and four degrees-of-freedom, a *Mathematica* code has been developed and a large number of cases have been examined to reproduce the equilibrium condition *exactly*.

In this paper, the stochastic counterpart $\tilde{X}$ of a deterministic variable X will be denoted with an overhead *tilde*. In the finite element formulation the *stochastic shape functions* yield the stochastic strain- displacement transformation matrix $[\tilde{\mathcal{B}}]$. For stochastic constitutive matrix $[\tilde{\mathcal{D}}]$ the matrix $[\tilde{\mathcal{B}}]$ is computed from:

$$[\tilde{\mathcal{B}}] = [\tilde{\mathcal{D}}]^{-1}[\mathcal{B}]$$

vide Dasgupta (1989) for details.

It was the objective of an earlier publication, Gyebi and Dasgupta (1992), to construct via an integro-differential equation a relation for $[\mathcal{D}]$ in terms of a prescribed Q. Since the approximation should hold case by case hence the same computer code became reusable for generating $[\tilde{\mathcal{D}}]$ from $\tilde{Q}$, and $[\bar{\mathcal{D}}]$ from $\bar{Q}$.

In order to capture the radiation damping condition the outer region was modeled according to the boundary element method. Therein the stochastic Green's function plays the role of finite element shape functions. The crucial results from Dasgupta (1989) is summarized below.

The stochastic Green's function $[\tilde{\mathcal{G}}]$ is obtained from:

$$\tilde{\mathcal{G}} = \mathcal{G}_o \otimes ([\tilde{\mathcal{D}}]^{-1}\mathcal{G}_o), \quad \mathcal{G}_o = [\mathcal{D}]\mathcal{L}_o\mathcal{G}, \quad \tilde{\mathcal{L}} = \mathcal{L}_o[\tilde{\mathcal{D}}]\mathcal{L}_o$$

in which $\otimes$ indicates convolution. The deterministic Green's function $\mathcal{G}$ can be found standard text books of boundary element method.

4. Numerical Considerations

The radiation damping effects are captured by a boundary element scheme, ref.2, for the exterior. The stochastic material properties are computed at finite element mesh points and are interpolated in an algebraic form, by methods of ref.3, for the boundary element. Monte Carlo simulation, ref.4, seems to be the only feasible resort to estimate the response statistics.

For nonlinear finite element systems with deterministic *Q-factors* an integro-differential force-displacement relation has already been successfully implemented in ref.1 with Padé temporal interpolants. The resulting algebraic *Mathematica* expressions, ref.3, were recoded in procedural languages (C and FORTRAN). For each Monte Carlo sample such a strategy was implemented. The near-field stochastic finite element model demanded *stochastic shape functions*, ref.5. The exterior stochastic boundary element was analogously formulated with stochastic Green's functions. Spatially correlated random fields for *Q-factors* were generated according to the standard techniques reported in ref.3. An indirect iterative technique for solving a system of linear equations was accelerated by ordering the sample space of stochastic constitutive properties. Therein a previously computed result is employed as an effective initial guess for the current realization.

It should be mentioned here that the utilization of Padé interpolation to account for the second degree frequency effects in the numerical scheme was possible since the formulation was carried out symbolically. For example, for a frequency of excitation ω the basic form $\exp(-b\omega)$ for a constant b can be readily obtained from *Mathematica* in the following form:

$$\exp(-b\omega) \cong \frac{1 - b\frac{\omega}{2} + b^2\frac{\omega^2}{12}}{1 + b\frac{\omega}{2} + b^2\frac{\omega^2}{12}}$$

The aforementioned close-form expression was numerically coded for the C routine to achieve accuracy and efficiency in the proposed computation with Q factors.

For the stochastic case using the Monte Carlo simulation strategy the decay factor b in the above equation is simulated from a given correlated random field. Thus in general the stochastic counterpart:

$$\exp(-\tilde{b}\omega) \cong \frac{1 - \tilde{b}\frac{\omega}{2} + \tilde{b}^2\frac{\omega^2}{12}}{1 + \tilde{b}\frac{\omega}{2} + \tilde{b}^2\frac{\omega^2}{12}}$$

5. Numerical Example

Numerical calculations were carried out with a variety of spatial correlation functions. A weakly correlated field would yield a high degree of fluctuation, hence the robustness of the prescribed method was examined under such a condition where the given mean field was described in Fig.-1 and a represented Monte Carlo sample input field was realized as depicted in Fig.-2

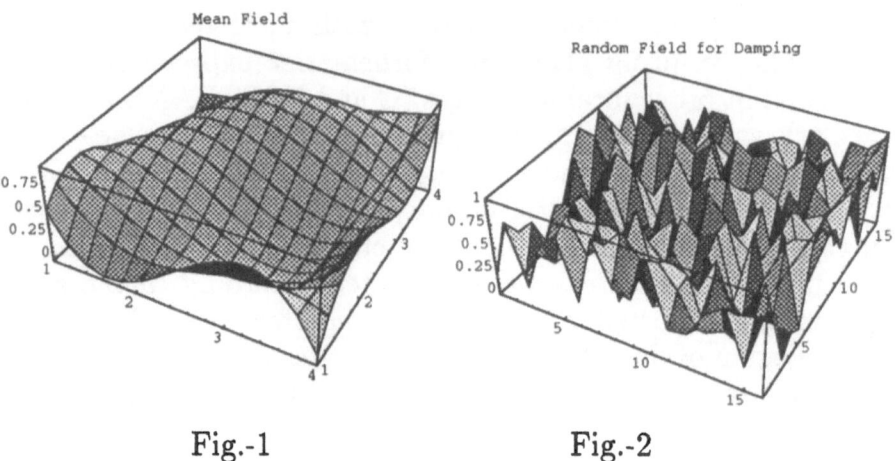

Fig.-1   Fig.-2

A field response with 100 Monte Carlo sample yielded average fields for mild and strong damping cases described in Fig.-3 and Fig.-4, respectively.

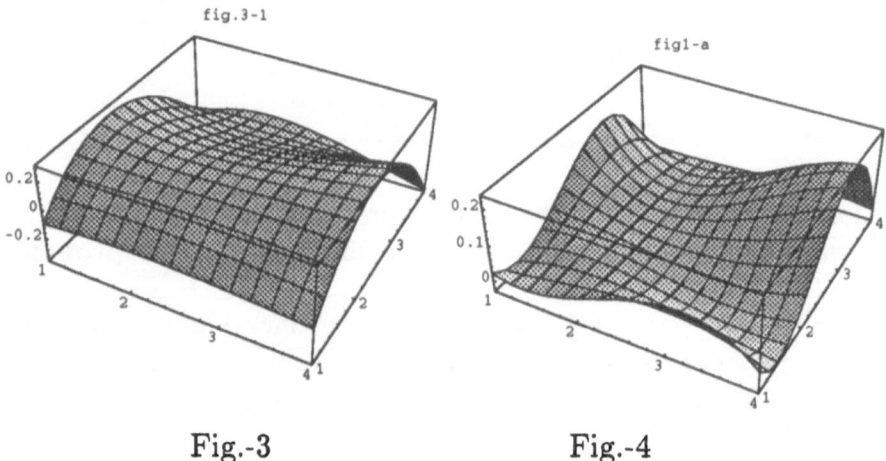

Fig.-3   Fig.-4

6. Computational Issues

Three numerical schemes, viz. finite difference, finite element, and boundary element, which constitute the discipline of computational mechanics, are implemented in computing responses for soil-structure systems. These discretization strategies have been successfully employed to solve stress analysis involving materials where the frequency independent viscous damping properties are spatially distributed as random fields. The difficulties faced in analytical methods due to arbitrary shape of soil-structure boundaries are alleviated with near-field finite elements. Variants of the Runge-Kutta time integration scheme are employed to capture the time history in congruence. Uncertainties in constitutive properties demanded *stochastic shape functions* which are required to be generated separately for each Monte Carlo realization. This part of the process is very computing intensive.

Nonlinear material properties strongly couple variability arising out of the system and the loading specification. Hence, it is possible to prescribe a single nonlinear strategy to address all questions related to the random aspect of the input dataset. There are distinct additional modeling requirements, due to stochastic considerations, by the aforementioned methods. A considerable volume of engineering analysis can be successfully carried out by restricting the stochastic effects, which are "small enough" to be adequately captured by the *perturbation method.* This elegant scheme allows most of the deterministic computer codes to be utilized with slight modification — by adding routines to evaluate the *sensitivity* parameters, which are essentially the partial derivatives of the response quantities with respect to the stochastic variables. Straightforward evaluation of such partial derivatives from first order finite differences very frequently introduces oscillatory errors of the same order as the fluctuations of responses due to the underlying correlated random fields. With the advent of symbolic computation software, it is possible to carry out *exactly* those first order (rather any order) Taylor derivatives. On the other hand, for *large stochasticity* — defined by the divergence of perturbation scheme, the Monte Carlo simulation seems to be the only alternative. Since the Neumann expansion of the stochastic operator still guarantees convergence, an ordering of the sample space accelerates indirect method of solving large stochastic system of equations. It is needless to state that such class of *large random dispersions* dominates practical applications.

7. Concluding Remarks

The basic development within the context of computational stochastic mechanics was the incorporation of random constitutive parameters in the solution algorithm for the *nonlinear* integro-differential equation governing the time history of viscoplastic responses. First, the issue of *nonlinearity* was critically examined for a beam on a Winkler foundation with spatially random $Q-$ parameter $\tilde{c}$. The explicit *nonlinear* foundation damping matrix was prescribed as $\tilde{c}[\{u\}\{u\}^T]$ due to transverse beam displacement $\{u\}$. The second problem involving the two-dimensional *potential* case was solved next. The stochastic Green's functions for the outer boundary element was calculated by recursively modifying an available deterministic Green's function. The final problem of embedment in a three-dimensional half space is currently under study when the midpoint displacement histories of a cubic embedment is being calculated. A wide range of average Q- *factors* of 100, 10, and 1 is being considered, the sample parameter $\tilde{c}$ being digitally simulated based on numerically prescribed generic spectra. Qualitative *nondimensional* plots are being studied as attention is focused on computationally intensive aspects. The *linear* model of the far-field, with stochastic Green's function, is numerically tested in order to furnish boundary conditions appropriate for the nonlinear soil-structure finite element system. In the interest of clarity, stochasticity is confined only to the $Q-$ parameter. The associated elasticity constants are treated to be deterministic quantities. Padé expansion for the decay is assumed with three sets of terms in the numerator and in the denominator.

Extensive computation with two-dimensional elements for the S_H (antiplane) and S_V (plane) cases has been completed. At this current state the three-dimensional program has proved to be very time consuming, since for analytical accuracy a large volume of numerics are carried out in the symbolic computing environment. The final routines in fully numerical form will be in C++ as currently developed in a `Macintosh`, using the `CodeWarrior` environment. In this paper, therefore, only some two-dimensional cases are reported. The numerics and graphics are from *Mathematica*.

A significant saving in Monte Carlo simulation procedure is due to the formulation of the spatial aspects of the response distribution. The finite element *stochastic shape functions* due to material damping decay are assumed in the Padé form. All rational polynomial expansions are currently carried out analytically.

Three-dimensional stochasticity in the experimental soil damping data demands more advancements to the existing finite element
technology than enhancement in area of stochastic mechanics. Since
spatial correlation influences only the initial data the subsequent
behavior in the nonlinear regime for the soil elements can be conveniently analyzed by conventional Monte Carlo schemes.

The standard problem of a square embedment under an impact loading of sufficient strength necessitates an averaging of the
propagation front in order not to smear the effects of discontinuity
propagation. The corresponding numerical results are summarized
in Fig.-5 when a realization indicated a steep rise in Q- over the
central element. Relevant <u>mean compressive</u> stresses are shown.

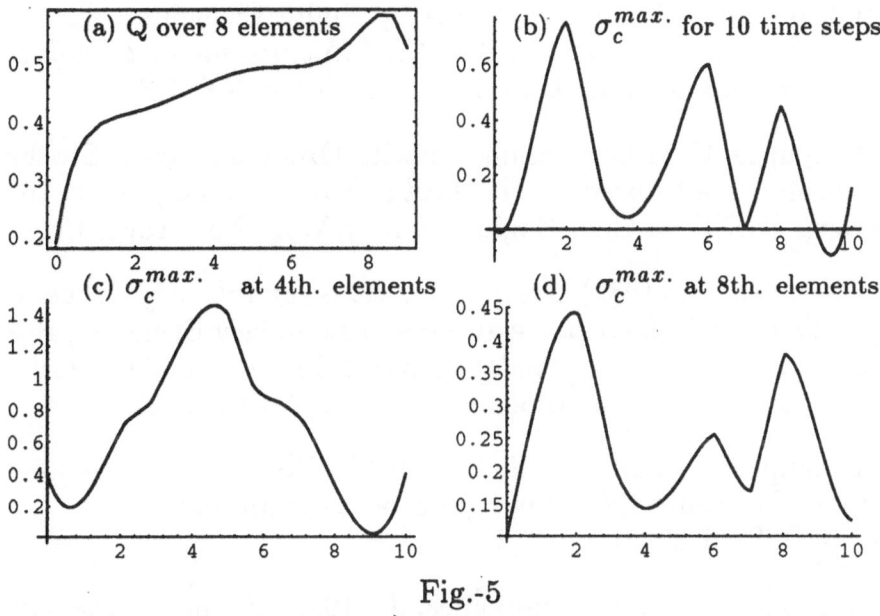

Fig.-5

The strength of the impact was adjusted to be weak enough not to
introduce any tensile stress beyond four elements down the exterior
region. The numerical solution was observed to be sufficiently stable.

The eighth element displayed a sharp peak at the sixth time
step as a result of discontinuity propagation. Attention is drawn
to the numerical result which could faithfully reproduce the effects
of a sharp front when the stochastic nonlinear material constitutive
model is admitted. The proposed integro-differential equation formulation with *stochastic shape functions* deemed adequate for predicting catastrophic failure at the soil-structure interface region.

8. References

Keller, J. B. (1962) *Wave propagation in random media,* SIAM Proc. **13**, American Mathematical Society, Providence, RI, 227-246.

Day, S. M. and Bernard Minster, J. (1984) "Numerical simulation of attenuated wavefields using a Padé approximant method," *Geophysics, J. Royal Astr. Soc.* **78**, 105-118.

Shinozuka, M., (1986) *Stochastic Mechanics*, vols. **1,2,3**, *Department of Civil Engineering and Engineering Mechanics,* Columbia University, New York, NY, USA.

Dasgupta, G. (1987) A Computational Scheme to Analyze Nonlinear Stochastic Dynamic Systems by Finite Elements. *Research Report: April 1987, Department of Civil Engineering and Engineering Mechanics,* Columbia University, New York, NY, USA.

Dasgupta, G. (1988) Simulation with Ordered Discrete Stochastic Samples. *Advancement in Probabilistic Methods,* May 1988, American Society of Civil Engineers, New York, New York, USA.

Dasgupta, G. (1989) *Green's Functions for Inhomogeneous Media for Boundary Elements.* Advances in Boundary Elements, Brebbia, C. A. and J. J. Connor (eds), Eleventh International Conference on Boundary Element Methods, Cambridge, Mass., **1**, 37-46.

Dasgupta, G. and Yip, S.-C., (1989) "Nondeterministic Shape Functions for Finite Elements with Stochastic Moduli," Proc. ICOSSAR 89, 1065-1072.

Brebbia, C. A. and Dominguez, J. (1992) *Boundary Elements,* McGraw-Hill.

Dasgupta, G. (1992) Approximate Dynamic Responses in Random Media. *Acta Mechanica, Springer Verlag,* Wien, Austria, **3**, 99-114.

Gyebi, O. K. and Dasgupta, G. (1992) "A Finite Element Model for Viscoplastic Soil with a Q-Factor," Int. J. Soil Dynamics and Earthquake Engineering, **11**, 187-192.

Wolfram, S., (1992) *Mathematica,* Addison-Wesley.

GAUSSIAN WHITE NOISE EXCITED ELASTO-PLASTIC OSCILLATOR OF SEVERAL DEGREES OF FREEDOM

OVE DITLEVSEN AND SØREN RANDRUP-THOMSEN
Department of Structural Engineering
Technical University of Denmark
Building 118, DK 2800 Lyngby, Denmark

Abstract. The Slepian model process method has turned out to be a powerful tool to obtain accurate approximations to the long run probability distributions of the plastic displacements of a one degree of freedom linear elastic-ideal plastic oscillator (EPO) subject to stationary Gaussian white noise excitation. The paper extends the method to the simplest EPO of more than one degree of freedom. This EPO contains only a single structural component that can yield. It is experienced by direct response simulation that the loss of the property of slowly varying amplitudes when extending to more that one degree of freedom tends to jeopardize the assumption that the position of the response amplitude at the end of any clump of plastic displacements has practically no influence on the distributional properties of the response just before the start of the next clump of plastic displacements. Therefore the results obtained herein strictly apply to the *first* clump and not to *any* clump of plastic displacements. Under this restriction the obtained Slepian model results fit well with the results obtained by direct response simulations. Also it is observed that the restriction gets less importance for decreasing intensity of the white noise excitation.

1. Introduction

Empirical distributions of the permanent displacement of an elasto-plastic oscillator (EPO) of one or more degrees of freedom subject to random process excitation can be obtained by direct simulation of the response vector of the oscillator given that realizations of the excitation can be simulated.

127

A. Naess and S. Krenk (eds.), IUTAM Symposium on Advances in Nonlinear Stochastic Mechanics, 127–142.
© *1996 Kluwer Academic Publishers.*

Apart from the computational efforts spent on direct response simulations, it may be difficult to analyse the details of the behavior of the oscillator solely studying this type of simulated empirical data. If for nothing else it is an intellectual challenge to try to apply approximate reasoning to establish random variable models that make it easier to appreciate the detailed features of the elasto-plastic response process and that allow the derivation of analytical or numerical approximations to the permanent displacement distributions.

For stationary Gaussian white noise excitation of a single degree of freedom EPO a tool for such approximate reasoning is the linear regression method used to formulate so-called Slepian model processes. With the dominating contribution to the randomness contained in a single Rayleigh distributed random variable Z the Slepian model process for given Z is an almost deterministic description of the response of the associated linear elastic oscillator (ALO) in the vicinity of an upcrossing of any given displacement level. The ALO is obtained from the EPO by letting the plasticity limits be $\pm\infty$ and changing nothing else. If the given displacement level corresponds to the yield limit, the Slepian model process for the ALO defines the initial conditions for the movement of the EPO after entrance to the plastic domain. The method has been used by Ditlevsen and Bognár (1993) to obtain closed form analytical approximations to the probability distributions of the single plastic displacement increments as well as to the accumulated plastic displacement and plastic work occurring in a clump of directly after each other following excursions into the plastic domain. Comparisons with direct simulation results as well as with alternative theoretical results obtained by stochastic averaging, Roberts (1980), show excellent agreement.

The following is about applications of the Slepian model process method to a stationary Gaussian white noise excited EPO of two or more degrees of freedom. The study is restricted to the simplest EPO of more than one degree of freedom that may reflect an idealized engineering application: A plane "shear wall" frame with n infinitely rigid floors (traverses) as sketched in Figure 1 is considered. The floors are numbered 1 to n from the top and solely floor n is subject to stationary Gaussian white noise. The damping is assumed to be modal. The movement of floor 2 is thought of as to represent the ground motion during the stationary part of an earthquake. Thus only floor 1 represents the "real" frame structure, while the floors $2, ..., n$ are model elements that simulate the ground behavior under an earthquake. By varying the number of floors and the individual mass, stiffness and damping properties, a certain family of different ground motion spectra can be generated.

The shear connection between floor 1 and floor 2 is assumed to resist the relative movement by a symmetric linear elastic-ideal plastic restoring force while all the other shear connections are assumed to be linear elastic.

This EPO also simulates the soil-structure interaction through the exchange of energy between the top floor and the elastic structure below the top floor.

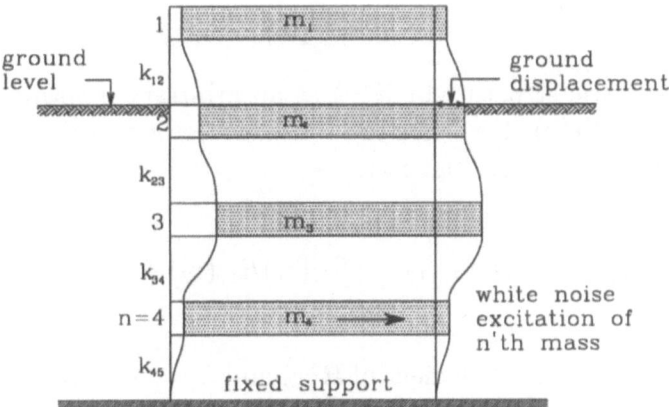

Figure 1. Linear $(n-1)$-mass earthquake simulator for one-floor building.

2. Slepian model vector processes

The time and the response of the oscillator are made dimensionless with the dominating harmonic component of floor 1 relative to floor 2 having the dimensionless period 2π. The relative dimensionless response $\mathbf{X}(\tau)$ is related to the absolute dimensionless response $\mathbf{Y}(\tau)$ by $X_1(\tau) = Y_1(\tau) - Y_2(\tau), ..., X_n(\tau) = Y_n(\tau)$ and $\mathbf{X}(\tau)$ is normalized such that $X_1(\tau)$ has unit variance (Appendix). The corresponding dimensionless yield limits of the EPO are u and $-u$.

Consider the points of upcrossing of level u of the relative displacement process $X_1(\tau)$. With the time origin chosen at an arbitrarily chosen upcrossing, the linear regression and the residual covariance function matrix together define the non-stationary Gaussian vector process

$$\mathbf{X}_u(\tau) = \mathbf{r}(\tau)u - \dot{\mathbf{r}}(\tau)Z/\sqrt{\lambda_2} + \mathbf{\Delta}(\tau) \tag{1}$$

where $[\mathbf{r}(\tau)]'$ (prime indicates transposition) is the first row in the covariance function matrix $\mathbf{R}(\tau)$ of the vector process $\mathbf{X}(\tau)$ as obtained from standard modal theory (Appendix), $Z = \dot{X}_1(0)/\sqrt{\lambda_2}$ is a standard Rayleigh variable, that is, Z has the density function $f_Z(z) = z\,e^{-z^2/2}$, $z > 0$, and $\mathbf{\Delta}(\tau)$ is the residual vector process. The vector process (1) is the so-called

Slepian model vector process for the probabilistic structure of the vector process $\mathbf{X}(\tau)$ at an arbitrary u-upcrossing of $X_1(\tau)$.

If instead of having an upcrossing between $X_1(\tau)$ and level u at time $\tau = 0$ we have that $X(0)$ is a local maximum of $X_1(\tau)$, that is, we have that $\dot{X}_1(0) = 0$, then the Slepian model vector process becomes

$$\mathbf{X}_{max}(\tau) = \mathbf{r}(\tau)X_1(0) + \mathbf{\Delta}(\tau) \tag{2}$$

for the probabilistic structure of $\mathbf{X}(\tau)$ at an arbitrary 0-downcrossing (or 0-upcrossing) of $\dot{X}_1(\tau)$. For both the residual vector processes in (1) and (2) the covariance function matrix is

$$\text{Cov}[\mathbf{\Delta}(\tau_1), \mathbf{\Delta}(\tau_2)'] = \mathbf{R}(\tau_2 - \tau_1) - \{R_{1i}(\tau_1)R_{1j}(\tau_2) + \dot{R}_{1i}(\tau_1)\dot{R}_{1j}(\tau_2)/\lambda_2\} \tag{3}$$

where $R_{ij}(\tau)$ is the generic element of $\mathbf{R}(\tau)$ and $\lambda_2 = -\ddot{R}_{11}(0)$ is the second spectral moment of the process $X_1(\tau)$.

Numerical studies reported in Randrup-Thomsen and Ditlevsen (1995) show that the first component of the linear regression part of (1) approximately takes the value M where

$$M^2 - u^2 = \frac{aZ^2 + b}{Z^2 + b}Z^2 \tag{4}$$

at the first local maximum after $\tau = 0$. The coefficients a and b are functions of the parameters of the oscillator (degrees of freedom, damping ratios, stiffness to mass ratios, and the level u). The numerical studies show that $b \simeq 0$ for $u = 0$ so that (4) in that case simplifies to $M \simeq \sqrt{a}Z$.

Ditlevsen and Bognár (1993) points out that the white noise excitation causes that the response curves have a large number of densely distributed small buckles of consecutive local maxima and minima situated within narrow intervals around the macro scale crests and troughs of the response curve. This property is crucial for the understanding that the Slepian model (2) applies for the macro variation at the crest with $X_1(0)$ replaced by the crest level M obtained from the linear regression part of the Slepian model (1) applicable at a u-upcrossing at $\tau = 0$ of $X_1(\tau)$. In fact, by the limit passage to infinity of the fourth spectral moment it follows as shown in Ditlevsen and Bognár (1993) that the local maxima on a sample curve of $X_1(\tau)$ gets exactly the same Gaussian distribution as the response process $X_1(\tau)$ itself. The clumping of the local maxima and minima makes this Gaussian distribution invalid as a distribution for the crest levels. Consequently the local maximal value $X_1(0)$ in (2) is replaced by the "crest" value M defined by (4) with Z being a standard Rayleigh variable.

3. Conditional distributions related to the Slepian model vector process $\mathbf{X}_{max}(\tau)$

In the following let $\mathbf{X}(\tau) = [X_1(\tau)...X_n(\tau)]'$ denote $\mathbf{X}_{max}(\tau)$ as defined by (2) and adopt the notation $\mathbf{X}_2(\tau) = [X_2(\tau)...X_n(\tau)]'$ for this and any other vector. Bayes' formula gives the conditional density

$$f_{\mathbf{X}(0),\dot{\mathbf{X}}_2(0)}(\mathbf{x}, \dot{\mathbf{x}}_2 \mid X_1(-\pi) > -u) \propto$$
$$P[X_1(-\pi) > -u \mid \mathbf{X}(0) = \mathbf{x}, \dot{\mathbf{X}}_2(0) = \dot{\mathbf{x}}_2] f_{\mathbf{X}(0),\dot{\mathbf{X}}_2(0)}(\mathbf{x}, \dot{\mathbf{x}}_2) \tag{5}$$

where the last factor is

$$f_{\mathbf{X}(0),\dot{\mathbf{X}}_2(0)}(\mathbf{x}, \dot{\mathbf{x}}_2) = f_M(x) f_{\mathbf{X}_2(0),\dot{\mathbf{X}}_2(0)}(\mathbf{x}_2, \dot{\mathbf{x}}_2 \mid X_1(0) = x) \tag{6}$$

with

$$x = \sqrt{u^2 + \frac{az^2 + b}{z^2 + b} z^2}, \quad f_M(x) \propto \frac{dz}{dx} z e^{-\frac{1}{2}z^2} \tag{7}$$

The first factor is

$$P(\mathbf{x}, \dot{\mathbf{x}}_2) = P[X_1(-\pi) > -u \mid \mathbf{X}(0) = \mathbf{x}, \dot{\mathbf{X}}_2(0) = \dot{\mathbf{x}}_2] =$$
$$\Phi\left(\frac{u + E[X_1(-\pi) \mid \mathbf{X}(0) = \mathbf{x}, \dot{\mathbf{X}}_2(0) = \dot{\mathbf{x}}_2]}{D[\Delta_1(-\pi) \mid \mathbf{\Delta}_2(0), \dot{\mathbf{\Delta}}_2(0)]}\right) \tag{8}$$

where $\Phi(\cdot)$ is the standard normal distribution function,

$$E[X_1(-\pi) \mid \mathbf{X}(0) = \mathbf{x}, \dot{\mathbf{X}}_2(0) = \dot{\mathbf{x}}_2] =$$
$$R_{11}(-\pi)x + E[\Delta_1(-\pi) \mid \mathbf{\Delta}_2(0) - \mathbf{x}_2 - r_2(0)x, \dot{\mathbf{\Delta}}_2(0) - \dot{\mathbf{x}}_2 - \dot{r}_2(0)x] \tag{9}$$

and $D[\Delta_1(-\pi) \mid \mathbf{\Delta}_2(0), \dot{\mathbf{\Delta}}_2(0)]$ is the residual standard deviation corresponding to the linear regression of $\Delta_1(-\pi)$ on $[\mathbf{\Delta}_2(0), \dot{\mathbf{\Delta}}_2(0)]$.

We also need the conditional density

$$f_M(x \mid \mathbf{X}(-\pi) = \boldsymbol{\xi}, \, \dot{\mathbf{X}}_2(-\pi) = \dot{\boldsymbol{\xi}}_2) \propto$$

$$f_{\mathbf{X}(-\pi), \dot{\mathbf{X}}_2(-\pi)}(\boldsymbol{\xi}, \dot{\boldsymbol{\xi}}_2 \mid X_1(0) = x) f_M(x) \qquad (10)$$

Considered as a function of x the first factor on the right side is proportional to the normal density $\varphi[(x - \mu)/\sigma]$ because it is proportional to

$$\exp\left\{-\frac{1}{2}([\boldsymbol{\xi}' \, \dot{\boldsymbol{\xi}}_2'] - x[\mathbf{r}(-\pi)' \, \dot{\mathbf{r}}_2(-\pi)])\mathbf{C}^{-1}\left(\begin{bmatrix} \boldsymbol{\xi} \\ \dot{\boldsymbol{\xi}}_2 \end{bmatrix} - x\begin{bmatrix} \mathbf{r}(-\pi) \\ \dot{\mathbf{r}}_2(-\pi) \end{bmatrix}\right)\right\} (11)$$

showing that the parameters σ and μ are given by

$$\sigma^2 = \left\{[\mathbf{r}(-\pi)' \, \dot{\mathbf{r}}_2(-\pi)']\mathbf{C}^{-1}\begin{bmatrix} \mathbf{r}(-\pi) \\ \dot{\mathbf{r}}_2(-\pi) \end{bmatrix}\right\}^{-1} \qquad (12)$$

$$\mu = \sigma^2[\mathbf{r}(-\pi)' \, \dot{\mathbf{r}}_2(-\pi)']\mathbf{C}^{-1}\begin{bmatrix} \boldsymbol{\xi} \\ \dot{\boldsymbol{\xi}}_2 \end{bmatrix} \qquad (13)$$

where

$$\mathbf{C} = \begin{bmatrix} \text{Cov}[\boldsymbol{\Delta}(-\pi), \, \boldsymbol{\Delta}(-\pi)'] & \text{Cov}[\boldsymbol{\Delta}(-\pi), \, \dot{\boldsymbol{\Delta}}_2(-\pi)'] \\ \text{Cov}[\dot{\boldsymbol{\Delta}}_2(-\pi), \, \boldsymbol{\Delta}(-\pi)'] & \text{Cov}[\dot{\boldsymbol{\Delta}}_2(-\pi), \, \dot{\boldsymbol{\Delta}}_2(-\pi)'] \end{bmatrix} \qquad (14)$$

As in Ditlevsen and Bognár (1993) we thus have the conditional density of M given $M > u = 0$ as

$$f_M(x \mid \mathbf{X}(-\pi) = \boldsymbol{\xi}, \, \dot{\mathbf{X}}_2(-\pi) = \dot{\boldsymbol{\xi}}_2) \propto \varphi\left(\frac{x - \mu}{\sigma}\right) x e^{-x^2/(2a)}, \quad x > 0 \qquad (15)$$

with the corresponding complementary distribution function

$$1 - F_M[X \mid \mathbf{X}(-\pi) = \boldsymbol{\xi}, \, \dot{\mathbf{X}}_2(-\pi) = \dot{\boldsymbol{\xi}}_2] =$$

$$\frac{\varphi\left(\dfrac{x - \mu\sqrt{a}}{\sigma}\right) + \dfrac{\mu\sqrt{a}}{\sigma}\Phi\left(-\dfrac{x - \mu\sqrt{a}}{\sigma}\right)}{\varphi\left(\dfrac{\mu\sqrt{a}}{\sigma}\right) + \dfrac{\mu\sqrt{a}}{\sigma}\Phi\left(\dfrac{\mu\sqrt{a}}{\sigma}\right)}, \, x > 0 \qquad (16)$$

Finally it is noted that the conditional density

$$f_{\mathbf{X}_2(0),\dot{\mathbf{X}}_2(0)}(\mathbf{x}_2, \dot{\mathbf{x}}_2 \mid \mathbf{X}(-\pi) = \boldsymbol{\xi}, \dot{\mathbf{X}}_2(-\pi) = \dot{\boldsymbol{\xi}}_2, X_1(0) = x) \qquad (17)$$

is the $(2n - 2)$-dimensional normal density with mean vector

$$\begin{bmatrix} r_2(0)x \\ \dot{r}_2(0)x \end{bmatrix} + \mathrm{Cov}\left[\begin{bmatrix} \boldsymbol{\Delta}_2(0) \\ \dot{\boldsymbol{\Delta}}_2(0) \end{bmatrix}, [\boldsymbol{\Delta}(-\pi)' \ \dot{\boldsymbol{\Delta}}_2(-\pi)'] \right] \mathbf{C}^{-1} \begin{bmatrix} \boldsymbol{\xi} - xr(-\pi) \\ \dot{\boldsymbol{\xi}}_2 - x\dot{r}_2(-\pi) \end{bmatrix} \quad (18)$$

and covariance matrix

$$\mathrm{Cov}\left[\begin{bmatrix} \boldsymbol{\Delta}_2(0) \\ \dot{\boldsymbol{\Delta}}_2(0) \end{bmatrix}, [\boldsymbol{\Delta}_2(0)' \ \dot{\boldsymbol{\Delta}}_2(0)'] \right] - \mathrm{Cov}\left[\begin{bmatrix} \boldsymbol{\Delta}_2(0) \\ \dot{\boldsymbol{\Delta}}_2(0) \end{bmatrix}, [\boldsymbol{\Delta}(-\pi)' \ \dot{\boldsymbol{\Delta}}_2(-\pi)'] \right] \cdot$$

$$\mathbf{C}^{-1} \mathrm{Cov}\left[\begin{bmatrix} \boldsymbol{\Delta}(-\pi) \\ \dot{\boldsymbol{\Delta}}_2(-\pi) \end{bmatrix}, [\boldsymbol{\Delta}_2(0)' \ \dot{\boldsymbol{\Delta}}_2(0)'] \right] \qquad (19)$$

4. Plastic displacements in a clump

Following the presentation in Ditlevsen and Bognár (1993) the excess elastic energy $(M^2 - u^2)/2$ of the ALO is available for the EPO to be dissipated into plastic work after the first yield limit crossing ($+u$, say) in a clump of successive crossings alternating between crossings of level $+u$ and level $-u$. This assumption was used for the EPO of one degree of freedom by Karnopp and Scharton (1966) and is also the basis for the theory in Ditlevsen and Bognár (1993). However, for $n > 1$ there is the possibility of having exchange of energy between the first floor and the floors beneath the first floor. Thus a modification could possibly be needed.

After an upcrossing of level u by the ALO the relative displacement of the first floor is approximately given by $X_1(\tau) = u \cos \tau + \kappa Z \sin \tau$, an expression that follows asymptotically from (1) when neglecting the residual process $\Delta_1(\tau)$ and letting the modal damping ratio for the dominating frequency approach zero. By introducing the correction factor κ on Z where $\kappa^2 = (aZ^2 + b)/(Z^2 + b)$, this displacement function of τ becomes consistent with the expression (4) for the first local maximum value M of $X_1(\tau)$ after $\tau = 0$. This maximum is obtained for $\tau = \tau_{max} = \arctan(\kappa Z/u)$. During yielding the restoring force on the first floor is equal to u for the EPO and equal to $X_1(\tau) \geq u$ for the ALO for $0 \leq \tau \leq \tau_{max}$. This means that the ALO restoring force on floor 2 should be corrected by a force equal to $u - X_1(\tau)$ for $0 \leq \tau \leq \tau_{max}$. By conservation of momentum the dimensionless impulse from this force causes a change of the dimensionless velocity $\dot{Y}_2(t)$ during the time of yielding. Assuming as an approximation that this change is instantaneous and takes place at time $\tau = 0$, it can be shown that the increment of the kinetic energy of the second floor approximately equals $m_2 \alpha_k^2 C_{11}(0) \dot{Y}_2(0) \Delta \dot{Y}_2$. Upon calculating the impulse by integrating the force $u - X_1(u)$ from 0 to τ_{max} it can finally be seen that the dimensionless plastic displacement D approximately becomes

$$D \simeq \frac{u}{2} \left[\left(\frac{M}{u} \right)^2 - 1 \right] - \left[\arctan \left(\sqrt{ \left(\frac{M}{u} \right)^2 - 1} \right) - \sqrt{ \left(\frac{M}{u} \right)^2 - 1} \right] \sum_{i=2}^{n} \dot{X}_i(0) \quad (20)$$

in which the first term corresponds to the excess elastic energy $(M^2 - u^2)/2$, and the last term is the impulse exchange correction term.

The characterization of a local crest of $X_1(\tau)$ at $\tau = 0$ to be the first crest above level u in a clump of local crests above level u is defined to be equivalent to the characterization $X_1(-\pi) > -u$. Conditional on this event the probability density of M is obtained by use of Bayes' formula. With the inclusion of the impulse exchange correction term in (20) it is necessary to apply a numerical method or simulation to obtain the distribution of the first plastic displacement D_1 in a clump of consecutive plastic displacements. Without taking the conditioning on the event $X(-\pi) > -u$ into account a sample value of D_1 defined by (20) is easily generated together with a sample vector $(\mathbf{x}, \dot{\mathbf{x}}_2)$ of $(\mathbf{X}(0), \dot{\mathbf{X}}_2(0))$ from the Slepian model (2) by use of (6). Due to the necessary conditioning on the event $X(-\pi) > -u$ the weight $P_1 = P(\mathbf{x}, \mathbf{x}_2)$ given by (8) is attached to this sample vector.

If the impulse exchange correction term in (20) is neglected, the distribution of D_1 can be derived analytically. For any $a > 0$ and $b \geq 0$ the density function of D_1 becomes

$$f_{D_1}(x) \propto \Phi \left[\frac{1}{\sigma}(u - \mu \sqrt{u^2 + 2ux}) \right] \frac{(z^2 + b)^2}{a(z^2 + b)^2 + (1 - a)b^2} e^{-\frac{1}{2}z^2} \quad (21)$$

in which $\mu = -R_{11}(-\pi)$, $\sigma = D[\Delta_1(-\pi)]$, and where

$$2az^2 = 2ux - b + \sqrt{[2ux + (2a - 1)b]^2 + 4a(1 - a)b^2}, \quad x > 0 \quad (22)$$

For $a = 1$ this formula is the same as in Ditlevsen and Bognár (1993).

To derive the second plastic displacement $-D_2 \leq 0$ in the clump it is in principle necessary to know the position $\mathbf{X}_2(0)$ and the velocity $\dot{\mathbf{X}}_2(0)$ besides that $X_1(0) = u$. These initial conditions define conditioning events for the Slepian model for the sample function behavior given that there is a crest at zero. Since the dominating period is 2π, the specification that there is a crest at $\tau = 0$ implies that almost deterministically there is a trough within a narrow neighbourhood of $\tau = \pi$. Using the previously simulated vector to define $\boldsymbol{\xi}_2 = -\mathbf{x}_2$ and $\dot{\boldsymbol{\xi}}_2 = -\dot{\mathbf{x}}_2 - [\Delta \dot{Y}_2 \, 0...0]'$ and setting $X_1(0) = -u$, $(\mathbf{X}_2(0), \dot{\mathbf{X}}_2(0)) = (\boldsymbol{\xi}_2, \dot{\boldsymbol{\xi}}_2)$ as conditions, we can simulate a conditional sample vector $(\mathbf{x}, \dot{\mathbf{x}}_2)$ of $(\mathbf{X}(\pi), \dot{\mathbf{X}}_2(\pi))$ by first simulating a realization x of M by use of the complementary distribution function (16). If $x < u$ the clump terminates after the first plastic excursion. Otherwise a realization of D_2 is calculated from (20), and a realization of $(\mathbf{X}_2(0), \dot{\mathbf{X}}_2(0))$ is simulated from the normal density (17). Thus we can go on to the third plastic displacement repeating the simulation procedure exactly as it was applied to obtain the realization of D_2 but of course with the conditioning values of $(\mathbf{X}_2(-\pi), \dot{\mathbf{X}}_2(-\pi))$ equal to those values (with opposite sign) obtained from the previous simulation after correction for the impulse exchange.

Following this principle recursively, a weighted sample of the sequence $D_1, D_2, ..., D_N$ is obtained with $N \in \{1, 2, ...\}$ as the random clump size. Repeating the recursive simulation independently m times we get a sample $\{\nu_i; d_{1i}, ..., d_{n_i i}; P_{1i}\}_{i \in \{1, ..., m\}}$ of $\{N; D_1, D_2, ..., D_N; P_1\}$.

Let g be some function of the sequence of plastic displacements such as the accumulated net plastic displacement $D_{net} = D_1 - D_2 + D_3 - ... + (-1)^{N-1} D_N$ or the accumulated absolute plastic displacement $D_{abs} = D_1 + D_2 + ... + D_N$, that is, let g be such that $g(D_1, ..., D_\nu, 0, D_{\nu+2}, ...)$ does not depend on $D_{\nu+i}$ for $i > 1$. As a function of x the probability

$$P[g(D_1, D_2, ..., D_N, 0, ...) \leq x] \simeq \left[\sum_{j=1}^{m} P_{1j} \mathbf{1}_{g(d_{1j}, ..., d_{qj}, 0, ...) \leq x} \right] \Big/ \sum_{j=1}^{m} P_{1j} \quad (23)$$

is the distribution function of the random variable $g(D_1, D_2, ..., D_N, 0, ...)$. Moreover

$$P(N = \nu) \simeq \left[\sum_{j=1}^{m} P_{1j} \mathbf{1}_{n_j = \nu} \right] \Big/ \sum_{j=1}^{m} P_{1j} \quad (24)$$

Without conditioning on $\mathbf{X}_2(0)$ and $\dot{\mathbf{X}}_2(0)$, and neglecting the impulse exchange correction term in (20), the density

$$f_{D_2}(x) \propto \varphi \left[\frac{1}{\sigma} (u - \mu \sqrt{u^2 + 2ux}) \right] \frac{(z^2 + b)^2}{a(z^2 + b)^2 + (1 - a)b^2} e^{-\frac{1}{2} z^2} \quad (25)$$

is obtained with z given by (22). For $a = 1$ this formula is the same as in Ditlevsen and Bognár (1993). Except for the approximation made by neglecting the impulse exchange, the density (25) is the marginal density of D_2 and of any of the following plastic displacements.

5. Examples

The specific results given in this section are obtained for the masses $m_1 = m_2 = \ldots = m_n = m$ and the stiffnesses $k_{12} = k$, $k_{23} = \ldots = k_{n,n+1} = 20k$, with the ratio $k/m = 152[\text{sec}^{-2}]$. The examples show the effect of varying the degree of freedom n, the common modal damping ratio ζ and the dimensionless yield level u.

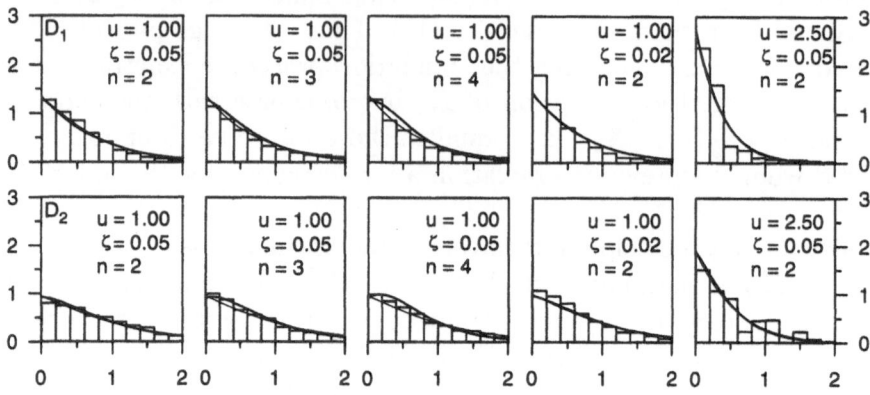

Figure 2. Marginal density functions for D_1 and D_2 given by (21) and (25), respectively, (heavy line), and for $a = 1$ (thin line $\sim$ one degree of freedom EPO). The histograms are obtained by direct simulation of the response of the EPO.

The heavy curves in Figure 2 show examples of the marginal density functions $f_{D_1}(x)$ and $f_{D_2}(x)$ defined by (21) and (25), respectively, with a and b as reported in Randrup-Thomsen and Ditlevsen (1995). The thin curves represent the density functions that correspond to a single degree of freedom EPO with the same values of μ and σ as for the considered EPO of n degrees of freedom. The histograms in Figure 2 are obtained by direct autoregressive simulation of the response of the EPO, Randrup-Thomsen and Ditlevsen (1995).

It is seen from Figure 2 that the distributions of the plastic displacements D_1 and D_2 are fairly well predicted by the Slepian model process method. To make the correct reading of sample values of $D_1, D_2, \ldots$ from the directly simulated response of the EPO it is emphasized that the plastic displacements $D_1, D_2, \ldots$ are accumulations of several small contributions coming from the micro variation generated by the white noise excitation.

For the particular parameter values that define the oscillator considered in this example it is seen that the density functions mostly only deviate modestly from the density functions for the white noise excited single degree of freedom EPO given that the values of μ and σ are those obtained from

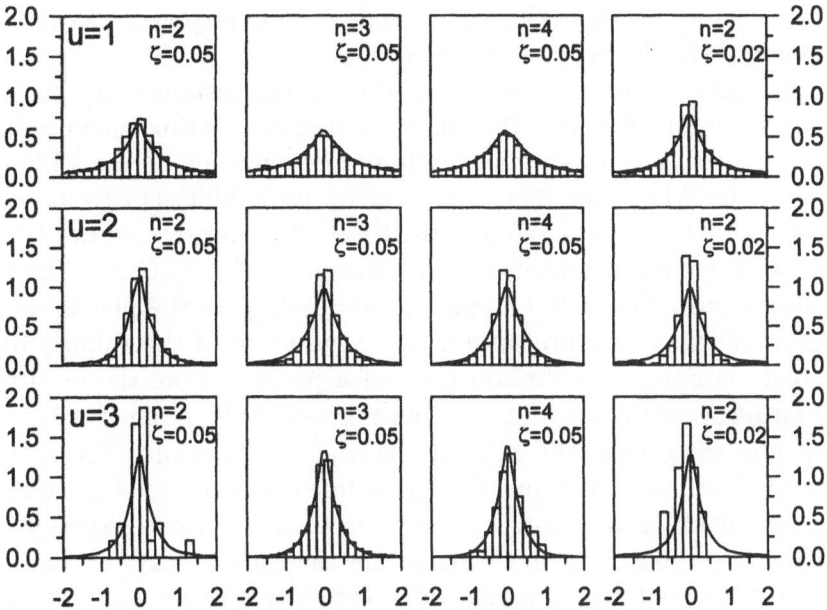

Figure 3. Density functions for $D_{net} = \pm[D_1 - D_2 + ...(-1)^{N-1}D_N]$ calculated by Slepian model simulations compared to histograms obtained by direct simulation of the response of the EPO.

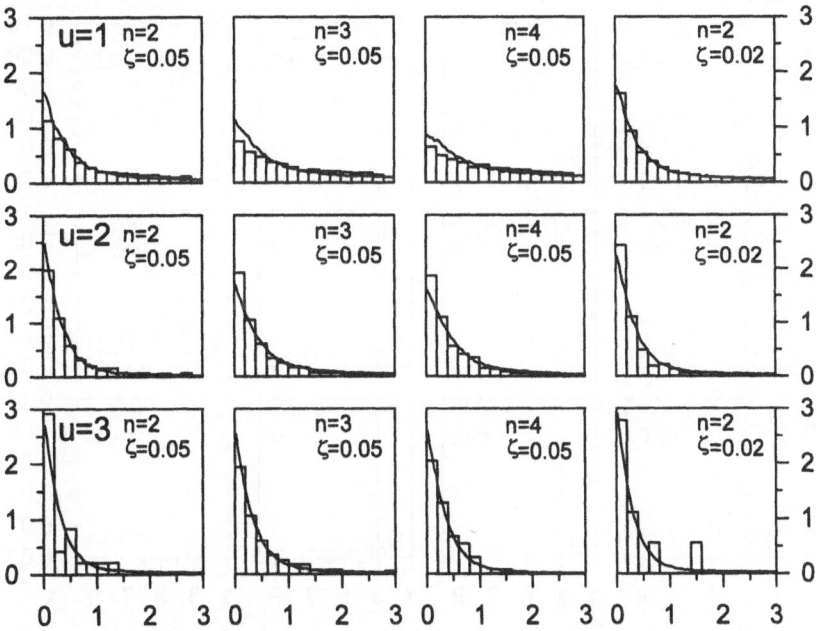

Figure 4. Density functions for $D_{abs} = D_1 + ...D_N$ calculated by Slepian model simulations compared to histograms obtained by direct simulation of the response of the EPO.

the Slepian model for the n degree of freedom ALO response considered at a maximum of the first component of the response.

The simulations are made such that they are consistent with the conditional Gaussian distribution (8). This is ensured by starting the response of the ALO according to the stationary initial distribution of the ALO. The response of the ALO may then with a given probability start out having amplitudes in the plastic domain of the EPO. However, the effect of this is eliminated by simulating such that the change of the ALO into the EPO is postponed until this initial clump of outcrossings to the plastic domain has ended. Without this precaution the assumption of the validity of the conditional Gaussian distribution (8) half a period before the occurrence of the first plastic displacement may not necessarily be as good an approximation as in the case of the EPO of one degree of freedom. The reason is that an ALO of more than one degree of freedom may not have a slowly varying amplitude as it is characteristic for an ALO of one degree of freedom. Therefore the fading of X_1 inside the yield limits may not have long enough duration not to be significantly influenced by the non-stationary start after the last visit to the plastic domain. In fact, if data for D_1 are obtained from the second and the following clumps of plastic excursions

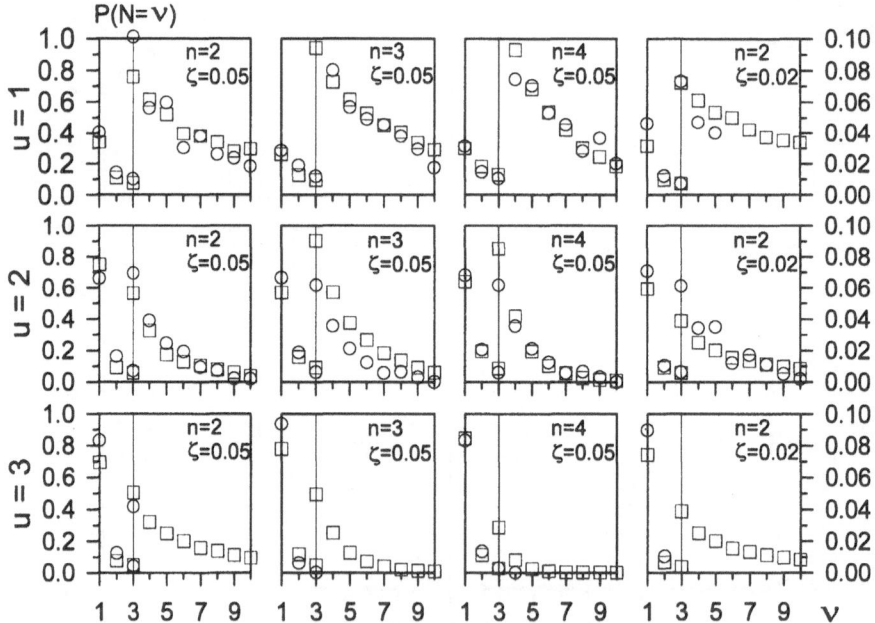

Figure 5. Clump size probabilities $P(N = \nu)$ calculated by Slepian model simulations (shown by □) compared to estimates obtained by direct simulation of the response of the EPO (shown by ○). (The points corresponding to $\nu \leq 3/\nu \geq 3$ refer to the left/right ordinate scale)

of a simulated response of the EPO, the fit to the marginal density of D_1 given by (21) underestimates the size of the plastic displacement. For the examples of Figure 2 it is found that D_1 in the mean is about 15% larger for $u = 1$ and about 1% larger for $u = 2.5$. Not surprising, for the second plastic displacement D_2 and the following plastic displacements in the clump the sensitivity to deviations from the assumption of (8) is less significant.

Figures 3 and 4 show the density functions for ID_{net} ($I = +1$ or -1 with equal probability) and D_{abs} obtained by the simulation of samples based on the Slepian model process method (100.000 clumps) together with histograms based on data from the direct simulation of the EPO response (5.000 clumps). It is seen that the fits between the two data sets are convincing.

Figure 5 shows a similar comparison for the clump size probabilities $P(N = \nu)$.

During the simulations the importance of the impulse exchange correction term in (20) has been evaluated in the form of the density functions of the ratio between the correction term and D. Figure 6 shows that these density functions are practically coincident for D_1 and D_2. Also it is seen that the importance decreases with the level u and increases with the number of degrees of freedom n.

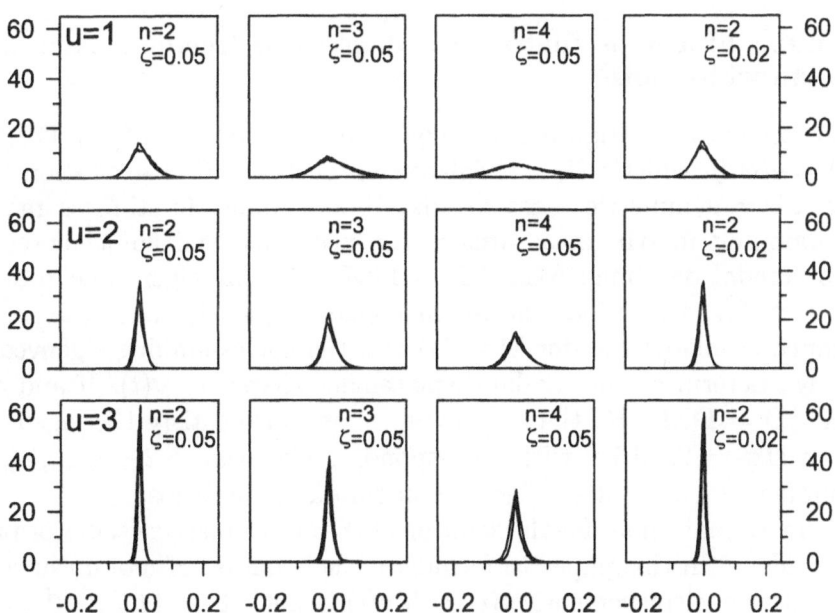

Figure 6. Practically coincident density functions of the ratio between the impulse exchange correction term and D in (20) for $D = D_1$ and $D = D_2$, respectively.

In particular the correction causes an increase of the relative number of small plastic displacements. As compared to the analytical density functions (21) and (25) the effect of the correction is most visible for (25) by moving the maximal density to zero displacement. As it is seen from the histograms in Figure 2 this brings the distributions obtained from the Slepian model to a better fit to the histograms.

Acknowledgement

This work has been financially supported by the Danish Technical Research Council.

References

Ditlevsen, O. and Bognár. L. (1993) Plastic displacement distributions of the Gaussian white noise excited elasto-plastic oscillator. *Probabilistic Engineering Mechanics*, **8**, 209-231.

Randrup-Thomsen, S. and Ditlevsen, O. (1995) One-floor building as elasto-plastic oscillator subject to and interacting with Gaussian base motion. *Third International Conference on Stochastic Structural Dynamics*. San Juan, Puerto Rico, Jan. 15-18, 1995.

Roberts, J.B. (1980) The yielding behaviour of a randomly excited elasto-plastic structure. *J. Sound Vib.*. **72**, 71-85.

Appendix: Linear oscillator of n degrees of freedom excited by Gaussian white noise

For the oscillator in Figure 1 the equation of motion is $\mathbf{M}\ddot{\mathbf{y}} + \mathbf{D}\dot{\mathbf{y}}(t) + \mathbf{K}\mathbf{y}(t) = \mathbf{e}(t)$ in which $\mathbf{y}(t)$ = displacement vector, $\mathbf{e}(t)$ = excitation force vector, $\mathbf{M}$ = symmetric mass matrix, $\mathbf{K}$ = symmetric stiffness matrix, $\mathbf{D}$ = damping matrix that satisfies the necessary and sufficient condition for modal damping: $\mathbf{M}\mathbf{K}^{-1}\mathbf{D} = \mathbf{D}\mathbf{M}^{-1}\mathbf{K}$. The eigenvalue problem $\mathbf{K}\mathbf{w} = \omega^2\mathbf{M}\mathbf{w}$ determines the angular eigenfrequencies $\omega_1, ..., \omega_n$ of the undamped oscillator (i.e. for $\mathbf{D} = \mathbf{O}$) and the corresponding eigenvectors $\mathbf{w}_1, ..., \mathbf{w}_n$ determine the modal displacement shapes of $\mathbf{y}(t)$. If and only if $\mathbf{M}\mathbf{K}^{-1}\mathbf{D} = \mathbf{D}\mathbf{M}^{-1}\mathbf{K}$, the vectors $\mathbf{w}_1, ..., \mathbf{w}_n$ also satisfy the eigenvalue problem $\mathbf{D}\mathbf{w} = 2\zeta\omega\mathbf{M}\mathbf{w}$ with corresponding eigenvalues $2\zeta_1\omega_1, ..., 2\zeta_n\omega_n$. The numbers $\zeta_1, ..., \zeta_n$ are called the modal damping ratios.

It is more convenient for the solution of the elasto-plastic oscillator problem considered in this paper to formulate the equation of motion in terms of the relative displacements $\mathbf{x}(t) = \mathbf{T}\mathbf{y}(t)$, where $\mathbf{T} = \{1_{i=j} - 1_{i=j-1}\}$, $\mathbf{T}^{-1} = \{1_{i\leq j}\}$, $1_A = 1$ if A true, $1_A = 0$ if A false. Then $\mathbf{x}(t)$ satisfies the equation of motion with $\mathbf{M}$, $\mathbf{D}$, $\mathbf{K}$ and $\mathbf{e}(t)$ replaced by $\mathbf{M}_r = (\mathbf{T}^{-1})'\mathbf{M}\mathbf{T}^{-1}$, $\mathbf{D}_r = (\mathbf{T}^{-1})'\mathbf{D}\mathbf{T}^{-1}$, $\mathbf{K}_r = (\mathbf{T}^{-1})'\mathbf{K}\mathbf{T}^{-1}$, and $(\mathbf{T}^{-1})'\mathbf{e}(t)$, respectively. The eigenfrequencies $\omega_1, ..., \omega_n$ and the modal damping ratios $\zeta_1, ..., \zeta_n$ remain invariant to this transformation and the eigenvectors $\mathbf{w}_1, ..., \mathbf{w}_n$ are transformed into $\mathbf{v}_1 = \mathbf{T}\mathbf{w}_1, ..., \mathbf{v}_n = \mathbf{T}\mathbf{w}_n$. Specifically we have $\mathbf{M} = \{m_i\mathbf{1}_{i=j}\}$, $\mathbf{M}_r = \{\sum_{k=1}^{\min\{i,j\}} m_k\}$, $\mathbf{K} = \{-k_{(i-1)i}\mathbf{1}_{i=j+1} + (k_{(i-1)i} + k_{i(i+1)})\mathbf{1}_{i=j} - k_{i(i+1)}\mathbf{1}_{i=j-1}\}$, $\mathbf{K}_r = \{k_{i(i+1)}\mathbf{1}_{i=j}\}$. The damping matrix $\mathbf{D}$ (or $\mathbf{D}_r$) is defined by $\mathbf{M}$, $\mathbf{K}$ and the damping ratios $\zeta_1, ..., \zeta_n$ according to the eigenvalue problem $\mathbf{D}\mathbf{w} = 2\zeta\omega\mathbf{M}\mathbf{w}$.

Let the eigenvectors $\mathbf{w}_1, ..., \mathbf{w}_n$ be normalized with respect to the mass matrix $\mathbf{M}$, that is, let $\mathbf{w}_r'\mathbf{M}\mathbf{w}_s = \delta_{rs}$ (Kronecker's delta $= \mathbf{1}_{r=s}$), so that $\mathbf{w}_r'\mathbf{D}\mathbf{w}_s = 2\zeta_r\omega_r\delta_{rs}$. Then the general solution for the relative displacements is

$$\mathbf{x}(t) = \sum_{r=1}^{n} T_r(t)\mathbf{v}_r \qquad (26)$$

where $T_r(t)$ satisfies the standard form $\ddot{T}(t) + 2\zeta_r\omega_r\dot{T}(t) + \omega_r^2 T(t) = \mathbf{w}_r'\mathbf{e}(t)$ of the equation of motion of the single degree of freedom linear oscillator. Under the assumption that $\mathbf{e}(t)$ is a stationary random vector process $\mathbf{E}(t)$, the stationary solution is

$$T_r(t) = \frac{1}{\alpha_r}\mathbf{w}_r'\int_{-\infty}^{t} \mathbf{E}(\tau)e^{-\zeta_r\omega_r(t-\tau)}\sin[\alpha_r(t-\tau)]d\tau \qquad (27)$$

in which $\alpha_r = \omega_r\sqrt{1-\zeta_r^2}$. Assume now that $\mathbf{E}(t) = N(t)\mathbf{e}_n$ where $N(t)$ is a stationary Gaussian white noise process of intensity πS and $\mathbf{e}_n' = [0...01]$. Then

$$\mathrm{Cov}[T_r(0), T_s(t)] = \frac{v_{rn}v_{sn}}{\alpha_r\alpha_s}\pi S\Bigg\{\int_{-\infty}^{0}\int_{-\infty}^{t}\delta(\tau_1-\tau_2)e^{\zeta_r\omega_r\tau_1}\sin(-\alpha_r\tau_1)$$

$$e^{-\zeta_s\omega_s(t-\tau_2)}\sin[\alpha_s(t-\tau_2)]d\tau_1 d\tau_2\Bigg\} \qquad (28)$$

where $\delta(\cdot)$ is Dirac's delta function and $v_{rn} = \mathbf{v}_r'\mathbf{e}_n = w_{rn}$, $v_{sn} = \mathbf{v}_s'\mathbf{e}_n = w_{sn}$. For $t \geq 0$, (28) reduces to

$$\mathrm{Cov}[T_r(0), T_s(t)] = \frac{v_{rn}v_{sn}}{2\alpha_r\alpha_s}\pi S e^{-\zeta_s\omega_s t}I_{rs}(t), \quad I_{rs}(t) = a_{rs}\cos\alpha_s t + b_{rs}\sin\alpha_s t \qquad (29,30)$$

$$a_{rs} = \frac{4\omega_r\omega_s(\zeta_r\omega_r + \zeta_s\omega_s)\sqrt{(1 - \zeta_r^2)(1 - \zeta_s^2)}}{(\omega_r^2 + \omega_s^2 + 2\omega_r\omega_s\zeta_r\zeta_s)^2 - 4\omega_r^2\omega_s^2(1 - \zeta_r^2)(1 - \zeta_s^2)} \qquad (31)$$

$$b_{rs} = \frac{2\omega_r\sqrt{1 - \zeta_r^2}[2\zeta_s\omega_s(\omega_r\zeta_r + \omega_s\zeta_s) + \omega_r^2 - \omega_s^2]}{(\omega_r^2 + \omega_s^2 + 2\omega_r\omega_s\zeta_r\zeta_s)^2 - 4\omega_r^2\omega_s^2(1 - \zeta_r^2)(1 - \zeta_s^2)} \qquad (32)$$

The covariance function matrix of the stationary response vector process $\mathbf{x}(t)$ as given by (26) is

$$\mathbf{C}(t) = \{C_{ij}(t)\} = \text{Cov}[\mathbf{x}(0), \mathbf{x}(t)'] = \sum_{r=1}^{n}\sum_{s=1}^{n}\mathbf{v}_r\mathbf{v}_s'\text{Cov}[T_r(0), T_s(t)] \quad (33)$$

where the values for $t < 0$ are obtained by using that $\mathbf{C}(t) = \mathbf{C}(-t)'$. In order to be able to make direct comparisons with the results for the one degree of freedom EPO reported by Ditlevsen and Bognár (1993), we define the dimensionless response vector process $\mathbf{X}(\tau) = \mathbf{x}(t)/\sqrt{C_{11}(0)}$ of relative displacements expressed as functions of the dimensionless time parameter $\tau = \alpha_k t$, $\alpha_k = \omega_k\sqrt{1 - \zeta_k^2}$, in which k is chosen such that $v_{k1}^2\text{Var}[T_k(0)] \geq v_{r1}^2\text{Var}[T_r(0)]$ for all $r = 1, ..., n$.

Thus the covariance function matrix of $\mathbf{X}(\tau)$ is obtained from (33) and (30) as

$$\mathbf{R}(\tau) = \{R_{ij}(\tau)\} = \text{Cov}[\mathbf{X}(0), \mathbf{X}(\tau)'] = \sum_{r=1}^{n}\sum_{s=1}^{n}\mathbf{A}_{rs}e^{-\zeta_s\omega_s t}I_{rs}(t) \quad (34)$$

$$\mathbf{A}_{rs} = \mathbf{v}_r\mathbf{v}_s'\frac{v_{rn}v_{sn}}{\alpha_r\alpha_s} \Big/ \sum_{i=1}^{n}\sum_{j=1}^{n}v_{i1}v_{j1}\frac{v_{in}v_{jn}}{\alpha_i\alpha_j}a_{ij} \qquad (35)$$

valid for $\tau \geq 0$. For $\tau < 0$ we have $\mathbf{R}(\tau) = \mathbf{R}(-\tau)'$, $\dot{\mathbf{R}}(\tau) = -\dot{\mathbf{R}}(-\tau)'$, $\ddot{\mathbf{R}}(\tau) = \ddot{\mathbf{R}}(-\tau)'$.

FATIGUE CRACK GROWTH UNDER STOCHASTIC LOADING

K. DOLIŃSKI[*], P. COLOMBI[**]
[*]Institute of Fundamental Technological Research, Centre of Mechanics
ul. Świętokrzyska 21, 00-049 Warsaw, POLAND
[**]Polytechnic of Milan, Department of Structural Engineering
Milan, ITALY

1. Introduction

Load sequence effects are observed in fatigue experiments under variable amplitude loading. The time, called the lifetime, of reaching a given critical length by a macro crack strongly depends on the arrangement of sequence of load maxima, e.g. [Schijve 1973]. Numerous experiments show rapid changes of the fatigue crack propagation rate after a load cycle with maximum of greater magnitude (overload) than the subsequent maxima. The usual change in the Mode I of crack growth is a transient diminution of the crack propagation rate after the overload. This phenomenon is called the retardation of the crack growth. The duration of the retardation phase and the magnitude of the retardation effect depend on many factors including specimen geometry, environmental effects, material properties, the magnitude of the overload and of subsequent extremes. The physical nature of this phenomenon has not been completely explained, yet. Among several mechanisms suggested in the literature the plasticity-induced fatigue crack closure is generally considered as a dominant cause of the retardation, [Shin & Fleck 1987]. Most of the models that are proposed in the literature to predict the fatigue crack growth with regard to the load sequence effects refer to the overload-induced plastic zone and a diminution of the effective stress intensity factor range after an overload, see e.g. [Wanhill & Schijve 1988]. Such a model will be also used in the present paper to assess some probabilistic characteristics of the structural lifetime when a critical fatigue macro crack length defines the structural failure due to stochastic loading.

2. Fatigue crack growth law

Depending on the method used in solution of the fatigue crack growth problem the fatigue crack growth laws are formulated in a form of continuous differential equations versus time or in a discrete incremental form where the crack length increments due to single stress cycles are considered. The latter will be considered in this paper. A lot of fatigue crack growth laws have been proposed in the literature, see e.g. [Kocańda 1978]. For a very wide class of fatigue crack growth laws the crack length increment, Δa_i, is described as a product of two functions

A. Naess and S. Krenk (eds.), IUTAM Symposium on Advances in Nonlinear Stochastic Mechanics, 143–152.

$$\Delta a_i = F\left(a_i, S_i^+, S_i^- | \mathbf{x}\right) = g(a_i | \mathbf{x}) \cdot \Xi\left(S_i^+, S_i^-\right) \tag{1}$$

where a_i denotes the crack length due to the i-th load cycle. The quantities S_i^+ and S_i^- denote, respectively, the stress maximum and minimum in the i-th cycle of the far-field stress applied to a cracked element. The components of the vector $\mathbf{x} = [x_1, x_2, ..., x_K]$ represent the material parameters. The parameters may be assumed to be some random variables. The vector $\mathbf{x}$ denotes a sample of the random vector $\mathbf{X} = [X_1, X_2, ..., X_K]$. Keeping in mind the discretization the subscript "i" will often be dropped.

The equation that is now and again used in application and modelling of fatigue crack growth under constant load amplitude conditions was proposed by Paris and Erdogan [1963] where the functions $g(\cdot | \cdot)$ and $\Xi(\cdot, \cdot)$ are assumed as follows

$$g(a|c) = c \cdot Y^m(a) \cdot \left(\sqrt{\pi \cdot a}\right)^m \qquad \Xi(S^+, S^-) = (S^+ - S^-)^m = \Delta S^m \tag{2}$$

with the coefficient c and the exponent m being some parameters that generally depend on material and load conditions. The dimensionless function Y(a) depends on the crack and specimen geometry.

Since Elber [1971] has noticed the crack closure phenomenon and pointed out its significance for fatigue crack growth the effective stress cycle amplitude

$$\Delta S_{eff} = S^+ - S_{eff}^- = S^+ \cdot (1 - q(R)) \tag{3}$$

where the ratio, $R = S^-/S^+$, defines the stress cycle asymmetry coefficient, is usually considered in fatigue crack equations instead of ΔS. In the literature there is no universal formula describing the relation (3). Most of the proposals are based on experimental data, see e.g. [Bulloch 1991]. The bilinear form, $q(R) = S_{op}/S^+ = \min\left\{q_0 \cdot \left(1 + R/|R_0|\right), R\right\}$ proposed by Veers [1987] with $q_0 \in [0.2, 0.5]$ and $R_0 \in [-5, -2]$ depending on the material properties was also used in [Doliński 1994] to identify some statistical properties of the random material parameters, $\mathbf{X} = [C_2, A_{th}, A_{fc}]$. The fatigue crack growth equation with $\Xi\left(S^+, S_{eff}^-\right) = \left(S^+ - S_{eff}^-\right)^2 = \Delta S_{eff}^2$ and g(a|x) = $= c_2 \cdot \left(Y^2(a) \cdot a - a_{th}\right) \big/ \left[\sigma_Y^2 \cdot \left(1 - Y^2(a) \cdot a/a_{fc}\right)\right]$ was there derived from some energy considerations.

The crack closure effect resulting from crack tip plasticity is usually modelled by referring to the plastic zone that develops at the crack tip due to the stress cycle maximum. The range of the plastic zone can be estimated [Irwin 1960] as $r_Y(a,S) = = \gamma \cdot \left(Y(a) \cdot S/\sigma_Y\right)^2 \cdot a$ where σ_Y denotes the yield stress of material. The coefficient $\gamma = 1$ for plane stress and $\gamma = (1 - 2\nu)^2$ for plane strain condition with ν as the Poisson coefficient. It is observed in fatigue experiments that the crack opening stress, S_{op}, varies under variable amplitude loading and depends on the last previous stress extremes. It lessens the effective stress amplitude, ΔS_{eff}, and the fatigue crack growth, eventually. In order to specify the retardation intensity Veers [1987] admitted the so-called reset stress, S_r. It determines the stress level that is necessary to reset the maximum extend of the plastic zone, $r_Y(a_{ol}, s_{ol})$, produced by the overload stress maximum, $S_{ol} = s_{ol}$, when the crack length $a = a_{ol}$. Applying the equality, $a_{ol} + r_Y(a_{ol}, s_{ol}) = a + r_Y(a, s_r)$, the reset stress at current a can be calculated as follows

$$S_r = S_r(a|a_{ol}, s_{ol}) = \sigma_Y \cdot \sqrt{\frac{\gamma \cdot s_{ol}^2}{\sigma_Y^2} \left(\frac{Y(a_{ol})}{Y(a)} \right)^2 \cdot \frac{a_{ol}}{a} - \frac{1}{Y^2(a)} \left(1 - \frac{a_{ol}}{a} \right)} \qquad (4)$$

Now, the effective minimum of the stress amplitude, $\overline{S}_{eff}$, takes the following form

$$\overline{S}_{eff} = \overline{S}_{eff}\left(a, S^-, S^+ | a_{ol}, s_{ol};\right) = \begin{cases} q \cdot S^+ & \text{if } S^+ > S_r \text{ and } S^- < q \cdot S^+ \\ q \cdot S_r(a|a_{ol}, s_{ol}) & \text{if } S_r > S^+ > q \cdot S_r \text{ and } S^- < q \cdot S_r \\ q \cdot S^+ & \text{if } S_r > S^+ > q \cdot S_r \text{ and } S^- \geq q \cdot S_r \\ S^+ & \text{if } S^+ \leq q \cdot S_r \end{cases} \qquad (5)$$

Substituting the real cycle minimum, S^-, in (1) with the effective minimum (5) we obtain the fatigue crack growth equation

$$\Delta a_i = g(a_i|x) \cdot \Xi\left[S_i^+, \overline{S}_{eff,i}\left(a_i, S_i^+, S_i^- | a_{ol}, s_{ol}\right) \right] \qquad (6)$$

which involves some load sequence effects due to the presence of parameters, a_{ol} and s_{ol}, associated with the last overload. It excludes the variable separation and requires a cycle-by-cycle summation in calculation of the fatigue crack length for deterministic variable amplitude loading.

3. Stochastic loading and retardation

For stochastic loading every maximum can likely be an overload. In our earlier paper [Colombi & Doliński 1994] some conditions for a maximum to be an overload, $S_k^+ = s_{ol}$, and for the subsequent maxima to be admissible to retain the crack to propagate in a retardation and then in a post-retardation phase were discussed. It appears that the whole fatigue crack propagation process alternately consists of retardation and post-retardation phases. The couples of these successive phases are considered as the blocks starting and terminating with overloads.

The stress extremes are the only load parameters involved in fatigue crack growth equations, c.f. (6). Recently, Frendahl & Rychlik [1992] have shown on a very wide numerical simulation basis that a homogeneous Markov chain is a good approximation of the random sequence of extremes of stationary Gaussian and some non-Gaussian processes with various spectral characteristics. In practical application the length of correlation of the sequence of extremes, n_{corr} = a dozen of cycles or so, appears much shorter than the lengths of blocks consisting of retardation+post-retardation phases, N_B = several dozen of cycles. Then the number of cycles to failure, N_F = several thousands cycles or more, is much longer than N_B's. This property, $n_{corr} \ll N_B \ll N_F$, originated an approach presented by the authors in their earlier papers [Colombi & Doliński 1994, 1995] where a continuous Markov approximation of the fatigue crack growth equation was used. A numerical procedure for determination of the drift and diffusion coefficients of the crack growth process has been described in [Colombi & Doliński 1994]. A numerical example carried out in that paper and some reliability calculation recently presented in [Colombi & Doliński 1995] show the method to be quite efficient in estimating the probability distribution of the mean number of load cycles to failure in the case of random material parameters, X, and random initial crack length, A_0. Some features of the considered problem, assumptions made in the load analysis and properties of the results observed on subsequent steps during the calculation allow us to propose some modifications of the approach presented in the subsequent section.

4. Discrete model of fatigue crack growth under stochastic stationary loading

The numerical procedure proposed in [Colombi & Doliński 1994] provides the joint probability distribution, $P_{B,N_B}(b_i, n | a_{ol}, x) = P[B(a_{ol}, x) = b_i \wedge N_B(a_{ol}, x) = n]$, of the length, $B(a_{ol}, x)$, of the retardation + post-retardation block, which has started at $a = a_{ol}$, and the number of cycles, $N_B(a_{ol}, x)$, within the block whereas $b_i = i \cdot \delta a$ and δa denotes a crack length increment used in the discrete numerical calculation scheme. The overload intensity, S_{ol}, is not involved anymore due to the averaging integration accounting for the probability distribution of overloads. The calculation confirms the strong inequalities $n_{corr} \ll N_B \ll N_F$ for both narrow- and wide-band spectra of stochastic load processes and shows that the length of a block $B(a_{ol}, x) \ll a_{ol}$ for most cases of practical engineering interest. Thus, the retardation + post-retardation blocks may be considered to constitute a sequence of random, approximately statistically independent, crack length increments.

After removing the overload condition the equation (6) can be transformed into the form

$$\frac{\Delta A_n}{g(A_n|x)} = \Xi\left[S_n^+, S_{\text{eff}}^-\left(A_n; S_n^-, S_n^+ | a_{ol}\right)\right] = \Delta\Gamma(A_n; S_n^-, S_n^+ | a_{ol}) \tag{7}$$

where the subscript "n", $n = 1, 2, .., N_B(a_{ol}, x)$, numbers the subsequent load cycle within a block, $B(a_{ol}, x)$, and all quantities with the subscript "n" are taken in the n-th cycle of the block. The left-hand side of (7) denotes an increment of the fatigue damage parameter due to the crack length increment ΔA_n. The right-hand side of (7) can be considered as an elementary increment of the load fatigue indicator corresponding with a single random load cycle within the block $B(a_{ol}, x)$. The sum of $N_B(a_{ol}, x)$ elementary $\Delta\Gamma$ increments yields an increment of the load fatigue indicator corresponding with the crack advance over the whole block of the length $B(a_{ol}, x)$, i.e.

$$\Gamma(N_B|a_{ol},\mathbf{x}) = \sum_{n=1}^{N_B(a_{ol},\mathbf{x})} \Delta\Gamma(A_n;S_n^-,S_n^+|a_{ol}) = \int_{a_{ol}}^{a_{ol}+B(a_{ol},\mathbf{x})} \frac{da}{g(a|\mathbf{x})} \qquad (8)$$

The equality (8) allows us to apply the joint probability distribution, $P_{B,I}(b_i,n|a_{ol},\mathbf{x})$, to calculate the statistical moments of $\Gamma(N_B|a_{ol},\mathbf{x})$, i.e. $\overline{\Gamma^k}(a_{ol},\mathbf{x}) = E\left[\Gamma^k(N_B|a_{ol},\mathbf{x})\right]$, from the following averaging equations

$$\overline{\Gamma^k}(a_{ol},\mathbf{x}) = \sum_{i=1}^{\infty}\sum_{n_{b_i}=1}^{\infty} \gamma^k(n_{b_i}|a_{ol},\mathbf{x}) \cdot P\left[B(a_{ol},\mathbf{x}) = b_i \wedge N_B(a_{ol},\mathbf{x}) = n_{b_i}\right] =$$

$$= \sum_{i=1}^{\infty}\sum_{n_{b_i}=1}^{\infty} \left(\int_{a_{ol}}^{a_{ol}+b_i}\frac{da}{g(a|\mathbf{x})}\right)^k \cdot P\left[B(a_{ol},\mathbf{x}) = b_i \wedge N_B(a_{ol},\mathbf{x}) = n_{b_i}\right] \qquad (9)$$

where $\gamma(n_{b_i}|a_{ol},\mathbf{x})$ is a sample of the load fatigue indicator increment, $\Gamma(N_B|a_{ol},\mathbf{x})$, corresponding with the block of length b_i starting at $a = a_{ol}$ and with n_{b_i} stress cycles occurring within it. It is observed that the empirical moments (9), are practically independent of a_{ol}, and $\mathbf{x}$, i.e. $\overline{\Gamma^k}(a_{ol},\mathbf{x}) = \overline{\Gamma^k} = \text{const}$. This feature becomes obvious looking at the reset stress expression given in (4) and recalling the strong inequalities $B_m \ll a_{ol}$ for any a_{ol} from the admissible crack length interval $[a_0,a_F]$. Since the effect of the reset stress depends on the ratio a_{ol}/a rather than on the location of the block within the interval $[a_0,a_F]$ the statistical characteristics of increments of the load fatigue indicator over a block are almost independent of the block location, a_{ol}. Moreover, the reset stress does not depend on the material parameter vector, $\mathbf{x}$. Hence, the number of cycles within a block is independent of $\mathbf{x}$ as well, i.e. $N_B(a_{ol},\mathbf{x}) = N_B$. All of these make the fatigue load indicators over blocks, B_m, independent of a_{ol} and $\mathbf{x}$, i.e. $\Gamma\left(N_{B_m}|a_{ol,m},\mathbf{x}\right) = \Gamma_m$. It allows us to investigate the load fatigue indicator, Γ_M, after M blocks as a sum of independent random variables having the same probability distributions with the mean and variance, respectively, $\overline{\Gamma}_m = \overline{\Gamma}$ and $\text{Var}\left[\Gamma_m\right] = \overline{\Gamma^2} - \overline{\Gamma}^2 = \sigma_\Gamma^2$. At failure when the crack, initially of the length a_0, reaches its critical size, a_F, the load fatigue indicator takes a value $\gamma_F(\mathbf{x}) = \gamma(a_0,a_F|\mathbf{x})$. i.e.

$$\gamma_F(\mathbf{x}) = \gamma(a_0,a_F|\mathbf{x}) = \int_{a_0}^{a_F} \frac{da}{g(a|\mathbf{x})} \approx$$

$$\approx \sum_{m=1}^{M_F(\mathbf{x})}\Gamma(N_{B_m}) = \sum_{m=1}^{M_F(\mathbf{x})}\Gamma_m = \sum_{m=1}^{M_F(\mathbf{x})}\sum_{n=1}^{N_{B_m}}\Delta\Gamma(A_{m,n},S_{m,n}^+,S_{m,n}^-|A_{ol,m}) \qquad (10)$$

where the subscript "m", m = 1,2,...,M, numbers the retardation + post-retardation blocks, $B_m = B(A_{ol,m}, x)$, that start at the random crack lengths $A_{ol,1}$ and

$$A_{ol,m} = A_{ol,1} + \sum_{l=1}^{m-1} B_l \text{ , subsequently for m = 2,3,...,M. The approximate equality sign}$$

in (10) points out an error resulting from an inconsistency of the initial crack length, a_0, with the beginning of the first block, $A_{ol,1}$, and of the critical crack length, a_F, with the crack length at the end of the last M-th block. The strong inequality $B_m \ll (a_F - a_0)$ assures this error to be negligible.

The load fatigue indicator, $\Gamma_F(M|x)$, is the random variable depending on the number, $M(x)$, of the retardation + post-retardation blocks, B_m. Until the fatigue failure the load fatigue indicator should be less than a critical value of the fatigue damage parameter, i.e. $\Gamma_F(M_F|x) < \gamma_F(x)$, corresponding with the critical crack length, a_F, given the material parameter vector $X = x$. Thus, the probability of failure given $X = x$ is defined as

$$P_F(x) = P[\Gamma_F(M_F|x) > \gamma_F(x)] = P\left[\sum_{m=1}^{M_F(x)} \Gamma_m > \gamma_F(x)\right] \tag{11}$$

where $M_F(x)$ denotes the random number of blocks to failure given $X = x$. It is well-known that the mean, $\overline{M_F}(x)$, of the random number, $M_F(x)$, of random variables, Γ_m, in the sum in (11) is equal to

$$\overline{M_F}(x) = \gamma(x)/\overline{\Gamma} \tag{12}$$

For a great $\overline{M_F}$ value the probability distribution of the random number of independent random variables in such a sum can be approximated by the inverse Gaussian probability distribution, see [Johnson & Kotz 1970], with the variance $\sigma^2_{M_F}(x) = \sigma^2_\Gamma \cdot \gamma_F(x)/\overline{\Gamma}^3 = v^2_\Gamma \cdot \overline{M_F}(x)$.

The total number of cycles to failure, $N_F(x)$, given $X = x$ is then given as the sum

$$N_F(x) = \sum_{m=1}^{M_F(x)} N_{B_m} \text{ with } N_{B_m} \text{ denoting the random number of cycles within the m-th}$$

block. The numbers N_{B_m} are statistically independent random variables and their moments are easily obtained from the empirical probability distribution $P_{B,l}(b_i,n|a_{ol},x)$. As discussed above they are also independent of a_{ol} and x and equal to each other. The means and variances of N_{B_m} will be denoted, respectively, as $E[N_{B_m}] = \overline{N_B}$ and $Var[N_{B_m}] = \sigma^2_{N_B}$. The number of blocks to failure, $M_F(x)$, is usually sufficiently great to apply the central limit theorem modified for a sum of a random number of random variables by Renyi [1962]. Thus, the probability distribution of the number of cycles to failure, $N_F(x)$, can be approximated by the Gaussian probability distribution

$$F_{N_F}(n;x) = P[N_F(x) \leq n] \approx \Phi\left[\frac{n - \overline{M_F(x)} \cdot \overline{N_B}}{\sigma_{N_F}(x)}\right] \tag{13}$$

where the variance of $N_F(x)$ is given as follows

$$\sigma^2_{N_F}(x) = \overline{M_F(x)} \cdot \sigma^2_{N_B} + \sigma^2_{M_F}(x) \cdot \overline{N_B}^2 = \frac{\gamma(x)}{\overline{\Gamma}} \cdot \left(\sigma^2_{N_B} + v^2_\Gamma(x) \cdot \overline{N_B}^2\right) \tag{14}$$

A general relation between the random number, N_t, of cycles of stochastic process and the corresponding time interval, T_n, results from the renewal theory

$$F_{N_t}(n|t) = P[N_t \leq n|t] = 1 - P[T_n \leq t|n] = 1 - F_{T_n}(t|n) \tag{15}$$

and involves the conditional probability distribution, $F_{N_t}(n|t)$, of the random number, N_t, of cycles that occur within a time interval $[0,t]$ and the conditional probability distribution, $F_{T_n}(t|n)$, of the random time interval $[0,T_n]$ that contains a given number of cycles, n. For long time intervals and some usually satisfied assumptions concerning the rate of decay of correlation function Malevich [1962] proofed the limit theorem assuring the probability distribution of the number of zeros of a zero mean stationary Gaussian stochastic to be approximately normal as $t \rightarrow \infty$, i.e.

$$F_{N_t}(n|t) = P[N_t \leq n|t] \approx \Phi\left[\frac{n - \overline{N_t}(t)}{\sigma_{N_t}(t)}\right] \tag{16}$$

where $\overline{N_t}(t)$ and $\sigma^2_{N_t}(t)$ denote the mean and variance of the number of zeros being some linear functions of the time interval range. In the fatigue analysis we have to count the maxima of the process. The number of maxima, $N_t^+(t)$, is equivalent to the number of zero upcrossings of the first derivative of the process. Thus, substituting in (16) $\overline{N_t}(t)$ and $\sigma^2_{N_t}(t)$ with the mean, $\overline{N_t^+}(t) = v^+ \cdot t$, and variance, $\sigma^2_{N_t^+}(t) = S_N^+ \cdot t + S_N^+$, of maxima the probability distribution, $F_{N_t^+}(n|t)$, of the number of maxima, $N_t^+(t)$, within the time interval $[0,t]$ is obtained whereas $v^+ = \sqrt{\lambda_4/\lambda_2}/(2\pi)$ denotes the mean rate of maxima while λ_i's are the spectral moments of the i-th order. The coefficients S_N^+ and S_N^+ depend on the correlation function of the stress process and their analytical derivation is given in [Tichonov 1970], say, and applied to fatigue analysis in [Doliński 1986]. The coefficients can be also estimated from the stress path sample used in calculation of the transition probability matrices in the Markov approximation of the sequence of extremes.

Equation (15) and the probability distributions (13) and (16) allow us to derive the conditional probability distribution of the fatigue lifetime T_F given $X = x$ and write it down in the following form

$$F_{T_F}(t|x) = P[T_F \le t|x] = \int_{-\infty}^{\infty} F_{T_n}(t|n) \cdot f_{N_F}(n|x) \, dn = 1 - \int_{-\infty}^{\infty} F_{N_t}(n|t) \cdot f_{N_F}(n|x) \, dn \approx$$

$$\approx 1 - \int_{-\infty}^{\infty} \Phi\left[\frac{n - \overline{N_t}(t)}{\sigma_{N_t}(t)}\right] \cdot \frac{1}{\sigma_{N_F}(x)} \cdot \varphi\left[\frac{n - \overline{M_F}(x) \cdot \overline{N_B}}{\sigma_{N_F}(x)}\right] dn = \Phi\left[\frac{\overline{N_t}(t) - \overline{M_F}(x) \cdot \overline{N_B}}{\sqrt{\sigma_{N_t}^2(t) + \sigma_{N_F}^2(x)}}\right] \quad (17)$$

similar to the so-called Birnbaum-Saunders probability distribution [1969]. Depending on the dimension, K, of the material parameter vector $\mathbf{X} = [X_1, X_2, ..., X_K]$ the unconditional probability distribution of the lifetime can be calculated by a direct integration

$$F_{T_F}(t) = \int_{-\infty}^{\infty} F_{T_F}(t|x) \cdot f_X(x) \, dx \qquad (18)$$

where $f_X(x)$ denotes the probability density function of the parameter vector $\mathbf{X}$ or by employing some approximate methods of reliability analysis involving a search for design points, first and second order reliability method, importance sampling, say.

5. Example

The fatigue crack propagation in a metal sheet in a region of stress concentration is considered. The Paris-Erdogan equation (2) with $c = 4.1 \cdot 10^{-13}$, $m = 3.5$ and $Y(a) = (K_t - 1) \cdot \exp(-\beta \cdot a^\delta) + 1$ is assumed to describe the crack growth, where $K_t = 3.3$, $\beta = 1.0648$ and $\delta = 0.44$. The external loading is modelled as a stationary zero mean Gaussian stochastic process with the following bimodal one side power spectrum density often used as stress spectrum in joints of offshore structures [Wirsching & Light 1980]

$$G(\omega) = \sigma^2 \cdot \frac{\kappa \cdot \exp\left[-1050 / (\omega \cdot T_D)^4\right]}{\omega^5 \cdot T_D^4 \cdot \left\{\left[1 - (\omega/1.797)^2\right]^2 + (0.041 \cdot \omega/1.797)^2\right\}} \qquad (19)$$

where $\omega = 2 \cdot \pi \cdot f$ is the frequency in radians/sec with f in 1/sec, κ is the normalizing coefficient, T_D is the dominant wave period and σ denotes the standard variation of the stress process admitted to be equal to 75 Mpa in the example. The nine values of T_D, namely T_D = 46.15, 27.91, 21.33, 17.50, 14.77, 12.54, 10.48, 8.18, were considered resulting, respectively, in nine values of the band-width parameter $\alpha = \lambda_2 / \sqrt{\lambda_0 \cdot \lambda_4}$ = 0.2, 0.3, 0.4, 0.5, 0.6, 0.7, 0.8, 0.9. For $\alpha = 0$ the process is the ideally broad band one and for $\alpha = 1$ it is the ideally narrow band one.

The stress amplitude in (2) is assumed to be the effective one with $S^- = S_{eff}^-$ as in (5) with the parameters $q_0 = 0.496$, $R_0 = -5$ and $\sigma_Y = 350$ MPa that are involved in definitions of S_{eff}^-, eq. (5), and S_r, eq. (4). For the initial crack length, $a_0 = 0.15$ mm, and critical crack length, $a_F = 4$ mm, the means of lifetimes, $\overline{T}_F$ in hours, standard deviations, σ_{T_F}, and coefficients of variations, COV, resulting from the crack growth models *with*, denoted by "R", and *without*, denoted by "C", retardation effects are compared for varying band-width parameter α of the stress power density spectrum (19). The results and the ratios between both solutions are plotted in Fig.1.

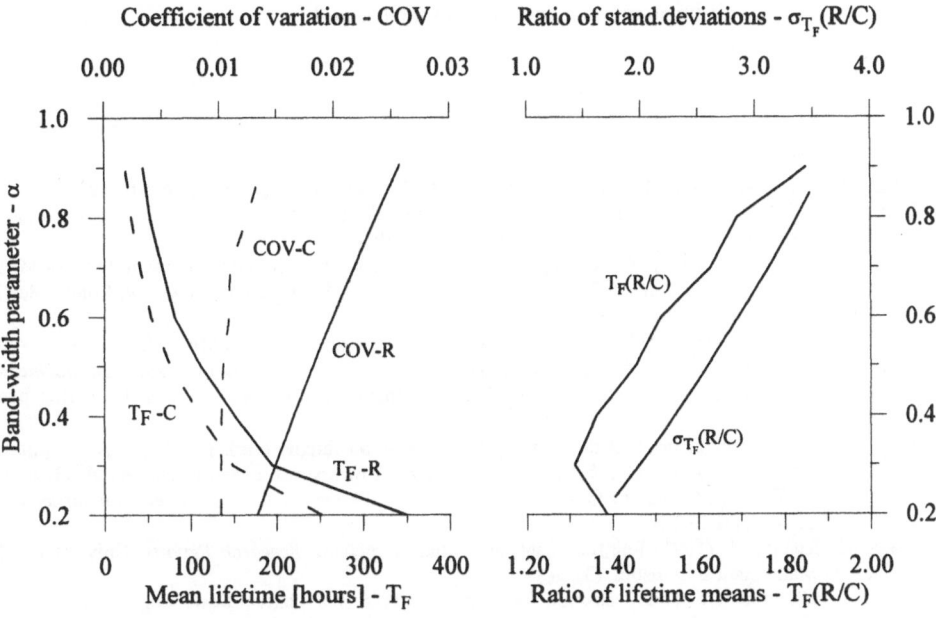

Fig. 1. a) mean lifetimes, T_F, and coefficients of variation, COV, for retarded, "R", and unretarded, "C", crack growth models; b) mean lifetime ratio, $T_F(R/C)$, and standard deviation ratio between retarded and unretarded crack growth models.

6. Concluding remarks

Considering the forms of the mean (12) and variance (14) it is seen that the lifetime probability function (17) depends directly and analytically on the following parameters:
• mean and variance of the number of stress cycles, N_B, within a retardation+post-retardation block, B, of the fatigue crack growth process;
• mean and variance of the load fatigue indicator, Γ_{N_B}, corresponding with that block;

- coefficients, S_N^+ and $\overline{S_N^+}$, of the linear approximation of the variance,

$$\sigma_{N_t^+}^2 (t) = S_N^+ \cdot t + \overline{S_N^+}$$

- load fatigue indicator, $\gamma_F(x)$, corresponding with the critical crack length interval $[a_0, a_F]$

The division of the fatigue crack growth process into the blocks and Markov assumption about the sequence of extremes of the stress process shortens the simulation procedure providing all necessary parameters involved in the lifetime probability distribution. Many retardation models and most of the fatigue crack growth equations can be easily and well implemented into the numerical code.

7. References

Birnbaum, Z.W., Saunders, S.C. (1969) A new family of life distributions, *J.Appl.Prob.*, **6**, 319-327.

Bulloch, J.H. (1991) The influence of mean stress or R-ratio on the fatigue crack threshold characteristics of steels - A review, *Int.J.Pres. Ves. & Piping*, **47**, (3), 263-292.

Colombi, P., Doliński, K. (1994) Markov approach to fatigue crack growth under stochastic load, in M.H. Aliabadi et al. (eds.), *Localized Damage III, Computer-Aided Assessment and Control*, Comp. Mech. Publications, Southhampton Boston, pp. 89-96.

Colombi, P., Doliński, K. (1995) Fatigue reliability under stochastic load, *Proc. OMAE'95* (in press).

Doliński, K. (1986) Retardation of fatigue crack growth under stochastic loading in: *Studies on Stochastic Fatigue Crack Growth*, Berichte der Zuverlässigkeitstheorie der Bauwerke, Techn.Univ.München, Heft 78;

Doliński, K. (1994) Random material non-homogeneity effects on fatigue crack growth, in A. Carpinteri (ed.), *Handbook of Fatigue Crack Propagation in Metallic Structures*, Elsevier Science Publs B.V..

Elber, W. (1971) The significance of fatigue crack closure, in *Damage Tolerance in Aircraft Structures*, ASTM STP 486, pp. 230-242;

Frendahl, M., Rychlik, I. (1992) Rainflow analysis - Markov method, *Research Reports*, Univ. of Lund, Dept. of Mathematical Statistics, (3), 1-57.

Irwin, G.R. (1960) Fracture mode transition for a crack traversing a plate, *J.Basic Engineering, Trans. ASME*, **82**.

Johnson, N.L., Kotz, S. (1970) *Distributions in Statistics: Continuous Univariate Distributions -1*, Houghton Mifflin Comp. Boston.

Kocańda, S. (1978) *Fatigue Failure of Metals*, Sijthoff & Noordhoff Int. Publ., Alphen aan den Rijn.

Malevich, T.L. (1969) Asymptotic normality of the number of crossings of the zero level by a Gaussian process, *Teoria veroiatnosti i ieio promienienia*, **14**, (2), 292-301, (in Russian).

Paris, P., Erdogan, F. (1963) A critical analysis of crack propagation laws, *Journal of Basic Engineering, Trans. ASME*, **85**, 528-534.

Renyi, A. (1962) *Wahrscheinlichkeitsrechnung mit einem Anhang über Informationstheorie*, VEB Deutscher Verlag der Wissenschaften, Berlin.

Schijve, J. (1973) Effect of load sequences on crack propagation under random and program loading, *Engng Fract.Mech.*, **5**, 269-280.

Shin, C.S., Fleck N.A. (1987) Overload retardation in a structural steel, *Fatigue Fract. Engng Mater.Struct.*, **9**, (5), 379-393.

Tichonov, V.I. (1970) *Excursion of Random Processes*, Nauka, Moskva (in Russian).

Veers, P.J. (1987) Fatigue crack growth due to random loading, *SAND87-2037*, Sandia National Laboratories.

Wanhill, R.J.H., Schijve, J. (1988) in J. Petit et al. (eds.), *Fatigue Crack Growth Under Variable Amplitude Loading*, Elsevier, London, p. 326.

Wirsching, A.M., Light, M.C. (1980) Fatigue under wide band random stresses, *Journal of Struct.Engng, ASCE*, **106**, (ST7), 1593-1607.

SMALL NOISE EXPANSION OF MOMENT LYAPUNOV EXPONENTS FOR TWO-DIMENSIONAL SYSTEMS

M. M. DOYLE AND N. SRI NAMACHCHIVAYA
Nonlinear Systems Group
Department of Aeronautical and Astronautical Engineering
University of Illinois at Urbana-Champaign
104 S. Wright Street, Urbana, IL 61801

AND

L. ARNOLD
Institute for Dynamical Systems
University of Bremen
Bibliothek strasse, Postfach 330440
2800 Bremen 33, GERMANY

1. Introduction

Sample or almost-sure stability of a stationary solution of a random dynamical system is of importance in the context of dynamical systems theory since it guarantees all samples except for a set of measure zero tend to the stationary solution as time goes to infinity. The almost-sure stability or instability of a dynamical system is indicated by the sign of the maximal Lyapunov exponent. However, from the applications viewpoint, one may not be satisfied with such guarantees since a sample stable process may still exceed some threshold values or may possess a slow rate of decay. Although sample solutions may be stable with probability one, the mean square response of the system for the same parameter values may grow exponentially. It is well known that there are parameter values at which the top Lyapunov exponent λ is negative, indicating that the system is sample stable, while the pth moments grow exponentially for large p indicating the pth mean response is unstable. This implies that, although the system response $|x(t; x_0)| \to 0$ as $t \to \infty$ with probability one at an exponential rate λ, there is a small probability that $|x(t; x_0)|$ is large. This makes the

153

A. Naess and S. Krenk (eds.), IUTAM Symposium on Advances in Nonlinear Stochastic Mechanics, 153–168.
© 1996 *Kluwer Academic Publishers.*

expected value of this rare event large for large values of p and results in the pth mean instability.

The connection between moment and almost-sure stability of the trivial solution of a linear Itô equation was investigated by Khas'minskii [11] and later by Kozin and Sugimoto [12]. It was shown by Khas'minskii that the asymptotic stability in probability is based on $E\|x(t; x_0)\|^p$ for small $p > 0$ while Kozin and Sugimoto showed that the almost-sure stability region in some parameter space is the limit of the regions of the pth moment stability as $p \downarrow 0$. This implies that the sample stability criteria will include samples that are stable in some p^{th} moment no matter how small p may be.

It is well known that for real noise situations, where the noise $\xi(t)$ is some stationary process with associated generator G, the moment stability is harder to examine since the moments are not closed. The correlation between the state vector $x(t)$ and the real noise $\xi(t)$ does not permit an exact set of equations for the moments of $x(t)$. One can intuitively see that $x(t) \cdot \xi(t)$ behaves like a nonlinear term, hence the closure problem. These situations were handled, for example, by stochastic averaging under suitable conditions on the noise to obtain a set of approximate Itô equations for which all moments are obtained thereby allowing moment stability to be obtained in a straight forward manner.

The connection between moment stability and almost-sure stability for an undamped linear oscillator under real noise excitation was established for the first time by Molčanov [13]. These results were extended for an arbitrary d-dimensional system by Arnold [2] where a concise formulation of the relation between almost-sure sample stability and the pth mean stability is presented. The complete set of results on the so-called moment Lyapunov exponent, its properties and generators is obtained in two consecutive papers by Arnold et al. [4, 5] for real and white noise situations, respectively. More extensive coverage of the moment Lyapunov exponents under white and real noise assumptions is given by Arnold and Wihstutz [6]. The aim of this paper is to use these theoretical results to obtain an asymptotic expansion of the moment Lyapunov exponents for arbitrary two-dimensional systems with small noise.

2. General Theory on Moment Lyapunov Exponents

In the study of stability of solutions of random dynamical systems, both almost-sure stability and p^{th} moment stability have been widely used. The top Lyapunov exponent describes the almost-sure stability. The exponential growth rate of $E \| x(t; x_0) \|^p$ is provided by the moment Lyapunov exponent defined as

$$g(p; x_0) = \lim_{t \to \infty} \frac{1}{t} \log E \parallel x(t; x_0) \parallel^p$$

where $x(t; x_0)$ is the solution process of a linear random dynamical system. If $g(p; x_0) < 0$, then, by definition, $E \parallel x(t; x_0) \parallel^p \to 0$ as $t \to \infty$ and this is referred to as p^{th} moment stability. The concept of moment Lyapunov exponents was introduced for linear random dynamical systems by Arnold [2] and completely discussed in the subsequent works by Arnold *et al.* [3, 4, 5] for the white noise and real noise cases. The connection between the moment Lyapunov exponent and the associated large deviation problem for linear systems perturbed by white noise was established by Stroock [18]. Arnold and Kliemann [3] provide a unified treatment of the large deviation problem considering a combination of real and white noise perturbations. Baxendale [7] provides an alternative approach to the study of moment Lyapunov exponents and large deviations of linear stochastic differential equations.

In this section, we present a summary of essential results for moment Lyapunov exponents. We shall not produce these theorems in detail in this paper but rather give an outline of some of the central ideas involved in the arguments and refer the reader to Arnold *et al.* [4, 5] for details of the white and real noise situations. We shall, however, provide a complete set of equations that is essential in calculating the moment Lyapunov exponents.

For the real noise case, we consider

$$\dot{x} = A(\xi(t)) x, \quad x \in \mathbf{R}^d \tag{1}$$

$$d\xi = X_0(\xi) + \sum_{i=1}^{r} X_i(\xi) \circ dW_i, \quad \xi \in M$$

In order to ensure that there is a unique smooth and positive invariant density ν on the compact manifold M, we assume $\xi(t)$ is strongly elliptic in the sense that $\dim LA(X_1, \cdots, X_r)(\xi) = \dim M$ for all $\xi \in M$ where $LA(Z)$ denotes the Lie algebra generated by the set Z of vector fields. Introducing polar coordinates in $\mathbf{R}^d$ through the Khas'minskii transformation [11]

$$s = \frac{x}{\parallel x \parallel} \in S^{d-1} \quad \text{and} \quad \parallel x \parallel \in \mathbf{R}^+$$

we obtain $\parallel x(t; x_0) \parallel = \parallel x_0 \parallel \exp\left\{\int_0^t q(\tau)d\tau\right\}$ and $\dot{s} = h(\xi(t), s)$ where $q(\tau) = q(\xi(\tau), s) = s^T A(\xi(\tau))s$, $h(\xi(t), s) = (A - qI) s$. and (ξ, s) together form a diffusion process on $M \times \mathbf{P}^{d-1}$ (obtained from S^{d-1} by identifying s and $-s$). The generator of this process is given by

$$\mathcal{L} = G + h\frac{\partial}{\partial s}$$

where $G = X_0 + \frac{1}{2}\sum_{i=1}^{r} X_i^2$ is the generator of ξ written in Hörmander form. For a fixed $\xi \in M$, $h(\xi, \cdot)$ is a vector field on the projective space. To avoid degenerate situations for $\mathcal{L}$, we impose the following ellipticity condition given by Arnold and Kliemann [3]:

$$(HR) \quad \dim LA\left(X_0 + h + \frac{\partial}{\partial t}, X_1, \cdots, X_r\right)(\xi, s, t) = \dim M + d$$

$$\forall\, (\xi, s, t) \in M \times \mathbf{P}^{d-1} \times \mathbf{R}$$

Thus, combining the above results with the definition of moment Lyapunov exponents yields

$$g(p, x_0) = \lim_{t \to \infty} \frac{1}{t} \log E\left[\exp\left\{p \int_0^t q(\tau)d\tau\right\}\right] \text{ for } p \in \mathbf{R} \text{ and } x_0 \in \mathbf{R}^d - \{0\}$$

The following was proven by Arnold *et al.* [3, 4]

Theorem 1 *Assume (HR)*

1. *Let $\lambda = \int_M \int_{\mathbf{P}^{d-1}} q(\xi, s)d\mu$ where μ is the unique invariant probability measure of (ξ, s) on $M \times \mathbf{P}^{d-1}$. Then λ is the maximal Lyapunov exponent for (1), i.e. for $x_0 \neq 0$*

$$\lim_{t \to \infty} \frac{1}{t} \log \| x(t; x_0) \| = \lambda \quad almost - surely$$

2. *For $p \in \mathbf{R}$, let $g(p)$ be the principal eigenvalue of $L(p) = \mathcal{L} + pq(\xi, s)$ acting on $C(M \times \mathbf{P}^{d-1})$. Then $g(p)$ is the pth moment Lyapunov exponent for (1), i.e. for $x_0 \neq 0$*

$$\lim_{t \to \infty} \frac{1}{t} \log E \| x(t; x_0) \|^p = g(p)$$

Moreover $g : \mathbf{R} \to \mathbf{R}$ is convex and analytic, $\frac{g(p)}{p}$ is increasing, $g(0) = 0$, $g'(0) = \lambda$, and the corresponding eigenfunction of $g(p)$ is non-negative. Furthermore, $g''(0)$ is the variance in the central limit theorem, i.e.

$$\frac{1}{\sqrt{t}}\left(\log \|x(t; x_0)\| - \lambda t\right) \Rightarrow \mathcal{N}(0, g''(0)) \ (t \to \infty) \text{ for any } x_0 \neq 0$$

where $\mathcal{N}$ is the normal distribution and $\Rightarrow$ denotes weak convergence.

In general, the solution of $g(p)$ to all orders of p is very difficult to obtain. However, using the convex and analytic properties of $g(p)$, one can write

$$g(p) = g(0) + pg'(0) + \frac{p^2}{2}g''(0) + o(p^2) = p\lambda + \frac{p^2}{2}g''(0) + o(p^2) \quad (2)$$

Consider the operator $L(p)$ and its adjoint $L^*(p)$. By Theorem 1, $g(p)$ is an isolated simple eigenvalue of $L(p)$ with non-negative eigenfunction $\phi(p)$ such that $\| \phi(p) \| = 1$. The adjoint operator $L^*(p)$ has an eigenfunction $\nu(p)$ corresponding to $g(p)$ which is unique and has the property $\langle \phi(p), \nu(p) \rangle = 1$, i.e.

$$L(p)\phi(p) = g(p)\phi(p) , \quad \langle \phi(p), \nu(p) \rangle = 1 \quad \forall p \in \mathbf{R} \tag{3}$$

It is clear that for $p = 0$ (since $g(0) = 0$) we have

$$
\begin{aligned}
L(0)\phi(0) &= \mathcal{L}\phi(0) = 0 \quad \text{with } \phi(0) = 1 \\
L^*(0)\nu(0) &= \mathcal{L}^*\nu(0) = 0 \quad \text{with } \nu(0) = \mu(\xi, s)
\end{aligned}
$$

where $\mu(\xi, s)$ is the density of the unique ergodic invariant measure associated with the generator $\mathcal{L}$. As in Arnold *et al.* [4], we make use of the analytical dependence of these equations on p. Differentiating both sides of the eigenvalue problem, Eq. (3), and taking the appropriate scalar product with $\nu(p)$, we arrive at the following set of equations and associated solvability conditions for $\psi = \left.\frac{\partial \phi}{\partial p}\right|_{p=0}$ and $\eta = \left.\frac{\partial^2 \phi}{\partial p^2}\right|_{p=0}$:

$$
\begin{aligned}
\mathcal{L}\psi &= \lambda - q(\xi, s) \\
&\quad \text{solvability}: \quad \lambda = \langle \mu(\xi, s), q(\xi, s) \rangle \\
\mathcal{L}\eta &= g''(0) - 2\left(q(\xi, s) - \lambda\right)\psi \\
&\quad \text{solvability}: \quad g''(0) = 2\langle \mu(\xi, s), (q(\xi, s) - \lambda)\psi \rangle
\end{aligned}
\tag{4}
$$

3. Asymptotic Analysis

The analysis in this paper is only valid for small intensity noise perturbations. To make this smallness explicit, we rescale the noise in Eq. (1) by a factor of ε ($\varepsilon \ll 1$). The equations of motion for the real noise case become

$$
\begin{aligned}
\dot{x} &= Ax + \varepsilon f(\xi(t))Bx , \quad x \in \mathbf{R}^2 \\
d\xi &= X_0(\xi)dt + \sum_{i=1}^{r} X_i(\xi) \circ dW_i
\end{aligned}
\tag{5}
$$

where $B = [b_{jk}]$, $f : M \to \mathbf{R}$ is a smooth function with zero mean. The dynamical system perturbed by small white noise is governed by the following equation

$$dx = Ax\, dt + \varepsilon \sum_{i=1}^{m} B_i x \circ dW_i(t) , \quad x \in \mathbf{R}^2 \tag{6}$$

in which $B_i = \left[b_{jk}^{(i)} \right]$ and W_i are independent unit variance Wiener processes.

Equation (2) represents the Taylor expansion of $g(p)$ around $p = 0$. Using this equation, and the identities $g(0) = 0$ and $g'(0) = \lambda$, we can approximate the moment Lyapunov exponent up to $O(p^2)$ by first obtaining asymptotic expansions, in orders of ε, for λ and $g''(0)$.

We denote the asymptotic expansion for λ by λ^ε and the asymptotic expansion for $g''(0)$ by V^ε. As shown in Eq. (4), the expression for $g''(0)$ contains λ. Hence, in order to calculate the terms in V^ε, it is necessary to first compute successive orders of λ^ε and feed these quantities into the calculation of V^ε at each level.

In this section, sufficient conditions for the validity of the asymptotic expansions for λ and $g''(0)$ are given for the real noise case. Although the general theory given in the previous section and the subsequent theorems providing valid asymptotic expansions for λ^ε and $V^\varepsilon = g''(0)$ are for an arbitrary d-dimensional system, the explicit asymptotic results are obtained only for two-dimensional systems in Section 4.

A proof that the expressions for λ and $g''(0)$ are valid asymptotic expansions begins with the construction of a formal expansion of the invariant measure, i.e. $\mu^\varepsilon(\xi) = \mu_0(\xi) + \varepsilon\mu_1(\xi) + \varepsilon^2\mu_2(\xi) + \cdots$, and of the generator $\mathcal{L}$, i.e. $\mathcal{L}^\varepsilon = \mathcal{L}^0 + \varepsilon\mathcal{L}^1 + \varepsilon^2\mathcal{L}^2$.

Consider the problem

$$\mathcal{L}^\varepsilon \psi^\varepsilon = \lambda^\varepsilon - Q^\varepsilon$$

where $\psi^\varepsilon = \psi_0 + \varepsilon\psi_1 + \varepsilon^2\psi_2 + \cdots$ and $Q^\varepsilon = Q^0 + \varepsilon Q^1 + \varepsilon^2 Q^2$. For ψ^ε to exist, it is necessary that $\langle \lambda^\varepsilon - Q^\varepsilon, \mu^\varepsilon \rangle = 0$. Solving for λ^ε yields $\lambda^\varepsilon = \langle Q^\varepsilon, \mu^\varepsilon \rangle$.

Theorem 2 *Suppose that the functions ψ_i and μ^ε are such that all inner products are well defined and assume that*

$$\sup_{\theta,\xi} \left| \mathcal{L}^1\psi_N + \mathcal{L}^2\psi_{N-1} \right| \leq K_1 < \infty \quad \text{and} \quad \sup_{\theta,\xi} \left| \mathcal{L}^2\psi_N \right| \leq K_2 < \infty$$

Applying the above estimate, it is clear that an asymptotic expansion for λ^ε for a fixed $N \geq 0$ is given by

$$\lambda^\varepsilon = \lambda_0 + \cdots + \varepsilon^N\lambda_N + \varepsilon^{N+1}\langle \mathcal{L}^1\psi_N + \mathcal{L}^2\psi_{N-1}, \mu^\varepsilon \rangle + \varepsilon^{N+2}\langle \mathcal{L}^2\psi_N, \mu^\varepsilon \rangle \quad (7)$$

Next consider the expansion for $V^\varepsilon = g''_\varepsilon(0)$. The pertinent equation is

$$\mathcal{L}^\varepsilon \eta^\varepsilon = V^\varepsilon - 2\left(Q^\varepsilon - \lambda^\varepsilon\right)\psi^\varepsilon \qquad (8)$$

where $\eta^\varepsilon = \eta_0 + \varepsilon\eta_1 + \varepsilon^2\eta_2 + \cdots$ and ψ^ε is determined from $\mathcal{L}^\varepsilon\psi^\varepsilon = \lambda^\varepsilon - Q^\varepsilon$. Now make the ansatz $V^\varepsilon = V_0 + \varepsilon V_1 + \varepsilon V_2 + \cdots$

Theorem 3 *Suppose that the functions η_i and μ^ε are such that all inner products are well defined and assume that*

$$\sup_{\theta,\xi} \left| \mathcal{L}^1 \eta_N + \mathcal{L}^2 \eta_{N-1} \right| \leq K_3 < \infty \quad \text{and} \quad \sup_{\theta,\xi} \left| \mathcal{L}^2 \eta_N \right| \leq K_4 < \infty$$

Then an asymptotic expansion for V^ε for all $N \geq 0$ is given by

$$V^\varepsilon = V_0 + \cdots + \varepsilon^N V_N + \varepsilon^{N+1} \langle \mathcal{L}^1 \eta_N + \mathcal{L}^2 \eta_{N-1}, \mu^\varepsilon \rangle$$
$$\varepsilon^{N+2} \langle \mathcal{L}^2 \eta_N, \mu^\varepsilon \rangle + \varepsilon^{N+1} \langle \tilde{f}_N, \mu^\varepsilon \rangle + O(\varepsilon^{N+3}) \tag{9}$$

For details of the derivation of Theorems 2 and 3, the reader is refered to Doyle *et al.* [10].

4. Moment Lyapunov Exponents of Two-Dimensional Systems

The following cases for the matrix A are considered: (a) A has a pair of complex eigenvalues and (b) A has two distinct real eigenvalues. Since the maximal and moment Lyapunov exponents do not depend on the choice of coordinate system, consider

$$\text{case(a)} \quad A = \begin{bmatrix} \delta & -\omega \\ \omega & \delta \end{bmatrix}$$

$$\text{case(b)} \quad A = \begin{bmatrix} a_1 & 0 \\ 0 & a_2 \end{bmatrix}, \quad a_1 > a_2$$

In this section, the moment Lyapunov exponents for systems subjected to small real noise excitation are approximated using the procedure outlined in Section 3.

Consider Eq. (5) in which the system is excited by a real noise process whose generator is specified by the model of the noise. Projecting the equations of motion onto the circle yields

$$\dot{\rho} = q(\theta) = q_0(\theta) + \varepsilon f(\xi) q_1(\theta)$$
$$\dot{\theta} = h_0 + \varepsilon f(\xi) h_1(\theta) \tag{10}$$
$$d\xi = X_0(\xi) dt + \sum_{i=1}^{r} X_i(\xi) \circ dW_i$$

Comparing this to the notation of Section 3, $Q^0 = q_0(\theta)$, $Q^1 = f(\xi) q_1(\theta)$, $Q^2 = 0$ and

$$\mathcal{L}^0 = G(\xi) + h_0(\theta) \frac{\partial}{\partial \theta}, \quad \mathcal{L}^1 = f(\xi) h_1(\theta) \frac{\partial}{\partial \theta}, \quad \mathcal{L}^2 = 0$$

This procedure for calculating the moment Lyapunov exponent is analogous to the perturbation method presented by Sri Namachchivaya and Van Roessel [17] and Doyle and Sri Namachchivaya [9] in calculating the maximal Lyapunov exponent. In [17] and [9], however, the adjoint equations were used.

Constructing formal expansions for ψ^ε and η^ε and expanding the equations for ψ^ε and η^ε in powers of ε up to $O(\varepsilon^2)$ yields the following sets of Poisson equations

$$\mathcal{L}^0 \psi_0 = \lambda_0 - q_0 \tag{11}$$

$$\mathcal{L}^0 \psi_1 = \lambda_1 - q_1 f(\xi) - \mathcal{L}^1 \psi_0 \tag{12}$$

$$\mathcal{L}^0 \psi_2 = \lambda_2 - \mathcal{L}^1 \psi_1 \tag{13}$$

and

$$\mathcal{L}^0 \eta_0 = V_0 - 2(q_0 - \lambda_0)\psi_0 \tag{14}$$

$$\mathcal{L}^0 \eta_1 = V_1 - 2\left[(q_0 - \lambda_0)\psi_1 - (\lambda_1 - q_1 f(\xi))\psi_0\right] - \mathcal{L}^1 \eta_0 \tag{15}$$

$$\mathcal{L}^0 \eta_2 = V_2 - 2\left[(q_0 - \lambda_0)\psi_2 - (\lambda_1 - q_1 f(\xi))\psi_1 - \lambda_2 \psi_0\right] - \mathcal{L}^1 \eta_1 \tag{16}$$

The quantities V_0, V_1 and V_2 are the terms in the asymptotic expansion for V^ε, i.e. $V^\varepsilon = V_0 + \varepsilon V_1 + \varepsilon^2 V_2 + \cdots$.

Case (a): A has a pair of complex eigenvalues
Here

$$\dot{x} = \begin{bmatrix} \delta & -\omega \\ \omega & \delta \end{bmatrix} x + \varepsilon f(\xi(t)) \begin{bmatrix} b_{11} & b_{12} \\ b_{21} & b_{22} \end{bmatrix} x \tag{17}$$

In the absence of noise, this represents an oscillator with frequency ω and damping -2δ. The expressions appearing in Eq. (10) are $q_0(\theta) = \delta$, $h_0(\theta) = \omega$ and

$$q_1 = \frac{1}{2}(b_{11} + b_{22}) + \frac{1}{2}(b_{12} + b_{21})\sin(2\theta) + \frac{1}{2}(b_{11} - b_{22})\cos(2\theta)$$

$$h_1 = \frac{1}{2}(b_{21} - b_{12}) + \frac{1}{2}(b_{22} - b_{11})\sin(2\theta) + \frac{1}{2}(b_{21} + b_{12})\cos(2\theta)$$

Recall that our aim is to expand $g_\varepsilon(p)$ about $\varepsilon = 0$ and $p = 0$. Using the fact that $q_0 = \delta$, one can easily verify that $\lambda_0 = \delta$. Employing this result in Eq. (11) implies $\psi_0 = 1$. Due to the periodicity of θ and the zero mean assumption on the noise process, the solvability condition for Eq. (12) yields $\lambda_1 = 0$.

Consider Eq. (12) for $\psi_1(\theta, \xi)$. In order to make use of the transient density $\mathcal{G}(\eta, T; \xi, 0)$ which solves

$$\frac{\partial \mathcal{G}}{\partial t} = G\mathcal{G}$$

we introduce an auxiliary time t' and the transformation

$$t = \frac{1}{2}\left(t' - \frac{\theta}{\omega}\right), \qquad s = \frac{1}{2}\left(t' + \frac{\theta}{\omega}\right)$$ (18)

This yields

$$\left(\frac{\partial}{\partial t} - G\right)\psi_{1_t}(\theta(s,t),\xi) = q_1(\theta(s,t))f(\xi)$$ (19)

where $\theta(s,t) = \omega(s-t)$. Equation (19) may be solved using Duhamel's principle. Taking the limit as $t \to \infty$ yields

$$
\begin{aligned}
\psi_1(\theta,\xi) \\
= \quad & \frac{1}{2}\left[(b_{12}+b_{21})\sin 2\theta + (b_{11}-b_{22})\cos 2\theta\right]\int_0^\infty K(\xi,T)\cos(2\omega T)dT \\
+ \quad & \frac{1}{2}\left[(b_{12}+b_{21})\cos 2\theta - (b_{11}-b_{22})\sin 2\theta\right]\int_0^\infty K(\xi,T)\sin(2\omega T)dT \\
+ \quad & \frac{1}{2}(b_{11}+b_{22})\int_0^\infty K(\xi,T)dT
\end{aligned}
$$

where $K(\xi,T) = \int_M f(\eta)\mathcal{G}(\eta,T;\xi,0)d\eta$.

The final term in the expansion for the maximal Lyapunov exponent can now be calculated and is given by

$$\lambda_2 = \langle \mathcal{L}^1\psi_1, \mu_0\rangle$$

This solvability condition gives the following expression:

$$\lambda_2 = \frac{1}{8}\left[(b_{12}+b_{21})^2 + (b_{22}-b_{11})^2\right]S(2\omega)$$

where $S(\omega)$, the spectral density of $f(\xi)$, is given in terms of the correlation $R(T)$ as

$$S(\omega) = 2\int_0^\infty R(T)\cos\omega T dT \quad \text{with} \quad R(T) = \int_M \nu(\xi)f(\xi)K(\xi,T)d\xi$$

Define

$$\beta = (b_{12}+b_{21})^2 + (b_{22}-b_{11})^2 \quad \text{and} \quad \kappa = 2(b_{22}+b_{11})^2$$

Then $\lambda_2 = \frac{\beta}{8}S(2\omega)$ and, finally,

$$\lambda^\varepsilon = \delta + \frac{\varepsilon^2}{8}\beta S(2\omega) + o(\varepsilon^2)$$

We thus have recovered a well-known formula (see, for example, [16, 14]). We are now in a position to expand V^ε.

Since $\lambda_0 = q_0$, it can easily be shown that $V_0 = 0$ and, in turn, $\eta_0 =$ constant. The solvability condition for Eq. (15) simplifies to

$$V_1 = 2\langle q_1(\theta) f(\xi), \mu_0\rangle \tag{20}$$

Due to the zero mean assumption on $f(\xi)$, $V_1 = 0$. This yields $V^\varepsilon = \varepsilon^2 V_2 + \cdots$ where V_2 is obtained from the solvability of Eq. (16), i.e.

$$V_2 = 2\langle q_1 f(\xi)\psi_1 - \lambda_2, \mu_0\rangle + \langle \mathcal{L}^1 \eta_1, \mu_0\rangle \tag{21}$$

In order to solve for V_2, it is necessary to solve for η_1 from Eq. (15). Upon substitution of the appropriate expressions, Eq. (15) reduces to

$$\mathcal{L}^0 \eta_1 = -2q_1 f(\xi) \quad \text{implying} \quad \eta_1 = 2\psi_1$$

Then

$$V_2 = 2\langle q_1 f(\xi)\psi_1(\xi, \theta)\rangle = \frac{\kappa}{8}S(0) + \frac{\beta}{8}S(2\omega)$$

Theorem 4 *In case (a), the asymptotic expansion of the moment Lyapunov exponent of Eq. (17) is given by*

$$g_\varepsilon(p) = p\left(\delta + \frac{\varepsilon^2}{8}\beta S(2\omega)\right) + \frac{\varepsilon^2 p^2}{16}\left(\kappa S(0) + \beta S(2\omega)\right) + o(\varepsilon^2 p^2)$$

where $\beta = (b_{12} + b_{21})^2 + (b_{22} - b_{11})^2$ and $\kappa = 2(b_{22} + b_{11})^2$

Case (b): A has two distinct real eigenvalues
Now consider

$$\dot{x} = \begin{bmatrix} a_1 & 0 \\ 0 & a_2 \end{bmatrix} x + \varepsilon f(\xi(t)) \begin{bmatrix} b_{11} & b_{12} \\ b_{21} & b_{22} \end{bmatrix} x, \quad a_1 > a_2 \tag{22}$$

The expressions appearing in Eq. (10) are

$$q_0 = a_1 \cos^2\theta + a_2 \sin^2\theta \quad \text{and} \quad h_0 = (a_2 - a_1)\cos\theta \sin\theta$$

and q_1 and h_1 are given in case (a). Hence

$$\mathcal{L} = \left(G + (a_2 - a_1)\cos\theta \sin\theta \frac{\partial}{\partial\theta}\right) + \varepsilon \left(f(\xi)h_1(\theta)\frac{\partial}{\partial\theta}\right) = \mathcal{L}^0 + \varepsilon \mathcal{L}^1$$

To determine the invariant density, consider $\mathcal{L}^{0^*}\mu_0(\xi, \theta) = 0$ and assume $\mu_0(\xi, \theta) = \nu(\xi)z(\theta)$. We restrict G such that $G = G^*$. The eigenfunctions

$\Phi_k(\xi)$ and eigenvalues γ_k of the generator satisfy $G\Phi_k(\xi) = -\gamma_k\Phi_k(\xi)$ with $G1 = 0$, i.e. $\gamma_1 = 0$ is associated with the steady state (see Pinsky [15]). The remaining eigenfunctions are then normalized such that $\langle \Phi_k, \Phi_j \rangle = \delta_{kl}$. The function $\nu(\xi)$ can then be expanded in terms of eigenfunctions as

$$\nu(\xi) = \sum_{k=2}^{N} \nu_k \Phi_k(\xi)$$

The coefficients ν_k are chosen such that $\int_M \left(\sum_{k=2}^{N} \nu_k \Phi_k(\xi)\right) d\xi = 1$. For $\varepsilon = 0$, the θ equation has stable (attracting) equilibrium points at 0 and π and unstable (repelling) points at $\pi/2$ and $3\pi/2$. Therefore, as in [14], $z(\theta)$ can be chosen such that $z(\theta) = \frac{1}{2}(\delta_0 + \delta_\pi)$ where δ_0 and δ_π denote Dirac measures at 0 and π, respectively.

Following the procedure outlined for case (a), yields

Theorem 5 *In case (b), the asymptotic expansion of the moment Lyapunov exponent of Eq. (22) is given by*

$$g_\varepsilon(p) = p\left(a_1 + \frac{\varepsilon^2}{2}b_{12}b_{21}S_{hyp}(a_1 - a_2)\right) + \frac{\varepsilon^2 p^2}{2}b_{11}^2 S_{hyp}(0) + o(\varepsilon^2 p^2)$$

where $S_{hyp}(a) = 2\int_0^\infty R(T)e^{-aT}\,dT$, $R(T) = \int_M \nu(\xi)f(\xi)K(\xi, T)d\eta$

5. Results for Small White Noise

Next consider the asymptotic behavior of a linear Stratonovich stochastic differential equation in $\mathbf{R}^d$

$$dx = Ax\,dt + \varepsilon\sum_{i=1}^{m} B_i x \circ dW_i(t)\,, \quad x \in \mathbf{R}^2 \tag{23}$$

Projecting the first order equations onto the circle yields the following equations of motion

$$\dot{\rho} = q_0(\theta) + \varepsilon\sum_{i=1}^{m}\sigma_i q_i(\theta) \circ dW_i(t)$$

$$\dot{\theta} = h_0(\theta) + \varepsilon\sum_{i=1}^{m}\sigma_i h_i(\theta) \circ dW_i(t) \tag{24}$$

where, for $i = 1, \cdots, m$,

$$q_i = \frac{1}{2}\left(b_{11}^i + b_{22}^i\right) + \frac{1}{2}\left(b_{12}^i + b_{21}^i\right)\sin(2\theta) + \frac{1}{2}\left(b_{11}^i - b_{22}^i\right)\cos(2\theta)$$

$$h_i = \frac{1}{2}\left(b_{21}^i - b_{12}^i\right) + \frac{1}{2}\left(b_{22}^i - b_{11}^i\right)\sin(2\theta) + \frac{1}{2}\left(b_{21}^i + b_{12}^i\right)\cos(2\theta)$$

with $\mathcal{L}^0 = h_0 \frac{\partial}{\partial \theta}$, $\mathcal{L}^1 = 0$ and $\mathcal{L}^2 = \frac{1}{2} \left(h_1(\theta) \frac{\partial}{\partial \theta} \right)^2$, and $q_0(\theta)$ and $h_0(\theta)$ depend on the structure of the matrix A.

The asymptotic expansions of $g_\varepsilon(p)$ for cases (a) and (b) are summarized in the following theorems:

Theorem 6 *In case (a), the asymptotic expansion for the moment Lyapunov exponent is*

$$g_\varepsilon(p) = p\lambda^\varepsilon + \frac{p^2}{2} V^\varepsilon + o(p^2) = p \left(\delta + \frac{\varepsilon^2}{8} \gamma_1 \right) + \frac{\varepsilon^2 p^2}{16} (\gamma_1 + \gamma_2) + o(\varepsilon^2 p^2) \quad (25)$$

where

$$\beta^i = \left(b_{12}^i + b_{21}^i \right)^2 + \left(b_{22}^i - b_{11}^i \right)^2 , \quad \kappa^i = 2 \left(b_{22}^i + b_{11}^i \right)^2$$

$\gamma_1 = \sum_{i=1}^m \beta^i$ and $\gamma_2 = \sum_{i=1}^m \kappa^i$.

Theorem 7 *In case (b), the asymptotic expansion for the moment Lyapunov exponent is*

$$g_\varepsilon(p) = p\lambda^\varepsilon + \frac{p^2}{2} V^\varepsilon + o(p^2) = p \left(a_1 + \frac{\varepsilon^2}{2} \gamma_1 \right) + \frac{\varepsilon^2 p^2}{2} \gamma_2 + o(\varepsilon^2 p^2)$$

where $\gamma_1 = \sum_{i=1}^m b_{12}^i b_{21}^i$ *and* $\gamma_2 = \sum_{i=1}^m \left(b_{11}^i \right)^2$.

6. Application to Stochastic Bifurcation Theory

In this section, the asymptotic results obtained in this paper are applied to concepts in stochastic bifurcation theory. It will be shown that the moment Lyapunov exponent gives important information regarding characteristics of the probability density of the response of the system.

Given the nonlinear stochastic system

$$dx = f_0^\alpha(x) dt + \sum_{i=1}^m f_i^\alpha(x) \circ dW_t^i , \quad x \in \mathbf{R}^d , \quad f_i^\alpha(0) = 0 \ \forall \alpha \in \mathbf{R} \quad (26)$$

Let $\varphi_\alpha(t, \cdot, x)$ denote the (possibly local) stochastic flow generated by Eq. (26). A dynamic bifurcation (D-Bifurcation) is related to the stability of $\varphi_\alpha(t, \cdot, x)$ with associated invariant measure μ^α through the Lyapunov exponent $\lambda_i(\mu^\alpha)$ of the linearized system $D\varphi_\alpha(t, \omega, x)$ (see Arnold [1]). The parameter value at which the top Lyapunov exponent becomes zero is denoted by α_1. Due to the assumption $f_i^\alpha(0) = 0$, $\mu^\alpha = \delta_0$, the Dirac measure at $x = 0$, is always an invariant measure which we call the 'trivial solution'.

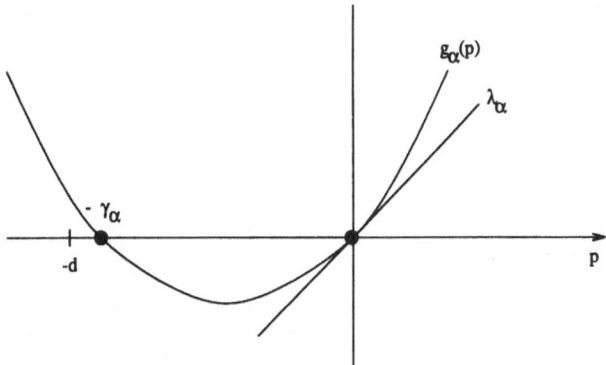

Figure 1. Moment and maximal Lyapunov exponents. For $\gamma_\alpha = d$, the probability density function attains a finite maximum at $x = 0$.

Baxendale [8] has proven that if λ_α of the trivial solution becomes positive, the Fokker-Planck equation has a new nontrivial solution with density $p_\alpha(x)$. He has also shown that

$$\frac{\int_{|x| \leq r} p_\alpha(x)\,dx}{\text{vol}(|x| \leq r)} \sim Cr^{\gamma_\alpha - d}$$

where $-\gamma_\alpha$ is the second zero of $g_\alpha(p)$. Figure 1 depicts a typical plot of the moment Lyapunov exponent as a function of p as well as the maximal Lyapunov exponent, $\lambda_\alpha = g'_\alpha(0)$. Hence p_α has a pole at $x = 0$ for $\gamma_\alpha < d$, and a zero at $x = 0$ for $\gamma_\alpha > d$. This qualitative change in the probability density function as the bifurcation parameter γ_α is varied is shown in Figure 2. This explains and locates a phenomenological bifurcation, or P-bifurcation, (qualitative change of density) at α_0 for which $\gamma_{\alpha_0} = d$. The bifurcation of the response amplitude and the corresponding changes in the density function at the points of D-bifurcation and P-bifurcation are displayed in Figure 3.

For the small noise case with dimension $d = 2$, we can now determine α_1 as well as α_0. Consider, for example, case (a) when the noise is white. In this case, we have

$$dx = \begin{bmatrix} \delta & -\omega \\ \omega & \delta \end{bmatrix} x\,dt + \varepsilon \sum_{i=1}^{m} B_i x \circ dW_i(t)$$

where $-\delta = \alpha$ is our bifurcation parameter. Using the previous results, we can write the expansion for λ^ε and V^ε as

$$\lambda^\varepsilon(\delta) = \delta + \frac{\varepsilon^2}{8}\gamma_1 + \cdots, \quad V^\varepsilon(\delta) = \frac{\varepsilon^2}{8}(\gamma_1 + \gamma_2) + \cdots$$

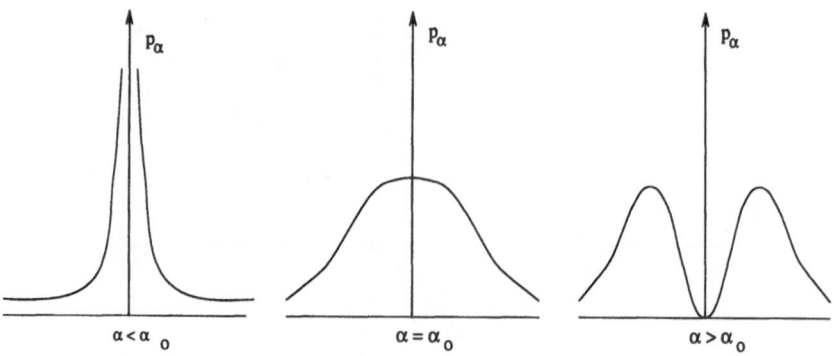

Figure 2.   Qualitative change in the probability density function.

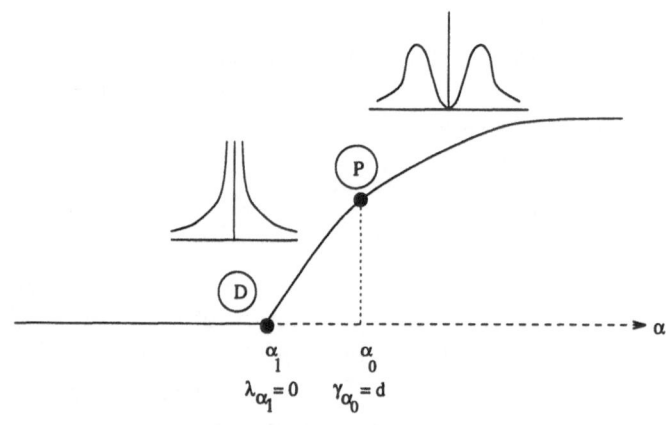

Figure 3.   P-bifurcation and D-bifurcation as α varies.

In addition, it is clear from the trace formula that the second exponent is $\lambda_2^\varepsilon(\delta) = \delta - \frac{\varepsilon^2}{8}\gamma_1 + \cdots$. Thus the moment Lyapunov exponent is

$$g_\varepsilon^\delta = \lambda^\varepsilon(\delta)p + \frac{V^\varepsilon(\delta)}{2}p^2 + \cdots$$

Denoting the parameter values δ at which the first and second Lyapunov exponents become zero by δ_1 and δ_2, respectively, we obtain

$$\lambda^\varepsilon(\delta_1) = 0 \quad \text{implies} \quad \delta_1 = -\frac{\varepsilon^2}{8}\gamma_1$$

$$\lambda_2^\varepsilon(\delta_2) = 0 \quad \text{implies} \quad \delta_2 = \frac{\varepsilon^2}{8}\gamma_1$$

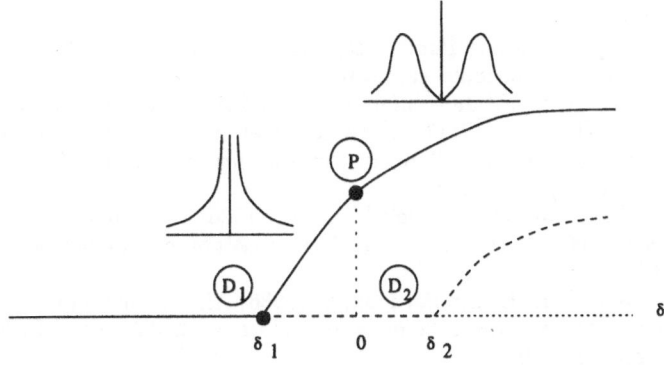

Figure 4.   Duffing-Van der Pol oscillator: D- and P-bifurcations as δ varies.

The parameter value δ_0 at which $g_\varepsilon^\delta(-2) = 0$ can be calculated as follows:

$$\gamma_\delta^\varepsilon = -\frac{2V^\varepsilon(\delta)}{\lambda^\varepsilon(\delta)} = -2 \quad \Leftrightarrow \quad V^\varepsilon(\delta) = \lambda^\varepsilon(\delta)$$

$$\Leftrightarrow \quad \frac{\varepsilon^2}{8}(\gamma_1 + \gamma_2) = \delta + \frac{\varepsilon^2}{8}\gamma_1$$

$$\Leftrightarrow \quad \delta_0 = \frac{\varepsilon^2}{8}\gamma_2$$

Note that $\delta_1 \leq 0$, $\delta_2 \geq 0$, $\delta_0 \geq 0$, but both $\delta_2 > \delta_0$ and $\delta_2 < \delta_0$ are possible depending on the B_i's.

As an example, consider the Duffing-Van der Pol oscillator at the point of deterministic Hopf bifurcation. The linearization at $x = 0$ is

$$dx = \begin{bmatrix} \delta & 1 \\ -1 & \delta \end{bmatrix} x\, dt + \varepsilon \begin{bmatrix} 0 & 0 \\ 1 & 0 \end{bmatrix} x \circ dW$$

Here, $\gamma_1 = 1$ and $\gamma_2 = 0$. Then $\delta_1 = -\frac{\varepsilon^2}{8}$, $\delta_0 = 0$ and $\delta_2 = \frac{\varepsilon^2}{8}$. The D- and P-bifurcations and corresponding density functions for this example are shown in Figure 4.

Acknowledgments

The first two authors would like to acknowledge the support of the Air Force Office of Scientific Research through grant 93-0063 monitored by Dr. Spencer Wu.

References

1. Arnold, L. (In preparation) *Randon Dynamical Systems*.
2. Arnold, L. (1984) A formula connecting sample and moment stability of linear stochastic systems, *SIAM Journal of Applied Mathematics*, **44**(4), pp. 793-802.
3. Arnold, L. and Kliemann, W. (1987) Large deviations of linear stochastic differential equations, vol. 96 of *Lecture Notes in Control and Information Sciences*, Springer-Verlag, New York, pp. 117-151.
4. Arnold, L., Kliemann, W. and Oeljeklaus, E. (1986) Lyapunov exponents of linear stochastic systems, vol. 1186 of *Lecture Notes in Mathematics*, Springer-Verlag, New York, pp. 85-125.
5. Arnold, L., Oeljeklaus, E. and Pardoux, E. (1986) Almost sure and moment stability for linear Itô equations, vol. 1186 of *Lecture Notes in Mathematics*, Springer-Verlag, New York, pp. 129-159.
6. Arnold, L. and Wihstutz, V. (1986) "Lyapunov exponents", vol. 1186 of *Lecture Notes in Mathematics*, Springer-Verlag, New York.
7. Baxendale, P. H. (1987) Moment stability and large deviations for linear stochastic differential equations, in *Taniguchi Symposium PMPP*.
8. Baxendale, P. H. (1990) Invariant measures for nonlinear stochastic differential equations, vol. 1486 of *Lecture Notes in Mathematics*, Springer-Verlag, New York, pp. 123-140.
9. Doyle, M. M. and N. Sri Namachchivaya (1994) Almost-sure asymptotic stability of a general four dimensional dynamical system driven by real noise, *Journal of Statistical Physics*, **75**(3/4), pp. 525-552.
10. Doyle, M. M., N. Sri Namachchivaya and L. Arnold (Submitted for publication) Small noise expansion of moment Lyapunov exponents for general two dimensional systems.
11. Khas'minskii, R. Z. (1980) *Stochastic Stability of Differential Equations*, Sijthoff and Noordhoff, Alphen aan den Rijn. (Translation of the Russian edition, Nauka, Moscow, 1969)
12. Kozin, F. and S. Sugimoto (1977) Relations between sample and moment stability for linear stochastic differential equations, in J. David Mason (ed.), *Proceedings of Conference on Stochastic Differential Equations and Applications*, Academic Press, pp. 145-162.
13. Molčanov, S. A. (1978) The structure of eigenfunctions of one-dimensional unordered structures, *Math. USSR Izvestija*, **12**(1), pp. 69-101.
14. Pardoux, E. and V. Wihstutz (1988) Lyapunov exponent and rotation number of two-dimensional linear stochastic aystems with amall diffusion, *SIAM Journal of Applied Mathematics*, **48**(2), pp. 442-457.
15. Pinsky, M. A. (1992) Lyapunov exponent and rotation number of the linear harmonic oscillator, in M. A. Pinsky and V. Wihstutz (eds.), *Diffusion Processes and Related Problems in Analysis, Volume II (Stochastic Flows)* Birkhäuser, Boston", pp. 257-267.
16. Sri Namachchivaya, N. and Ariaratnam, S. T. (1987) Stochastically perturbed Hopf bifurcation *International Journal of Non-Linear Mechanics* **22**(5), pp. 363-372 (see also Stochastically perturbed Hopf bifurcation, in F. M. A. Salam and M. L. Levi (eds.), *Dynamical Systems Approaches to Nonlinear Problems in Systems and Circuits*, SIAM, pp. 39-52, 1988).
17. Sri Namachchivaya, N. and H. J. Van Roessel, Maximal Lyapunov exponent and rotation numbers for two coupled oscillators driven by real noise, *Journal of Statistical Physics*, **71**(3/4), pp. 549-567.
18. Stroock, D. W. (1986) On the rate at which a homogeneous diffusion approaches a limit, an application of the large deviation theory of certain stochastic integrals, *Annals of Probability*, **14**, pp. 840-859.

NON-PERTURBATIVE FEM FOR DETERMINISTIC AND STOCHASTIC BEAMS THROUGH INVERSE OF STIFFNESS MATRIX

I. ELISHAKOFF AND Y.J. REN
Department of Mechanical Engineering, Florida Atlantic University
Boca Raton, FL 33431-0991

M. SHINOZUKA
Department of Civil Engineering, University of Southern California
University Park, Los Angeles, CA 90089-0242

1. Abstract

This paper proposes an alternative way of constructing the global stiffness matrix in the finite element analysis of bending beams, which involve spatially deterministic or stochastical bending stiffness. Originating from Fuchs' idea of decoupling the shear and bending components in the bending beam, the element level stiffness matrix is diagonalized. The generalized stress-strain, strain-displacement and equilibrium relationships are assembled, respectively, and then are combined to form the global stiffness matrix. The advantage of the new formulation is that the bending stiffness explicitly appears in the global stiffness matrix, which can be inverted exactly without application of perturbation based expansion. The mean vector and correlation matrix of the displacement of the beam are then obtained in terms of probabilistic characteristics of the uncertain bending stiffness. The example is given to illustrate the efficacy of the new formulation and its application to bending of stochastic beams.

2. Introduction

The finite element method for stochastic problems (FEMSP) has gained significant interest among researchers in the past decade. The literature includes numerous papers and three monographs published since 1985 (Nakagiri and Hisada 1985; Ghanem and Spanos 1991; Kleiber and Hien 1993). Tthe existing FEMSP falls into the category of perturbation and/or series expansion methods. It is basically a combination of the deterministic finite element method, perturbation technique and probabilistic analysis. The perturbed components of the response are obtained in terms of the perturbed components of the uncertain parameters. The perturbation-based FEMSP has several disadvantages. For example, the accuracy of low-order perturbation method is acceptable only for the problems whose uncertain parameters have only small deviations, whereas the algorithm

A. Naess and S. Krenk (eds.), IUTAM Symposium on Advances in Nonlinear Stochastic Mechanics, 169–178.

to obtain high-order perturbations may turn out to be extremely time-consuming.

In this paper, we propose a new formulation of the finite element method; the global stiffness matrix is constructed by combining the global equations of generalized stress-strain relationship, strain-displacement relationship and equilibrium condition. The idea is originated from Fuchs' approach (Fuchs 1991,1992), which was designated to solve deterministic *optimal design* of structures. Through the application of the physical idea, Fuchs decoupled the shear and bending components in the generalized stress-strain relationship and obtained the final stiffness matrix explicitly in terms of the bending stiffness. The latter constitutes a design parameter in the optimization analysis. Present study applies the *generalized* procedure to isolate the bending stiffness in the global stiffness matrix so that the response can explicitly be expressed in terms of the stochastic bending stiffness. To achieve this, an *orthogonal transformation* is used to diagonalize the stiffness matrix at the element level. The element-level generalized stress-strain relationship, strain-displacement relationship and static equilibrium equation are then assembled into the global ones, respectively, and then are combined into the global stiffness matrix. The fact that the global generalized stress-strain matrix is diagonal allows to express the response symbolically in terms of the stochastic bending stiffness. The mean vector and variance-covariance matrix of the response are consequently obtained based on the probabilistic information of the beam's stiffness.

3. New Formulation of FE Stiffness Matrix

Consider an elastic beam with spatially varying bending stiffness $D(x)=E(x)I(x)$, where $E(x)$ is the Young's modulus and $I(x)$ is the inertia moment of the section of the beam. The beam is equally divided into N elements each with length a. The finite element equilibrium equation for the i-th element of the beam is derived as

$$\frac{D_i}{a^3}\begin{bmatrix} 12 & 6a & -12 & 6a \\ 6a & 4a^2 & -6a & 2a^2 \\ -12 & -6a & 12 & -6a \\ 6a & 2a^2 & -6a & 4a^2 \end{bmatrix}\begin{Bmatrix} w_{i1} \\ \theta_{i1} \\ w_{i2} \\ \theta_{i2} \end{Bmatrix} = \begin{Bmatrix} Q_{i1} \\ M_{i1} \\ Q_{i2} \\ M_{i2} \end{Bmatrix} \tag{1}$$

where the subscripts " *1* " and " *2* " represent the left and right nodes of the element, respectively, w=displacement of the beam, $\theta=dw/dx$=slope of the beam, Q and M are generalized nodal forces representing shear force and bending moment, respectively, D_i is the representative of the bending stiffness $D(x)$ on the i-th element. Assume that the finite element mesh is uniform, i.e., all elements have the same length a, eq.(1) can be simply rewritten as

$$\frac{D_i}{a^3}\boldsymbol{K}_i\boldsymbol{q}_i = \boldsymbol{F}_i \tag{2}$$

where $q_i = [w_{i1} \ a\theta_{i1} \ w_{i2} \ a\theta_{i2}]^T$, $F_i = [Q_{i1} \ M_{i1}/a \ Q_{i2} \ M_{i2}/a]^T$ and

$$K_i = \begin{bmatrix} 12 & 6 & -12 & 6 \\ 6 & 4 & -6 & 2 \\ -12 & -6 & 12 & -6 \\ 6 & 2 & -6 & 4 \end{bmatrix} \qquad (3)$$

Since K_i is a full matrix, four simultaneous equations expressed by eq.(2) are *coupled* with each other. Orthogonal transformation of variables can be applied to *decouple* eq.(2). The eigenvalue matrix and eigenvector matrices of K_i are, respectively

$$\lambda = diag[0,0,2,30] \qquad (4)$$

$$V = \begin{bmatrix} -1 & 1 & 0 & 2 \\ 1 & 0 & -1 & 1 \\ 0 & 1 & 0 & -2 \\ 1 & 0 & 1 & 1 \end{bmatrix} \qquad (5)$$

Let $q_i = V^T e_i$ and pre-multiply eq.(2) by V, we get

$$\frac{D_i}{a^3} diag[0,0,4,300]e_i = t_i \qquad (6)$$

where $t_i = VF_i$. Eq.(6) constitutes four equations. The first two equations are $t_{i1} = t_{i2} = 0$. The third and fourth equations read

$$\frac{D_i}{a^3}\begin{bmatrix} 4 & 0 \\ 0 & 300 \end{bmatrix}\begin{Bmatrix} e_{i3} \\ e_{i4} \end{Bmatrix} = \begin{Bmatrix} t_{i3} \\ t_{i4} \end{Bmatrix} \qquad (7)$$

Equation (7) consists of two uncoupled equations, which, in actuality, constitute the generalized strain-stress law in terms of generalized strain e and generalized strain t. The advantage of uncoupledness of strain-stress law in eq.(7) will be used later to obtain the explicit expression of the displacement in terms of the space-wise varying bending stiffness vector with elements D_i ($i=1,2,...,N$). Bearing in mind that $t_{i1} = t_{i2} = 0$, we have

$$F_i = V^{-1}t_i = \frac{1}{5}\begin{bmatrix} 0 & -\frac{5}{2} & 0 & \frac{5}{2} \\ 1 & \frac{1}{2} & -1 & \frac{1}{2} \end{bmatrix}^T \begin{Bmatrix} t_{i3} \\ t_{i4} \end{Bmatrix} \tag{8}$$

From definition, we deduce $e_i = V^{-T}q_i$. Its third and fourth components of e_i can be explicitly obtained as

$$\begin{Bmatrix} e_{i3} \\ e_{i4} \end{Bmatrix} = \frac{1}{5}\begin{bmatrix} 0 & -\frac{5}{2} & 0 & \frac{5}{2} \\ 1 & \frac{1}{2} & -1 & \frac{1}{2} \end{bmatrix} q_i \tag{9}$$

Eqs.(7,8,9) can respectively be viewed as the generalized constitutive law, equilibrium condition and kinematic equation of the i-th element, if t and e are considered to be generalized stresses and strains. Assembling eq.(7), eq.(8) and eq.(9) over all elements, respectively, we get the global-level constitutive law, equilibrium condition and kinematic equation as follows

$$T = S\varepsilon \tag{10}$$

$$F = QT \tag{11}$$

$$\varepsilon = Ru \tag{12}$$

where

$$u = [w_1, a\theta_1, w_2, a\theta_2, \dots, w_{N+1}, a\theta_{N+1}]^T$$

$$F = [Q_1, M_1/a, Q_2, M_2/a, \dots, Q_{N+1}, M_{N+1}/a]^T$$

$$\varepsilon = [e_{13}, e_{14}, e_{23}, e_{24}, \dots, e_{N3}, e_{N4}]^T \tag{13}$$

$$T = [t_{13}, t_{14}, t_{23}, t_{24}, \dots, t_{N3}, t_{N4}]^T$$

and

$$Q = R^T = \frac{1}{5}\begin{bmatrix} 0 & 1 & 0 & 0 & 0 & 0 & \cdots & 0 & 0 \\ -\frac{5}{2} & \frac{1}{2} & 0 & 0 & 0 & 0 & \cdots & 0 & 0 \\ 0 & -1 & 0 & 1 & 0 & 0 & \cdots & 0 & 0 \\ \frac{5}{2} & \frac{1}{2} & -\frac{5}{2} & \frac{1}{2} & 0 & 0 & \cdots & 0 & 0 \\ & & \cdots & & & \cdots & & & \\ 0 & 0 & 0 & 0 & 0 & 0 & \cdots & 0 & -1 \\ 0 & 0 & 0 & 0 & 0 & 0 & \cdots & \frac{5}{2} & \frac{1}{2} \end{bmatrix} \tag{14}$$

Moreover,

$$S = \frac{1}{a^3} diag[4D_1, 300D_1, 4D_2, 300D_2, \cdots, 4D_N, 300D_N] \tag{15}$$

Combining eqs.(10-12) results in the global finite element equilibrium equation

$$Ku = F \tag{16}$$

where $K = QSR$ is the global finite element stiffness matrix. We digress that Q is a $2(N+1)$ by $2N$ matrix of constants, and $R = Q^T$ is a $2N$ by $2(N+1)$ matrix of constants. The bending stiffness D_i ($i=1,2, \ldots ,N$) appears only in the $2N$ by $2N$ diagonal matrix S. This feature makes it possible to obtain the solution for the displacement explicitly with respect to the discretized bending stiffness of the beam.

It is worthy to remark that there is no necessity that the finite element mesh should be uniform. Analogous analysis can be carried out for the non-uniform mesh, with a more complicated transformation matrix V.

4. Imposition of Displacement Constraints

To obtain the explicit displacement vector u, one needs to invert the global stiffness matrix K. However, the matrix K is singular without the incorporation of the displacement boundary conditions or constraints; moreover, Q and R are non-square matrices.

We herein propose a way to impose the displacement boundary conditions or constraints. Assume that the beam is subjected to M displacement constraints including boundary conditions and/or intermediate supports. We first apply only two displacement constraints onto eq.(16), as a result of that the stiffness matrix K becomes non-singular, with the attendant change of the matrices Q and R into square ones. This can be achieved simply by discarding two rows in the matrix Q, which are related to the constrained nodal displacement components, together with two corresponding columns in the matrix R. With the incorporation of the first two displacement constraints, eqs.(16) reduce to

$$K_1 u_1 = F_1 \qquad (17)$$

where u_1 is obtained from u by cancelling the two constrained displacements, F_1 is obtained from F by cancelling the corresponding two forces; $K_1 = Q_1 S R_1$, in which Q_1 is obtained from Q without two rows corresponding to constrained displacements and $R_1 = Q_1^T$. All matrices Q_1, R_1, K_1 and S are $2N$ by $2N$ square matrices. The inverse matrix Z of K_1 is then

$$Z = K_1^{-1} = R_1^{-1} S^{-1} Q_1^{-1} = G^T S^{-1} G \qquad (18)$$

where the matrix $G = Q_1^{-1}$ (or $G^T = Q_1^{-T} = R_1^{-T}$) can be numerically obtained since Q_1 (or R_1) is the matrix of constants. The inverse of S is straightforwardly obtained

$$S^{-1} = a^3 diag[\frac{1}{4D_1}, \frac{1}{300D_1}, \frac{1}{4D_2}, \frac{1}{300D_2}, \cdots, \frac{1}{4D_N}, \frac{1}{300D_N}] \qquad (19)$$

For statically determinate beams, the displacement solution is obtained immediately from eq.(17)

$$u_1 = K_1^{-1} F_1 = Q_1^{-T} S^{-1} Q_1^{-1} F_1 \qquad (20)$$

For statically indeterminate beams, additional displacement constraints exist and should be incorporated into the equilibrium equation (17). Without loss of generality, suppose that u_1 is composed of two parts, namely,

$$u_1 = [u_1^{cT} \quad u_1^{uT}]^T \qquad (21)$$

where u_1^c=vector of constrained nodal displacements excluding two previously-imposed constraints, namely, $u_1^c = 0$; u_1^u=vector of unconstrained nodal displacements. Eq.(20) becomes

$$\begin{Bmatrix} u_1^c \\ u_1^u \end{Bmatrix} = \begin{bmatrix} Z_{cc} & Z_{cu} \\ Z_{uc} & Z_{uu} \end{bmatrix} \begin{Bmatrix} F_1^c \\ F_1^u \end{Bmatrix} \qquad (22)$$

where F_1^c=vector of unknown nodal forces relevant to constrained nodal displacements and F_1^c=vector of equivalent nodal forces of applied loads relevant to unconstrained nodal displacements. The matrices Z_{cc}, Z_{cu}, Z_{uc} and Z_{uu} are submatrices of Z. Solving out the unknown reactions F_1^c from the first equation of eq.(31) and applying the condition $u_1^c{=}0$, we obtain

$$F_1^c = Z_{cc}^{-1} Z_{cu} F_1^u \tag{23}$$

Substituting eq.(23) back into the second equation of eq.(22), we arrive at the solution for the unknown nodal displacement vector u_1^u

$$u_1^u = (Z_{uu} - Z_{uc} Z_{cc}^{-1} Z_{cu}) F_1^u \tag{24}$$

Rewriting $G = [G_c , G_u]$, where G_c and G_u are submatrices of G, we have

$$u_1^u = [G_u^T S^{-1} G_u - G_u^T S^{-1} G_c (G_c^T S^{-1} G_c)^{-1} G_c^T S^{-1} G_u] F_1^u \tag{25}$$

5. Mean and Correlation Functions of the Displacement

For statically determinate beams, the mean vector and variance-covariance matrix of the nodal displacement vector u_1 are, respectively,

$$E[u_1] = Q_1^{-T} E[S^{-1}] Q_1^{-1} F_1 \tag{26}$$

$$R(u_1, u_1^T) = Q_1^{-T} E[S^{-1} Q_1^{-1} F_1 F_1^T Q_1^{-T} S^{-1}] Q_1^{-1} \tag{27}$$

where $E[\cdot]$ is the expectation operator. For statically undeterminate beams, the mean vector and variance-covariance matrix of the nodal displacement vector u_1^u are, respectively,

$$E[u_1^u] = G_u^T E[S^{-1} - H] G_u F_1^u \tag{28}$$

$$R(u_1^u, (u_1^u)^T) = G_u^T E[(S^{-1} - H) G_u F_1^u (F_1^u)^T G_u^T (S^{-1} - H)] G_u \tag{29}$$

where

$$H = S^{-1} G_c (G_c^T S^{-1} G_c)^{-1} G_c^T S^{-1} \tag{30}$$

6. Example: Clamped-Hinged Beam Under Uniform Load

Consider a clamped-hinged beam subjected to a uniformly distributed deterministic load q. The beam is comprised of three equal length segments. The two side-segments have the stiffness $D_1 = D_0(1 + k\alpha)$ and the mid-segment has the stiffness $D_2 = D_0(1 + k\beta)$, where α and β are two random variables and k=constant. The normalized random variables α and

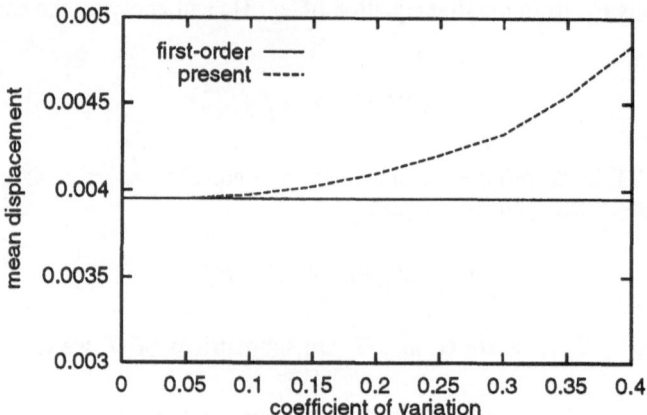

Figure 1: The mean value of mid-span displacement of clamped-hinged beam versus coefficient of variation

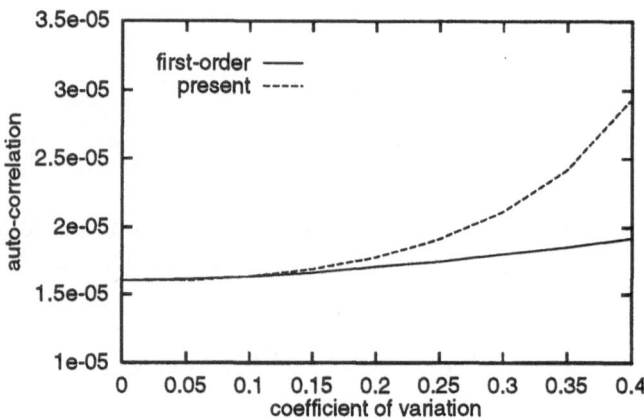

Figure 2: The auto-correlation of mid-span displacement of clamped-hinged beam versus coefficient of variation

β are assumed to possess the joint Pearson Type II distribution with the following density function (Chmielewski 1981 and Johnson 1987)

$$p_{\alpha\beta}(x,y) = \frac{1}{\pi\sqrt{1-\rho^2}} \quad , \quad for \quad x^2 - 2\rho xy + y^2 \leq 1 - \rho^2 \tag{31}$$

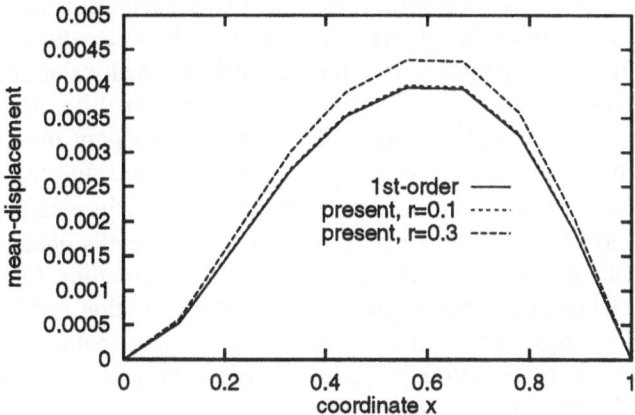

Figure 3: The mean displacement of clamped -hinged beam for different coefficients of variation r=0.1 and r=0.3 (1st-order stands for first-order perturbation based FEMSP)

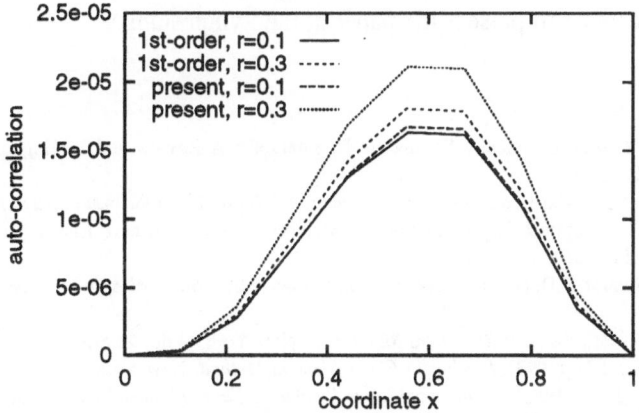

Figure 4: The auto-correlation of the displacement of clamped-hinged beam for r=0.1 and r=0.3

where ρ is the correlation coefficient between α and β and is fixed at $\rho=0.5$. The distribution in eq.(31) assures that the values of the bending stiffness are positive and bounded, in contrast to often used physically unjustifiable assumption of Gaussianity.

Fig.1 and Fig.2 portray the calculated results for the mean and auto-correlation of the displacement at the center of the beam, obtained via the new formulation, as well as by the first-order perturbation FEMSP, for various coefficients of variation $r=k/2$ of the stochastic stiffness. It is observed that the agreement between the new formulation solution and the first-order pertubation solution is good for small values of coefficient of variation. For larger values of coefficient of variation, the difference between two solutions increases. Fig.3 and Fig.4 depict the computed results of the mean value and auto-correlation of the displacement along the beam, for two different values of the coefficient of variation $r=0.1$ and $r=0.3$. For comparison, the results obtained by first-order perturbational FEMSP are also plotted. As we expect, the results obtained by the new FEMSP agree with the results obtained by first-order perturbation FEMSP for small coefficient of variation $r=0.1$. For a larger coefficient of variation $r=0.3$, the first-order perturbation based FEMSP solution underestimates the response autocorrelation by about 20%. In contrast to the perturbation analysis, the present approach can be applied for any coefficient of variation of the stochastic stiffness.

Acknoledgement

This work has been supported in part by the National Center for Earthquake Engineering Research (NCEER), through Grant 94-2306   (Professor G.Lee, Director). We thank Professor Arvid Naess for an invitation to present the lecture at the IUTAM Symposium on "Stochastic Mechanics". We express our sincere thanks to Professor H.Ugur Köylüoğlu for his kind agreement to present the paper at the Symposium.

References

Chmielewski, M.A. (1981). Elliptically Symmetric Distributions: A Review and Bibliography, *International Statistical Review*, **49**, 67-74.
Fuchs, M.B. (1991). Unimodal Beam Elements, *International Journal of Solids and Structures*, **27**, 533-545.
Fuchs, M.B. (1992). Analytical Representation of Member Forces in Linear Elastic Redundant Trusses, *International Journal of Solids and Structures*, **29**, 519-530.
Ghanem, R.G. and Spanos, P.D. (1991). *Stochastic Finite Elements: A Spectral Approach*. New York, Springer-Verlag.
Johnson, M.E. (1987). *Multivariate Statistical Simulation*. New York, Wiley & Sons.
Kleiber, M. and Hien, T.D. (1993). *Stochastic Finite Element Method*. New York, Wiley & Sons.
Nakagiri, S. and Hisada, T. (1985). *An Introduction to Stochastic Finite Element Method: Analysis of Uncertain Structures*. BaiFuKan, Tokyo, Japan (in Japanese).

STRENGTH AND SERVICEABILITY REQUIREMENTS IN SEISMIC DESIGN USING NONLINEAR SFEM

L. GAO, A. HALDAR, and N. SHOME
University of Arizona
Dept. of Civil Engineering and Engineering Mechanics,
Tucson, AZ 85721, U.S.A.

1. Introduction

The damage suffered by steel structures in a recent strong earthquake forced the profession to reevaluate issues related to the seismic design of steel structures. Moment-resisting steel frames or parts of a frame are being tested to study their behavior during an earthquake. In a recently published report, SAC (1995) pointed out that it is not possible to generalize observations from limited experimental results since experiments may not represent inplace conditions, it is not economically feasible to test commonly used structural members and connection details, and the uncertainty in the problem can not be considered in limited experiments. Mathematical models verified by experimental results are necessary for this purpose.

To quote the statements made in the Commentary of the NEHRP (1992) guidelines, "Since general damage control for economic reasons is not a goal of this document and since the state of the art is not well developed in this area, the drift limits have been established without regard to considerations such as present worth of future repairs versus additional structural costs to limit drift." The Director of the EERI stated "this quake (Northridge) could be an opportunity to correct "misperceptions" about building codes, which many people don't realize are based on life safety and not protecting structures. If [society] is not satisfied with life safety codes, then it may be time to change that."

In a typical deterministic design procedure, structural members are sized first considering strength, and then the deflection of the structure is checked. For reliability-based design, the element level reliability index is usually estimated for the strength limit state. However, the serviceability, e.g., deflection, is a structural limit state. According to NEHRP (1992), "Stress or strength limitations imposed by design level forces occasionally may provide adequate drift control. However, it is expected that the design of moment resisting frames, especially steel building frames, ... will be governed at least in part by drift consideration." The presence of PR connections will make the structure more flexible, producing more lateral drift. Since the presence of PR connections is usually neglected in both elastic and inelastic analyses, it can be

179

A. Naess and S. Krenk (eds.), IUTAM Symposium on Advances in Nonlinear Stochastic Mechanics, 179–188.
© 1996 Kluwer Academic Publishers.

concluded that for steel frames, the calculated drift is expected to be extremely unconservative.

2. A Unified Nonlinear Stochastic Finite Element Method

To evaluate the safety of nonlinear structures, a stochastic finite element method (SFEM) has been under development by the authors for a long time (Mahadevan and Haldar, 1991; Haldar and Zhou, 1992; Gao and Haldar, 1995). Due to severe page limitations, the details of the method can not be discussed here. Only some of its essential features are discussed very briefly below.

The method is based on the first-order reliability method (FORM) and necessitates the definition of a limit state function $G(x,u,s)$, where vector x denotes the set of basic random variables pertaining to a structure (e.g., loads, material properties and structural geometry), vector u denotes the set of displacements involved in the limit state function, and vector s denotes the set of load effects (except the displacement) involved in the limit state function (e.g., stresses, internal forces). The displacement u can be expressed as $u = QD$, where D is the global displacement vector and Q is a transformation matrix. In general, x, u and s are related in an algorithmic sense, e.g., a finite element code. In the SFEM developed by the authors, nonlinearities due to geometric, material, and partially restrained (PR) connections are incorporated. The algorithm evaluates the performance function deterministicaly, with the corresponding gradients at each iteration point. It converges to the most probable failure point or checking point and calculates the corresponding reliability index ß. The following iteration scheme can be used for finding the checking point:

$$y_{i+1} = \left[y_i^t \alpha_i + \frac{G(y_i)}{|\nabla G(y_i)|} \right] \alpha_i \tag{1}$$

where

$$\nabla G(y) = \left[\frac{\partial G(y)}{\partial y_1}, \ldots, \frac{\partial G(y)}{\partial y_n} \right]^t \tag{2}$$

$$\alpha_i = - \frac{\nabla G(y_i)}{|\nabla G(y_i)|} \tag{3}$$

$$\nabla G = \left[\frac{\partial G}{\partial s} J_{s,x} + \left(Q \frac{\partial G}{\partial u} + \frac{\partial G}{\partial s} J_{s,D} \right) J_{D,x} + \frac{\partial G}{\partial x} \right] J_{y,x}^{-1} \tag{4}$$

where $J_{i,j}$ are the Jacobians of transformation and y_i's are statistically independent random variables in the standard normal space. The evaluation of the quantities in Eq. 4 will depend on the problem under consideration (linear or nonlinear, 2-D or 3-D, etc.) and the performance functions used. These quantities need to be derived as described briefly below.

2.1 STRENGTH PERFORMANCE FUNCTIONS

For reliability evaluation, the strength performance function can be expressed as (AISC, 1994):

$$G(x,u,s) = 1.0 - \frac{P_u}{P_n} - \frac{8}{9}\left[\frac{M_{ux}}{M_{nx}} + \frac{M_{uy}}{M_{ny}}\right]; \quad if \quad \frac{P_u}{\phi P_n} \geq 0.2 \tag{5}$$

$$G(x,u,s) = 1.0 - \left[\frac{P_u}{2P_n} + \frac{M_{ux}}{M_{nx}} + \frac{M_{uy}}{M_{ny}}\right]; \quad if \quad \frac{P_u}{\phi P_n} < 0.2 \tag{6}$$

where ϕ is the resistance factor; P_u is the required tensile/compressive strength; P_n is the nominal tensile/compressive strength; M_{ux} and M_{uy} are the required flexural strength with respect to the x-axis and y-axis, respectively; and M_{nx} and M_{ny} are the nominal flexural strength with respect to the x-axis and y-axis, respectively. As stated in the LRFD manual (AISC, 1994), P_n can be evaluated as $P_n = AF_{cr}$ (compression) or $P_n = AF_y$ (tension), $M_{nx} = Z_x F_y$ and $M_{ny} = Z_y F_y$. They cannot be discussed further due to lack of space.

One can derive $\partial G/\partial x$, $\partial G/\partial u$ and $\partial G/\partial s$ from Eqs. 5 and 6. Since both G functions do not contain any displacement component explicitly, $\partial G/\partial u = 0$. In order to calculate $\partial G/\partial x$, the basic random variables are defined in this study as the Young's modulus E, area A, yield stress F_y, plastic modulus Z_x and Z_y, and the moments of inertia of a cross section J, I_x and I_y along with the external load F. Therefore, the following expression can be obtained:

$$\frac{\partial G}{\partial x} = \left[\frac{\partial G}{\partial E} \frac{\partial G}{\partial A} \frac{\partial G}{\partial J} \frac{\partial G}{\partial I_x} \frac{\partial G}{\partial I_y} \frac{\partial G}{\partial Z_x} \frac{\partial G}{\partial Z_y} \frac{\partial G}{\partial F_y}\right] \tag{7}$$

The following expression is used to calculate the partial differential $\partial G/\partial s$:

$$\frac{\partial G}{\partial s} = \left[\frac{\partial G}{\partial P_u} \frac{\partial G}{\partial M_{ux}} \frac{\partial G}{\partial M_{uy}}\right] \tag{8}$$

Considering either Eq. 5 or 6, the corresponding terms in Eqs. 7 and 8 can be calculated.

2.2 SERVICEABILITY PERFORMANCE FUNCTION

For the serviceability criterion, the following performance function is used:

$$G(x, u, s) = 1.0 - \delta / \delta_{limit} \qquad (9)$$

where δ is the calculated displacement component at a particular point in the frame and δ_{limit} is the allowable maximum value of the displacement component. From Eq. 9, one obtains:

$$\frac{\partial G}{\partial x} = \frac{\partial G}{\partial s} = 0 \quad and \quad \frac{\partial G}{\partial u} = \left[\frac{\partial G}{\partial \delta} \; 0 \right] \qquad (10)$$

where $\partial G / \partial \delta = -1/\delta_{limit}$. The actual δ_{limit} would depend on the structure under consideration. In this study, the vertical deflection at the midspan of the beam under live load is considered to be less than or equal to 1/360 of the span of the beam. For the side-sway, lateral movement at the top of the frame is considered to be less than or equal to 1/400 of the height of the frame.

2.3 EVALUATION OF JACOBIANS AND ADJOINT VARIABLE METHOD

To evaluate the gradient ∇G, the four Jacobians in Eq. 4 need to be computed properly. Because of the triangular nature of the transformation, $J_{y,x}$ and its inverse are easy to compute. Since s is not an explicit function of the basic random variables x, $J_{s,x} = 0$. $J_{s,D}$ and $J_{D,x}$, however, are not easy to compute since s, D and x are implicit functions of each other. The adjoint variable method (Gao and Haldar, 1995) is used in this study to compute the product of the second term in Eq. 4 directly rather than evaluating its constituent parts. It can be shown that:

$$\frac{\partial G}{\partial D} J_{D,x} = \left[Q \frac{\partial G}{\partial u} + \frac{\partial G}{\partial s} J_{s,D} \right] J_{D,x} \qquad (11)$$

The equilibrium equation of a structural system can be expressed as:

$$\psi (x, D) = 0 \qquad (12)$$

For nonlinear problems, it becomes

$$\psi (x, D) = K^{s} (x, D) D - F(x) = 0 \qquad (13)$$

where $K^{s}(x, D)$ is the secant stiffness matrix, which is a function of basic variables x and displacement vector D, and $F(x)$ is the same as $F^{(n)}$ i.e., the external load vector at the nth iteration. Differentiating Eq. 12 with respect to x, one obtains:

$$\left[\frac{\partial\psi}{\partial D}\right] J_{D,x} = -\left[\frac{\partial\psi}{\partial x}\right]$$ (14)

Substituting Eq. 13 into Eq. 14 and rearranging, one obtains:

$$[K_s + \bar{K}]\, J_{D,x} = \bar{R}$$ (15)

where

$$\bar{K} = \frac{\partial[K^s(x,D)\check{D}]}{\partial D}$$ (16)

and

$$\bar{R} \equiv -\frac{\partial\psi}{\partial x} = \frac{\partial F(x)}{\partial x} - \frac{\partial[K^s(x,\check{D})\check{D}]}{\partial x}$$ (17)

where " $\sim$ " over a variable indicates that it is to be held constant during partial differentiations.

A linear iterative strategy is used for solving nonlinear structural problems. In order to solve Eq. 11, an adjoint vector λ is introduced. Using Eq. 15 and some mathematical manipulations, it can be shown that:

$$\lambda^t K^{(n)} J_{D,x} = \left[Q\frac{\partial G}{\partial u} + \frac{\partial G}{\partial s} J_{s,D}\right] J_{D,x} = \lambda^t\left[\frac{\partial F}{\partial x} - \frac{\partial R^{n-1}}{\partial x}\right]$$ (18)

where $\partial F/\partial x$ can be calculated since explicit dependence of F on basic variables is known and the superscript n represents the total number of iterations.

Generally, when the strength performance functions are considered, the internal force vector σ is the only contribution of the load effect s and can be expressed as $s = A\sigma$, where A is the transformation matrix with constant elements. Thus, $J_{s,D}$ in Eq. 18 can be obtained as:

$$J_{s,D} = \frac{\partial s}{\partial D} = A\frac{\partial\sigma}{\partial D} = A\left[\frac{\partial\sigma}{\partial d}\ 0\right]$$ (19)

where d is the nodal displacement vector in the global coordinate for the element. For a three-dimensional problem it can be defined as:

$$d = [{}^1u_1 \quad {}^1u_2 \quad {}^1u_3 \quad {}^1\theta_1 \quad {}^1\theta_2 \quad {}^1\theta_3 \quad {}^2u_1 \quad {}^2u_2 \quad {}^2u_3 \quad {}^2\theta_1 \quad {}^2\theta_2 \quad {}^2\theta_3]^t$$ (20)

and

$$\sigma = [n \quad m_1 \quad {}^1m_2 \quad {}^2m_2 \quad {}^1m_3 \quad {}^2m_3]^t \tag{21}$$

Thus, the derivative of the internal forces with respect to the displacements becomes:

$$\frac{\partial \sigma}{\partial d} = \left[\frac{\partial n}{\partial d} \quad \frac{\partial m_1}{\partial d} \quad \frac{\partial^1 m_2}{\partial d} \quad \frac{\partial^2 m_2}{\partial d} \quad \frac{\partial^1 m_3}{\partial d} \quad \frac{\partial^2 m_3}{\partial d} \right]^t \tag{22}$$

To calculate $\partial R^{(n-1)}/\partial x$ in Eq. 18, it can be shown that (Gao, 1994):

$$R = -A_{ado}^t \, R_{A\sigma} + R_{do} \tag{23}$$

where $R_{A\sigma}$ can be expressed as:

$$R_{A\sigma} = \left[n - \frac{EAH}{l} \quad m_1 - \frac{GJ}{l}({}^2\theta_1^* - {}^1\theta_1^*) \quad {}^1m_2 + \frac{2EI_2}{l}(2{}^1\theta_2^* + {}^2\theta_2^*)\,{}^2m_2 - \frac{2EI_2}{l} \right.$$
$$\left. ({}^1\theta_2^* + 2{}^2\theta_2^*)\,{}^1m_3 + \frac{2EI_3}{l}(2{}^1\theta_3^* + {}^2\theta_3^*) \quad {}^2m_3 - \frac{2EI_3}{l}({}^1\theta_3^* + 2{}^2\theta_3^*) \right]^t \tag{24}$$

Since R_{do} and A_{ado}^t are not functions of basic variables, the derivative of R with respect to x can be expressed as:

$$\frac{\partial R}{\partial x}\Big|_{D,\sigma} = -A_{\sigma do}^t \frac{\partial R_{A\sigma}}{\partial x} \tag{25}$$

As mentioned before, the evaluation of Eq. 25 depends on how the basic variables are selected. It will become a non-zero matrix if the Young's modulus E, area A, and the moments of inertia of a cross section J, I_2, I_3 are considered to be basic random variables. In this case, $\partial R_{A\sigma}/\partial x$ for a 3-D beam-column element can be shown to be:

$$\frac{\partial R_{A\sigma}}{\partial x} = \left[\frac{\partial R_{A\sigma}}{\partial E} \quad \frac{\partial R_{A\sigma}}{\partial A} \quad \frac{\partial R_{A\sigma}}{\partial J} \quad \frac{\partial R_{A\sigma}}{\partial I_2} \quad \frac{\partial R_{A\sigma}}{\partial I_3} \quad 0 \right] \tag{26}$$

2.4 ELASTO-PLASTIC NONLINEAR CASE

The plastic hinge model is used in this study. When the combined axial force and bending moments of an elastic-perfectly-plastic material model satisfy a prescribed yield function at a node of an element, a plastic hinge is assumed to occur instantaneously at that location. Therefore, it can be shown that (Gao, 1994):

$$\frac{\partial R}{\partial x} = 0 \qquad\qquad (27)$$

2.5 PARTIALLY RESTRAINED CONNECTIONS

The AISC LRFD (1994) specifications explicitly recognize two types of steel construction: fully restrained (FR) and partially restrained (PR) or semi-rigid. However, despite these classifications, almost all steel connections used in practice are in fact essentially PR connections with different rigidities. For seismic response analysis, proper consideration of the rigidity of connections is essential.

The flexible behavior of a connection is generally described by the relationship between the moment M transmitted by the connection and the relative angle of rotation between connecting members Θ. Among many alternatives, Richard's four parameter M-Θ curves are used to represent partially restrained behavior of a connection. The Richard model can be expressed as:

$$M = \frac{(k-k_p)\,\theta}{\left[1+\left|\frac{(k-k_p)\theta}{M_o}\right|^N\right]^{\frac{1}{N}}} + k_p\theta \qquad\qquad (28)$$

where M is the connection moment; θ is the connection rotation; k is the initial or elastic stiffness; k_p is the plastic stiffness; M_o is the reference moment and N is the curve shape parameter. For numerical analyses, an ordinary beam-column element is used to represent a flexible connection. However, the element's stiffness needs to be continually updated since the stiffness and the current Young's modulus of an element are functions of the relative rotation θ.

2.6 UNCERTAINTY IN FLEXIBLE CONNECTIONS

A considerable amount of uncertainty is expected in M-θ curves. In this study, a stochastic computational model is presented to account for the scatter in the connection behavior by considering the four parameters in the Richard model to be random variables.

For 3-D frames, the out-of-plane behavior of the connections needs to be considered properly. Therefore, both major- and minor-axis rotational flexibility of connections are considered in the present work. Denoting k, k_p, M_o and N as the four parameters representing the major-axis rotational flexibility of a connection element, and k', k'_p, M'_o and N' as the four parameters representing the minor-axis rotational flexibility of the connection element, the stochastic information on both major and minor axes can be included in Eq. 26 as follows:

$$\frac{\partial R_{A\sigma}}{\partial x} = \left[\frac{\partial R_{A\sigma}}{\partial k} \quad \frac{\partial R_{A\sigma}}{\partial k_p} \quad \frac{\partial R_{A\sigma}}{\partial M_o} \quad \frac{\partial R_{A\sigma}}{\partial N} \quad \frac{\partial R_{A\sigma}}{\partial k'} \quad \frac{\partial R_{A\sigma}}{\partial k'_p} \quad \frac{\partial R_{A\sigma}}{\partial M'_o} \quad \frac{\partial R_{A\sigma}}{\partial N'} \quad 0 \right] \quad (29)$$

The components in the major-axis direction in Eq. 29 can be shown to be

$$\frac{\partial R_{A\sigma}}{\partial \xi_i} = 2\frac{\partial K(\theta)}{\partial \xi_i}[0 \quad 0 \quad (2^1\theta_2^* + {}^2\theta_2^*) - ({}^1\theta_2^* + 2^2\theta_2^*) \quad 0 \quad 0]^r \quad (30)$$

where $\xi = k$, k_p, M_o and N. The corresponding components in the minor-axis direction are:

$$\frac{\partial R_{A\sigma}}{\partial \xi'_i} = 2\frac{\partial K(\theta)}{\partial \xi'_i} [0 \quad 0 \quad 0 \quad 0 \quad (2^1\theta_2^* + {}^2\theta_2^*) - ({}^1\theta_2^* + 2^2\theta_2^*)]^r \quad (31)$$

where $\xi'_i = k'$, k'_p, M'_o, and N', and

$$\frac{\partial K(\theta)}{\partial k} = \frac{1 - Na^N}{[1+a^N]^{(2N+1)/N}} \quad (32)$$

$$\frac{\partial K(\theta)}{\partial k_p} = \frac{-(N+2)a^N - 1}{[1+a^N]^{(2N+1)/N}} + 1 \quad (33)$$

$$\frac{\partial K(\theta)}{\partial M_o} = \frac{(N+1)a^N}{[1+a^N]^{(2N+1)/N}} \frac{(k-k_p)}{M_o} \quad (34)$$

$$\frac{\partial K(\theta)}{\partial N} = \left[-\frac{a^N(1+N)\log a}{N(1+a^N)^{(1+2N)/N}} + \frac{-N+(1+N)\log(1+a^N)}{N^2(1+a^N)^{(1+N)/N}} \right](k-k_p) \quad (35)$$

where

$$a = \left| \frac{(k-k_p)\theta}{M_o} \right| \quad (36)$$

Once Eq. 29 is evaluated, the rest of the steps are same as those for an ordinary beam-column element.

All the quantities required for the computation of $\nabla G(y)$ in Eq. 4 for a 3-D frame are now available in a simple explicit form considering geometric nonlinearity, elasto-plastic material behavior, and partially restrained connections.

3. Example

A two-story three-dimensional space frame structure shown in Fig. 1 is considered to show the application potential of the proposed method. The geometric dimensions of the frame and the sizes of the beams and columns are shown in Fig. 1. The probabilistic descriptions of basic variables required for the analysis of the structure are given in Table 1. M-Θ curves for the major and minor axes are assumed to be the same for all connections. They are shown in Fig. 1. The probabilistic descriptions of the four parameters of the Richard model are given in Table 2.

Assuming all the members are connected rigidly, the reliability indexes for the strength and serviceability limit states, discussed earlier, are calculated using the proposed SFEM. Reliability indexes for beam 1 and column 2, shown in Fig. 1, are calculated and shown in Table 3. Reliability indexes for the drift at the top of column at location A and vertical deflection at the mid-span of the top floor beam at location B are also given in Table 3.

The connections are then assumed to be PR-type, as shown in Fig. 1, and the corresponding reliabilities for the strength and serviceability limit states are evaluated; the results are summarized in Table 3 for comparison. The implications of considering PR connections can not be presented here due to lack of space, but will be discussed in detail during the presentation.

4. Acknowledgment

This paper is based upon work partly supported by the National Science Foundation under Grant No. MSM-8896267. Any opinions, findings, conclusions, or recommendations expressed in this publication are those of the authors and do not necessarily reflect the views of the sponsor.

5. References

American Institute of Steel Construction (1994) *Manual of Steel Construction: Load and Resistance Factor Design.*

Gao, L. (1994) *Stochastic Finite Element Method for the Reliability Analysis of Nonlinear Frames with PR Connections,* University of Arizona, Ph.D thesis.

Gao, L. and Haldar, A. (1995) Safety evaluation of frames with PR connections, *J. Structural Engineering, ASCE,* 121(7), 1101-1109.

Haldar, A. and Zhou, Y. (1992) Reliability of geometrically nonlinear PR frames, *J. Engineering Mechanics, ASCE,* 118(10), 2148-2155.

Mahadevan, S. and Haldar, A. (1991) Stochastic FEM-based validation of LRFD, *J. Structural Engineering, ASCE,* 117(5), 1393-1412.

NEHRP Recommended Provisions for the Development of Seismic Regulations for New Buildings, FEMA 223, 1992.

SAC (1995) *Steel Moment Frame Advisory No. 3,* Report No. SAC-95-01.

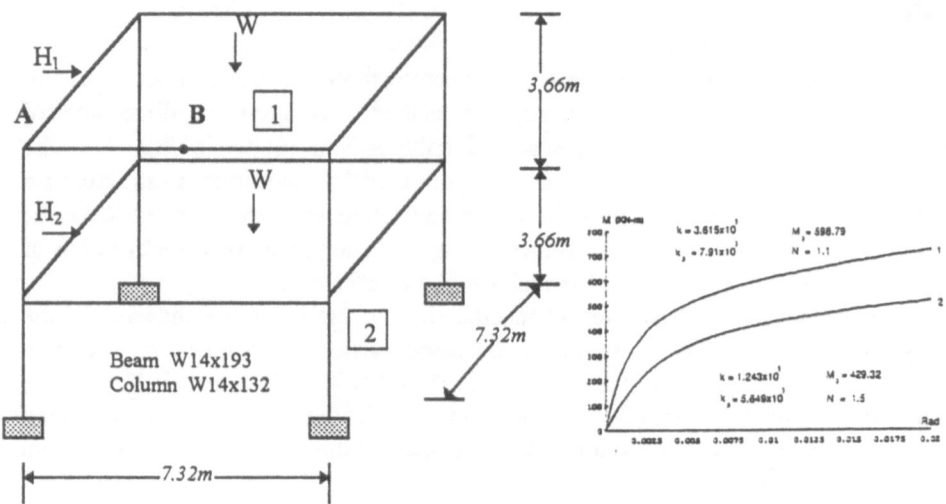

Figure 1. Geometry of the 3-D structure and M-θ curve of the connection

TABLE 1   Description of Basic Random Variables

Random	Beam			Column		
Variable	Mean	COV	Dist.	Mean	COV	Dist.
$A(\times10^2 \text{ mm}^2)$	36.64	0.05	Ln	25.03	0.05	Ln
$J(\times10^6 \text{ mm}^4)$	14.480	0.05	Ln	5.120	0.05	Ln
$I_2(\times10^8 \text{ mm}^4)$	9.989	0.05	Ln	6.368	0.05	Ln
$I_1(\times10^8 \text{ mm}^4)$	3.875	0.05	Ln	2.281	0.05	Ln
$Z_X(\times10^6 \text{ mm}^3)$	5.817	0.05	Ln	3.835	0.05	Ln
$Z_Y(\times10^6 \text{ mm}^3)$	2.950	0.05	Ln	1.852	0.05	Ln
$F_Y(\text{MPa})$	260.6	0.10	Ln	260.6	0.10	Ln
$E(\text{MPa})$	2×10^5	0.06	Ln	2×10^5	0.06	Ln
$W(\text{kN/m}^2)$	6.225	0.20	Ln			
H_1 (kN)	81.398	0.30	Type 1			
H_2 (kN)	40.477	0.30	Type 1			

TABLE 2   Description of Stocastic Parameter of Connection Element

Random	Major Axis			Minor Axis		
Variable	Mean	COV	Dist.	Mean	COV	Dist.
$K(\times10^5 \text{ KN-m/rad})$	3.616	0.15	Normal	1.243	0.15	Normal
$K_P(\times10^3 \text{ KN-m/rad})$	7.910	0.15	Normal	5.650	0.15	Normal
$M_O(\text{KN-m})$	598.9	0.15	Normal	429.4	0.15	Normal
N	1.1	0.05	Normal	1.5	0.05	Normal

TABLE 3   Reliability Indexes of the Frame

Limit State	Location	β(FR)	β(PR)	Limit State	Location	β(FR)	β(PR)
Strength	Beam[1]	4.26	3.78	Serviceability	Drift[A]	5.84	4.15
	Column[2]	3.35	3.62		Deflection[B]	6.29	5.54

PARAMETRIC MODELS AND STOCHASTIC INTEGRALS

M. GRIGORIU
Cornell University
Ithaca, NY 14853, USA

1. Introduction

Solutions of linear and nonlinear stochastic differential equations describing the response of dynamic systems with random input can be expressed as stochastic integrals involving the input and the system state vector. The stochastic integrals can be defined in the Itô or the Stratonovich sense [16]. These integrals become ordinary Stieltjes integrals in some cases if the input can be described by parametric models. Parametric models can also be used efficiently in Monte Carlo simulation studies [6, 14].

The objectives of this study are to (1) define and outline properties of several Gaussian and non-Gaussian parametric models with continuous and discrete time and (2) develop methods for finding the response of non-linear dynamic systems to Lévy white noise. The Lévy white noise can be interpreted as the formal derivative of the α-stable Lévy motion process, $0 < \alpha \leq 2$, and includes the Brownian motion as a special case ($\alpha = 2$) [6]. The Itô calculus cannot be extended directly to solve nonlinear systems with Lévy white noise because this noise has no moments of order two and higher for $0 < \alpha < 2$. However, Itô-type stochastic integrals can be defined and used for solution.

2. Parametric models

Consider a real-valued random process $\{X(t),\ t \geq 0\}$ and let $\{Y_n(t),\ t \geq 0\}$ be a parametric model of $\{X(t),\ t \geq 0\}$ depending on n random variables and approaching $\{X(t),\ t \geq 0\}$ in some sense as $n \to \infty$. Most available parametric models are for Gaussian processes and can be used to specify a class of non-Gaussian processes, defined by memoryless transformations of Gaussian processes and referred to as translation processes [6].

189

A. Naess and S. Krenk (eds.), IUTAM Symposium on Advances in Nonlinear Stochastic Mechanics, 189–212.
© *1996 Kluwer Academic Publishers.*

The parametric models can be used efficiently to generate realizations of random processes because they depend on a finite or an at most countable number of random variables. These models can also be applied to determine the response of physical systems to random input.

2.1. CONTINUOUS TIME GAUSSIAN PROCESSES

Three parametric models are defined and illustrated for stationary and nonstationary Gaussian processes.

2.1.1. *Spectral representation method*
Suppose that $\{X(t),\ t \geq 0\}$ is a stationary Gaussian processes with mean zero, covariance function $c(\tau) = E\,X(t)X(t + \tau)$, and one-sided spectral density $g(\omega)$. Let

$$Y_n(t) = \sum_{r=1}^{n}(A_r \cos \omega_r t + B_r \sin \omega_r t) \tag{1}$$

be a parametric representation of $\{X(t),\ t \geq 0\}$, in which (α_{r-1}, α_r), $r = 1, 2, \ldots, n$, $\alpha_0 = 0$, $\alpha_n = \omega^*$, is a partition of the frequency band $(0, \omega^*)$, $0 < \omega^* < \infty$, of $\{X(t),\ t \geq 0\}$ and the frequencies $\{\omega_r\}$, $r = 1, 2, \ldots, n$ are the midpoints of these intervals. If the frequency band of $\{X(t),\ t \geq 0\}$ is not finite, it is common to select a cutoff frequency $\omega^* < \infty$ such that most of the power of the process is in $(0, \omega^*)$. This approximation is satisfactory in applications because physical signals do not have harmonics of infinite frequency. The coefficients $\{A_r, B_r\}$, $r = 1, \ldots, n$, are zero-mean Gaussian variables with covariances

$$E A_r A_p = E B_r B_p = \delta_{rp} \int_{\alpha_{r-1}}^{\alpha_r} g(\omega)\,d\omega \simeq \delta_{rp}\sigma_r^2 \tag{2}$$

for $r, p = 1, \ldots, q$, where $\Delta\omega_r = \alpha_r - \alpha_{r-1}$.

The model of Eq. 1 is a zero-mean Gaussian process for any partition of the frequency range, has covariance function

$$c_n(\tau) = E\,Y_n(t)Y_n(t + \tau) = \sum_{r=1}^{n} \sigma_r^2 \cos \omega_k \tau \tag{3}$$

approaching $c(\tau)$ as $n \to \infty$ and $\max_{r=1,2,\ldots,n} \Delta\omega_r \to 0$, and satisfies some asymptotic ergodic properties [4].

2.1.2. *Sampling theorem method*
Let $\{X(t),\ t \geq 0\}$ be a zero-mean stationary Gaussian process with covariance function $c(\tau) = E\,X(t)X(t + \tau)$ and frequency band $(0, \omega^*)$, $\omega^* < \infty$.

Consider the parametric model

$$Y_{m^*}(t) = \sum_{u=m_t-m^*}^{m_t+m^*+1} X_u \, \alpha_u(t;T), \quad t \in [m_t,(m_t+1)T] \tag{4}$$

in which $T = \pi/\omega^*$, m^* is a positive integer defining the window size, m_t denotes the largest integer smaller that t/T, and

$$\alpha_u(t;T) = \frac{\sin[\pi(t-uT)/T]}{\pi(t-uT)/T} . \tag{5}$$

The model is a zero-mean Gaussian process, has a covariance function $c_{m^*}(t,t+\tau) = E Y_{m^*}(t) Y_{m^*}(t+\tau)$ approaching $c(\tau)$ as $m^* \to \infty$, can be extended to vector and/or nonstationary Gaussian processes, and can be used to develop efficient algorithms for generating samples of Gaussian processes [5, 6].

Let $X(t)$ be a zero-mean stationary Gaussian process with spectral density $s(\omega) = 1/(2\pi)$ for $\omega \in (-\pi,\pi)$ and zero otherwise. Figures 1(a), (b),

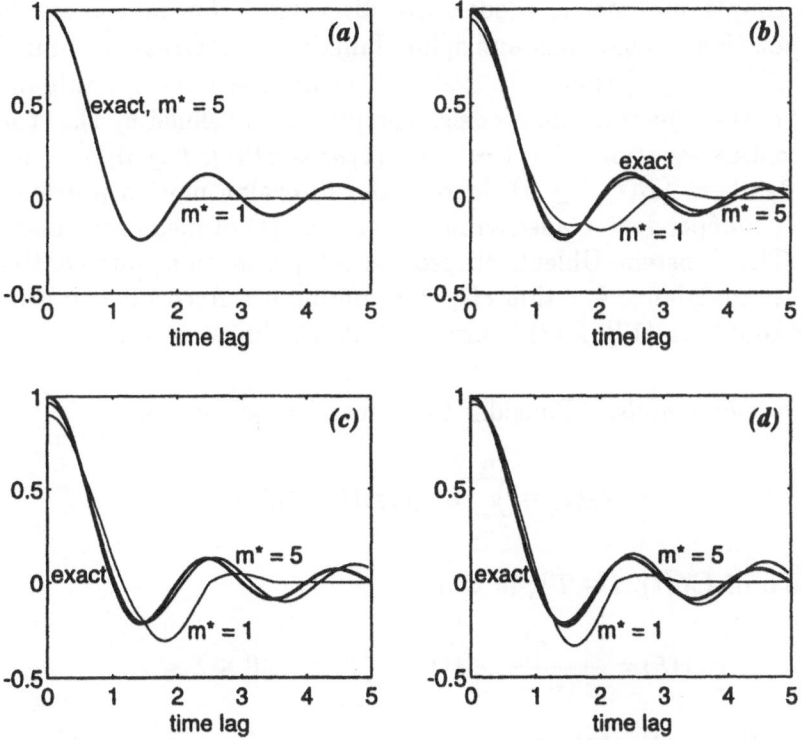

Figure 1. Covariance functions $c(\tau)$ and $c_{m^*}(t,t+\tau)$ of a zero-mean band-limited white noise process $X(t)$ and of model $Y_{m^*}(t)$.

(c), and (d) show the dependence of $c_{m^*}(t, t + \tau)$ on the time lag τ for $t/T = l, l + 1/4, l + 1/2$, and $l + 3/4$, where l is an arbitrary integer. These covariance functions are not stationary but approach $c(\tau)$ rapidly as m^* increases. Therefore, $Y_{m^*}(t)$ becomes stationary and provides an accurate approximation of $X(t)$ with increasing m^*.

2.1.3. *Random polynomial method*

Consider a zero-mean real-valued Gaussian process $X(t)$ with covariance function $c(t, s) = E\, X(t)X(s)$. The domain of definition of the process can be the real line or intervals T of this line bounded to the left, right, or both ends. Let

$$Y_n(t) = \sum_{k \in I_n} X_k\, \varphi_k(t), \quad t \in T, \tag{6}$$

be a parametric representation of $\{X(t),\, t \in T\}$, in which I_n is an index set with a finite or countable number of elements, $\{X_k\}$ are random variables depending on $\{X(t),\, t \in T\}$, and $\{\varphi_k(t)\}$ represent specified deterministic functions. If $X_n(t)$ approximates $X(t)$ satisfactorily and I_n is finite, the parametric representation of Eq. 10 can be applied to generate realizations of $\{X(t),\, t \in T\}$.

Two simple parametric models are discussed. The models are based on interpolation polynomials and spline functions. An Ornstein–Uhlenbeck process $\{X(t),\, t \geq 0\}$ is used to evaluate the accuracy and the rate of convergence of these parametric models. The process is defined by the stochastic differential equation $dX(t) = -\lambda\, X(t)\, dt + dB(t),\ t \geq 0,\ \lambda > 0$, and $X(0) = 0$, where $\{B(t), t \geq 0\}$ denotes the Brownian motion process with stationary independent Gaussian increments $dB(t)$ of mean zero and variance dt. The Ornstein–Uhlenbeck process $X(t)$ is mean-square continuous because its covariance function $c(t, s)$ is continuous. According to the Kolmogorov condition [17], $X(t)$ is also continuous almost surely.

Bernstein polynomials.   Consider the Bernstein polynomial

$$Y_n(t) = \sum_{k=0}^{n} X_k\, p_{n,k}((t - a)/\tau) \tag{7}$$

of degree n of $\{X(t),\, t \in T\}$, in which

$$p_{n,k}(\xi) = \frac{n!}{k!(n - k)!}\, \xi^k (1 - \xi)^{n-k}, \quad 0 \leq \xi \leq 1, \tag{8}$$

$\tau = b - a$ is the length of the time interval T, $T = \tau/n$, and $X_k = X(a + kT)$, $k = 0, 1, \ldots, n$. The model depends on the values of the process at $n + 1$ times, referred to as nodal points. It can be shown that the covariance

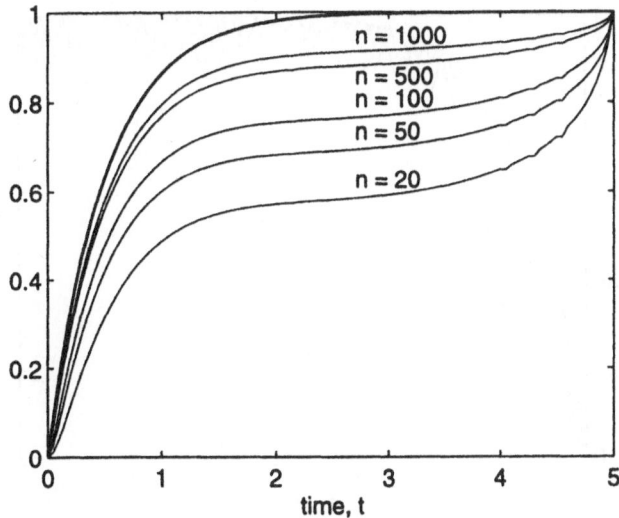

Figure 2.   Exact and approximate variances for Bernstein polynomials.

function of $Y_n(t)$ approaches $c(t, s)$ as $n \to \infty$. Because $X(t)$ is a Gaussian process, $Y_n(t)$ is Gaussian for any value of n and becomes a version of $X(t)$ asymptotically as n increases.

The random Bernstein polynomials are attractive because their defini-tion does not require any calculations. A drawback of these polynomials is the slow rate at which $Y_n(t)$ approaches $X(t)$ as n increases. Figure 3 shows the variation in time of the exact and approximate variances of the Ornstein–Uhlenbeck process $X(t)$ with $\lambda = 1$, $a = 0$, and $\tau = b = 5$. The maximum of the absolute value of the difference between the covari-ance functions of $X(t)$ and $Y_n(t)$ on $[0, \tau] \times [0, \tau]$ is 0.4122, 0.3007, 0.2297, 0.1149, and 0.0838 for $n = 20$, 50, 100, 500, and 1000, respectively.

Spline functions.   Spline functions of various orders can be used to approx-imate random processes. Because the Ornstein–Uhlenbeck $X(t)$ process is continuous but not differentiable, linear spline functions are considered. The approximation of $X(t)$ based on linear spline functions is

$$Y_n(t) = (1 - \xi(t)) X_{n_t} + \xi(t) X_{n_t+1} \tag{9}$$

for $t \in [a + n_t T, a + (n_t + 1)T]$, where n_t is the largest integer smaller than $(t - a)/T$ and $\xi(t) = (t - a - n_t T)/T$. Figure 4 shows the variation in time of the variances of $X(t)$ and $Y_n(t)$. The variance of $Y_n(t)$ converges to $c(t, t) = E X(t)^2$ much faster than the variance of the approximation of $X(t)$ based on Bernstein polynomials. The maximum of the absolute value of the difference between the covariance functions of $X(t)$ with $\lambda = 1$ and

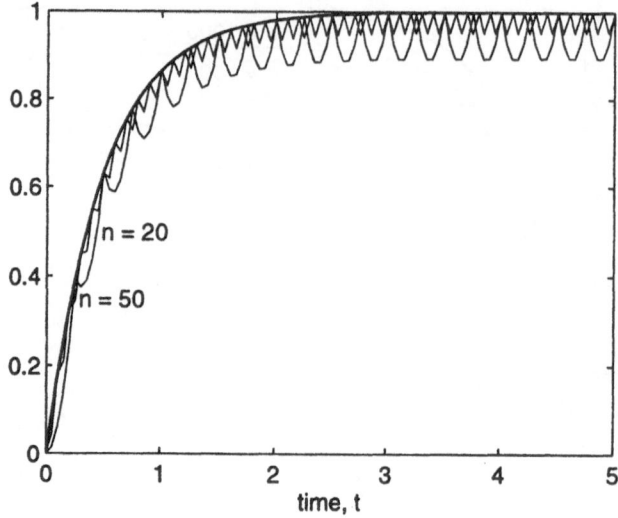

Figure 3.   Exact and approximate variances for linear spline functions.

$Y_n(t)$ on $[0, \tau] \times [0, \tau]$, $\tau = 5$, is 0.1183, 0.0498, and 1.3323×10^{-15} for $n = 20$, 50, and 100, respectively.

The random spline functions $Y_n(t)$ can be used to approximate $X(t)$ because (1) the covariance function $E\,Y_n(t)Y_n(s)$ of the random spline function $Y_n(t)$ approaches the covariance function $c(t, s) = E\,X(t)X(s)$ with increasing values of n, (2) $Y_n(t)$ converges in the mean-square sense to $X(t)$ as $n \to \infty$ at any time $t \in T$, and (3) $Y_n(t)$ converges almost surely to $X(t)$ as $n \to \infty$ at any time $t \in T$ [6].

2.2. CONTINUOUS TIME α-STABLE PROCESSES

Let X be a random variable following an α-stable distribution with scale $\sigma > 0$, location $\mu \in \Re$, and skewness parameter $\beta \in \Re$. The class of these variables, denoted by $S_\alpha(\sigma, \beta, \mu)$, is defined for values of α in the range $(0; 2]$ and has the characteristic function

$$\varphi(u) = E \exp iuX = \exp\{i\mu u - \sigma^\alpha |u|^\alpha [1 + i\beta \, \text{sign}(u)\, \omega(|u|, \alpha)]\} \quad (10)$$

where

$$\omega(|u|, \alpha) = \begin{cases} -\tan(\pi\alpha/2), & \alpha \neq 1 \\ (2/\pi)\log|u|, & \alpha = 1. \end{cases} \quad (11)$$

An α-stable distribution with $\beta = 0$ is symmetric about μ. Characteristic functions of α-stable random variables with various parameters are shown in Fig. 5. If the density of these variables is symmetric about the origin, that is, $\beta = 0$ and $\mu = 0$, their characteristic functions are real-valued.

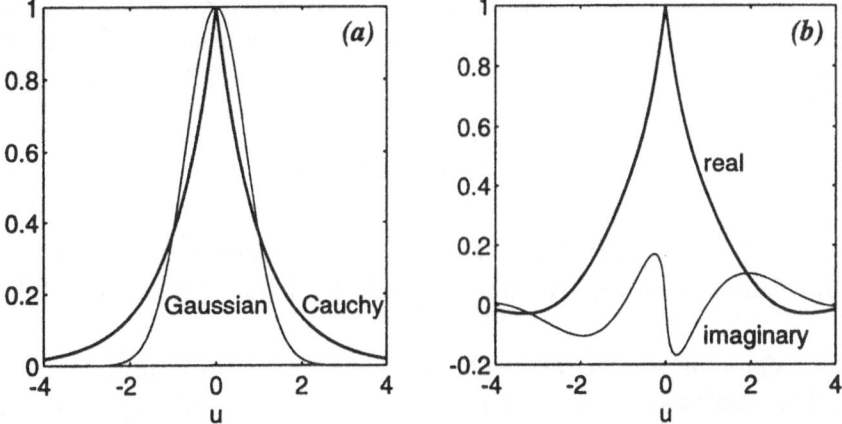

Figure 4.   Characteristic function of α-stable random variables: (a) the Gaussian $S_2(1,0,0)$ and the Cauchy $S_1(1,0,0)$ variables and (b) the skewed variable $S_1(1,-1,0)$.

Such characteristic functions are shown in Fig. $5(a)$ and correspond to the α-stable variables $S_2(1,0,0)$ and $S_1(1,0,0)$ or the Gaussian and Cauchy variables, respectively. The complex-valued characteristic function shown in Fig. $5(b)$ is for $S_1(1,-1,0)$.

The α-stable variables are closed to linear transformations, a property resembling the behavior of Gaussian variables, and have no moments $E|X|^p$ for $p \geq \alpha$. Samples of these variables can be generated simply [6, 13].

α-Stable processes.   There are several ways in which an α-stable process $\{X(t),\ t \in T\}$ can be defined [6, 13]. The integral representation of such a process is

$$\{X(t),\ t \in T\} \overset{d}{=} \left\{ \int_S f(t,s)\, M(ds),\ t \in T \right\} \tag{12}$$

in which $\alpha \in (0,2]$, $(S, \mathcal{S}, m)$ denotes a σ-finite measure space, M is an α-stable random measure with control measure m on S and skewness intensity β, and $\{f(t,\cdot),\ t \in T\}$ denotes a family of measurable functions on S satisfying the condition $f(t,\cdot) \in L^\alpha(m)$, or $\int_S |f(t,s)|^\alpha\, m(ds) < \infty$, and also $f(t,\cdot)\,\beta(\cdot)\log|f(t,\cdot)| \in L'(m)$ for $\alpha = 1$, for every $t \in T$. It can be shown that practically all α-stable processes have such an integral representation [13]. The random variable $X(t)$ is α-stable with parameters $S_\alpha(\sigma(t), \beta(t), 0)$, in which

$$\sigma(t) = \left[\int_S |f(t,s)|^\alpha\, m(ds) \right]^{1/\alpha}$$

$$\beta(t) = \sigma(t)^{-\alpha} \int_S |f(t,s)|^\alpha\, \beta(s)\, \text{sign}(f(t,s))\, m(ds). \tag{13}$$

The integral representation of Eq. 16 resembles the spectral representation of wide sense stationary processes and can be used to develop parametric representations of $\{X(t),\ t \in T\}$. For example, $X(t)$ can be approximated by

$$Y_n(t) = \sum_{k=1}^{n} f(t, s_k)\, M(S_k) \qquad (14)$$

in which $\{S_k\}$, is a partition of S and s_k interior points of S_k, $k = 1, \ldots, n$. The approximation of Eq. 18 approaches $X(t)$ of Eq. 16 as $n \to \infty$ [6].

α-Stable Lévy motion process.   The α-stable Lévy motion process $\{L_\alpha(t), t \geq 0\}$ has stationary independent increments $L_\alpha(t) - L_\alpha(s)$, $t > s$, following the α-stable distribution $S_\alpha((t-s)^{1/\alpha}, 0, 0)$ and starts at zero. The characteristic function of an arbitrary increment $L_\alpha(t) - L_\alpha(s)$, $t > s$, of $\{L_\alpha(t), t \geq 0\}$ is

$$E \exp iu[L_\alpha(t) - L_\alpha(s)] = \exp[-(t-s)|u|^\alpha], \quad t > s, \qquad (15)$$

showing that $L_2(t)$ has the same finite dimensional distributions as the scaled Brownian motion $\sqrt{2}\, B(t)$.

The samples of the Lévy motion process can exhibit very different features depending on the value of α (Fig. 6). The features of these samples

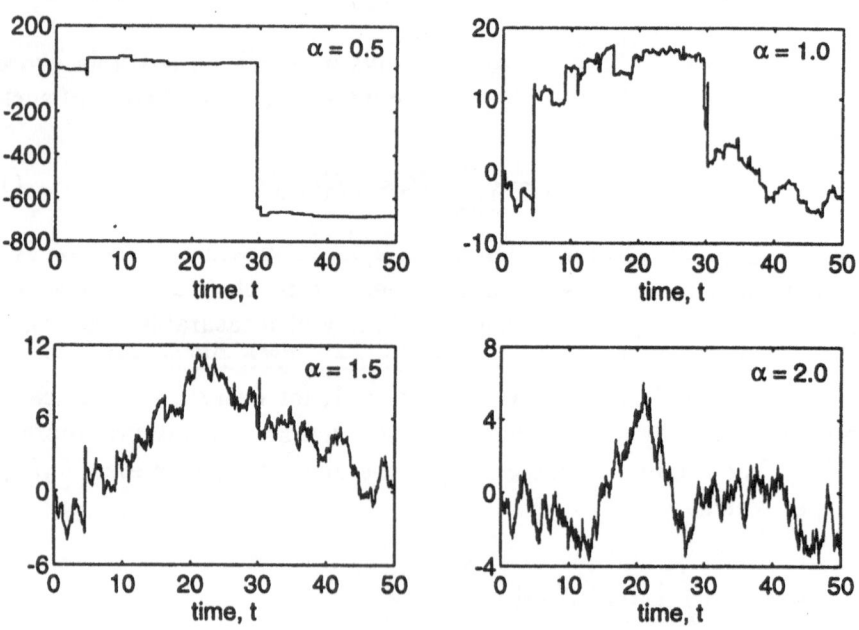

Figure 5.   Realizations of the α-stable Lévy motion process.

are consistent with the decomposition

$$L_\alpha(t) = \int_{|x|<1} x\,[N_t(dx) - t\,\nu(dx)] + \sum_{0<s\leq t} \Delta X(s)\,1_{\{|\Delta X(s)|\geq 1\}} \qquad (16)$$

where $N_t^\Lambda = \int_\Lambda N_t(dx)$ is a Poisson process of intensity $\nu(\Lambda)$ and $\nu(dx)$ is a measure on $\Re \setminus \{0\}$ such that $\int \min(1, x^2)\,\nu(dx) < \infty$ [11]. This decomposition, referred to as the Lévy decomposition, shows that the Lévy motion process consists of a compound Poisson process with jumps $\Delta X(s)$ of magnitude larger than unity and a compensated compound Poisson process $N_t(dx) - t\,\nu(dx)$ associated with infinitely many jumps of magnitudes smaller than unity.

The α-stable Lévy process can be used to define a white noise process because $\{L_\alpha(t),\ t \geq 0\}$ has independent stationary increments. By analogy with the interpretation of the Gaussian white noise process, the α-stable Lévy white noise is

$$W_\alpha(t) = \frac{dL_\alpha(t)}{dt} \qquad (17)$$

and represents the formal derivative of $\{L_\alpha(t),\ t \geq 0\}$.

2.3. DISCRETE TIME NON-GAUSSIAN PROCESSES

Two non-Gaussian time series are examined. The series are defined by linear models with random coefficients and ARMA process driven by α-stable noise. These time series can be used efficiently to determine the response of linear systems to non-Gaussian inputs as well as in simulation studies of linear and nonlinear systems with random noise.

2.3.1. *Linear model with random coefficients*
Consider the series $\{X_k\}$ defined by the recurrence formula

$$X_k = V_k X_{k-1} + (1 - V_k)W_k, \quad k = 1, 2, \dots. \qquad (18)$$

in which $\{V_k\}$ are independent identically distributed variables equal to 1 and 0 with probabilities ρ and $1-\rho$, $\{W_k\}$ another sequence of independent variables following an arbitrary distribution F, X_0 a random variable, and $\{V_k\}, \{W_k\}$, and X_0 are independent [6]. Thus, X_k is equal to X_{k-1} and W_k with probabilities ρ and $(1 - \rho)$. Figure 7 shows realizations of X_k for several values of ρ when X_0 and W_k follow the exponential distribution $F(x) = 1 - e^{-\beta x}$, $x > 0$, $\beta = 1$. The realizations demonstrate that $\{X_k\}$ follows a non-Gaussian distribution and has a longer memory for larger values of ρ.

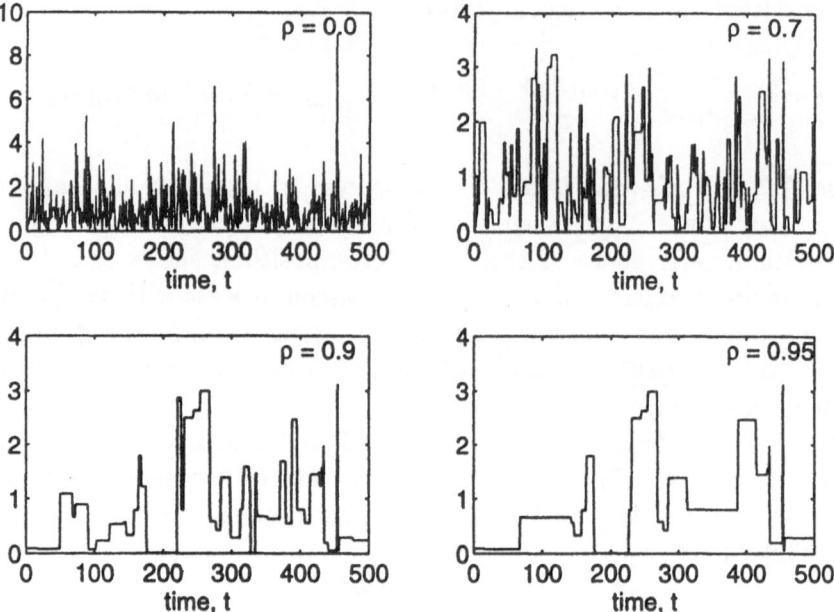

Figure 6.   Realizations of an exponential time series $\{X_k\}$ with marginal density $f(x) = \beta \exp(-\beta x)$, $\beta = 1$, defined by Eq. 23.

The marginal distribution of the series can be calculated from the recurrence formula

$$P(X_k \leq x) = \rho\, P(X_{k-1} \leq x) + (1 - \rho)\, P(W_k \leq x)\,, \quad k = 1, 2, \ldots, \quad (19)$$

because X_k is equal to X_{k-1} and W_k with probabilities ρ and $(1 - \rho)$, respectively. If X_0 follows the distribution F, the probability $P(X_1 \leq x)$ is $\rho\, P(X_0 \leq x) + (1 - \rho)\, P(W_1 \leq x) = F(x)$. The recurrence formula of Eq. 26 shows that $P(X_k \leq x) = F(x)$ for all values of k. If X_0 follows an arbitrary distribution, the probability of X_k, $k \geq 1$, is

$$P(X_k \leq x) = P(X_0 \leq x)\, \rho^k + (1 - \rho)\, F(x) \sum_{i=1}^{k} \rho^{i-1} \qquad (20)$$

and depends on k until the stationary solution is reached. This probability approaches $F(x)$ because $\rho^k \to 0$ and $(1 - \rho)\sum_{i=1}^{k} \rho^{i-1} = 1 - \rho^k \to 1$ as $k \to \infty$. If $P(X_0 \leq x) = F(x)$, the series is stationary to the second order and has the covariance function

$$c(q) = \sigma^2 \rho^q\,. \qquad (21)$$

The series $\{X_k\}$ of Eq. 23 can follow any marginal distribution F. However, the correlation structure of $\{X_k\}$ is restricted to the functional form of Eq. 28. This restriction may be significant in applications and can be alleviated by extending the memory of the series [6].

2.3.2. *Non-Gaussian ARMA models*

Consider a general autoregressive moving average or ARMA series

$$X_k + \sum_{i=1}^{p} a_i X_{k-i} = \sum_{i=0}^{r} c_i W_{k-i} \qquad (22)$$

in which $\{W_i\}$ are uncorrelated random variables and $c_0 = 1$. We say that the series is causal if there exists a sequence of constants $\{\zeta_j\}$ such that $\sum_{j=0}^{\infty} |\zeta_j| < \infty$ and

$$X_k = \sum_{j=0}^{\infty} \zeta_j W_{k-j}, \qquad k = 0, \pm 1, \ldots. \qquad (23)$$

It can be shown that $\{X_k\}$ is causal if, and only if, the polynomial $1 + \sum_{i=1}^{p} a_i z^i$ has no roots in the unit disk of the complex plane. The coefficients ζ_j can be obtained from the recurrence formulas

$$\zeta_j = \begin{cases} -\sum_{k=1}^{j} a_k \zeta_{j-k} + c_j, & 0 \le j < \max\{p, r+1\} \\ -\sum_{k=1}^{p} a_k \zeta_{j-k}, & j \ge \max\{p, r+1\} \end{cases} \qquad (24)$$

and $\zeta_0 = 1$ for causal ARMA models, where $a_j = 0$ and $c_j = 0$ if $j > r$ and $j > p$, respectively [1].

The ARMA series has been used extensively in applications to model Gaussian data. The model can be generalized and used to describe non-Gaussian phenomena. Memoryless transformations of Gaussian ARMA series can be used to generate non-Gaussian series. Alternatively, non-Gaussian models can be obtained by driving ARMA series with non-Gaussian noises of distributions that are closed to linear transformations. For example, suppose that the noise follows a multivariate Cauchy distribution with scale $a > 0$, density [8]

$$f(\boldsymbol{w}) = a\pi^{-(n+1)/2} \, \Gamma((n+1)/2) \, (a^2 + \boldsymbol{w}'\boldsymbol{w})^{-(n+1)/2}, \qquad (25)$$

and characteristic function

$$\varphi(\boldsymbol{u}) = E \exp(i\boldsymbol{u}'\boldsymbol{W}) = \exp(-a \, \|\boldsymbol{u}\|). \qquad (26)$$

It can be shown by direct calculations that X_k is Cauchy for any value of k and has scale $\tilde{a} = a(\sum_{j=0}^{\infty} \zeta_j^2)^{1/2}$ as $n \to \infty$. The ARMA model of Eqs. 29–33 belongs to the class of α-stable ARMA models defined by Eq. 29 in which the noise $\{W_r\}$ follows an α-stable distribution.

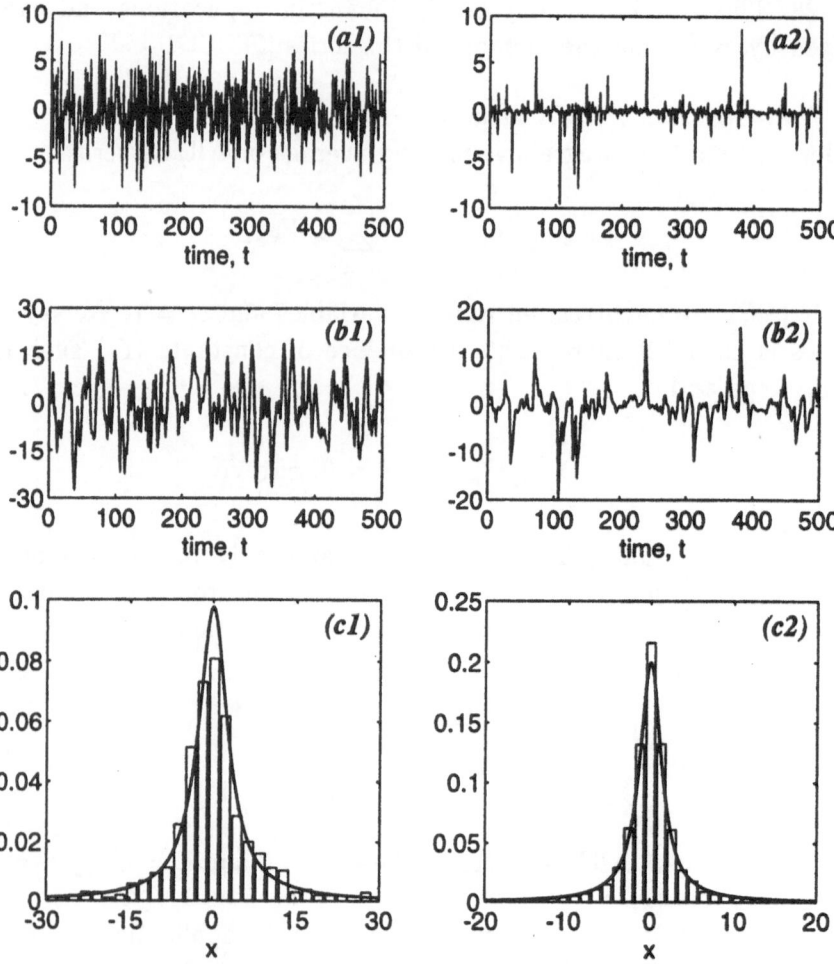

Figure 7.   Cauchy ARMA series with parameters $p = 2$, $r = 1$, $a_1 = -1$, $a_2 = 0.25$, $c_0 = 1$, and $c_1 = 1$: (*a*) realizations of the Cauchy white noise with scale $a = 1$, (*b*) realizations of the Cauchy ARMA series, and (*c*) histograms and densities of the Cauchy ARMA series.

Figures 8(*a*), (*b*), and (*c*) show realizations of $\{W_k\}$, $\{X_k\}$, and histograms and densities of $\{X_k\}$, respectively, for $p = 2$, $r = 1$, $a_1 = -1$, $a_2 = 1/4$, $c_0 = 1$, and $c_1 = 1$. Two Cauchy noises are considered: (1) a noise W with dependent components having the density and characteristic functions of Eqs. 32–33, Figs. 8(*a1*), (*b1*), (*c1*), and (2) a noise W with independent Cauchy components of scale a, Figs. 8(*a2*), (*b2*), (*c2*). The distribution and characteristic function of these variables can be obtained from Eqs. 32–33 with $n = 1$. The ARMA series is causal so that it admits the representation of Eq. 30 in which $\zeta_0 = 1$, $\zeta_1 = 2$, $\zeta_j = -a_1\zeta_{j-1} - a_2\zeta_{j-2}$, $j \geq 2$. The marginal density of the Cauchy ARMA series $\{X_k\}$ is $f(x) = \tilde{a}/[\pi(\tilde{a}^2+x^2)]$, in which $\tilde{a}$ is $a(\sum_{j=0}^{\infty} \zeta_j^2)^{1/2}$ and $a \sum_{j=0}^{\infty} |\zeta_j|$ for the dependent and independent Cauchy noise, respectively. The samples of $\{X_k\}$ depend strongly on the type of Cauchy noise. The histograms of the series agree with the theoretical densities and are characterized by heavy tails. It is not possible to compare these results directly with Gaussian densities because the Cauchy series $\{X_k\}$ has no moments.

3. Stochastic integrals and differential equations

Consider a d-dimensional process $\{X(t), \ t \geq 0\}$ defined by the stochastic differential equation

$$dX(t) = a(X(t), t) \, dt + b(X(t), t) \, dL_\alpha(t), \quad t \geq 0, \qquad (27)$$

where the vector process $L_\alpha(t)$ consists of m independent Lévy motion processes with stationary independent increments of characteristic function given by Eq. 20. The integral form of this equation is

$$X(t) = X(0) + \int_0^t a(X(s), s) \, ds + \int_0^t b(X(s), s) \, dL_\alpha(s). \qquad (28)$$

The integral $\int_0^t a(X(s), s) \, ds$ can be interpreted in the Riemann sense but the definition of $\int_0^t b(X(s), s) \, dL_\alpha(s)$ requires special attention. The second integral of Eq. 36 is of the type $\int_0^t G(s) \, dL_\alpha(s)$, where $\{G(s), \ s \geq 0\}$ is a nonanticipating process, and can be defined as an Itô-type stochastic integral [7, 12, 11].

The objective is to find probabilistic characteristics of the state vector $X(t)$. Suppose that the α-stable driving noise of Eq. 36 is replaced by any of the parametric models of Sec. 2. If Eq. 36 is linear, that is the drift and diffusion coefficients $a(X(t), t)$ and $b(X(t), t)$ depend linearly on and are independent of $X(t)$, respectively, simple analytical solutions are available for $X(t)$. In this case, the stochastic integral $\int_0^t b(X(s), s) \, dL_\alpha(s)$ becomes a linear combination of deterministic Riemann integrals with random coefficients. If Eq. 36 is nonlinear, efficient Monte Carlo simulation algorithms can be developed for generating samples of $X(t)$ [6, 16].

Suppose now that the state $X(t)$ is defined by Eqs. 35–36 with α-stable noise and assume that these equations are linear. If $\alpha = 2$, the input noise is Gaussian and the mean and covariance functions of $X(t)$ define completely this process. Efficient methods of linear random vibration are available for calculating these functions [16]. If $\alpha \neq 2$, $X(t)$ is an α-stable process that can be given in a form similar to the integral representation of Eq. 16. The properties of this representation are known [6, 13]. If the input is a Poisson white noise, the output $X(t)$ is a filtered Poisson process with parameters that can be calculated simply [6, 10]. The Poisson white noise is a component of the α-stable Lévy white noise (Eq. 22) and can be interpreted as a sequence of independent identically distributed pulses arriving in time according to a homogeneous Poisson process [3].

Consider now the general form of Eqs. 35 and 36. If $\alpha = 2$, the driving noise is Gaussian and methods of the nonlinear random vibration can be applied to find probabilistic descriptors of $X(t)$ [15, 16]. These methods can be extended to characterize $X(t)$ for the case in which the input is a Poisson white noise [2]. However, such extensions are not possible for α-stable white noise inputs with $\alpha < 2$ because the methods of the nonlinear random vibration generally assume the existence of the first two moments and the Lévy motion process $L_\alpha(t)$ does not have moments of order two and higher. Alternative methods are needed for solution.

Two methods are developed for solving Eqs. 35 and 36. These methods are based on an Itô-type definition of stochastic integrals with Lévy measures and a finite difference approximation of Eq. 35 converging to $X(t)$. The proposed solutions of Eqs. 35 and 36 are based on (1) differential equations for the characteristic function of $X(t)$ and (2) an extension of the path integral method.

3.1. STOCHASTIC INTEGRALS

Consider a probability space $(\Omega, \mathcal{F}, \mathcal{P})$, a filtration $\{\mathcal{F}_t\}$, $t \geq 0$, that is an increasing family of σ-fields included in $\mathcal{F}$, a nonanticipating process $G(s)$ such that $G(s, \omega)$ is $\mathcal{B}([0, \infty)) \times \mathcal{F}_s$-measurable for every $s \geq 0$, and the stochastic integral

$$X(t) = \int_0^t G(s) \, M(ds). \tag{29}$$

depending on the random measure

$$M(ds) = dL_\alpha(s) \tag{30}$$

assumed to be the Lévy measure. If $\alpha = 2$, this measure coincides with the increment of the Brownian motion in $(s, s + ds)$ and the integral can be defined in the classical Itô sense. This integral exists in the mean square sense if $\int |G(s)|^2 \, ds < \infty$ almost surely [2].

Consider the stochastic integral of Eq. 37 with the Lévy random measure $M(dt) = dL_\alpha(t)$. If $\int_0^t |G(s)|^\alpha \, ds < \infty$ with probability one, the integral of Eq. 37 can be defined such that it (1) is linear in G, (2) satisfies the condition $\int_0^t 1_{[a,b]}(s) \, dL_\alpha(s) = L_\alpha(b) - L_\alpha(a)$, $0 \leq a < b \leq t$, and (3) is bounded in probability. The proof of this statement involves two major steps that are summarized briefly.

First, consider a nonanticipating step function $G_m(s) = G(t_k)$ for $s \in [t_k, t_{k+1})$ satisfying the condition $\sum_{k=0}^{m-1} |G(t_k)|^\alpha (t_{k+1} - t_k) < \infty$ with probability one, where $k = 0, 1, \ldots, m-1$ and $0 = t_0 < t_1 < \cdots < t_m = t$. The stochastic integral of Eq. 37 is

$$\int_0^t G_m(s) \, dL_\alpha(s) = \sum_{k=0}^{m-1} G(t_k) [L_\alpha(t_{k+1}) - L_\alpha(t_k)] \qquad (31)$$

for this function. It is simple to show that this definition satisfies properties (1) and (2). To show that the definition is also consistent with the third property, consider the associated step function $G_{\eta,m}(s)$ that coincides with $G_m(s)$ on $[t_k, t_{k+1})$ if $\sum_{k=0}^{k} |G(t_k)|^\alpha (t_{k+1} - t_k) \leq \eta$ and is zero otherwise. Therefore, there is a random integer $Q = 0, 1, \ldots, m-1$ such that $G_{\eta,m}(s) = G_m(s)$ for $s < t_{Q+1}$ and $G_{\eta,m}(s) = 0$ for $s \geq t_{Q+1}$, $\int_0^t |G_{\eta,m}(s)|^\alpha \, ds = \sum_{j=0}^{Q} |G(t_j)|^\alpha (t_{j+1} - t_j) \leq \eta$, and $G_{\eta,m}(s) = G_m(s)$ for all $s \in [0, t]$ if $\int_0^t |G_m(s)|^\alpha \, ds \leq \eta$. The probability of the random variable $\int_0^t G_m(s) \, dL_\alpha(s)$ can be bounded by

$$P\left(\left|\int_0^t G_m(s) \, dL_\alpha(s)\right| > \lambda\right) \leq P\left(\left|\int_0^t G_{\eta,m}(s) \, dL_\alpha(s)\right| > \lambda\right)$$
$$+ P\left(\int_0^t |G_m(s)|^\alpha \, ds > \eta\right). \qquad (32)$$

The first probability of the upper bound on $P(|\int_0^t G_m(s) \, dL_\alpha(s)| > \lambda)$ approaches zero as $\lambda \to \infty$ because $\sum_{j=0}^{Q} |G(t_j)|^\alpha (t_{j+1} - t_j) < \infty$ with probability one by hypothesis so that $\int_0^t G_{\eta,m}(s) \, dL_\alpha(s)$ is the α-stable variable $S_\alpha((\sum_{j=0}^{Q} |G(t_j)|^\alpha (t_{j+1} - t_j))^{1/\alpha}, 0, 0)$ for every sample of the step function and the tail probability of an α-stable variable Z with parameters $S_\alpha(\sigma, 0, 0)$ is $P(|Z| > \lambda) \sim 2c_\alpha \sigma^\alpha \lambda^{-\alpha}$ as $\lambda \to \infty$, where c_α is a real constant [13]. The second integral approaches zero as $\eta \to \infty$ with probability one by hypothesis. These considerations and the inequality of Eq. 40 show that the stochastic integral $\int_0^t G_m(s) \, dL_\alpha(s)$ is bounded in probability.

Second, let G be a general nonanticipating function. It is necessary to show that there is sequence of step functions $\{G_m\}$ converging to G in some sense and that the sequence of integrals $\int_0^t G_m(s) \, dL_\alpha(s)$ has a unique limit independent of the particular sequence $\{G_m\}$ considered in the analysis. If G is continuous on $[0, t]$, the sequence of step functions $G_m(s) = G(t_k)$, $t_k \leq s < t_{k+1}$, $0 = t_0 < t_1 < \cdots < t_m = t$, can be used to approximate G on $[0, t]$. If the sequence $\{G_m\}$ satisfies the condition $\int |G_m(s) - G_n(s)|^\alpha \, ds \to 0$ in probability as $m, n \to \infty$, the use of Eq. 40 with the step function $G_m - G_n$ shows that $P(|\int_0^t G_m(s) \, dL_\alpha(s) - G_n(s) \, dL_\alpha(s)| > \varepsilon) = P(|\int_0^t [G_m(s) - G_n(s)] \, dL_\alpha(s)| > \varepsilon)$ can be made as small as desired for arbitrary $\varepsilon > 0$. Therefore, $\int_0^t G_m(s)\,dL_\alpha(s)$ is a fundamental series in the sense of convergence in probability so that it converges to some limit that is independent of the particular sequence of step functions $\{G_m\}$. This limit is by definition the stochastic integral of Eq. 37.

3.2. CHARACTERISTIC FUNCTION EQUATION

Let

$$\varphi(u,t) = E \exp iu'X(t), \quad u \in \Re^d, \tag{33}$$

be the characteristic function of the solution $X(t)$ of Eqs. 35–36. It is assumed that $X(t)$ exists, is unique, and has a density with no atoms. Under these assumptions, the characteristic function $\varphi(u,t)$ satisfies partial differential equations depending on the functions $\{a,b\}$ and the value of parameter α. The derivation of differential equations for $\varphi(u,t)$ is based on the Euler approximation of Eq. 35 corresponding to some discrete times $0 = t_0 < t_1 < \cdots < t_n = t$. If the functions $a(x,s)$ and $b(x,s)$ are measurable, Lipschitz with respect to x, and finite for bounded values of x and s and if $a(x_r,s_r)$ and $b(x_r,s_r)$ converge to $a(x,s)$ and $b(x,s)$, respectively, for every sequence $\{x_r,s_r\}$ approaching $\{x,s\}$, the solution of the Euler recurrence formula converges in law to the solution $X(t)$ as $\max(t_{k+1} - t_k) \to 0$ [7]. Therefore, the Euler recurrence formula can be used to approximate $X(t)$ and generate realizations of this process for sufficiently small time steps $t_{k+1} - t_k$.

General case. If $X(t)$ is the solution of Eqs. 35–36 with initial state $X(0) = X_0$, the characteristic function $\varphi(u,t)$ satisfies the differential equation

$$\frac{\partial \varphi(u,t)}{\partial t} = iu' E\{a(X(t),t) \exp iu'X(t)\}$$
$$- E\left\{ \sum_{p=1}^m |v_p(X(t),t)|^\alpha \exp iu'X(t) \right\} \tag{34}$$

in which $v(X(t),t) = b(X(t),t)'u$. The initial condition of this differential equation is $\varphi(u,0) = E \exp(iu'X_0)$. The characteristic function $\varphi(u,t)$ has also to satisfy the additional conditions (1) $|\varphi(u,t)| \le 1$, (2) $\varphi(0,t) = 1$, and (3) $\lim_{|u_k| \to +\infty} \varphi(u,t) = 0$, $k = 1,\ldots,n$. The first two conditions follow from the definition of the characteristic function. The last condition indicates that the density of the state $X(t)$ has no atoms.

If the state vector $X(t)$ becomes stationary as $t \to \infty$, the characteristic function $\varphi(u,t)$ converges to the time-invariant function $\varphi(u) =$

$\lim_{t\to\infty} \varphi(u, t)$ satisfying the differential equation

$$iu'\, E\{a(X(t))\exp iu'X(t)\} - E\left\{\sum_{p=1}^{m} |v_p(X(t))|^\alpha \exp iu'X(t)\right\} = 0. \quad (35)$$

Necessary conditions for this asymptotic result are that matrices a and b do not depend on time explicitly.

To prove the equality of Eq. 42, consider the characteristic function of $X(t + \Delta t)$, $\Delta t > 0$, that can be approximated by

$$\varphi(u, t + \Delta t) \sim E\{\exp iu'[X(t) + a(X(t), t)\,\Delta t + b(X(t), t)\,\Delta L_\alpha(t)]\} \quad (36)$$

as Δt decreases to zero. An alternative form of this asymptotic result is

$$\varphi(u, t)$$
$$\sim E\{\exp iu'\,[X(t) + a(X(t), t)\,\Delta t]\,E[\exp(b(X(t), t)\,\Delta L_\alpha(t)) \mid X(t)]\}$$
$$\sim E\left\{\exp iu'\,[X(t) + a(X(t), t)\,\Delta t]\exp\left(-\Delta t \sum_{p=1}^{m} |v_p(X(t), t)|^\alpha\right)\right\} \quad (37)$$

as $\Delta t \to 0$. The second approximate equality holds because the noise increment in $(t, t + \Delta t)$ and the state $X(t)$ are independent. Expanding the exponential of the last equality of Eq. 45 and retaining terms of order one and Δt, the characteristic function $\varphi(u, t + \Delta t)$ becomes

$$\varphi(u, t + \Delta t)$$
$$\sim E\left\{\exp[iu'X(t)]\,[1 + iu'a(X(t), t)\,\Delta t]\left[1 - \Delta t \sum_{p=1}^{m} |v_p(X(t), t)|^\alpha\right]\right\}$$
$$\sim E\left\{\exp[iu'X(t)]\left[1 + iu'a(X(t), t)\,\Delta t - \Delta t \sum_{p=1}^{m} |v_p(X(t), t)|^\alpha\right]\right\}$$
$$\sim \varphi(u, t) + i\,\Delta t\, u'E\{a(X(t), t)\exp iu'X(t)\}$$
$$- \Delta t\, E\left\{\sum_{p=1}^{m} |v_p(X(t), t)|^\alpha \exp iu'X(t)\right\} \quad (38)$$

as $\Delta t \to 0$. The differential equation of Eq. 42 follows from Eqs. 44–46 by moving $\varphi(u, t)$ on the left side of Eq. 46, dividing by Δt, and letting $\Delta t \to 0$.

If the diffusion matrix $b(X(t), t) = b(t)$ of Eqs. 35–36 does not depend on the state vector, Eq. 42 simplifies to

$$\frac{\partial \varphi(u, t)}{\partial t} = iu'\, E\{a(X(t), t)\exp iu'X(t)\} - \varphi(u, t)\sum_{k=1}^{m} |v_p(t)|^\alpha. \quad (39)$$

The evolution of the characteristic function $\varphi(u,t)$ cannot be determined form Eqs. 42 and 47 because the right hand side of these equations contains expectations of functions of the state vector that are not known. However, if $a(X(t),t)$ and $\sum_{p=1}^{m}|v_p(X(t),t)|^{\alpha}$ are polynomials of the components of $X(t)$, the expectations $E\{a(X(t),t)\exp iu'X(t)\}$ and $E\{\sum_{p=1}^{m}|v_p(X(t),t)|^{\alpha}\exp iu'X(t)\}$ are proportional with partial derivatives of $\varphi(u,t)$ relative to components of u so that Eqs. 42 and 47 become partial differential equations for the characteristic function $\varphi(u,t)$. These equations can be used to calculate the characteristic function $\varphi(u,t)$ of $X(t)$.

Polynomial drift and additive noise. Consider the special case in which $b(X(t),t) = b(t)$ and $a(X(t),t)$ is the polynomial

$$a(X(t),t) = \sum_{q_1,\ldots,q_n} a_{q_1,\ldots,q_n}(t) \prod_{k=1}^{n} X_k(t)^{q_k} \tag{40}$$

with coefficients $a_{q_1,\ldots,q_n}(t)$ that may depend on time. The characteristic function of $X(t)$ is the solution of Eq. 47 in which the expectations $E\{a(X(t),t)\exp iu'X(t)\}$ are

$$E\{a(X(t),t)\exp iu'X(t)\} = \sum_{q_1,\ldots,q_n} a_{q_1,\ldots,q_n}(t) E\left\{\prod_{k=1}^{n} X_k(t)^{q_k} \exp iu'X(t)\right\} \tag{41}$$

or

$$E\{a(X(t),t)\exp iu'X(t)\} = \sum_{q_1,\ldots,q_n} i^{-\left(\sum_{k=1}^{n} q_k\right)} a_{q_1,\ldots,q_n}(t) \frac{\partial^{q_1+\cdots+q_n}\varphi(u,t)}{\partial u_1^{q_1}\cdots\partial u_n^{q_n}} \tag{42}$$

according to the definition of the characteristic function. The corresponding partial differential equation of the characteristic function is

$$\frac{\partial\varphi(u,t)}{\partial t} = iu' \sum_{q_1,\ldots,q_n} i^{-\left(\sum_{k=1}^{n} q_k\right)} a_{q_1,\ldots,q_n}(t) \frac{\partial^{q_1+\cdots+q_n}\varphi(u,t)}{\partial u_1^{q_1}\cdots\partial u_n^{q_n}}$$
$$- \varphi(u,t) \sum_{k=1}^{m} |v_p(t)|^{\alpha} \tag{43}$$

and can be used to determine $\varphi(u,t)$.

Consider another special case in which $n = m = 1$ and $a(X(t),t) = a(X(t),t) = \sum_{q=0}^{r} a_q X(t)^q$ is polynomial of degree r with time invariant coefficients. The characteristic function $\varphi(u,t)$ of $X(t)$ is the solution of

$$\frac{\partial\varphi(u,t)}{\partial t} = u \sum_{q=0}^{r} (-1)^q i^{q+1} a_q \frac{\partial^q\varphi(u,t)}{\partial u^q} - |u|^{\alpha} \varphi(u,t) \tag{44}$$

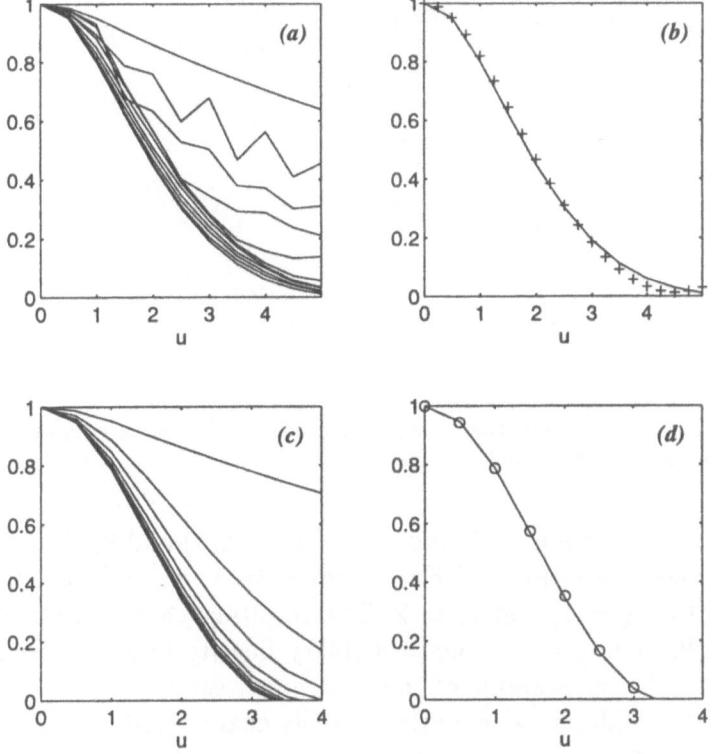

Figure 8. Characteristic function of Eq. 54 with $r = 3$, $a_0 = 0$, $a_1 = -1$, $a_2 = 0$, and $a_3 = -1$ for $\alpha = 1$ (a, b) and $\alpha = 2$ (c, d). Transient and stationary characteristic functions are in (a, c) and (b, d). Symbols "+" and "o" denote simulation and exact values of $\varphi(u, t)$, respectively.

with the convention $\partial^0 \varphi(u, t)/\partial u^0 = \varphi(u, t)$.

Figure 9 shows the evolution in time of the characteristic function $\varphi(u, t)$ and its limit $\varphi(u) = \lim_{t \to \infty} \varphi(u, t)$ for $n = m = 1$, initial condition $\varphi(u, 0) = \exp(-0.1|u|)$, $u \geq 0$, drift coefficient $a(X(t), t) = \sum_{q=0}^{r} a_q X(t)^q$ with $r = 3$, $a_0 = 0$, $a_1 = -1$, $a_2 = 0$, and $a_3 = -1$, diffusion coefficient $b(t) = b(t) = 1$, and two values of α. The solid lines of the figure have been obtained from a finite difference approximation of Eq. 54. The plots of $\varphi(u, t)$ in Figs. 9(a) and 9(c) correspond to times $t = 100 \, k \, \Delta t$, $k = 0, 1, 2, \ldots, 5$, $\Delta t = 0.002$. The asymptotic characteristic functions $\varphi(u)$ of $X(t)$ obtained from Eq. 54 by numerical integration are shown with solid lines in Figs. 9(b) and 9(d) for $\alpha = 1$ and $\alpha = 2$, respectively. These asymptotic results practically coincides with estimates of $\varphi(u)$ determined form realizations of $X(t)$ generated by Monte Carlo simulation for $\alpha = 1$ and the Fourier transform of the stationary density of $X(t)$, that is available in closed form for $\alpha = 2$ [16]. The estimates and the exact values of $\varphi(u)$

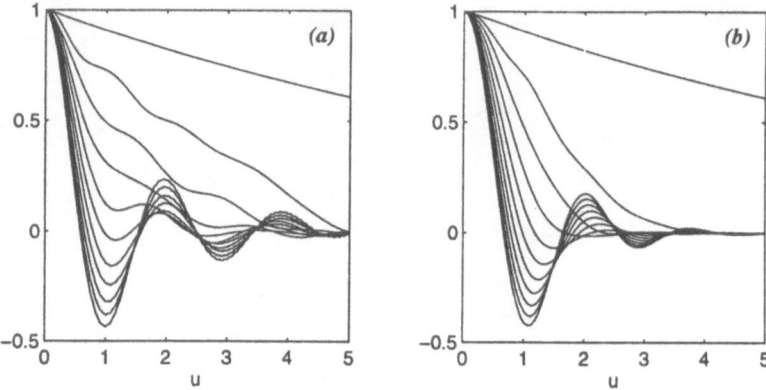

Figure 9. Characteristic function given by Eq. 54 with $r = 3$, $a_0 = 0$, $a_1 = 1$, $a_2 = 0$, and $a_3 = -0.1$ for $\alpha = 1$ (a) and $\alpha = 2$ (b).

are shown with symbols "+" and "o" in Figs. 9(*b*) and 9(*d*), respectively. Figure 10 shows solutions of Eq. 54 for $r = 3$, $a_0 = 0$, $a_1 = 1$, $a_2 = 0$, $a_3 = -0.1$, $\alpha = 1$, and $\alpha = 2$. The resulting characteristic functions exhibit oscillations because the probability density of $X(t)$ is bimodal, as demonstrated by an example of the subsequent section.

The partial differential equations of the characteristic function $\varphi(u, t)$ of the state vector incorporate the implicit assumption that $\varphi(u, t)$ is differentiable relative to both u and t for all values of these arguments. However, the characteristic function may not be differentiable with respect to u everywhere. For example, the derivative of the characteristic functions φ satisfying Eq. 54 may not exist at $u = 0$ so that these equations do not hold for this value of u. However, Eq. 54 can be solved for $u > 0$ and $u < 0$.

Gaussian white noise. If $\alpha = 2$, Eqs. 42 and 47 of $\varphi(u, t)$ can be found by the Itô differentiation rule because $dL_2(t) = \sqrt{2}\, dB(t)$, where $\{B(t), \ t \geq 0\}$ consists of m independent Brownian motion processes. In this case, $X(t)$ is the solution of the Itô stochastic differential equation

$$dX(t) = a(X(t), t)\, dt + \sqrt{2}\, b(X(t), t)\, dB(t), \quad t \geq 0. \qquad (45)$$

According to the Itô differentiation formula [16], the characteristic function $\varphi(u, t) = E \exp iu' X(t)$ can be obtained from

$$\frac{\partial}{\partial t}\, \varphi(u, t) = E\left[\sum_{k=1}^{n} a_k(X(t), t)(iu_k) \exp iu' X(t) \right.$$

$$\left. + \frac{1}{2} \sum_{k,l=1}^{m} d_{k,l}(X(t), t)(iu_k)(iu_l) \exp iu' X(t) \right] \qquad (46)$$

where $d_{k,l}(x, t) = \{2\, b(x, t)\, b(x, t)'\}_{k,l}$. This result coincides with Eq. 42.

A generalized version of the Itô formula can be applied to derive Eq. 42 for $\alpha \neq 2$. This version of the Itô formula is associated with stochastic integrals of the type $\int_0^t G(s)\, dM(s)$, in which $\{M(s), s \geq 0\}$ is a semimartingale [11].

3.3. PATH INTEGRAL METHOD

Let $\{X(t),\, t \geq 0\}$ be the solution of Eqs. 35–36. The state vector at a time $t + \Delta t$, $\Delta t > 0$, can be related to $X(t)$ by the recurrence formula

$$X(t + \Delta t) \sim X(t) + a(X(t), t)\, \Delta t + b(X(t), t)\, \Delta L_\alpha(t) \qquad (47)$$

as $\Delta t \to 0$. The series $\{X_k = X(k\, \Delta t)\}$, $k = 0, 1, 2, \ldots$, constitutes a discrete-time Markov chain with time step Δt because future and past values of the state vector are independent conditional on the present value of this vector. The conditional vector $\hat{X}(t + \Delta t) = X(t + \Delta t) \mid X(t) = x$ is α-stable with the characteristic function

$$\hat{\varphi}(u, t) \sim \exp\{iu'\, [x + a(x, t)\, \Delta t]\} \exp\Big\{-\Delta t \sum_{p=1}^m |v_p(x, t)|^\alpha\Big\}, \quad \Delta t \to 0,$$

$$(48)$$

because $\hat{\varphi}(u, t) \sim E \exp\{iu'\, [x + a(x, t)\, \Delta t + b(x, t)\, \Delta L_\alpha(t)]\}$ as $\Delta t \to 0$, the increments $\{\Delta L_{\alpha,p}\}$, $p = 1, \ldots, m$, are independent and have the characteristic function $E \exp[iu\, \Delta L_{\alpha,p}] = \exp[-\Delta t\, |u|^\alpha]$, and the expectation $E \exp[iu'\, b(x, t)\, \Delta L_\alpha(t)] = E \sum_{p=1}^m i\, v_p(x, t)\, \Delta L_{\alpha,p}(t)$ can be calculated from $\prod_{p=1}^m \exp[-\Delta t\, |v_p(x, t)|^\alpha]$ [6, 13].

It is convenient for calculations to discretize the values of the state vector. Let $D \subseteq \Re^d$ be a bounded domain containing all possible values of $\{X(t),\, t \geq 0\}$ with nearly unit probability. The domain can be divided in cells $\{C_s\}$, $s = 1, \ldots, r$, such that $C_s \cap C_{s'} = \emptyset$, $s \neq s'$, and $\cup_{s=1}^r C_s = D$. Let x_s be an arbitrary point of C_s. For example, x_s can be the mid point of cell s. The time and space discretization can be used to approximate $X(t)$ by a Markov chain with time step Δt and taking a finite number of values $\{x_1, \ldots, x_r\}$. Transitions of the state vector from a cell C_s at time t to an arbitrary cell $C_{s'}$ at a later time $t + \Delta t$ occur with a probability that can be approximated by

$$p(s, s'; t, t + \Delta t) = \int_{C_{s'}} \hat{f}(y \mid x_s)\, dy \qquad (49)$$

for $s, s' = 1, \ldots, r$, where $\hat{f}(y \mid x_s)$ is the density of $X(t + \Delta t) \mid X(t) = x_s$ at y. Let $P(t) = \{P(X(t) \in C_1), \ldots, P(X(t) \in C_r)\}'$, $t = k\, \Delta t$, $k = 0, 1, 2, \ldots$, be a vector of probabilities characterizing the system state at

time t. The initial value $P(0)$ of this vector can be obtained from the
definition of X_0. Consider the matrix $p(t, t + \Delta t)$ of transition probabilities
$p(s, s'; t, t + \Delta t)$. The s'th row of this transition matrix is $\{p(s, 1; t, t + \Delta t), \ldots, p(s, r; t, t + \Delta t)\}$. The evolution of the probability vector $P(t)$ can
be determined from the recurrence formula

$$P(t + \Delta t) = p(t, t + \Delta t) \, P(t), \quad t = 0, \Delta t, 2 \, \Delta t, \ldots, \tag{50}$$

with the initial condition $P(0)$. If $p(t, t + \Delta t) = p(\Delta t)$ is time invariant
and the eigenvalues of this matrix are in the unit disk of the complex plane,
$P(t)$ approaches a stationary value P as time t increases. This stationary
probability vector is the solution of the linear homogeneous system

$$P = p(\Delta t) \, P \tag{51}$$

and has a unique solution because $\sum_{s=1}^{r} P_s = 1$, where $P_s = \lim_{t \to \infty} P(X(t) \in C_s)$. The time and space discretization must satisfy some conditions to
assure that the approach of Eqs. 59 and 60 converges [9].

Consider for illustration a half nonlinear oscillator with displacement
$X(t)$ satisfying the equation $dX(t) = (a \, X(t) + b \, X(t)^3) \, dt + dL_\alpha(t)$, $t \geq 0$.
The conditional process $\hat{X}(t + \Delta t) = X(t + \Delta t) \mid X(t) = x$ is α-stable
with location $x + (ax + bx^3) \, \Delta t$, scale $(\Delta t)^{1/\alpha}$, and characteristic function
$\hat{\varphi}(u) = E \exp iu \, \hat{X}(t + \Delta t) = \exp\{iu[x + (ax + bx^3) \, \Delta t] - \Delta t \, |u|^\alpha\}$. This
characteristic function can be inverted to find the density of $\hat{X}(t + \Delta t)$.
There are only three types of α-stable variables for which the density is
available in closed form: the Gaussian, the Cauchy, and the Lévy variables
corresponding to $\alpha = 2$, $\alpha = 1$, and $\alpha = 0.5$, respectively.

Figures 11(a, b) show stationary densities and histograms of $X(t)$ obtained by the path integral and Monte Carlo simulation methods for $\alpha = 1$.
Stationary densities calculated by the path integral method and exact densities of $X(t)$ for $\alpha = 2$ are in Figs. 11(c, d). The unimodal and bimodal
densities and histograms of Figs. 11(a, c) and Figs. 11(b, d) correspond to
$(a = -1, b = -0.1)$ and $(a = 1, b = -0.1)$, respectively. The exact stationary density of $X(t)$ is proportional with $\exp(ax^2/2 + bx^4/4)$ for $\alpha = 2$
[16]. Results of the path integral, exact, and simulation methods shown in
Fig. 11 are in good agreement.

4. Conclusions

Parametric models of random processes and solutions of stochastic differential equations with Lévy random measures were examined. It was
shown that parametric models (1) can represent a large class of continu-
ous/discrete time Gaussian/non-Gaussian processes $X(t)$, (2) become versions of $X(t)$ asymptotically as the representation is refined, and (3) can

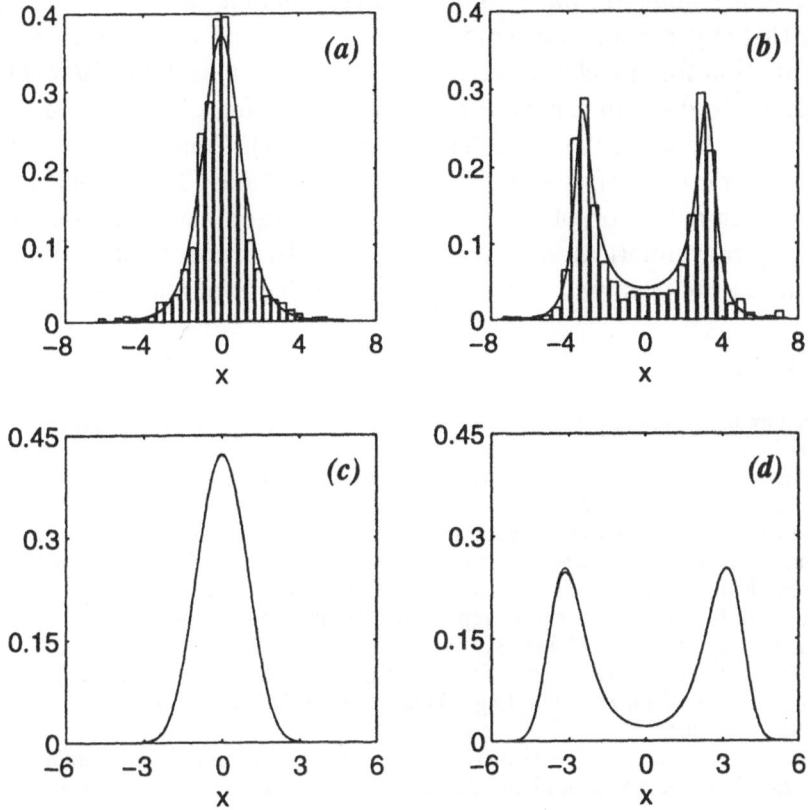

Figure 10. Stationary probability density function of $X(t)$ for $\alpha = 1$ (a, b), $\alpha = 2$ (c, d), $(a = -1, b = -0.1)$ (a, c), and $(a = 1, b = -0.1)$ (b, d).

be used to facilitate the analysis of stochastic differential equations. If the input of these equations is represented by parametric models, simple analytical solutions and efficient Monte Carlo simulation algorithms are available for finding probabilistic characteristics of $X(t)$.

The solution $X(t)$ of stochastic differential equations with white noise input requires different techniques depending on the type of the noise and the differential equation. For linear differential equation with Gaussian, Poisson, and α-stable Lévy white noise input $X(t)$ is a Gaussian, filtered Poisson, and α-stable process, respectively. The probability law of $X(t)$ can be obtained simply in this case. If $X(t)$ satisfies a nonlinear differential equation, the determination of probabilistic characteristics of this process poses significant difficulties. The random vibration theory provides exact and approximate methods for characterizing $X(t)$ corresponding to Gaussian and Poisson white noise inputs but does not consider nonlinear stochastic differential equations driven by α-stable white noise. Two methods were developed for solving stochastic differential equations with α-stable white noise. The methods are based on an Itô-type definition of stochastic integrals with Lévy random measures. The first method is based on a differential equation for the characteristic function $\varphi(u, t) = E \exp iuX(t)$. It is not possible to develop a partial differential equation of $\varphi(u, t)$ for general nonlinear stochastic differential equations of $X(t)$. Conditions were establish for $\varphi(u, t)$ to satisfy partial differential equations. The second method is based on an extension of the path integral method developed for Itô processes. Several numerical examples were used to illustrate the application and evaluate the accuracy of the proposed methods of analysis.

References

1. Brockwell, P.J. and Davis, R.A.: *Time Series: Theory and Methods*, Springer-Verlag, New York, 1987.
2. Gilhman, I.I. and Skorohod, A.V.: *Stochastic Differential Equations*, Springer-Verlag, New York, 1972.
3. Grigoriu, M.: White Noise Processes in Random Vibration, *International Symposium on Nonlinear Dynamics and Stochastic Mechanics*, The Fields Institute for Research in Mathematical Sciences, Waterloo, Ontario, Canada, 1993.
4. Grigoriu, M.: On the spectral representation method in simulation, *Probabilistic Engineering Mechanics* 8 (1993), 75–90.
5. Grigoriu, M. and Balopoulou, S.: A Simulation method for Stationary Gaussian Functions Based on the Sampling Theorem, *Probabilistic Engineering Mechanics* 8 (1993), 239–254.
6. Grigoriu, M.: *Applied Non-Gaussian Processes. Example, Theory, Simulation, Linear Random Vibration, and MATLAB Solutions*, Prentice-Hall, Englewood Cliffs, N.J., 1995.
7. Janicki, A., Michna, Z., and Weron, A.: Approximation of Stochastic Differential Equations Driven by α-Stable Lévy Motion, *Report, Hugo Steinhaus Center for Stochastic Methods*, Wroclaw, Poland, 1995.
8. Johnson, N.L. and Kotz, S.: *Distributions in Statistics: Continuous Multivariate Distributions*, John Wiley, New York, N.Y., 1972.
9. Naess, A. and Johnsen, J.M.: The Path Integral Solution Technique Applied to the Random Vibration of Hysteretic Systems, in P.D. Spanos and C.A. Brebbia (eds.), *Computational Stochastic Mechanics*, Elsevier Applied Science, London, 1991.
10. Parzen, E.: *Stochastic Processes*, Holden-Day, San Francisco, 1962.
11. Protter, P.: *Stochastic Integration and Differential Equation*, Springer-Verlag, New York, N. Y., 1990.
12. Rosiński, J. and Woyczyński, W.A.: On Itô Stochastic Integration with respect to p-Stable Motion: Inner Clock, Integrability of Sample Paths, Double and Multiple Integration, *The Annals of Probability* 14 (1986), 271–286.
13. Samorodnitsky, G. and Taqqu, M.S.: *Stable Non-Gaussian Random Processes. Stochastic Models with Infinite Variance*, Chapman and Hall, New York, N.Y., 1994.
14. Shinozuka, M. and Deodatis, G.: Simulation of Stochastic Processes by Spectral Representation, *Applied Mechanics Reviews* 44 (1991), 191–203.
15. Soize, C.: *The Fokker-Planck Equation for Stochastic Dynamical Systems and its Explicit Steady State Solutions*, World Scientific, Singapore, 1994.
16. Soong, T.T. and Grigoriu, M.: *Random Vibration of Mechanical and Structural Systems*, Prentice-Hall, Englewood Cliffs, N.J., 1993.
17. Wong, E. and Hajek, B.: *Stochastic Processes in Engineering Systems*, Springer-Verlag, New York, N.Y., 1985.

STOCHASTIC RESPONSE OF IRREGULAR TRACKS UNDER MOVING VEHICLES

R N IYENGAR and O R JAISWAL
Central Building Research Institute
Roorkee 247 667, India

1 Introduction

Profile irregularity is invariably present in any railway track. Among the various causes which influence the dynamics of track-vehicle system, track irregularity is perhaps the most important one. Profile irregularity make the moving vehicle vibrate, which in turn, induces dynamic forces on track. These forces depend also on the dynamic characteristics of moving vehicles. Under the influence of these forces, supporting subgrade yields in the regions of high stresses. This leads to further changes of the track profile. Thus dynamic forces on track and nonlinear behviour of the subgrade are the basic causes of track deterioration. Furthermore irregularities are stochastic in nature and so are the track forces. For proper understanding of track behaviour it is necessary to include vehicle characteristics and the stochastic nature of irregularities in to the analysis. In the past most of the studies have either ignored or have oversimplified track-vehicle interaction. Also there are very few studies on track deterioration. Patil (1988) and Duffy (1990) have considered the vehicle mass in the dynamic analysis of railway tracks. They studied the effect of vehicle mass on the resonant track frequency and the critical velocity. Further, it can be mentioned here that subgrade properties vary along the track and these variations are stochastic. Fryba *et al.* (1991) have studied the effect of these variations on the dynamic response of tracks. Most of the studies on tracks with nonlinear foundations have considered track response to a single moving force. Mulachy (1973) analysed the response of a beam on elasto-plastic foundation subjected to a moving force. Choros and Adams (1979) have studied the response of a beam on tensionless foundation under a moving force. Frederick (1978) calculated dynamic forces on the track due to stochastic vertical profile using a lumped parameter model. Based on the maximum value of this force, track settlement was obtained using experimentally determined settlement laws.

In the present paper, analysis of tracks with stochastic vertical profile is presented. First dynamic analysis of track-vehicle system is presented. Track is modelled as an infinitely long beam on Winkler foundation. A moving vehicle is represented as a spring-mass-damper system. Irregularity in vertical profile is modelled as a stationary gaussian random process. Theory of linear systems is used to obtain statistics of track and vehicle response. Track deterioration is studied using a simple quasi-static model. Effects of vehicle characteristics and vertical profile are retained in this model. Nonlinear behaviour of the subgrade is included in this model. Equivalent linearisation technique is used to obtain results on track settlement.

2 Stochastic Track Subjected to a Moving Vehicle

With reference to figure 1 track is modelled as an infinitely long beam with irregular vertical profile, $w(x)$, resting on an elastic foundation and subjected to a vehicle moving with velocity v. Here $w(x)$ is the track profile measured from the X-axis under no load condition. The elastic foundation consists of closely spaced linear independent springs and dash-pots. The vehicle is modelled as a spring-mass-

A. Naess and S. Krenk (eds.), IUTAM Symposium on Advances in Nonlinear Stochastic Mechanics, 213–224.
© 1996 *Kluwer Academic Publishers.*

damper system with mass ρ, stiffness K_v and viscous damping coefficient C_v. In the moving co-ordinate $\xi = x - vt$ the equations of motion for the track deflection $y(\xi, t)$ measured from the vertical profile $w(x)$ and vehicle deflection $u(t)$ are given by

$$EI\frac{\partial^4 y}{\partial \xi^4} + Q\frac{\partial^2 y}{\partial x^2} + m\{\frac{\partial^2 y}{\partial t^2} - 2v\frac{\partial^2 y}{\partial \xi \partial t} + v^2\frac{\partial^2 y}{\partial \xi^2}\} + C_t\{\frac{\partial y}{\partial t} - v\frac{\partial y}{\partial \xi}\} + K_t y =$$

$$\{K_v(u - y - w) + C_v(\frac{du}{dt} - \frac{\partial y}{\partial t} - \frac{dw}{dt})\}\delta(\xi) \qquad (1)$$

$$\rho\frac{d^2 u}{dt^2} + C_v\frac{du}{dt} + K_v u = \{K_v(y + w) + C_v(\frac{\partial y}{\partial t} + \frac{dw}{dt})\}|_{\xi=0} \qquad (2)$$

Here, EI is the bending rigidity of the track; Q is axial force on the track; m is mass per unit length of the track; C_t is the foundation viscous damping coefficient and K_t is the foundation modulus. The vertical profile, $w(x)$ is considered as a Gaussian stationary random field with power spectral density (PSD) $\phi_{ww}(f)$, where, f is the spatial frequency in cycles/meter. Eqs. (1) and (2) are coupled linear differential equations under stochastic input. In these equations $w(t)$ is $w(x)$ evaluated at $x = vt$. The PSD of track response $y(\xi, t)$ can be obtained using linear system theory. Following the frequency domain approach PSD of track deflection, $\phi_{yy}(\omega, \xi)$ can be obtained as

$$\phi_{yy}(\omega, \xi) = H_t(\omega, \xi)H_t^*(\omega, \xi)\phi_{ww}(\omega) \qquad (3)$$

$$\phi_{ww}(\omega) = \phi_{ww}(f)/(2\pi v) \qquad (4)$$

where, $H_t(\omega, \xi)$ is track frequency response function and $\omega = 2\pi f v$ is the temporal frequency in radians/sec. In this case track frequency response function is the response of a track with harmonic vertical profile. Similarly, PSD of vehicle acceleration is given by

$$\phi_{\ddot{u}\ddot{u}}(\omega) = \omega^4 H_v(\omega)H_v^*(\omega)\phi_{ww}(\omega) \qquad (5)$$

where, $H_v(\omega)$ is vehicle frequency response function.

2.1 TRACK FREQUENCY RESPONSE FUNCTION

Consider a track with harmonic vertical profile $w(x) = e^{2\pi i f x}$. For this case eqs. (1) and (2) can be written as

$$EI\frac{\partial^4 y}{\partial \xi^4} + Q\frac{\partial^2 y}{\partial x^2} + m\{\frac{\partial^2 y}{\partial t^2} - 2v\frac{\partial^2 y}{\partial \xi \partial t} + v^2\frac{\partial^2 y}{\partial \xi^2}\} + C_t\{\frac{\partial y}{\partial t} - v\frac{\partial y}{\partial \xi}\} + K_t y =$$

$$\{K_v(u - y - e^{i\omega t}) + C_v(\frac{du}{dt} - \frac{\partial y}{\partial t} - i\omega e^{i\omega t})\}\delta(\xi) \qquad (6)$$

$$\rho\frac{d^2 u}{dt^2} + C_v\frac{du}{dt} + K_v u = \{K_v(y + e^{i\omega t}) + C_v(\frac{\partial y}{\partial t} + i\omega e^{i\omega t})\}|_{\xi=0} \qquad (7)$$

Here, $\omega = 2\pi f v$. In the steady state the solution can be written as

$$y(\xi, t) = z(\xi)e^{i\omega t} \qquad (8)$$

Substituting this solution in eqs. (6) and (7) one gets,

$$EI\frac{d^4 z_j}{d\xi^4}+(Q+mv^2)\frac{d^2 z_j}{d\xi^2}-v(2miw+C_t)\frac{dz_j}{d\xi}+(K_t-mw^2+iwC_v)z_j = 0 \quad j=1,2 \quad (9)$$

where, $z = z_1$ for $\xi \geq 0$ and $z = z_2$ for $\xi < 0$. The boundary and continuity conditions to be satisfied are

$$z_1(\xi \to \infty) = 0; \quad z_2(\xi \to -\infty) = 0 \tag{10}$$

$$z_1(\xi = 0) - z_2(\xi = 0) = 0 \; ; \; z'_1(\xi = 0) - z'_2(\xi = 0) = 0 \; ; \; z''_1(\xi = 0) - z''_2(\xi = 0) = 0$$
$$EI\, z'''_1(\xi = 0) - EI\, z'''_2(\xi = 0) = \{\rho w^2(K_v + iwC_v)\}\{K_v - \rho w^2 + iwC_v\}^{-1}(z+1)|_{\xi=0}(11)$$

Here the primes denote derivatives with respect to ξ. Taking the solution as

$$z_j = e^{s\xi} \tag{12}$$

from eq. (9) one gets the following characteristic equation

$$s^4 + (Q + mv^2)s^2 - v(C_t + 2miw)s + (K_t - mw^2 + C_t iw) = 0 \tag{13}$$

Let d_k ($k = 1, 2, 3, 4$) be the roots of this equation such that the real parts of d_1, d_2 are negative and that of d_3, d_4 are positive. In view of the boundary conditions in eq. (10), the solution for $z_j(j = 1, 2)$ can be written as

$$z_1(\xi) = C_1 e^{d_1\xi} + C_2 e^{d_2\xi} \; ; \; z_2(\xi) = C_3 e^{d_3\xi} + C_4 e^{d_4\xi} \tag{14}$$

Constants $C_k(k = 1, 2, 3, 4)$ can be obtained form the continuity conditions given in eq. (11). Thus, track frequency response function is given by

$$\begin{aligned} H_t(w, \xi) &= C_1 e^{d_1\xi} + C_2 e^{d_2\xi} \quad \text{for } \xi > 0 \\ &= C_3 e^{d_3\xi} + C_4 e^{d_4\xi} \quad \text{for } \xi \leq 0 \end{aligned} \tag{15}$$

The vehicle frequency response function is given by

$$H_v(w) = \{\rho w^2(K_v + iwC_v)\}\{K_v - \rho w^2 + iwC_v\}^{-1}(1 + C_1 + C_2) \tag{16}$$

It is to be noted here that track frequency response function is a function of $\xi = x - vt$. Thus, track response at any fixed point x, will be non-stationary. Since vehicle is always at $\xi = 0$, vehicle frequency response is not a function of ξ and vehicle response is stationary.

3 Numerical Results

Numerical results are presented for the following values of various track parameters : $E = 2 \times 10^{10}$ kg/m^2, $I = 5 \times 10^{-5}$ m^4, $m = 70$ kg-sec^2/m^2, $K_t = 1 \times 10^6$ kg/m^2, $C_v = 2 \times 10^3$ kg-sec/m^2. Vehicle parameters are selected as $\rho = 6 \times 10^3$ kg-sec^2/m, $K_v = 6 \times 10^5$ kg/m, $C_v = 8 \times 10^3$ kg-sec/m. These parameters correspond to a typical ICF passenger coach of Indian Railways. Sample PSD obtained from recorded data of vertical profile from a typical Indian railway track is used for computation. This profile PSD is shown in figure 2.

PSD of track response given by eq. (3) is a function of ξ. In figure 3 PSD of track response at $\xi = 0$ is shown. Track PSD shows a predominant peak at 1.7 Hz which is the natural frequency of the vehicle. In figure 4 PSD of vehicle acceleration obtained from eq. (5) is shown. Also plotted in this figure is the PSD obtained from the recorded data of vertical acceleration of the ICF coach. It is seen that PSD obtained from eq. (5) compares well with the observed data. PSD of vehicle acceleration also shows a predominant peak at 1.7 Hz. In figure 5 variation of standard deviation, σ_y, of track response with ξ is plotted. Since $\xi = x - vt$, results of figure 5 also indicate the variation of σ_y with time at a fixed value of x.

4 Quasi-static Model

In previous sections dynamic analysis of stochastic tracks under a moving vehicle is presented. It is seen that track-vehicle response is governed by coupled partial differential equations. As pointed out earlier, for track deterioration studies it is necessary to consider the nonlinear behaviour of the subgrade. If this nonlinear behaviour is included in the dynamic analysis, then it will be almost impossible to solve track-vehicle equations. Thus, to obtain useful engineering results on track settlement, it is necessary to suitably simplify track-vehicle equations. Iyengar and Jaiswal (1994) have developed a simple quasi-static model for this purpose. In this model a moving train is considered as an infinitely long flow of closely spaced vehicles. Analysis of such a track-train model showed that during the passage of a long train, track is subjected to spatially varying quasi-static forces. These forces can be estimated by considering a moving train over a rigid track. For a track with stochastic profile $w(x)$, the PSD of quasi-static force, $p(x)$, is given by

$$\phi_{pp}(f) = \frac{[\bar{\rho}(2\pi v f)^2]^2 [\bar{K}_v^2 + (\bar{C}_v 2\pi v f)^2]}{[\bar{K}_v^2 - \bar{\rho}(2\pi v f)^2]^2 + (\bar{C}_v 2\pi v f)^2} \phi_{ww}(f) \tag{17}$$

Here, $\bar{\rho}$, $\bar{K}_v$ and $\bar{C}_v$ are respectively the mass, stiffness and damping coefficients per unit length of the train. Track response to this quasi-static force is given by

$$EI\frac{d^4 y}{dx^4} + K_t y = p(x) \tag{18}$$

PSD of track deflection is obtained as

$$\phi_{yy}(f) = \phi_{pp}(f)[EI(2\pi f)^4 + K_t]^{-2} \tag{19}$$

Thus, quasi-static approach considerably simplifies the analysis of track response to a moving train.

5 Nonlinear Track

Due to differential yielding of subgrade, track accumulates damage over a period of time. Such behaviour in track can be simulated by incorporating nonlinearity in the track foundation. As a simple model, here the flexible foundation is considered to be made of elasto-plastic tensionless springs. The displacement of the track is given by

$$EI\frac{d^4 y}{dx^4} + g(y) = p_0 + p(x) \tag{20}$$

Here, $g(y) = K_t[yU(y) + (y_c - y)U(y - y_c)]$ is the nonlinear track spring force and U is the unit step function. As seen from figure 6, the track yields in compression when the displacement exceeds a critical value y_c and the spring does not take any tensile force. In eq. (20) p_0 is the uniform mean force due to self-weight of the train and $p(x)$ is the spatially varying force calculated by considering a moving train over a rigid profile.

5.1 HARMONIC FORCES ON NONLINEAR TRACK

Before one considers stochastic forces on nonlinear tracks, it would be informative to study the deflection of the track under sinusoidal forces. Thus, first consider

$$EI\frac{d^4y}{dx^4} + g(y) = a + b\sin(2\pi fx) \tag{21}$$

With the non-dimensionalisation $y_1 = y/y_c$ and $X = x/L_c$, $L_c = (4EI/K_t)^{0.25}$ the above equation is transformed to

$$\frac{d^4y_1}{dX^4} + 4\bar{g}(y_1) = A + B\sin(2\pi\Omega X) \tag{22}$$

Where, $\Omega = L_c f$, $A = aL_c^4/(EIy_c)$ and $B = bL_c^4/(EIy_c)$.

This is a boundary value problem with boundaries at $\pm\infty$. Numerical treatment of such a problem poses considerable difficulties. For example, let us consider the linear equation

$$\frac{d^4y_1}{dX^4} + 4y_1 = A + B\sin(2\pi\Omega X) \tag{23}$$

The eigenvalues of the homogeneous part of this equation are $1 + i$, $1 - i$, $-1 + i$ and $-1 + i$. Two of the eigenvalues are having positive real parts and remaining two are with negative real parts. Now if a numerical technique like Runge-Kutta scheme is used, one will get unbounded solutions due to the presence of eigenvalues with positive real parts. Thus the eigenvalue structure of this boundary value problem does not allow direct use of well known numerical techniques. However, one can obtain the solution through an iterative scheme using the method of integral equations.

5.1.1 Integral Equation Method

The nonlinear term in eq. (22) can be expressed as

$$\bar{g}(y_1) = y_1 + q(y_1) \tag{24}$$

and the equation can be recast as an integral equation (Na, 1979).

$$y_1(X) = \int_{-\infty}^{\infty} \{A + B\sin(2\pi\Omega s) - q[y_1(s)]\}h(X - s)ds \tag{25}$$

Here, $h(s)$ the corresponding Green's function for the linear part is

$$h(s) = \{4\sqrt{2}\}^{-1}e^{-|s|}\sin(|s| + \pi/4) \tag{26}$$

Eq. (25) can be solved iteratively, starting with the known linear solution when $q = 0$. For obtaining the numerical solution, the limits of the integration in eq. (25) are set as $-400 \leq s \leq 400$ and the the convergence of the solution is verified in the interval $-50 \leq X \leq 50$. The values of A and B are taken as $A = B = 2.0$ and solutions are worked out for two values of Ω. The results of the convergence of the peak amplitude for these two cases are shown in Table 1. It is seen that the solution converges uniformly, however for $\Omega = 0.1$, the convergence is slow. The final solutions are shown in figures 7 and 8. These may be taken as the exact solutions for further comparison.

5.1.2 Equivalent Linearisation

The above iterative procedure for solving the nonlinear track problem is highly time consuming. A faster alternative is to use linearisation techniques widely adopted in the analysis of nonlinear oscillators. For this purpose we write $y_1(X) = R_0 + z_1(X)$ such that

$$\int_0^{1/\Omega} z_1(X)dX = 0 \tag{27}$$

Now, eq. (22) can be recast as

$$\frac{d^4 z_1}{dX^4} + 4\bar{g}(z_1 + R_0) = A + B\sin(2\pi\Omega X) \tag{28}$$

The nonlinear term in this equation is linearised as

$$\bar{g}(z_1 + R_0) \approx \alpha_1 z_1 + \alpha_2 \tag{29}$$

upon substituting eq. (29) in eq. (28) and making use of eq. (27) one gets

$$z_1(X) = R\sin(2\pi\Omega X) \tag{30}$$

where,

$$R = B\{(2\pi\Omega)^4 + 4\alpha_1\}^{-1} \tag{31}$$

From eq. (28) one also finds

$$A = 4\Omega \int_0^{1/\Omega} \bar{g}(R_0 + z_1)dX \tag{32}$$

Further, minimisation of the mean square error

$$\epsilon^2 = \int_0^{1/\Omega} [\bar{g}(R_0 + z_1) - \alpha_1 z_1 - \alpha_2]^2 dX \tag{33}$$

with respect to α_1 leads to

$$\alpha_1 = \int_0^{1/\Omega} z_1 \bar{g}(R_0 + z_1)dX / \int_0^{1/\Omega} z_1^2 dX \tag{34}$$

Using eq. (31) one finds that eqs. (32) and (34) are coupled transcendental equations in R_0 and R. These can be solved iteratively. These results are compared in figures 7 and 8 with solutions obtained using the integral equation formulation. It is seen that the approximate results compare well with the exact results. In figure 9 results on R_0 and R are plotted for various values of Ω.

5.1.3 Stability of Harmonic Solutions

Harmonic solutions obtained by ELT compared well with those obtained by integral equation method. However, it would be natural to find whether these harmonic solutions are also stable. Stability of the solution, $y_s = R_0 + R\sin(2\pi\Omega X)$ can be studied by considering the first order variational equation. The variational equation in this case is a fourth order ordinary linear differential equation with a periodic coefficient. To study the stability of y_s, it is required to evaluate the characteristic exponents of the variational equation. Earlier it is mentioned that for the linear equation given in eq. (23), two of the eigen values are with positive real parts and remaining two are having negative real parts. Based on this observation one can say that criterion for the stability of y_s is that real parts of two of the characteristic exponents should be positive and remaining two should be negative. The characteristic exponents are evaluated by constructing the fundamental solution matrix. These exponents are obtained for the harmonic solutions corresponding to $\Omega = 0.1$ and $\Omega = 0.15$. For these values of Ω the solutions are $R_0 = 0.93$, $R = 1.51$ for $\Omega = 0.1$ and $R_0 = 0.68$, $R = 0.56$ for $\Omega = 0.15$. The values of the real parts of the characteristic exponents, β_k, ($k = 1, 2, 3, 4$) for these two solutions are obtained as $\beta_{1,2} = \beta_{3,4} = \pm0.0288$ for $\Omega = 0.1$ and $\beta_{1,2} = \beta_{3,4} = \pm0.0488$ for $\Omega = 0.15$. Both the solutions are stable. It can be mentioned here that construction of the fundamental solution matrix requires numerical integration of the variational equation over one period of the periodic coefficient. Due to inherently unstable nature of the original system, numerical errors, during the process of integration get build up. This is particularly true for the cases of long period solutions or smaller values of Ω.

5.1.4 Tracks with Softening Foundation

Yielding behaviour of subgrade can also be simulated by considering the track foundation to be made up of softening springs. For such a model, track displacement is given by

$$EI\frac{d^4y}{dx^4} + K_t(y - \epsilon y^3) = a + b\sin(2\pi fx) \tag{35}$$

where, ϵ is the nonlinearity parameter. Solution of this equation obtained through integral equation method is shown in figure 10. This result correspond to $f = 0.1$ cycles/m and $\epsilon = 10^{-4}$ mm^{-2}. ELT can also be used to obtain approximate solution of eq. (35). Here, the nonlinear term is linearised as $\epsilon y^3 = \gamma y$. The solution is obtained as $y = C_0 + C_1\sin(2\pi fx)$ with $C_0 = a/[K_t(1 - \gamma)]$ and $C_1 = b/[EI(2\pi f)^4 + K_t(1 - \gamma)]$. Minimisation of mean square error over one period of the loading gives following transcendental equation for γ.

$$\gamma = \{\epsilon\{C_0^4 + 3C_0^2C_1^2 + 3C_1^4/8\}\}\{C_0^2 + 0.5C_1^2\}^{-1} \tag{36}$$

Solution obtained by solving this equation is copared in figure 10 with the solution obtained by integral equation method. The ELT solution compare well with the exact results. In figure 11 results on C_0 and C_1 are plotted for various values of f.

5.3 STOCHASTIC FORCES ON NONLINEAR TRACK

A track with stochastic irregularity is subjected to random forces during the passage of trains. For this case one needs to solve eq. (20), where $p(x)$ is stochastic. This stochastic force is obtained by considering a moving train on a rigid track. Here we consider tracks with elasto-plastic tensionless springs. Since an exact solution is not known, the equivalent linearisation technique as suggested by Roberts and Spanos (1990) for an unsymmetrical nonlinearity is followed. Accordingly, one takes $y(x) = m_y + z(x)$, where, $z(x)$ is a zero mean process and m_y is the constant mean of y. This transforms the track eq. (20) into

$$EI\frac{d^4z}{dx^4} + g(z + m_y) = p_0 + p(x) \tag{37}$$

The nonlinear term is linearised as

$$g(z + m_y) \approx k_1 z + k_2 \tag{38}$$

Taking expectation on eq. (38) one gets,

$$< g(z + m_y) > = p_0 \approx k_2 \tag{39}$$

Further, minimisation of mean square error leads to

$$k_1 = < zg(z + m_y) > / < z^2 > \tag{40}$$

The spatial fluctuation, $z(x)$ is a gaussian random field and hence eqs. (39) and (40) can be solved iteratively for k_1 and m_y after recognising that

$$< y^2 > = < z^2 > = \sigma^2 = \int_0^\infty \frac{\phi_{pp}(f)df}{(EIf^4 + k_1)^2} \tag{41}$$

Numerical results are obtained for $\bar{p} = 1.5 \times 10^3$ kg-sec^2/m^2, $\bar{K}_v = 3.5 \times 10^5$ kg/m^2 and $\bar{C}_v = 2 \times 10^3$ kg-sec /m^2. The value of y_c is taken to be 7 mm. The solutions for m_y and σ as functions of velocity are shown in figure 12. Here it is observed that the mean solution depends on the velocity in the case of nonlinear foundation. This is in contrast with a linear foundation wherein, the mean track deflection is independent of the speed of the train. Once the PSD of track deflection is known, results on the permanent settlement can easily be obtained. The quantities of interest are the number of regions where yield has occurred and the length of these regions. Since the track deflection is gaussian, the average spatial rate of upward crossings of the yield level y_c is given by

$$N^+(y_c) = \frac{\sigma_2}{2\pi\sigma} \exp[\frac{-(y_c - m_y)^2}{2\sigma^2}] \; ; \; \sigma_2^2 = < (dy/dx)^2 > \tag{42}$$

Similarly the spatial average fractional rate of the track which lies above the yield level is given by

$$D(y_c) = 0.5 \text{Erfc}[(\sqrt{2}\sigma)^{-1}(y_c - m_y)] \tag{43}$$

where, Erfc( ) is the complementary error function. At 100 kmph, the values of $N^+(y_c)$ and $D(y_c)$ are obtained as 0.02 and 0.05 respectively. This implies that in a track length of 100 meters the yield level will be crossed twice and average length of yield will be nearly five meters.

6 Summary and Conclusions

Inclusion of vehicle characteristics and vertical irregularity in the dynamic analysis of railway tracks leads to coupled partial differential equations with stochastic input. Solution of these equations can be obtained once track frequency response function is known. Here, track frequency response function is the response of a track with harmonic vertical profile under a moving vehicle. A new method to obtain this frequency response function is described. PSDs of track and vehicle response show considerable energy in the vicinity of vehicle natural frequency.

For track degradation studies it is necessary to consider yielding behviour of the subgrade. With the inclusion of subgrade nonlinearity it becomes very difficult to solve coupled track-vehicle equations. Quasi-static approach proposed by Iyengar and Jaiswal (1994) considerably simplifies the problem. In this, track model reduces to a beam on nonlinear foundation subjected to spatially varying stochastic static forces. This is a boundary value problem with boundaries at $\pm\infty$. Exact solution to this problem is still difficult. Here, for the case of harmonic loading, solution has been obtained using the method of integral equations. These results indicate that solutions no longer remain strictly harmonic. This hints at the possibility of existence of more complicated solutions. However, due to slow convergence of these solutions, integral equation method requires large computational effort. Equivalent linearisation as proposed here can lead to reasonable engineering approximation. However, other theoretical and numerical methods need to be developed for this nonlinear boundary value problem to obtain better solutions under deterministic and stochastic forces.

References

Choros, J., and Adams, G. G. (1979) "A steadily moving load on an elastic beam resting on a tensionless Winkler foundation", Journal of Applied Mechanics, ASME, Vol. 46, pp. 175-180.

Duffy, D. G. (1990) "The response of an infinite railroad track to a moving, vibrating mass", Journal of Applied Mechanics, ASME, Vol. 57, pp. 66-73.

Endo, M., and Inoue, M. (1990) "A fundamental study on the rolling noise of the wheel/rail system of a railway vehicle", Transactions of the Japan Society of Mechanical Engineers (In Japanese), Vol. 56, No. 256, Series C, pp. 70-75.

Frederick, C. O. (1978) "The effect of wheel and rail irregularities on the track", Proceedings of the first International Heavy Haul Conference, Perth, Australia, pp. 1-14.

Fryba, L., Nakagiri, S., and Yoshikawa, N. (1991) "Stochastic analysis of a beam on random foundation with uncertain damping subjected to a moving load", in *Nonlinear Stochastic Mechanics*, IUTAM Symposium, Turin (Editors: Bellomo, N., and Casciati, F.), Springer-Verlag, Berlin.

Iyengar, R N and Jaiswal O R (1994) "Stochastic Analysis of Railway Tracks." Proceedings of *Second International Conference on Computational and Stochastic Mechanics* held at Greece, June 13-15.

Mulcahy, T. M. (1973) "Steady state responses of a beam on idealised strain-hardening foundations for a moving load", Journal of Applied Mechanics, ASME, Vol. 40, pp. 1040-1044.

Na, T. Y. (1979) *"Computational Methods in Engineering Boundary Value Problems"*, Academic Press, New York.

Patil, S. P., (1988) "Response of infinite railroad track to vibrating mass", Journal of Engineering Mechanics, ASCE, Vol. 114, No. 4, pp. 688-703.

Roberts J B and Spanos P D (1990) *"Random Vibration and Statistical Linearisation"* John Wiley & Sons, West sussex, England.

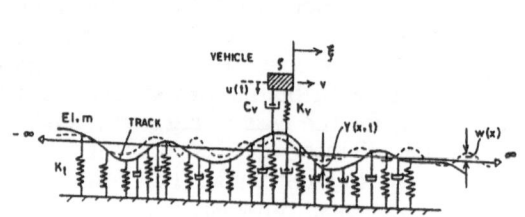

Fig 1 Track-vehicle system

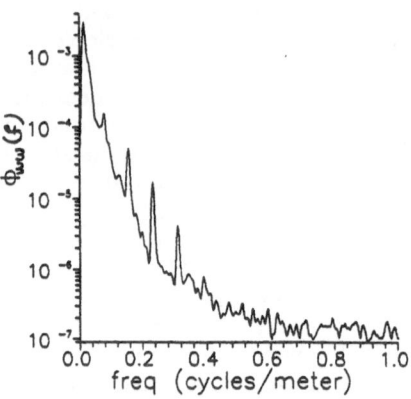

Fig 2 PSD of vertical profile

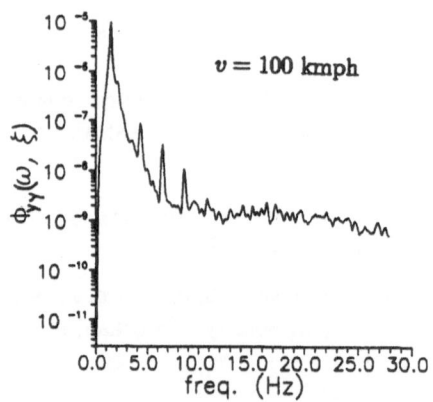

Fig 3 PSD of track response at $\xi = 0$

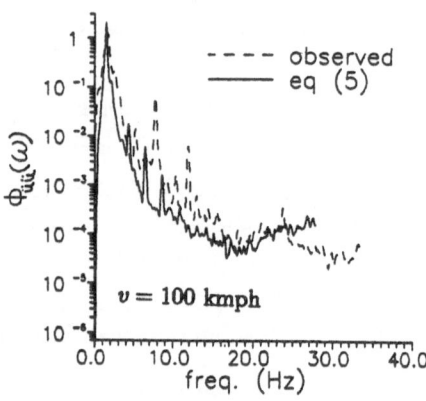

Fig 4 PSD of vehicle acceleration

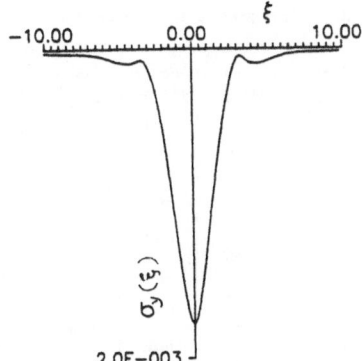

Fig 5 Variation if σ_y with ξ

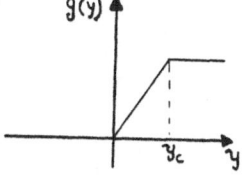

Fig 6 Nonlinear spring force

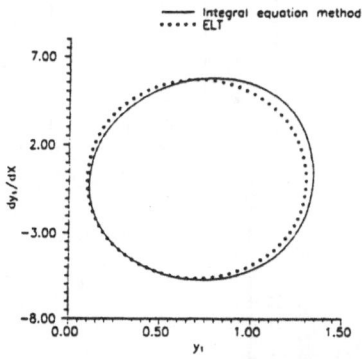

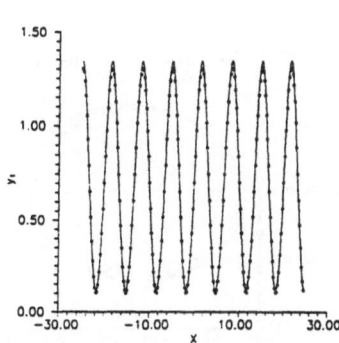

Fig 7 Solution under harmonic
loading; $A = B = 2.0$, $\Omega = 0.15$

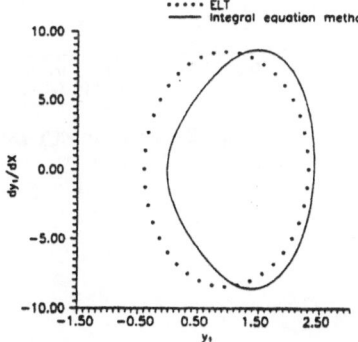

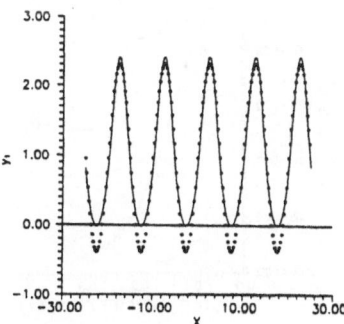

Fig 8 Solution under harmonic
loading; $A = B = 2.0$, $\Omega = 0.1$

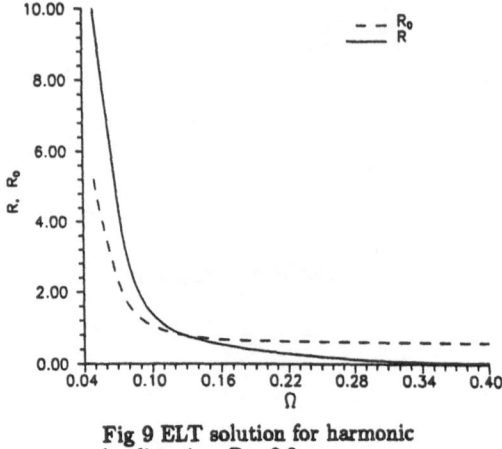

Fig 9 ELT solution for harmonic
loading; $A = B = 2.0$

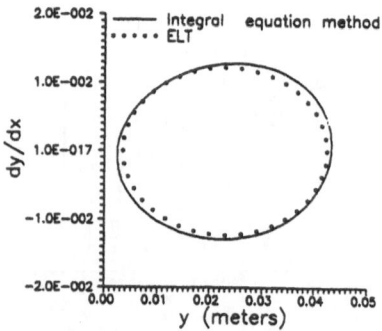

Fig 10 Solution under harmonic
loading; $a = b = 10000$ kg/m,
$\epsilon = 10^{-4}$ mm^{-2}, $f = 0.1$ cyc/m

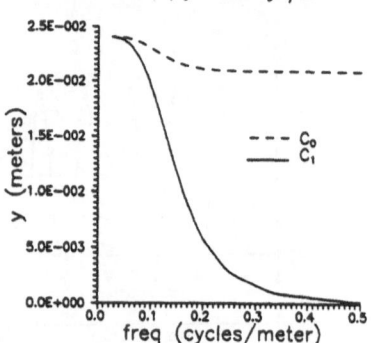

Fig 11 ELT solution;
$a = b = 10000$ kg/m, $\epsilon = 10^{-4}$ mm^{-2}

Table 1 Convergence of the peak amplitude, $A = B = 2.0$

Iteration No.	$\Omega = 0.15$ Peak amplitude	Successive difference	$\Omega = 0.1$ Peak amplitude	Successive difference
1	1.205		1.237	
		0.041		0.171
2	1.247		1.408	
		0.027		0.139
3	1.274		1.547	
		0.019		0.118
4	1.293		1.665	
		0.014		0.105
5	1.307		1.770	
		0.009		0.092
6	1.317		1.862	
		0.007		0.084
7	1.324		1.946	
		0.005		0.074
8	1.329		2.020	
		0.003		0.069
9	1.332		2.089	
		0.001		0.063
10	1.335	.	2.152	
				0.062
11	.	.	2.260	
				0.050
12	.	.	2.310	
				0.045
13	.	.	2.355	
				0.042
14	.	.	2.397	
				0.039
15	.	.	2.436	

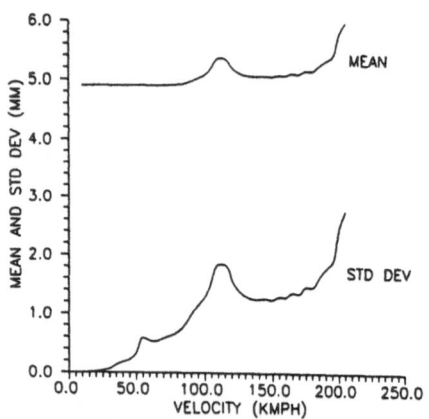

Fig 12 Stochastic ELT solution

MONTE CARLO SIMULATION OF DYNAMICAL SYSTEMS OF ENGINEERING INTEREST IN A MASSIVELY PARALLEL COMPUTING ENVIRONMENT: AN APPLICATION OF GENETIC ALGORITHMS

ERIK A. JOHNSON AND LAWRENCE A. BERGMAN
Department of Aeronautical and Astronautical Engineering
University of Illinois at Urbana-Champaign
Urbana, Illinois 61801 USA

DAVID E. GOLDBERG
Department of General Engineering
University of Illinois at Urbana-Champaign
Urbana, Illinois 61801 USA

SHIRLEY J. DYKE
Department of Civil Engineering and Geological Sciences
University of Notre Dame
Notre Dame, Indiana 46556 USA

1. Introduction

The evolution of a stochastic dynamical system is governed by a Fokker-Planck equation if its response process is Markovian. An analytical solution for nonstationary response does not exist for any but the simplest systems of engineering interest. The evolution of the transition probability density function over the phase space has been solved numerically for various two- and three-state systems subjected to additive and multiplicative white noise excitation using the finite element method [27,28]. Systems of higher order, however, can pose significant difficulty when using standard finite element formulations due to memory requirements and computational expense, leading to the use of various economization measures, a discussion of which lies beyond the scope of this paper.

Direct Monte Carlo simulation (MCS), while often regarded as less elegant than other methods, can indeed be used to solve transient problems of significant complexity. Low-order systems are often more efficiently investigated by other means (*e.g.*, stochastic averaging, path integral methods, finite element methods, etc.). However, a straightforward finite element solution with a grid of n points in each dimension requires one solution of n^d equations followed by forward and backward substitution at each time step for a d-dimensional problem. Thus, the required number of computations and memory allocation grow exponentially with the dimensionality of the problem. While these matrix equations are not fully populated, and in fact have relatively narrow bandwidth if node numbering is done optimally, the number of calculations required to solve them is at least n^d. Contrastingly, the number of computations required by Monte Carlo simulation is proportional to d and to the number of realizations. Furthermore, the accuracy of the Monte Carlo simulation is not dependent on the dimensionality of the system, but

225

A. Naess and S. Krenk (eds.), IUTAM Symposium on Advances in Nonlinear Stochastic Mechanics, 225–234.
© 1996 *Kluwer Academic Publishers.*

rather on the magnitude of the density function in areas of interest and on the number of realizations used to characterize the system [21].

The number of realizations required to accurately produce the transition probability density function over the entire phase space, especially in the tails, is large, but the Monte Carlo simulation is easily and efficiently adapted to massively-parallel computation, because each realization is entirely independent of the others. Furthermore, selective simulation methods have been and are currently being developed to refine the computation in the tail region, thus requiring fewer realizations for equivalent accuracy. Consequently, Monte Carlo simulation may be more efficient for higher-dimensional systems than other solution methods currently in use.

The purpose of this paper, then, is twofold. First, we will examine the above observations and lay the foundation for a new class of selective simulation procedures incorporating many of the ideas and strengths from the field of *Genetic Algorithms* (GAs), touching upon the performance of MCS on various platforms, including a massively-parallel supercomputer and distributed-network workstations. Second, we will examine several systems drawn from previous work by several of the authors [13] in order to show how the response of higher dimensional systems might be effectively visualized.

2. Application of Genetic Algorithms to Monte Carlo Simulation

In performing Monte Carlo simulation of large scale, stochastically excited, dynamical systems in a massively-parallel computing environment, it has been observed that the number of realizations required for good accuracy in the tails is formidable. Therefore, a selective form of Monte Carlo simulation known as *Double and Clump* (D&C) [21] has been adopted. However, it became apparent, as this procedure was implemented, that it acts much like a very basic genetic algorithm, and that a greater degree of sophistication may be employed with a GA.

2.1. WHY IS THIS IMPORTANT?

Recent efforts directed toward analyzing *large-scale* dynamical systems, with random coefficients and forcing, reinforce the prevailing wisdom that Monte Carlo simulation is the analysis method of choice at this time. However, these methods can be very time and resource consuming when the behavior of the response process in the tails of the probability distribution must be estimated (*e.g.*, to examine the reliability of the system) or when the distribution itself is complex and multimodal. Under those circumstances, the choice is straightforward: either employ millions of individual realizations or develop economization techniques to limit the number of realizations required to estimate a given level of probability with sufficient accuracy. The potential savings in resources as well as the impact upon the field of stochastic dynamical systems warrant this development.

2.2. BRIEF DESCRIPTION OF THE CURRENT STATE-OF-THE-ART

In a recent review of methods currently available to analyze large-scale dynamical systems under stochastic loading, their merits, limitations, and potential utility in engineering practice, it was concluded that only numerical procedures such as Monte Carlo simulation are capable of treating both systems of high dimension and nonlinearities without restriction to obtain information regarding the tails of the response distribution [25]. Since system reliability is perhaps the most important result of an analysis of this type, small probabilities, sometimes vanishingly small, must be computed. In direct simulation, the magnitude of the probability being estimated governs the number of realizations of the response that is required. The average relative error from a MCS in areas of low probability is proportional to $1/\sqrt{np}$, where n is the number of realizations and p is the probability. Given failure probabilities in the range of 10^{-4} to 10^{-7}, the required number of realizations in a direct implementation must be in the order of hundreds of millions in order to ensure that sufficient samples fall into the failure domain to facilitate an accurate estimate of the probability of failure. Even when the probability density function is required on the interior of the domain, many thousands of realizations may be required to ensure that features such as multimodality are accurately captured and portrayed. This is, of course, computationally unattractive, as discussed in a number of recent papers [12,21,25,22].

This shortcoming leads directly to the notion of selective simulation. One frequently cited example of this is importance sampling, wherein the original distribution, reflecting direct Monte Carlo simulation, is modified through the use of a sampling distribution [24]. The expected distortion of the original distribution is offset by modifying the weight of each realization appropriately. This method is of limited utility when attempting to estimate the tails of the distribution as the weights can become extremely small and difficult to determine accurately. An alternative to importance sampling has recently been proposed [21]. In order to fully describe the concept, we begin with a short discussion of the underlying problem. Consider, for example, a stable dynamical system excited by a continuous Gaussian white noise loading. The probability distribution governing response, beginning with the initial distribution and proceeding to stationarity, constitutes a cloud orbiting in the phase space, with the center of the distribution circling toward the origin and the cloud itself diffusing or coalescing (see, for example, [27]). From the standpoint of reliability, the flow of probability into the low probability region is of paramount importance. To approximate the flow, a procedure called *Double and Clump* is utilized. As described by Pradlwarter, *et al.* [21], certain trajectories are doubled at appropriate times, assuming they meet specific criteria based on energy and weight. Other trajectories, failing to meet these criteria, are clumped, thus preserving the total number of realizations. In order to not distort the TPDF, the *weight* of each realization is modified: halved if it was doubled, combined if clumped. Doubling (and halving the weight) has no effect on the instantaneous TPDF, but clumping does have such an effect (though first moments are preserved). In order to minimize the change, only realizations close to each other in the phase space are allowed to clump. This proximity requirement, however, causes serious problems with parallel implementation due to the need to search all other realizations for a sufficiently close partner with which to clump, thus requiring many slow inter-processor communications.

In order to alleviate this barricade to high parallel efficiency, Pradlwarter, *et al.* [22] recently reported a similar selective Monte Carlo method called *Splitting and Russian Roulette* (S&RR). This newer method operates similar to D&C, but rather than search for two close realizations suitable for clumping, it simply kills off one suitable realization and normalizes all other weights as necessary.

The *Double and Clump* and *Splitting and Russian Roulette* Monte Carlo procedures have been shown to give excellent results, even for dynamical systems of relatively large dimension, which are inaccessible by other methods. However, the authors realized that these procedures resemble a rudimentary genetic algorithm and have had several exchanges of correspondence [23,1] discussing the possibility of a joint development program with the Institute for Mechanics at the University of Innsbruck. The authors are, thus, presently engaged in incorporating some of the robust machinery from the field of genetic algorithms into the framework of Monte Carlo simulation.

2.3. IMPLEMENTATION ISSUES

The current study focuses on adapting work in genetic algorithms pioneered in part by one of the authors (Goldberg) [4] to improve the quality, speed, and reliability of large-scale Monte Carlo computations. The following activities have been identified as critical and are currently being evaluated:
- Map the GA to the D&C/S&RR Monte Carlo format most effectively
- Consider the implications of the fact that meiotic (haploid) reproduction is to a GA what mitotic (single-cell) reproduction is to S&RR
- Investigate the importance of niching, sharing, and crowding
- Improve sampling trade-off with theory and confirming experiments
- Consider overall computational efficiency on serial and parallel machines

In the remainder of this section, each of these is discussed in somewhat greater detail.

To map the genetic algorithm to the Monte Carlo computation effectively requires that we recognize the component parts, as noted in Table 1 showing a comparison of a traditional GA with the S&RR algorithm. In the "doubling" portion of the D&C (or "splitting" in the S&RR) algorithm, the states of a realization are replicated and permitted to evolve independently. Clearly, this is an analog of pure selection and mitotic reproduction. In the "clumping" portion of D&C, two realizations are combined using an averaging technique, and the single offspring is allowed to carry on. This is an analog of adverse selection and recombination. Likewise, the "Russian Roulette" portion in S&RR is adverse selection and death of unfit "genetic" material. The current study will perform similar operations; however, recognizing the connection to GAs permits the use of the current state of GA art to design more effective codes and operators.

One area of inquiry that should be particularly important is the use of niching or a similar operator. Since the probability density functions being investigated are often multimodal, it is important to spread one's sampling over a representative sampling of the most important peaks. The D&C and S&RR schemes, by preferring high energy and high weight realizations, have something like a niching effect, but care is required since competitive population schemes are notorious for converging or drifting to local solutions, even when there is no difference between alternative solutions [4]. One of the authors (Goldberg) has been involved in the development of the most widely used schemes, which will be adapted to this development, to stably select multiple optima, including sharing, crowding, tournament niching and others [8,19,17.].

TABLE 1. Similarities and differences between *Splitting and Russian Roulette* and *Genetic Algorithms*.

Simple Genetic Algorithm (sGA)	Splitting and Russian Roulette (S&RR)
selection based on fitness	selection based on energy/weight
two parents	one parent
meiotic (haploid) reproduction	mitotic (single-cell) replication
crossover	cloning
mutation	integration through next time steps

Monte Carlo simulation involves large quantities of computation, but oftentimes sensible sampling strategies can reduce the burden considerably. GA work in this area has been limited, and the most widely cited paper [11] is tantalizingly suggestive, but does not offer concrete heuristics or computational assistance for making the exploration-sampling trade-off. As part of this work, the population-sizing work pioneered by one of the authors (Goldberg) [7] will be used to optimize the number of samples given to each structure under a fixed clock-time constraint.

Such an optimization will be helpful, but it will also be important to optimize the overall setup so that a fixed amount of time on a given number of processors will yield the solution set of highest quality. Some preliminary calculations and experiments have been run in this area, and the results are promising.

Actually, results such as these were foreshadowed as early as 1989 [5], and shortly thereafter two independent applications studies [14,9] showed that small populations with restart could be effective in many problems. Subsequent analysis within either a fixed-reliability or fixed-computation-time framework have unified these results with population sizing based on sound statistical decision making [9].

In general, many very small populations beat a few larger populations when it comes to getting a quality solution in fixed time. The ramifications for this study are clear. Optimizing the algorithm for efficiency at the setup and the sampling levels will beat solution techniques that have not considered these factors. Furthermore, since the GA approach is inherently parallelizable, a natural synergy exists with prior work in that area [12].

An important and interesting issue [23,1] and one that is currently being addressed is the trade-off between using computational resources to parallelize the system finite element model versus using the resources to parallelize the MCS-GA. In the general case, efficient decomposition of a structure can be difficult, and speedups of parallelization at the finite element level are likely to be far from linear. As a result, we are proceeding under the assumption that parallelization at the MCS-GA level will be the more fruitful course. Recent analytical results [9] suggest that relatively small populations should communicate relatively infrequently to obtain the best answer given either a fixed-reliability or a fixed-computational-time constraint. This is good news and suggests that speedups should increase nearly linearly as the number of processors used increases.

2.4. DESCRIPTION OF EARLY EXPERIMENTS

A series of preliminary investigations have been completed and show significant promise, but also that further study is required. The primary thrust has been in developing a simple genetic algorithm within the Monte Carlo framework. Given that the "fitness" in a GA is somewhat arbitrary, several schemes were attempted to give the best "push" out into the probability tails.

Essentially, the MCS-GA functions identically to that of a normal MCS for a stochastic dynamical system, except that before every m^{th} time step integration, a number of "fit" realizations are randomly chosen and mated with other realizations in that group, producing one or more children each. In order to keep the population size constant, a like number of "unfit" realizations are removed from the population, i.e., killed. (Constant population is not necessarily a requirement in a GA, but great care must be taken otherwise to require that the population size not grow or decline overly much.) Several issues, such as how often to do the GA, how many should mate in a given generation, the definition of "fit" and "unfit", and the crossover method (i.e., determining the states and weight of a child from those of its parents, along with new weights for the parents), are crucial to the performance of the MCS-GA. If chosen poorly, these can cause the MCS-GA to perform terribly, with inaccuracies magnitudes larger than a normal MCS. Choosing wisely, however, gives the MCS-GA accuracy several orders of magnitude better than the MCS.

It was quickly observed, as is true with S&RR and D&C, that performing the GA recombination in every time step was not only wasteful in computational resources, but often detrimental to the accuracy of the solution; every three to five time steps seemed to work well. A similar phenomenon occurs with the fraction of the population that is allowed to mate. If it is too small, there is little difference from a normal MCS; too large, and it disrupts trajectories too much. Values on the order of 10-20% appear to work well.

The fitness definition is probably the most crucial, and the most difficult. Thus far, the best scheme appears to be similar to those used in S&RR and D&C, with the fitness given as

$$\text{fitness} = \left(\frac{\text{weight}}{\text{max weight}}\right)^{\alpha}\left(\frac{\text{energy}}{\text{max energy}}\right) \tag{1}$$

with α in the range 0.1 to 0.3.

Another operator with many variations is crossover . Many GAs use a binary crossover. Here, that would mean using the most significant bits of the states of one parent and the least significant from the other. This was seen to be of little help in pushing mass toward the probability tails, and in fact often had the opposite effect. Similarly, a crossover in Euclidean space is obtained by letting the child be at a random point on the line between the two parents in the phase space. This, too, often had a detrimental effect since the line between two "fit" parents may pass through regions of low energy. Thus the method adopted here was to place the child randomly on a line of minimal energy gradient from one parent to the other, but letting one state vary spatially, and the other be determined such that the child lies on such a line.

In order to minimize the bias caused by the crossover, it was found beneficial to restrict the distance a child can be from its parent. Further study is required to fully understand the effect of this restriction.

Figure 1 shows a comparison of normal MCS and two of the variants studied herein. With 1000 realizations, a normal MCS is unable to determine probabilities lower than 10^{-3}, but S&RR and MCS-GA are able to go a couple of orders of magnitude lower with an identical population size.

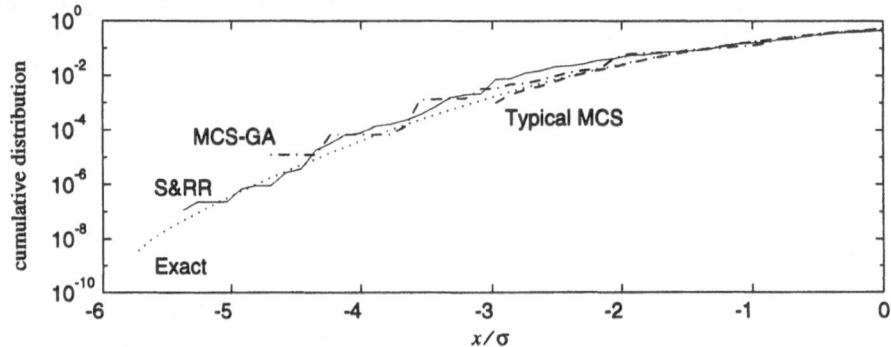

Figure 1. Cumulative distribution function for a linear oscillator excited by Gaussian white noise: exact, MCS, S&RR, and MCS-GA.

One modification to the *Splitting and Russian Roulette* scheme of Pradlwarter, *et al.* that was expected to be promising was allowing a parent to produce two or more identical children. This expectation was based upon the idea that one may define "fitness" in a biological sense as an organism's reproductive "power", *i.e.*, its ability have many thriving offspring. This modification did not, however, have any noticeable positive effect. One possible explanation is that those multiple children are in competition against each other in subsequent generations.

3. An Examination of Some Results

Three simple benchmark problems will be discussed herein:
- a linear oscillator subjected to a filtered white noise excitation (a simple model of a building excited by a Kanai-Tajimi earthquake excitation), resulting in a four-state system [12]
- a Duffing oscillator subject to a band-limited white noise
- a linear oscillator subjected to a low-pass filtered Gaussian white noise (3 states)

The first is included principally to provide some performance measures that can be used to assess the efficacy of the MCS-GA on serial and parallel computers. The second is a system for which there is no analytical solution, even at stationarity. The latter, for which an exact solution can be readily determined, provides an excellent vehicle for visualization.

3.1. THE FOUR-STATE LINEAR SYSTEM

A total of 210 simulations of the four-state oscillator system were performed on each of six computing platforms, varying the number of realizations and many other parameters. One characteristic set was used to compare computational efficiency; Figure 2 shows the floating-point efficiency and memory requirements on various platforms. (Advantages and disadvantages of the various platforms are discussed in [13].)

For a constant number of realizations, the effect on performance when adding S&RR or a simple GA for improving tail accuracy is minimal as long as the scheme does not hamper the parallelism inherent in MCS. But the improved accuracy will permit a greatly reduced number of realizations to achieve the same level of accuracy, providing tremen-

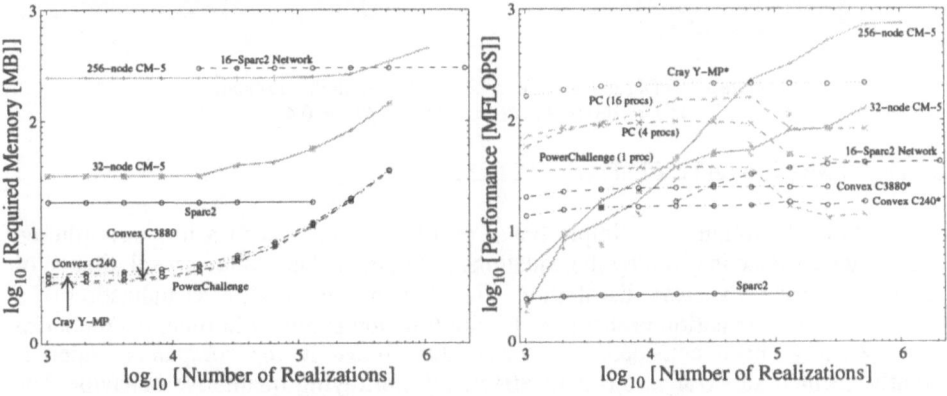

Figure 2. 4-D linear MCS performance on various platforms.

dous savings in computational resources. To date, the MCS-GA has not yet been tested on larger computational platforms or on more complex dynamical systems, but the effect on computational resource requirements is expected to be minimal.

3.2. THE DUFFING OSCILLATOR DRIVEN BY BAND-LIMITED WHITE NOISE

The Duffing oscillator has been shown to display chaotic behavior when driven by a sinusoidal force of certain amplitudes and frequencies [18]. Preliminary investigations of this system, with examination of the sensitivity of the density function to the excitation bandwidth, have been reported [13], but the visualization was incomplete and is concluded here.

The equation of motion for the system is $\ddot{X} + 2\zeta\dot{X} + (\varepsilon X^2 - 1)X = Z(t)$, where $Z(t)$ is a band-limited white noise simulated by the stochastic process (adapted from [26]),

$$Z(t) = Z_0 \sum_{k=0}^{m} \cos\left[\left(\omega_{center} - \Delta\omega/2 + k\Delta\omega/m\right)t + \Phi_k\right] \tag{2}$$

where the Φ_k are uniform random variables on $[0, 2\pi]$. Here, $\varepsilon = 0.1$, $\zeta = 0.2$, $\omega = 1$ rad/sec, $m = 100$, $\omega_{center} = 2$ rads/sec, and Z_0 is such that the energy in the signal is the same as the sinusoid $10\sin(\omega_{center}t)$.

One interesting observation about this system is that for a very narrow band, there appears to be no stationary solution, whereas wider bands tend toward the well-known bimodal distribution caused by pure white noise excitation. Furthermore, the very narrow band forces most of the probability mass into a periodic trajectory; the mass near that trajectory diminishes rapidly as the band is widened.

Figure 3 shows two final frames of the visualization of the evolutionary TPDF of this system on a companion video. One of the interesting differences is that wider bands go to a stationary solution whereas narrower bands do not.

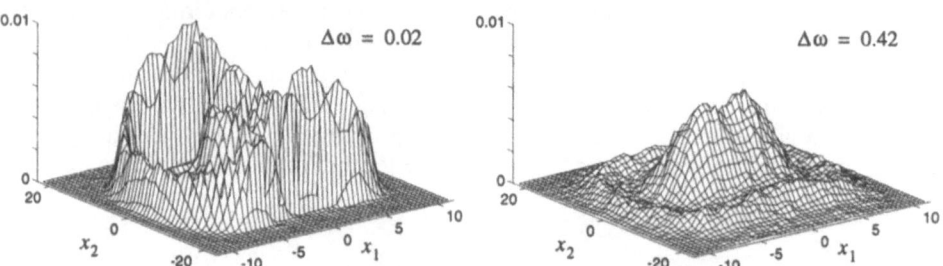

Figure 3. TPDFs of Duffing oscillator with band-limited excitation
at $t = 160$ secs : $\Delta\omega = 0.02$ and $\Delta\omega = 0.42$.

3.3. THE THREE-STATE LINEAR SYSTEM

The equation of motion of a simple linear oscillator subjected to a low-pass filtered Gaussian white noise is given by the equations in Figure 4. The stationary solution is, of course, a trivariate Gaussian distribution. The challenge here is the visualization of a three-dimensional transition probability density function evolving in time. A number of approaches have been considered; however, that shown in the companion video is currently deemed superior in terms of effectively portraying qualitative behavior. The final frame of the simulation is represented in Figure 4.

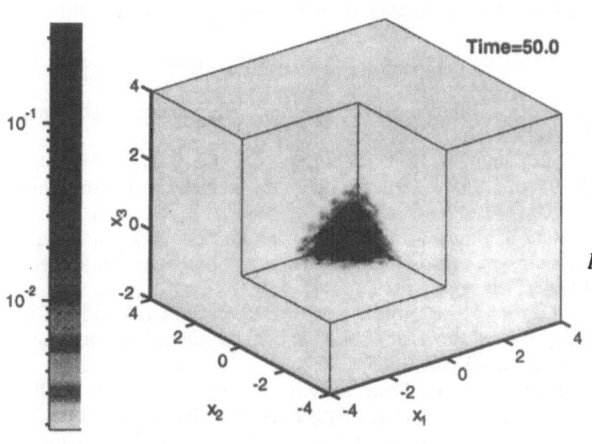

$$\dot{X}_1 = X_2$$
$$\dot{X}_2 = -\omega^2 X_1 - 2\zeta\omega X_2 + X_3$$
$$\dot{X}_3 = -\alpha X_3 + W(t)$$

$W(t)$ is Gaussian white noise.
$$E[W(t)] = 0$$
$$E[W(t)W(t-\tau)] = 4\zeta\delta(\tau)$$
$\omega = 1$ rad/sec, $\zeta = 0.2$, $\alpha = 1$

Figure 4. Visualization of a 3-D transition probability density function: the stationary distribution of a 3-D linear system driven by Gaussian white noise. (The cutout is centered at the mean. The colorbar shows the magnitude of the TPDF.)

4. Conclusions

A potentially useful relationship between genetic algorithms and Monte Carlo simulation has been identified. Building upon the *Double and Clump* and *Splitting and Russian Roulette* strategies of Pradlwarter, *et al.*, the MCS-GA shows promise, but is not yet fully developed. Further work is required, especially in determining the effect of bias and loss of independence caused by sampling schemes. Furthermore, the use of niching or similar operators must be examined for their potential in constructing complex distributions.

5. Acknowledgments

This project has been supported in part by National Science Foundation contracts ECS-9224828, CEE-92000-4N, and MSS-94000-1N, the latter two through the National Center for Supercomputing Applications at the University of Illinois at Urbana-Champaign (Bergman and Johnson) and AFOSR Grant No. F49620-94-1-0103 (Goldberg).

6. References

1. Bergman, L.A. and Goldberg, D.E.: *Personal Communications to G.I. Schueller*, 1994-1995.
2. Bergman, L.A., Spencer, B.F., Jr., Wojtkiewicz, S.F., and Johnson, E.A.: Robust Numerical Solution of the Fokker-Planck Equation for Second Order Dynamical Systems Under Parametric and External White Noise Excitations, in W. Langford, W. Kliemann and N. Sri Namachchivay (eds.), *Fields Institute Communications, Nonlinear Dynamics and Stochastic Mechanics*, American Mathematical Society, in press. Presented at The Fields Institute for Research in Mathematical Sciences, Waterloo, Ontario, CANADA, August 28-September 1, 1993.
3. Deb, K., and Goldberg, D.E.: An Investigation of Niche and Species Formation in Genetic Function Optimization, in J.D. Schaffer (ed.), *Proceedings of the Third International Conference on Genetic Algorithms*, Kaufmann, San Mateo, Calif., 1989, 42-50.
4. Goldberg, D.E.: *Genetic Algorithms in Search, Optimization, and Machine Learning*, Addison-Wesley, Reading, Massachusetts, 1989.
5. Goldberg, D.E.: Sizing Populations for Serial and Parallel Genetic Algorithms, in J.D. Schaffer (ed.), *Proceedings of the Third International Conference on Genetic Algorithms*, Kaufmann, San Mateo, Calif., 1989, 70-79.

6. Goldberg, D.E.: Making Genetic Algorithms Fly: A Lesson from the Wright Brothers, *Advanced Technology for Developers* **2** (1993), 1-8.
7. Goldberg, D.E., Deb, K., and Clark, J.H.: Genetic Algorithms, Noise, and the Sizing of Populations, *Complex Systems* **6** (1992), 333-362.
8. Goldberg, D.E., Deb, K., and Horn, J.: Massive Multimodality, Deception, and Genetic Algorithms, *Parallel Problem Solving from Nature* **2** (1992), 37-46.
9. Goldberg, D.E., Kargupta, H., Horn, J., and Cantu-Paz, E.: *Critical Deme Size for Serial and Parallel Genetic Algorithms* (IlliGAL Report No. 95002), Illinois Genetic Algorithms Laboratory, University of Illinois, Urbana, Illinois, 1995.
10. Goldberg, D.E., and Richardson, J.: Genetic Algorithms with Sharing for Multimodal Function Optimization, in J.J. Grefenstette (ed.), *Proceedings of the Second International Conference on Genetic Algorithms*, L. Erlbaum Associates, Hillsdale, New Jersey, 1989, 41-49.
11. Grefenstette, J.J. and Fitzpatrick, J.M.: Genetic Search with Approximate Function Evaluations, in J.J. Grefenstette (ed.), *Proceedings of an International Conference on Genetic Algorithms and Their Application*, L. Erlbaum Associates, Hillsdale, New Jersey, 1985, 20-27.
12. Johnson, E.A., Wojtkiewicz, S.F., and Bergman, L.A.: Some Experiments with Massively Parallel Computation for Monte Carlo Simulation of Stochastic Dynamical Systems, in P.D. Spanos (ed.), *Computational Stochastic Mechanics*: Proceedings of the Second International Conference on Computational Stochastic Mechanics, Athens, Greece, June 12–15, 1994, Balkema, Rotterdam, 325-336.
13. Johnson, E.A., Wojtkiewicz, S.F., Bergman, L.A., and Spencer, B.F., Jr.: Observations with Regard to Massively Parallel Computation for Monte Carlo Simulation of Stochastic Dynamical Systems, presented at the *Third International Conference on Stochastic Structural Dynamics*, San Juan, Puerto Rico, January 15-18, 1995.
14. Karr, C.L.: *Analysis and Optimization of an Air-injected Hydrocyclone* (TCGA Report No. 90001 and Doctoral Dissertation), The Clearinghouse for Genetic Algorithms, University of Alabama, Tuscaloosa, Alabama, 1990.
15. Krishnakumar, K.: Microgenetic Algorithms for Stationary and Non-stationary Function Optimization, *SPIE Proceedings on Intelligent Control and Adaptive Systems, 1196*, 1989, 289-296.
16. Kunert, A. and Pfeiffer, F.: Description of Chaotic Motion by an Invariant Probability Density, *Nonlinear Dynamics* **2** (1991), 291-304.
17. Mahfoud, S.W.: Crowding and Preselection Revisited, in R. Manner and B. Manderick (ed.), *Parallel Problem Solving from Nature: 2*, Elsevier, The Hague, 1992, 27-36.
18. Moon, F.C.: *Chaotic Vibrations: An Introduction for Applied Scientists and Engineers*, Wiley, New York, 1987.
19. Oei, C.K., Goldberg, D.E., and Chang, S.-J.: *Tournament Selection, Niching, and the Preservation of Diversity* (IlliGAL Report No. 91011), Illinois Genetic Algorithms Laboratory, University of Illinois, Urbana, Illinois, 1991.
20. Pradlwarter, H.J.: A Selective MC Simulation Technique for Non-linear Structural Reliability, *Proceedings of the 6th ASCE Specialty Conference on Probabilistic Mechanics and Structural and Geotechnical Reliability*, Denver, CO, July 8-10, 1992, 69-72.
21. Pradlwarter, H.J., Schuëller, G.I., and Melnik-Melnikov, P.G.: Reliability of MDOF-Systems, *Journal of Probabilistic Engineering Mechanics* **9** (1994), 235-244.
22. Pradlwarter, H.J. and Schuëller, G.I.: On Advanced Monte Carlo Simulation Procedures in Stochastic Structural Dynamics, presented at the *Third International Conference on Stochastic Structural Dynamics*, San Juan, Puerto Rico, January 15-18, 1995.
23. Schuëller, G.I.: *Personal Communications to L.A. Bergman*, 1994-1995.
24. Schuëller, G.I, Bucher, C.G., Bourgund, U., and Ouypornpraser, W.: On Efficient Computational Schemes to Calculate Structural Failure Probabilities, *Probabilistic Engineering Mechanics*, **4** (1989), 10-18.
25. Schuëller, G.I., Pradlwarter, H.J., and Pandey, M.D.: Methods for Reliability Assessment of Nonlinear System under Stochastic Dynamic Loading — a review, *Proceedings of the EURODYN '93*, Balkema, Rotterdam, 1993, 751-759.
26. Soong, T. and Grigoriu, M.: *Random Vibration of Mechanical and Structural Systems*, Prentice Hall, Englewood Cliffs, New Jersey, 1993.
27. Spencer, B.F., Jr. and Bergman, L.A.: On the Numerical Solution of the Fokker–Planck Equation for Nonlinear Stochastic Systems, *Nonlinear Dynamics* **4** (1993), 357-372.
28. Wojtkiewicz, S.F., Bergman, L.A., and Spencer, B.F., Jr.: New Insights on the Application of Moment Closure Methods to Nonlinear Stochastic Systems, presented at the *IUTAM '95 Symposium on Advances in Nonlinear Stochastic Mechanics*, Trondheim, Norway, July 3-7, 1995.

STOCHASTIC RESPONSE OF COUPLED PLATFORM-TETHER SYSTEM UNDER MULTI-DIRECTIONAL SEAS

AHSAN KAREEM
University of Notre Dame
Notre Dame, IN 46556-0767

XIAOBING SONG[1]
University of Houston

1. Introduction

A tension leg platform consists of a floating top platform which is kept in position by vertical tensioned cables referred to as tethers. The overall platform motions can be divided into two categories, e.g., vertical plane and horizontal plane motions. The horizonal plane motions, namely, surge, sway and yaw have a very low frequency which classifies these motions as compliant in nature, whereas, the vertical plane motions, pitch, roll and heave are relatively high frequency. By virtue of their compliant behavior in the horizontal plane, TLPs are less sensitive to loads at the wave-frequency, but their sensitivity to low frequency loads (e.g., wind and slowly varying second-order wave loads) increases. The level of low-frequency loads, though small in comparison with the wave-frequency loads, contributes significantly to the platform response. In a similar manner the vertical plane motions are more sensitive to second- and higher-order effects of waves than the first-order effects and result in spring and ringing types of motion.

A typical compliant offshore platform is exposed to a combination of environmental loads consisting of wind, waves and currents. The wind induced effects are described in terms of mean and fluctuating components based on the space-time structure of the wind field and the aerodynamic characteristics of the platform configuration. Fluctuations in the wave surface profile introduce hydrodynamic loads on a structure at the wave, and at low and high frequencies. The hydrodynamic loads of both viscous and potential origins exist, where the viscous effects are described by drag related and the potential type loads represent diffraction of waves and associated effects. The potential effects normally dominate when structural dimensions are comparable to typical wave lengths (Sarpkaya & Isaacson, 1981).

The tethers of a TLP are generally modelled by linear/nonlinear massless springs. The restoring force depends on the platform displacement in a nonlinear manner. Kareem & Li (1988) expanded the restoring force using Taylor series expansion in terms of the mean, first-order and second-order restoring force components. The modelling of tethers by equivalent springs has been shown to work reasonably well for shallow to moderate

---

1. Present affiliation, Ludwig Buildings Inc., New Orleans.

A. Naess and S. Krenk (eds.), IUTAM Symposium on Advances in Nonlinear Stochastic Mechanics, 235–246.
© 1996 *Kluwer Academic Publishers.*

water depths. However, the relative importance of tether dynamics becomes more significant with an increase in the water depth, as the natural period of tether system falls into the
energetic portion of the wave spectrum. The dynamics of tethers become more significant
as water depth increases primarily due to the tether curvature under distributed loads,
tether distributed mass and damping, and the top-platform setdown. This makes it essential to analyze the top-platform and tethers as a combined system exposed to environmental loads. The coupled analysis of platform-tethers system has been addressed in the
literature with different levels of simplifications in either modelling the coupled behavior
or environmental loading (Patel & Lynch, 1983; Davis and Mungall, 1991; Sircar et al.,
1993 and Kareem & Song, 1993).

   The problem in hand involves a coupled nonlinear system exposed to stochastic
environmental loads, i.e., wind and directional waves. The analysis in this study is
performed in both time and frequency domains employing substructure modelling of the
top-platform and tethers. The combined response analysis is conducted utilizing a parallel
computer to benefit from the parallel architecture which permits computations of different
parts of the system concurrently. The following sections describe the modelling of the load
effects, structural system and estimation of response statistics. Some of the techniques
utilized in this study to make the analysis of this complex problem computationally efficient are also described.

2. Environmental Loads

In Fig. 1, a schematic diagram of environmental loads and their effects on a TLP are
described. Under the influence of wind, the above-water portion of a TLP is subjected to
aerodynamic drag force. To formulate the total fluctuating wind load effects on a TLP, it is
necessary to establish multi-point statistics of the stochastic wind field in terms of spectra
and coherence. Aerodynamic transfer functions are then introduced to relate wind-velocity
fluctuations to the corresponding wind loads. Details of this formulation can be found in
Gurley and Kareem (1993).

   A hybrid combination of wave-induced viscous and potential effects is utilized to
describe wave loads on TLPs. The viscous effects are obtained through the drag term of
the Morison equation, and the potential effects, i.e., diffraction and radiation forces, are
based on the diffraction theory (Sarpkaya and Isaacson, 1981). Details concerning modelling of wave load effects are omitted here for the sake of brevity and the reader is refered
to Salvesen, 1982, de Boom and Pinkster and Tan, 1984, Kareem and Li, 1993 and Li and
Kareem, 1993.

   The description of wave height fluctuations in a statistical framework is provided
by a wave energy spectrum. Under directional seas or short crested sea states the wave
height spectrum is given by

$$S_\eta (f, \theta) = S_\eta (f) \, \psi (\theta, f) \qquad (1)$$

where $S_\eta (f)$ is the unidirectional wave spectrum and $\psi (\theta, f)$ is directional spreading
function, which in some case may be independent of frequency. Typically, analysis in
offshore mechanics is conducted to include a spectrum of frequencies and superposition of
directional components. Naess (1992) has used the concept of a random orthogonal

measure to describe the directionality as a linear superposition of unidirectional Gaussian wave fields.

In this study a different approach has been employed to account for a lack of spatial correlation by using a coherence function approach. The lack of correlation in wave height fluctuations between two points r_1 and r_2 in short-crested seas can be expressed in terms of cross-spectrum

$$S_\eta(r_1, r_2, f) = S_\eta(f) e^{ik[(y_1 - y_2)\cos\bar\theta + (x_1 - x_2)\sin\bar\theta]} \cdot R(\Delta r, f)$$

$$R(\Delta r, f) = exp\left[-\beta\frac{C_n}{\lambda}(\Delta x + \Delta y + \Delta z)\right] \qquad (2)$$

Δx, Δy, Δz are separation distances, β and C_n are constants and λ is wavelength. Details can be found in Mitwally and Noval (1989) and Kareem and Song (1993).

3. TLP Modelling

3.1 TETHER MODELLING

The tether system is modelled using a finite element discretization. The tethers are assumed pinned at the sea bottom and are connected to the top platform by a flex-joint. The overall stiffness matrix of each element is expressed in terms of the following

$$[K] = [K_o] + [K_g] + [K_b] \qquad (3)$$

where $[K_o]$, $[K_g]$ and $[K_b]$ are the elastic, geometric and the initial displacement stiffness matrices. Details concerning the formulation of these matrices are very lengthy, which can be found in Kareem and Song (1993). For deep water situation, the tether becomes very long and very slender which leads to ill conditioning of the stiffness matrix. In this study a hybrid finite element formulation based on the work of McNamara et al. (1986) has been employed which avoids the problems of ill-conditioning. This method employs a Lagrangian constraint which assumes tethers to be axially inextensible. This leads to an additional set of equations which exactly negates the axial extensions of the tethers.

3.2 TOP PLATFORM

The top-platform is treated as a rigid body with six degrees of freedom, namely, surge, sway, yaw and pitch, roll and heave. The platform stiffness without the effects of tethers arises only from buoyancy forces (Kareem and Song, 1993).

4. Sources of Nonlinearity in Loading and Structural System

The overall sources of nonlinearity and their treatment in this problem are briefly discussed here. The sources of nonlinearity can be broadly classified into three categories, i.e., nonlinearities in the wave environment, in the hydrodynamic loads and structural system.

The nonlinearity in a wave environment is manifested through wave profile which has higher peaks and flattened troughs. For example, a second-order stokes wave exhibits these features. A second-order wave introduces higher harmonics which are bound to their basic linear waves and typically sharpen their crests and fatten their troughs. The short-crested seas or directional spreading further complicates the wave field environment.

The nonlinearities of hydrodynamic origin include nonlinearities in Bernoulli's equation, Morison equation, free-surface profile, dependence of loads on platform state, and nonlinear diffraction. The nonlinearity of structural system arises from the nonlinear stiffness that results from the hydrodynamic and elastic effects and the geometric position of the system.

In this paper, treatment of two of these nonlinearities for frequency domain analysis is briefly reported. Details can be found in Kareem and Song (1993).

4.1. NONLINEAR DRAG FORCE ANALYSIS

The drag force acting on a body can be described in terms of Morison equation

$$f_d(t) = 1/2 \rho_w C_d A V(t) | V(t) + 1/2 \rho_w C_d D V'(t) | V'(t) | \eta(t) \quad (4)$$

The first part concerns force contributed by integrating force to mean water level, whereas the second part follows the instaneous wave profile. In this equation ρ_w = water density, C_d = drag coefficient, A is structural area below mean water, and $V(t)$ is the relative structural velocity (Li and Kareem, 1993). The water particle velocity distribution in zone above mean water level is best represented by higher-order wave theories. Approximations are made by a number of theories based on linear wave theory. One of these theories has been employed here for analysis (Li and Kareem, 1993). The preceding equation is expanded into bi-variate and tri-variate Hermite polynomials. This expansion allows a frequency domain analysis of systems excited by such nonlinear loading functions. The force in x direction on a cylindrical body in a frame-invariant form is proportional to

$$V_x(t) | V'(t) | = \sum_{i=0}^{\infty} \sum_{j=0}^{\infty} H^{ij} \quad (5)$$

in which $V'(t) = \{ \begin{matrix} V_x(t) \\ V_y(t) \end{matrix} \}$ and H^{ij} is a bi-variate Hermite polynomial. Its coefficient

matrices are determined by a mean square minimization. Similarly the other component

can be expanded as

$$V_x(t) | V'(t) | \eta(t) = \sum_{i=0}^{\infty} \sum_{j=0}^{\infty} \sum_{k=0}^{\infty} H^{ijk} \quad (6)$$

in which H^{ijk} is a tri-variate Hermite polynomial. The conditions of orthogonality and derivation of coefficient matrices are detailed in Li and Kareem (1993). Following the expansions given in the proceeding equations, the first- and second-order drag forces are described in terms of linear and quadratic transformations of the water particle velocities of the incident wave and the platform motions.

4.2. COMPUTATION OF FORCES AT DISPLACED POSITION OF PLATFORM

Traditionally, the dynamic response analysis of conventional platforms is based on the assumption that the structural displacements are small. This is not valid for compliant structures due to their inherent flexibility. The wave-induced loads computed at the displaced position of a compliant platform may differ from those calculated at the undisplaced position. For example wave force acting on a platform without displacement if given by

$$F_W(t) = \int_{-\infty}^{\infty} F_W(f) \, exp \, (j2\pi f t) \, df \tag{7}$$

in which $F_W(f)$ is the Fourier transform of the wave force. The wave forces acting on a platform with displacement $\xi(t)$

$$F_W(t, \xi) = \int_{-\infty}^{\infty} F_W(f) \, exp \, \{j \, (2\pi f t - k\xi(t))\} \, dt \tag{8}$$

In the time domain analysis, one may use summation of trignometric series to simulate the wave induced force

$$F_W(t, \xi) = \sqrt{2} \sum_{p=0}^{N_f} \sqrt{G_F(f)\Delta f} (\cos (2\pi f_p t - k\xi + \Phi_p + \varepsilon_p)) . \tag{9}$$

However, the treatment in the frequency domain is not very straightforward. In this study computationally efficient techniques in both the time and frequency domains that permit inclusion of the time-dependent drift forces, introduced by the platform displacement, in terms of linear and nonlinear feedback contributions are presented. These time dependent feedback forces are expressed in terms of the applied wave loads by linear and quadratic transformations, e.g.,

$$F_W(t, \xi) = F_W(t, 0) + \theta_W^{(1)}(t)\xi + \theta_W^{(2)}(t)\xi^2 + \dots$$

$$\theta_W^{(n)} = \frac{1}{n!} + (-jk)^n F_W(f, 0) \tag{10}$$

in which $F_W(t, 0)$ describes wave force at zero displaced position and $\theta_w^{(n)}(\ )$ are the feedback coefficients. The response estimates demonstrate the relative significance of response due to feedback forces which contribute to drift response. Details of this scheme are given in Li and Kareem (1992).

5.   Response Analysis

The top-platform and tether system are treated as two substructures $S^{(1)}$ and $S^{(2)}$ with an arbitrary boundary B. If each substructure is modelled by using appropriate discretization, the mass, damping, stiffness matrix and force vector for each substructure can be assembled following a mass reduction transformation (Kareem and Song, 1993)

$$
\begin{bmatrix} M_{ss}^{(j)} & 0 \\ 0 & M_{bb}^{(j)} \end{bmatrix} \begin{Bmatrix} \ddot{X}_s^{(j)}(t) \\ \ddot{X}_b^{(j)}(t) \end{Bmatrix} + \begin{bmatrix} C_{ss}^{(j)} & 0 \\ 0 & C_{bb}^{(j)} \end{bmatrix} \begin{Bmatrix} \dot{X}_s^{(j)}(t) \\ \dot{X}_b^{(j)}(t) \end{Bmatrix}
$$

$$
+ \begin{bmatrix} K_{ss}^{(j)} & K_{sb}^{(j)} \\ K_{bs}^{(j)} & K_{bb}^{(j)} \end{bmatrix} \begin{Bmatrix} X_s^{(j)}(t) \\ X_b^{(j)}(t) \end{Bmatrix} = \begin{Bmatrix} F_s^{(j)}(t) \\ F_b^{(j)}(t) \end{Bmatrix}. \tag{11}
$$

where each element of the mass, damping and stiffness matrices is a submatrix (Kareem and Song, 1993), $j = 1, 2$ and $F_s^{(1)}$, is the force vector at the nodal points of $S^{(j)}$. One exception is the points on boundary B, where force vector $F_b^{(j)}$ can be decomposed to

$$
F_b^{(j)} = F_{be}^{(j)} + \tilde{F}_{bi}^{(j)}, \tag{12}
$$

in which $F_{be}^{(j)}$ is the externally applied nodal vector on boundary B for substructure $S^{(j)}$, while $\tilde{F}_{bi}^{(j)}$ is the internal nodal force vector due to the separation of two substructures. Typically, the loading terms include linear and quadratic terms concerning wind, wave induced viscous and potential forces including some terms from the boundary of substructure 2. There are different analysis schemes for the time and frequency domain treatment. Details are quite lengthy and the readers are referred to Kareem and Song (1993) for additional information.

5.1 PARAMETRIC MODELS FOR TIME DOMAIN ANALYSIS

In the time domain, analysis is carried out using parametric models (Li & Kareem, 1993, Mignolet & Spanos, 1992 and Naganuma et al., 1987). The models include autoregressive and moving averages (ARMA) algorithms for the simulation of multi-directional sea states, discrete convolution models for linear transformations of given time histories, discrete differentiation models for obtaining derivatives, discrete interpolation models and their hybrid combinations. Details concerning the selection of appropriate models and their orders is discussed in the context of their ability, accuracy and robustness can be found in Li and Kareem (1993).

The two-dimensional ARMA model $(p_t/p_r, q_t/q_r)$ is given by

$$
\eta(n\Delta t, l\Delta r) = -\sum_{p=1}^{p_t} \sum_{q=1}^{p_r} A_{pq}\eta_{pq}[(p-n)\Delta t, (q-1)\Delta r]
$$

$$+ \sum_{q=0}^{q_t} \sum_{q=1}^{q_r} B_{pq} \varepsilon_{pq} [(p-n) \Delta t, (q-1) \Delta r] \tag{13}$$

where A_{pq} and B_{pq} are coefficient matrices and $\varepsilon_{pq}(\ )$ is a two-dimensional white noise.

5.2 STOCHASTIC DECOMPOSITION FOR FREQUENCY DOMAIN ANALYSIS

The analysis of nested-cascade system with large degrees of freedom under multi-corre-lated input becomes computationally very intensive which lessens the attractiveness of the frequency domain approach. Typical examples of nested-cascade systems are found in the perturbation and iteration solution techniques. These approaches have been employed in this study. In a perturbation technique, the equations at different perturbation orders can be regarded as nested-cascade systems, in which the cross-spectral density functions between the responses at different orders are required for the final solution. This can become computationally very intensive for a large degrees of freedom system like the one being addressed here. In view of this shortcoming, a novel approach, referred to as stochastic decomposition, has been developed. This technique decomposes a set of random processes into component random processes, the relationship between any two of which is either fully coherent or noncoherent. Each component is described by a corresponding decomposed spectrum, which is related to conventional spectral description. The fully coherent or noncoherent processes alleviate the need for the computation of the cross-spectral density function, thus enhancing the computational efficiency of the frequency-domain approach. This concept is extended to second-order analysis for treating quadratic systems (Kareem and Li, 1988). Details of the stochastic decomposition approach can be found in Li & Kareem, 1995. In this reference details of a system subjected to wave loads are presented and the effectiveness of this approach is demonstrated through examples using conventional and decomposition approaches.

6. Example

The dynamic behavior of a typical TLP exposed to stochastic wind, wave and current induced loads is studied. The TLP has four columns, four pontoons and it is located in 542 meters deep water. There are sixteen tethers 510 meters in length each. The column and tether have an outer diameter of 18 and 0.27 meters, respectively. The tethers have a pretention of 10200 KN. The platform is exposed to wind field characterized by Kareem spectrum and wave field described by JONSWAP spectrum. The TLP above water and below water portions are discretized for computing wind and wave related forces and the tethers are modelled by a finite element discretization as discussed earlier. The calculated natural periods for surge, sway, heave, pitch, roll and yaw are 88.2, 88.2, 2.15, 3.05, 3.05 and 76.3 seconds, respectively.

The response of the TLP-tether system under different environmental conditions was estimated. Details of each condition are reported in Kareem & Song (1993). Here due to the length limitation only a part of the results is reported. First, the numerical simulation of directional sea wave surface elevation is given in fig. 2. The data represents spatial vari-ation of wave elevation at a fixed time. The comparison of simulated and target spectra showed very good agreement (Kareem & Song, 1993).

The TLP response based on time and frequency domain analysis is reported in Table 1. The response is separated into three frequency ranges, i.e., the low, medium and high frequency components. The low frequency response is near the platform natural frequency in the horizontal plane. In this range, the wind induced and the slowly varying drift provide the major make-up (below 0.02 Hz) of the response. The medium frequency response component consists of the wave frequency response that falls between 0.02 Hz and 0.2 Hz. The high frequency response is due to the resonant response of TLP in vertical modes. The results in Table 1 exhibit a very good agreement between the time and frequency domains, thus validating the frequency domain approach for such a complex system under nonlinear stochastic loads. In Table 2, the mean and standard deviation of coupled TLP-tether system is compared to conventional method that employs coupled equivalent springs to model tether dynamics. The difference between the results of the mean value obtained from the two methods in the horizontal plane motions (surge, sway and yaw) are found to be small, while the discrepancy increases significantly for the motions in the vertical plane. This difference in response due to different modelling procedures can be attributed to the following. The motions in the vertical plane are controlled by the elastic force in the tethers, therefore, these motions are more sensitive to the tether characteristics. In conventional method, the stiffness of a set of tethers is used as a constant and each tether is assumed to be a rigid element with no geometrical deformation and no spatial distribution of tether mass and damping effect. In reality, the stiffness of the TLP is influenced by the tether curvature, tether mass, and damping. These effects become more pronounced as the water depth increases.

In Table 3, a comparison of the mean and standard deviation of response under directional and uni-directional sea states is presented. The surge response is reduced in directional seas due to lack of coherence in the wave field introduced by directional spreading. However, the trend is reversed in the case of yaw response which is introduced by the imbalance of the approach wave field. Obviously, lack of coherence introduces more imbalance and hence higher yaw response.

In Fig. 3, the spectra of surge response is presented for the case of directional and uni-directional sea states. These spectral estimates reemphasize the comments made in the preceding section based on the mean and standard deviation response.

7. Concluding Remarks

In this paper a brief outline of efficient computational schemes for estimating the combined response of TLP-tether system under random wind and wave forces is presented. Both time and frequency domain procedures are included. The structural system involves a coupled TLP and tether subsystems in which the top platform is modelled as a rigid body and the tethers are modelled by a finite element discretization. The wind and wave loads are expanded in terms of linear and quadratic components. The directional sea states are modelled using a coherence function approach. A stochastic decomposition scheme is employed to expedite the frequency domain analysis. In the time domain, parametric models are used to simulate wind and wave fields and to evaluate response statistics. The coupled analysis scheme offers more accurate description of the tether-platform interaction in comparison with conventional equivalent spring approach.

The computational scheme presented here for the coupled system is especially suited for parallel computers. The analysis outlined here can be conveniently extended to the dynamic analysis of other combined systems that include weakly nonlinear subsystems under random loading.

8. Acknowledgements

The support for this research was provided in part by the Office of Naval Research Grant N00014-93-1-0761, the National Science Foundation Grant BCS90-96274, Texas Advanced Science & Technology Programs and several major oil companies.

9. References

Davis, K.B. and Mungall, J.G.H. (1991) Method for Coupled Analysis of TLPs, *OTC 6567*.

de Boom, W.C., Pinkster, W.C. and Tan, P.S.G. (1984) Motion and Tether Force Prediction for a TLP, *J. Wtrway, Port, Coast and Oc. Engrg.*, ASCE, 110(4).

Fylling, I.J. and Larson, C.M. (1987) TLP Tendon Analysis, in *Tension Leg Platform, State of the Art Review*, Ed. Aeki, Demirbilek, ASCE.

Gurley, K. and Kareem, A. (1993) Gust Loading Factors for Tension Leg Platforms, *Applied Ocean Research*, 15(3).

Kareem, A. and Li, Y. (1993) Wind-Excited Surge Response of Tension Leg Platform: Frequency-Domain Approach, *J. of Engrg. Mech., ASCE*, 119(1).

Kareem, A. and Li, Y. (1988) Stochastic Response of a Tension Leg Platform to Wind and Wave Loads, Tech. Rept. No. UHCE 88-18, Department of Civil Engineering, University of Houston, Houston, TX.

Kareem, A. and Song, X. (1993) Stochastic Response of Coupled TLP-Tether System Under Multi-Directional Seas, Department of Civil Engineering and Geological Sciences, University of Notre Dame, Tech. Rept. No. NDCE 93-003.

Li, Y. and Kareem, A. (1995) Stochastic Decomposition and Application to Probabilistic Dynamics, *J. of Engrg. Mech., ASCE*, 121(1).

Li, Y. and Kareem, A. (1993) Parametric Modelling of Stochastic Wave Effects on Offshore Platforms, *Applied Ocean Research*, 15.

Li, Y. and Kareem, A. (1993) Multivariate Hermite Expansion of Hydrodynamic Drag Loads on Tension Leg Platforms, *J. of Engrg. Mech., ASCE*, 119(1).

Li, Y. and Kareem, A. (1992) Computation of Wave-Induced Drift Forces Introduced by Displaced Position of Compliant Offshore Platforms, *J. of Off. Mech. & Arc. Engrg.*, ASME, 112.

McNamara, J.F., O'Brien, P.J. and Gilroy, S.G. (1998) Nonlinear Analysis of Flexible Risers Using Hybrid Finite Elements, *J. of Off. Mech. & Arc. Engrg. ASME*, 110.

Mignolet, M.P. and Spanos, P.D. (1992) Simulation of Homogeneous Two-dimensional Random Fields Part I-AR and ARMA Models, *J. of Applied Mechanics*, Transactions of the ASME, 59.

Mitwally, H. and Novak, M. (1989) Wave Force on Fixed Offshore Structures in Short-Crested Seas, *J. Eng. Mech.*, 115(3).

Naess, A. (1992) Statistical Analysis of Nonlinear, Second-order Forces and Motions of Offshore Structures in Short-Crested Random Seas, *Probabilistic Engineering Mechanics*, 5(2).

Naganoma, T., Deodatis, G. and Shirozuka, M. (1987) ARMA Representation of Random Processes, *J. of Engineering Mechanics*, 113(2).

Patel, M.H. and Lynch, E.J. (1983) Coupled Dynamics of Tensioned Bouyant Platform and Mooring Tethers, *Engr. Struct.*, 4(37).

Salvensen, N. (1982) Computation of Nonlinear Surge Motion of Tension Leg Platform, *Proceedings Offshore Tech. Conf. OTC 4394*, Houston, TX.

Sarpkaya, T. and Isaacson, M. (1981) *Mechanics of Wave Forces on Offshore Structures*, Van Nostrand Reinhold, New York, NY.

Sircar, S., Kleinhams, J.W. and Prasad, J. (1993) Impact of Coupled Analysis on Global Performance of Deep Water TLPs, *OTC 7145*, pp. 103-115.

Song, X and Kareem, A. (1994) Combined System Analysis of Tension Leg Platforms: A Parallel Computation Scheme, *Proceedings of the Off. Mech. and Arc. Engrg., OMAE, ASME*, **1**, Offshore Technology, Houston, TX.

Table 1: Comparison of the mean and standard Deviation of response using time domain and frequency domain approaches.

D.O.F	Unit	Mean		Standard Deviation			
				<0.02 Hz		0.02~0.2 Hz	
		Freq	Time	Freq	Time	Freq	Time
Surge	m	11.80	11.76	0.99	0.77	0.25	0.26
Sway	m	1.57	1.60	0.20	0.25	0.11	0.11
Heave	cm	-1.87	-1.82	0.16	0.13	0.06	0.06
Roll	$\times 10^{-4}$ rad	1.88	1.90	0.077	0.076	0.196	0.211
Pitch	$\times 10^{-3}$ rad	1.20	1.73	0.130	0.094	0.200	0.210
Yaw	$\times 10^{-2}$ rad	4.69	4.39	0.51	0.34	0.35	0.33

Note: Wave peak frequency $f_o = 0.1$ Hz and wave incident angle $\theta = 22.5°$.

Table 2: Comparison of the mean and standard deviation of response using the coupled and conventional methods.

D.O.F	Unit	Mean		Standard Deviation			
				<0.02 Hz		0.02~0.2 Hz	
		Coupl	Conv	Coupl	Conv	Coupl	Conv
Surge	m	11.80	11.50	0.99	0.98	0.25	0.25
Sway	m	1.57	1.62	0.20	0.19	0.11	0.11
Heave	cm	-1.87	-1.25	0.16	0.22	0.06	0.091
Roll	$\times 10^{-4}$ rad	1.88	2.70	0.077	0.083	0.196	0.173
Pitch	$\times 10^{-3}$ rad	1.20	2.87	0.130	0.092	0.200	0.169
Yaw	$\times 10^{-2}$ rad	4.69	5.09	0.51	0.42	0.35	0.35

Note: Wave peak frequency $f_s = 0.1$ Hz and wave incident angle $\theta = 22.5$.

Table 3: Comparison of the mean and standard deviation of response under directional and unidirectional sea states.

D.O.F	Unit	Mean		Standard Deviation			
				<0.02 Hz		0.02~0.2 Hz	
		Dir	Unidir	Dir	Unidir	Dir	Unidir
Surge	m	11.7	13.80	1.11	2.34	0.17	0.36
Sway	m	2.44	2.32	0.106	0.098	0.057	0.056
Heave	cm	-1.30	-1.33	0.265	0.27	0.060	0.061
Roll	$\times 10^{-4}$ rad	0.13	0.11	0.13	0.10	0.48	0.46
Pitch	$\times 10^{-3}$ rad	6.63	7.04	0.53	0.62	0.69	1.48
Yaw	$\times 10^{-2}$ rad	4.51	1.19	0.552	0.398	0.424	0.316

Note: Decay factor C = 5.0 and wave peak f_o = 0.06.

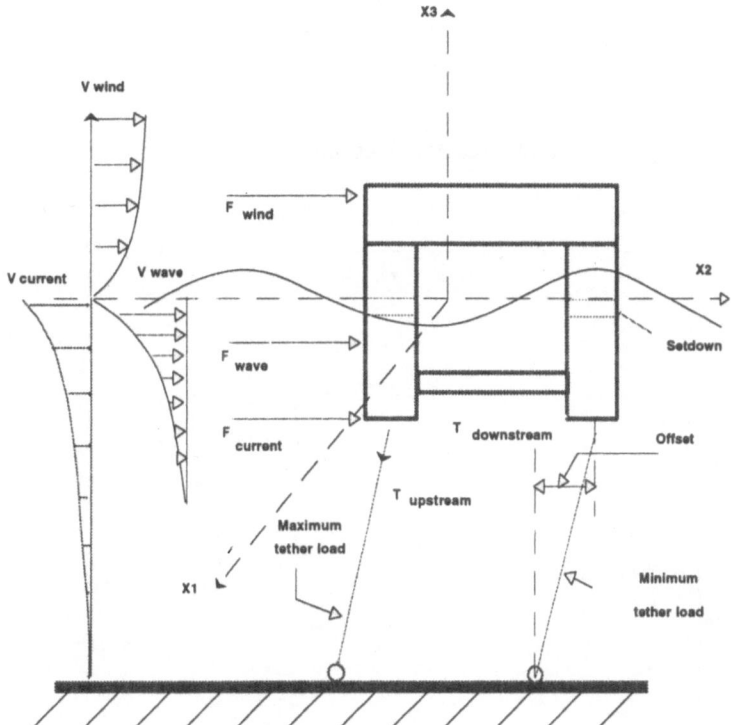

Fig. 1 Definition of load effects on TLP

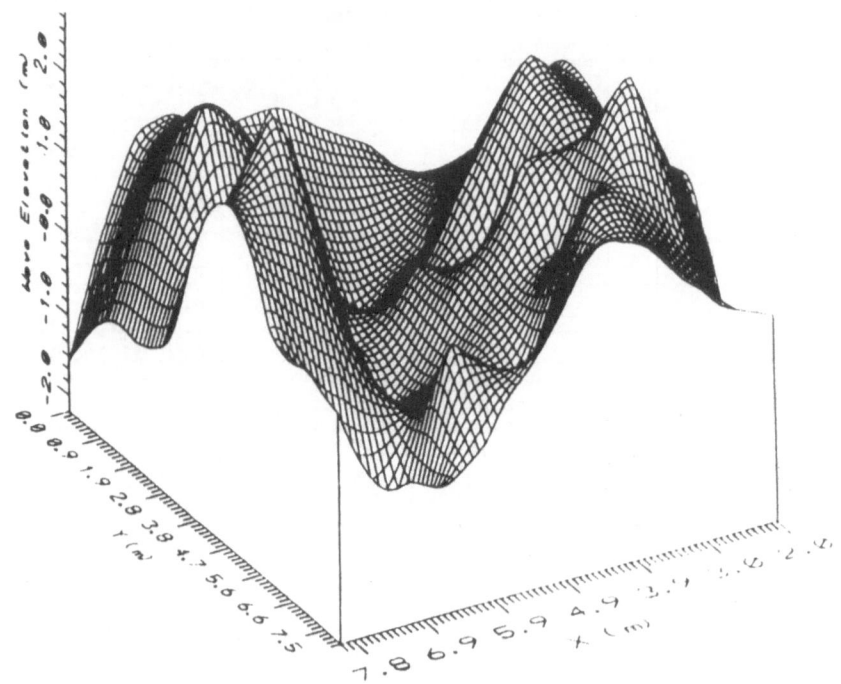

Fig. 2 Time history of directional sea at a fixed time

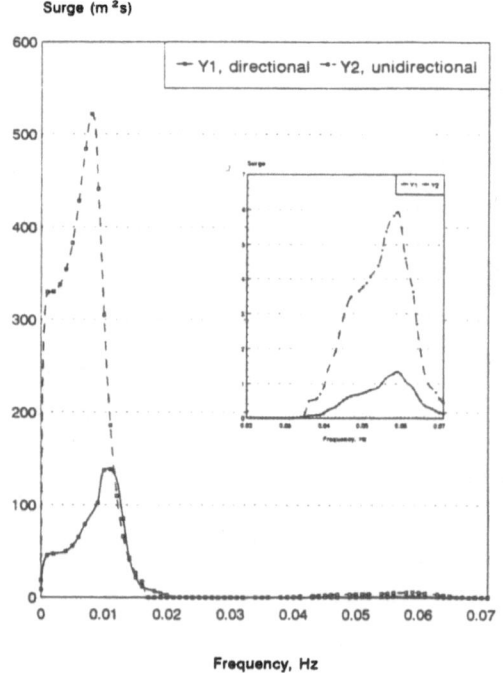

Fig. 3 PSD of surge response

OPTIMAL CONTROL PROBLEMS FOR NONLINEAR OSCILLATORY SYSTEMS WITH RANDOM PERTURBATIONS

A.S. KOVALEVA
Mechanical Engineering Research Institute
Russian Academy of Sciences
Griboedov Str 4, Moscow 101830 Russia

The paper is concerned with problems of optimal control for a class of dynamical systems with slow and fast variables and random perturbations. In general, real noises in physical systems are not white noises or Markov processes, and the well known dynamic programming equations can not be directly written. In this paper we develop special perturbation techniques for approximative solutions of the problems of interest.

1. Introduction

Consider a typical model of a weakly perturbed and weakly controlled oscillatory system [1], [2], [5]

$$\dot{x} = \varepsilon X(x,\theta,\xi(t),u,\varepsilon), \quad x = x^u(t,\varepsilon) \in S \subset R^n, \quad u \in U \subset R^d \qquad (1.1)$$

$$\dot{\theta} = \Omega(x) + \varepsilon Y(x,\theta,\xi(t),u,\varepsilon), \quad \theta = \theta^u(t,\varepsilon) \in T^m \subset R^m$$

Here S is a bounded open ball, $T^m = (2\pi)^m$, U is a compact in R^d, u is a control, $\xi(t)$ is a vector right-continuous random process, $0 < \varepsilon \ll 1$ is a small parameter. For oscillatory systems we suppose that the right sides of (1.1) are 2π-periodic in each component of θ and smooth enough and the system is nonresonant.

The objective is to find an optimal control u_* from the criterion

$$J(u_*) = \min_{u \subset U} J(u), \quad J(u) = E_{\tau,x} \Phi(T_\varepsilon, x^u(T_\varepsilon,\varepsilon)) \qquad (1.2)$$

where Φ is a smooth scalar function and $T_\varepsilon < \infty$. Detailed assumptions are

A. Naess and S. Krenk (eds.), IUTAM Symposium on Advances in Nonlinear Stochastic Mechanics, 247–256.
© 1996 *Kluwer Academic Publishers.*

discussed below. Here we only emphasize that the expectation $EX(x,\theta,\xi(t),u,0) = 0$ uniformly on x, θ, u. It implies [5], [8] that the natural time scaling for x is $\tau = \varepsilon^2 t$, and we can write

$$x = x^u(t,\varepsilon) = x^u_\varepsilon(\tau), \ \theta = \theta^u(t,\varepsilon) = \theta^u_\varepsilon(\tau), \ \tau \in [0,T], \ T_\varepsilon = T/\varepsilon^2$$

The deterministic analog of (1.1) has been exhaustively studied [1], [5]. In the stochastic situation only controlled periodic diffusions have been thoroughly studied [2].

In this paper we discuss optimal control problems for a more general class of perturbed systems when the dynamical programming equations can not be directly written.

As in the deterministic case, the slow variable $x(t,\varepsilon)$ describes a "macroscopic" evolution of the system, and we aim for eliminating the fast variables for studying the behaviour of the system "in the large". At the same time, the noise $\xi(t)$ is the fast mixing process as compared with the slow variable $x(t,\varepsilon)$. Hence, diffusion approximations and averaging ideas may be applied for the separation of variables. Diffusion approximations of real processes is very appealing because diffusions have a much simpler mathematical and physical structure than real processes, and a number of well developed methods can be applied to them.

The diffusion approximations approach has been successfully used in analysis problems for oscillatory systems with slow and fast variables [5], [6], [7] and for a simpler class of controlled systems

$$\dot{x} = \varepsilon X(x,\xi(t),u) \tag{1.3}$$

The main idea is to approximate the slow process $x(t,\varepsilon)$ by a diffusion, then to construct an optimal control for the diffusion and to prove the suboptimality of the constructed control for the original system [5],[9]

In this paper the diffusion approximations ideas are applied to a more general class of nonlinear controlled systems (1.1). The weak convergence of solutions of (1.1) to a controlled periodic diffusion is stated in Section 2. In Section 3 an optimal control problem is set up, and the main convergence theorem for optimal control problems is proved. The dynamic programming equations for this nearly optimal control problem are discussed in Section 4. Section 5 treats an example.

To simplify and shorten the exposition, we consider the system with a scalar phase ($m=1$). Generalization to multifrequency nonresonant systems can be easily obtained.

2. Diffusion Approximations

2.1. MAIN DEFINITIONS AND ASSUMPTIONS

A feedback control $u = u(\tau, x, \theta) \in \mathcal{U}^\varepsilon$ is said to be *an admissible control*
for (1.1) if u is periodic in θ, takes values in U when $\tau \in [0, T]$ and
(1.1) has a unique solution (in the weak sense) for $u = u(\tau, x, \theta)$.
     Set

$$X(x, \theta, \xi(t), u, \varepsilon) = F(x, \theta, \xi(t)) + \varepsilon G^u(x, \theta, \tau) \qquad (2.1)$$

$$Y(x, \theta, \xi(t), u, \varepsilon) = K(x, \theta, \xi(t)) + \varepsilon D^u(x, \theta, \tau)$$

$$F(x, \theta, \xi(t)) = F_o(x, \theta)\xi(t), \quad K(x, \theta, \xi(t)) = K_o(x, \theta)\xi(t) \qquad (2.2)$$

and

$$G^u(x, \theta, \tau) = G(x, \theta, u), \quad D^u(x, \theta, \tau) = D(x, \theta, u) \qquad (2.3)$$

Define the domain $Q = S \times (0, 2\pi) \times \mathcal{U}^\varepsilon$. Throughout the paper, we suppose that
the functions (2.1) -(2.3) are smooth enough in Q, and $\xi(t)$ is a zero-
mean right-continuous mixing stationary process [3], [5]-[9]. Note that
the required mixing conditions hold, in particular, for stationary
Gauss - Markov or strong mixing processes.

2.2. DIFFUSION APPROXIMATIONS OF PERTURBED PROCESSES

It was proved [6], [7] that under these broad assumptions, the slow
variable $x_\varepsilon^u(\tau)$ converges weakly [3], as $\varepsilon \to 0$ and $\tau \in [0, T]$, to   the
diffusion $x_o^u(\tau)$ with the differential operator $\mathcal{A}$

$$\mathcal{A}V = b'(x, \tau)V_x + 1/2 \, Tra(x)V_{xx} \qquad (2.4)$$

The drift coefficient $b = b_1 + b_2 + g$ is calculated by [6]

$$b_1(x) = \frac{1}{2\pi}\int_0^{2\pi}d\theta \int_0^\infty EF_x(x, \theta+\Omega(x)s, \xi(t+s))F(x, \theta, \xi(t))ds \qquad (2.5)$$

$$b_2(x) = \frac{1}{2\pi}\int_0^{2\pi}d\theta \int_0^\infty EF_\theta(x, \theta+\Omega(x)s, \xi(t+s))H(x, \theta, \xi(t))ds$$

$$g(x,\tau) = \frac{1}{2\pi}\int_0^{2\pi} G^u(x,\theta,\tau)d\theta$$

The ij-th component of the matrix $a(x)$ is defined by [6]

$$a(x) = \frac{1}{2\pi}\int_0^{2\pi}d\theta \int_{-\infty}^{\infty} EF^i(x,\theta+\Omega(x)s,\xi(t+s))F^j(x,\theta,\xi(t))ds \qquad (2.6)$$

where F^i denotes the i-th components of F. When Ω is independent of x, (2.4) is like the generator for quasilinear systems [5].

For a simple example that illustrates the above results, let

$$G^u = D^u = 0, \quad F_o(x,\theta) = f(x)sin\theta, \quad H_o(x,\theta) = h(x)cos\theta$$

and x, θ be scalars. Then

$$\dot{x} = [f(x)sin\theta]\xi(t), \quad \dot{\theta} = [h(x)sin\theta]\xi(t) \qquad (2.7)$$

By (2.6), we have

$$a(x) = \sigma^2(x) = \frac{f^2(x)}{2\pi}\int_0^{2\pi}sin\theta d\theta \int_{-\infty}^{\infty}K(u)sin(\theta+\Omega(x)u)du =$$

$$= \frac{f^2(x)}{2}\int_{-\infty}^{\infty}K(u)cos(\Omega(x)u)du = \frac{f^2(x)}{2}S(\Omega(x)) \qquad (2.8)$$

where $K(u)$ and $S(\Omega)$ are the covariance and the spectral density of the scalar perturbation $\xi(t)$.

Write b_1, b_2. By (2.5), we have $b_1(x) = \beta_1(x) + \beta_2(x)$, where

$$\beta_1(x) = \Omega_x(x)\frac{f^2(x)}{2\pi}\int_0^{2\pi}sin\theta d\theta\int_0^{\infty}K(u)cos(\theta+\Omega(x)u)du =$$

$$= \frac{f^2(x)}{4}\frac{dS(\Omega)}{d\Omega}\Omega_x(x) = \frac{f^2(x)}{4}\frac{dS(\Omega)}{dx} \qquad (2.9)$$

$$\beta_2(x) = \frac{f(x)f_x(x)2\pi}{2\pi}\int_0^{\infty}sin\theta d\theta\int_0^{\infty}K(u)sin(\theta+\Omega(x)u)du = \frac{f(x)f_x(x)}{4}S(\Omega(x))$$

By the same procedure

$$b_2(x) = f(x)h(x)S(\Omega(x))/4 \qquad (2.10)$$

Hence, the drift coefficient b depends on the derivative $dS/d\Omega$ if Ω depends on x. If $\Omega = \Omega_o = const$, then $\beta_2 = 0$, and we have the well known diffusion approximation results for the simplest quasilinear problem [8]. If $\Omega = \Omega(x)$, but $\xi(t)$ is a white noise and $S(\Omega) = S = const$, then also $\beta_2 = 0$, and we have asimptotic results for a periodic diffusion [2].

To clarify the applications of the weak convergence ideas to control problems, consider the Ito equation corresponding to the generator (2.4)

$$\dot{x} = b(x,\tau) + \sigma(x)\dot{w}(\tau), \quad x = x_o^u(\tau) \qquad (2.11)$$

Here $w(\tau)$ is a standard vector Wiener process, $\sigma\sigma' = a$, and the dimensions of the vectors b, w and the matrix σ are compatible. Note that (2.11) has a unique solution when the coefficients are smooth and u is an admissible control [3].

Employing the weak convergence, we can write [3]

$$\left| E_{\tau,x} \Phi(T,x_\varepsilon^u(T) - E_{\tau,x}\Phi(T,x_o^u(T)) \right| \to 0, \quad \varepsilon \to 0 \qquad (2.12)$$

for any admissible control $u \in U^\varepsilon$ and $T < \infty$.

2.2.1 *Corollary*
Consider the system

$$\dot{z} = \beta(z) + G^u(z,\psi,\tau) + \sigma(z)\dot{w}(\tau), \quad \dot{\psi} = \varepsilon^{-2}\Omega(z) \qquad (2.13)$$

$$\beta(z) = b_1(z) + b_2(z)$$

where the drift and diffusion coefficients are defined by (2.5) - (2.6). Let a class U of admissible controls for (2.13) be defined in the same way as that for (1.1). Obviously $U \subset U^\varepsilon$.

Let the above-mentioned assumptions hold. Then for each $u \in U$ and $\tau \in [0,T]$, the solution $x_\varepsilon^u(\tau)$ of (1.1) converges weakly as $\varepsilon \to 0$ to the slow component $z^u(\tau)$ of (2.13), so that

$$\left| E_{\tau,x} \Phi(T,x_\varepsilon^u(T) - E_{\tau,x}\Phi(T,z_\varepsilon^u(T)) \right| \to 0, \quad \varepsilon \to 0 \qquad (2.14)$$

for any admissible control $u \in \mathcal{U}$.

Proof. The weak convergence $z_\varepsilon^u(\tau) \to x_0^u(\tau)$ [10] implies

$$\left| E_{\tau,x} \Phi(T, z_\varepsilon^u(T)) - E_{\tau,x} \Phi(T, x_0^u(T)) \right| \to 0, \quad \varepsilon \to 0 \tag{2.15}$$

and (2.14) follows directly from (2.12), (2.15).

3. Convergence of Optimal Solutions

In this Section we apply the asymptotic results for the optimal control problem (1.1), (1.2). Let an optimal control $u_* \in \mathcal{U}$ exist, and write

$$J(u_*) = \min_{u \in U} J(u), \quad J(u) = E_{\tau,x} \Phi(T, x_\varepsilon^u(T)) \tag{3.1}$$

Consider a parallel optimal control problem for the diffusion (2.13) with the criterion

$$I(u_\varepsilon) = \min_{u \in U} I(u), \quad I(u) = E_{\tau,x} \Phi(T, z_\varepsilon^u(T)) \tag{3.2}$$

and suppose that an optimal control $u_\varepsilon \in \mathcal{U}$ for (2.13), (3.2) exists.

From (2.14), (3.1), (3.2) we have

$$\left| J(u) - I(u) \right| \to 0 \tag{3.3}$$

for each admissible control $u \in \mathcal{U}$. We state now the convergence theorem.

THEOREM 1. Let (2.12), (2.14) hold for both $u_*, u_\varepsilon \in \mathcal{U}$. Then u_ε is a nearly optimal control for the original problem (1.1), (1.2) and

$$0 \leqslant J(u_\varepsilon) - J(u_*) \leqslant C(\varepsilon) \to 0, \quad \varepsilon \to 0 \tag{3.4}$$

Proof follows directly from the weak convergence of $x_\varepsilon^u(\tau)$ to $z_\varepsilon^u(\tau)$ and is given in [5] for a general class of optimal control problems and in [9] for controlled systems of the type (1.3).

Following Theorem 1 we can replace the original complicated system (1.1) by the diffusion (2.13). Methods of constructing optimal controls for diffusions are well studied; technical problems may be overcome by using the well developed averaging and homogenization techniques [2].

4. Dynamic Programming Conditions

First we consider the Bellman equations for the diffusion (2.13) with the cost (3.2). We write the Bellman function [4]

$$V^{\varepsilon}(\tau,x,\psi) = \min_{u \in U} E_{\tau,x,\psi} \Phi(T, z_{\varepsilon}^{u}(\tau)) \tag{4.1}$$

where T is the exit time on the boundary Γ of the domain S. We write the Bellman equation satisfied by (4.1). If we set

$$h(x,\psi,p) = \min_{u \in U} p' G(x,\psi,u) \tag{4.2}$$

then [4] V^{ε} is a unique solution of

$$V_{\tau}^{\varepsilon} + \varepsilon^{-2}\Omega(x)V_{\psi}^{\varepsilon} + LV^{\varepsilon} + h(x,\psi,V_x^{\varepsilon}) = 0, \quad \tau,x,\psi \in (0,T) \times S \times (0,2\pi)$$

$$V^{\varepsilon}(\tau,x,\psi) = \Phi(\tau,x), \quad \tau = T, \; x \in \Gamma \tag{4.3}$$

where

$$LV^{\varepsilon} = \beta'(x)V_x^{\varepsilon} + 1/2 \; Tra(x)V_{xx}^{\varepsilon} \tag{4.4}$$

If the coefficients of (4.3), (4.4) are smooth, (4.3) has a smooth solution periodic in ψ [2]. Moreover, there exists a feedback

$$u_{\varepsilon}(\tau,x,\psi) = U(x,\psi,p^{\varepsilon}) \quad p^{\varepsilon} = V_x^{\varepsilon}(\tau,x,\psi) \tag{4.5}$$

which realizes the minimum in (4.2) [2], [4]. Thus, (4.5) is an optimal control for (2.13), (3.2). By Theorem 1, u_{ε} is a nearly optimal control for (1.1), (1.2), and (3.4) holds for the control (4.5).

Consider now the limit behaviour of V^{ε} and u_{ε}, as $\varepsilon \to 0$. Let V^{O} be a solution of the averaged equation

$$V_{\tau}^{O} + LV^{O} + h^{O}(x,V_x^{O}) = 0, \quad \tau, x \in (0,T) \times S \tag{4.6}$$

$$V^{O}(\tau,x) = \Phi(\tau,x), \quad \tau = T, \; x \in \Gamma$$

where

$$h^O(x,p) = \frac{1}{2\pi}\int_0^{2\pi} h(x,\psi,p)d\psi \qquad (4.7)$$

and (4.6) has a unique solution if the coefficients are smooth enough [4].

We write the feedback control

$$u_O(\tau,x,\psi) = U(x,\psi,p^O), \quad p^O = V_x^O(\tau,x) \qquad (4.8)$$

It is proved [2] that (4.8) is nearly optimal for (2.13) and

$$0 \leqslant I(u_O) - I(u_\varepsilon) \leqslant C(\varepsilon) \to 0, \quad \left|I(u_\varepsilon) - V^O(\tau,x)\right| \to 0 \qquad (4.9)$$

as $\varepsilon \to 0$, for each $x \in S$ uniformly in $\psi \in (0,2\pi)$.

We state now the main convergence theorem.

THEOREM 2. Let Theorem 1 hold, and the Bellman function $V^O(\tau,x)$ and the feedback control $u_O(\tau,x,\psi)$ be defined by (4.6), (4.8). Then u_O is a nearly optimal control for the original problem (1.1), (1.2), and

$$0 \leqslant J(u_O) - J(u_*) \leqslant C(\varepsilon) \to 0, \quad \left|J(u_*) - V^O(\tau,x)\right| \to 0 \qquad (4.10)$$

Proof. Write

$$J(u_*) - J(u_O) = [J(u_*) - J(u_\varepsilon)] + [J(u_\varepsilon) - I(u_\varepsilon)] +$$

$$+ [I(u_\varepsilon) - I(u_O)] + [I(u_O) - J(u_O)] \qquad (4.11)$$

By inserting (3.3), (3.4), (4.9) into the right side of (4.11) and recalling that u_* is optimal, we deduce the first assertion of (4.10).

In the same way we have

$$J(u_*)-V^O(\tau,x) = [J(u_*)-J(u_\varepsilon)]+[J(u_\varepsilon)-I(u_\varepsilon)] + \qquad (4.12)$$

$$+ [I(u_\varepsilon)-V^O(\tau,x)]$$

The second assertion of (4.10) follows from (3.3), (3.4) and (4.12).

5. Example

For concreteness in the illustration of the main idea, we employ a simple mechanical model. Consider fast weakly controlled rotation of the pendulum in the vertical plane. Random vertical vibrations of the base are considered as weak perturbations. Then, introducing the small parameter ε, we write the equation of the fast rotation in the form [1], [6]

$$\dot{x} = \varepsilon^2 k s in\theta + \varepsilon \varkappa \xi (t)sin\theta + \varepsilon^2 u, \quad \dot{\theta} = x, \quad x(0) = \Omega \qquad (5.1)$$

Here θ is an angle of rotation, k, $\varkappa$ are parameters of the pendulum, $\xi(t)$ is a random acceleration of the base, a zero-mean normal stationary process with the spectral density $S(\omega)$, the forcing moment u is a control, $|u| \leqslant U$.

We now set up the optimal speed reducing problem. The aim is to find a control u_* minimizing the expectation

$$J(u) = E(T/\varepsilon^2)// x(T/\varepsilon^2) = \omega, \quad |u| \leqslant U \qquad (5.2)$$

where $\omega < \Omega$ is a fixed angular velocity. Thus, we consider the minimum time problem.

Theorem 2 yields that $T \to T^o$, where $T^o = V^o(\Omega)$ and $V^o(x)$ is the minimum positive solution of the averaged Bellman equation for the minimum time problem. From (4.6), (2.8), (2.9) we have

$$\frac{1}{4}\frac{d}{dx}\left[S(x)\frac{dV^o}{dx}\right] + \min_{|u|\leqslant U} u\frac{dV^o}{dx} = -1, \quad V^o(\omega) = 0 \qquad (5.3)$$

It implies that

$$u = -U sign V_x^o(x) = -U \qquad (5.4)$$

and (5.3) can be written in the form

$$\frac{1}{4}\frac{d}{dx}\left[S(x)\frac{dV^o}{dx}\right] - U\frac{dV^o}{dx} = -1, \quad V^o(\omega) = 0 \qquad (5.5)$$

It follows that

$$V^o(x) = -4\int_x^\omega S^{-1}(r)dr \int_0^r exp\left[4\int_z^r US^{-1}(y)dy\right]dz \qquad (5.6)$$

and the solution strongly depends on the behaviour of $S(x)$. If $\zeta(t)$ is a white noise and $S(x) = S$, then $V^o(x) = U^{-1}(x-\omega)$, $T^o = U^{-1}(\Omega-\omega)$. In this case the solution does not depend on the perturbations.

6. Acknowledgements

This work is supported by the International Science Foundation, grant N19000, and the Russian Science Foundation, grant 93-01-00874.

7. References

1. Akulenko, L.D.: *Problems and Methods of Optimal Control*. Kluwer Academic Publishers, Dordrecht, 1994.
2. Bensoussan, A.: *Perturbation Methods in Optimal Control*. Springer Verlag, New York, 1987.
3. Ethier, S.N., Kurtz T.G.: *Markov Processes: Characterization and Convergence*, Jhon Wiley and Sons, New York, 1986.
4. Fleming W., Rishel, R.: *Optimal Deterministic and Stochastic Control*, Springer, New York, 1978.
5. Kovaleva, A.S.: *Control for Oscillatory and Vibroimpact Systems*. Nauka, Moscow, 1990 (in Russian).
6. Kovaleva, A.S.: Decomposition of motions in non-linear systems with a spinning phase under random perturbations. *J.Appl.Math. Mech.* 57 (1993), 237-255.
7. Kovaleva, A.S.: On the averaging for multifrequency nonlinear systems with random perturbations. *J.Appl.Math.Mech.* 59 (1995),#1.
8. Kushner, H.J.: *Approximation and Weak Convergence Methods for Random Processes; with Applications to Stochastic Systems Theory*, Cambridge, MA: MIT Press 1984.
9. Kushner,H.J.: *Weak Convergence Methods and Singularly Perturbed Stochastic Control and Filtering Problems*. Birkhaeuser, Boston, 1990.
10. Ventzel A.D., Freidlin M.I. *Random Perturbations of Dynamical Systems*. Springer Verlag, New York, 1984.

STOCHASTIC DYNAMICS OF NONLINEAR STRUCTURES WITH RANDOM PROPERTIES SUBJECT TO STATIONARY RANDOM EXCITATION

H. Uğur Köylüoğlu
Koç University, 80860 İstinye, İstanbul, Turkey

Søren R. K. Nielsen
University of Aalborg, DK-9000 Aalborg, Denmark

Ahmet Ş. Çakmak
Princeton University, Princeton, NJ 08544, USA

A nonlinear stochastic finite element formulation for the stochastic response analysis of geometrically nonlinear, elastic 2-dimensional frames with random stiffness and damping properties subject to stationary random excitation is derived utilizing deterministic shape functions and random nodal displacements. The discretized second order nonlinear stochastic differential equations with random coefficients are solved applying the total probability theorem with a mean-centered second order perturbation method in the frequency domain to evaluate the unconditional statistics of the response. Zeroth, first and second order perturbations are computed using a spectral approach in which a system reduction scheme to the modal subspace expanded by the deterministic linear eigenmodes and equivalent linearization with Gaussian closure are applied. Sample frames are solved and the results are compared with the ones obtained from extensive Monte Carlo simulations.

1. Introduction

The structural response processes will be stochastic when uncertainties in initial conditions and/or external loads and/or parameters of the constitutive relations are modeled using random variables or random fields. Stochastic Finite Element Method (SFEM) deals with the analysis of MDOF large scaled structural systems using discrete approximations. The developments in SFEM are reviewed in [1-6], some studies that address dynamic problems are listed in the references [7-14].

For the full quantification of uncertainty using random variables and/or random fields, joint probability density function (jpdf) should be assigned. Estimations of the jpdf for the random models are based on experimental tests, observations and engineering judgement. It is very difficult, usually impossible, to quantify the uncertainty in terms of the jpdf. In practice, only the first and second order statistical moments can be estimated accurately. This study considers that the first and second order statistics are known. By the application of SFEM with deterministic shape functions and random nodel displacements, the nonlinear stochastic partial differential equations are reduced to stochastic ordinary differential equations (sode) with random coefficients. Here, consistency in the discretization process is achieved as both the continuous deterministic part and the random field of each element are discretized in the same manner. The first and second order stationary statistical moments of the stochastic response are then calculated by solving the resulting nonlinear sode with random coefficients approximately in the frequency domain with a second order pertubation approach. The perturbation solutions in the time domain cannot be applied for the nonstationary solution of sode with random coefficients. Because when time is involved, partial derivatives of the response with respect to random quantities become proportional to time, thus there are divergent secular terms and solutions may blow up with time. These secular terms can be cured in the frequency domain, [6, 12, 13]. The nonlinearities considered are small to moderate and variabilities of the random parameters are limited to a coefficient of variations of 25-30 percent.

A. Naess and S. Krenk (eds.), IUTAM Symposium on Advances in Nonlinear Stochastic Mechanics, 257–268.
© *1996 Kluwer Academic Publishers.*

Sample frames are studied to illustrate the effects of the variability of random damping ratios and random bending rigidity field as well as the correlation length of the random bending rigidity field.

2. Nonlinear Stochastic Finite Element Formulation

The dynamic behaviour of a d-dimensional nonlinear, nonhysteretic structure with random stiffness and damping properties subject to random excitations can be approximated by SFEM to the following coupled second order sode.

$$\mathbf{M}\tilde{\mathbf{V}}(\mathbf{X},t) + \mathbf{C}(\mathbf{X})\dot{\mathbf{V}}(\mathbf{X},t) + \mathbf{K}(\mathbf{X})\mathbf{V}(\mathbf{X},t) + \mathbf{H}\big(\mathbf{V}(\mathbf{X},t),\mathbf{X}\big) = \mathbf{F}(t) \tag{1}$$

where $\mathbf{X}$ is the vector of basic random variables specifying the randomness of the structural system. $\mathbf{M}$, $\mathbf{C}(\mathbf{X})$ and $\mathbf{K}(\mathbf{X})$ are the linear deterministic consistent mass, random proportional viscous damping and linear random stiffness matrices of the discretized system, respectively. $\mathbf{V}(\mathbf{X},t)$ is the displacement vector of global degrees of freedom (dof), $\mathbf{H}\big(\mathbf{V}(\mathbf{X},t),\mathbf{X}\big)$ is the nonlinear random vector function of the displacement vector, carrying the related nonlinear restoring terms which are due to geometrical nonlinearities, and $\cdot$ signifies differentiation with respect to time. $\mathbf{F}(t)$ is the random nodal loading vector assumed as $\mathbf{F}(t) = \mathbf{F}_0 f(t)$, where $\mathbf{F}_0$ is a constant real amplitude vector and $f(t)$ is a random stationary excitation with an auto-spectral density of $S_{ff}(\omega)$.

In what follows, a nonlinear stochastic Galerkin SFEM for elastic 2-dimensional frame structures is derived from the differential equations of the nonlinear Bernoulli-Euler beam theory using a potential energy approach. Hence, a physical meaning is attained for the random variables of the linear random stiffness matrix and the nonlinear random vector function of equation (1). Geometrical nonlinearity and physical linear elasticity case is covered. For simplicity, only bending deformations are considered, and the bending stiffness field of all beam elements are modeled as homogeneous random fields. It is assumed that the stochastic viscous damping matrix decouples with respect to the shape function of the linear elastic deterministic structure, and modal damping ratios of the equivalent linear structure are assigned as random variables.

A global (X,Y,Z) coordinate system is defined with (X,Y) plane coincident with the plane of the frame structure. For each element a referential (x,y,z) coordinate system is introduced with the origin placed at one end of the beam element and the x-axis along the beam element in the statical equilibrium state, Figure 1. The global Z and local z-axes are unidirectional. Γ_e is the angle from the global X-axis to the local x-axis in the anti-clockwise direction. The curvature $\kappa(x,t)$ of a material point at x is related to the displacement field in the local y-direction $v(x,t)$ as, e.g. [16],

$$\kappa(x,t) = \frac{v''(x,t)}{\left(1 - v'(x,t)^2\right)^{\frac{1}{2}}} \tag{2}$$

where $'$ denotes partial differentiation with respect to the Lagrangean coordinate x. In (2), the beam is assumed to be inextensible along the beam axis so that the length of the central line is always the same. Due to such a deformation, a shortening, $\Delta_e(t)$, along the chord line takes place.

$$\Delta_e(t) = \int_0^{L_e} \left[1 - \left(1 - v'(x,t)^2\right)^{\frac{1}{2}}\right] dx \tag{3}$$

L_e is the length of the eth beam element. Based on assuming that plane sections remain plane after the deformation and are perpendicular to the deformed beam axis, the elastic linear constitutive equation between the curvature $\kappa(x,t)$ and the bending moment $M(x,t)$ is

$$M(x,t) = -(EI)_e(x)\kappa(x,t) \quad (4) \qquad (EI)_e(x) = \overline{(EI)_e}\left(1 + f_e(x)\right) \tag{5}$$

$(EI)_e(x)$ represents the bending stiffness field of the beam element. It is assumed that the assumptions of the nonlinear Bernoulli-Euler beam theory are valid even when μ_e, $(EI)_e(x)$ are modeled as random fields. For simplicity the mass per unit length μ_e is considered to be constant and deterministic, whereas the bending stiffness of each element is assumed to be a sum separable random field as in (5) where $\overline{(EI)_e}$ is the mean value of the bending stiffness of element e, and

$f_e(x)$ is a one-dimensional, zero-mean, homogeneous random field of the eth element. Since, the bending stiffness cannot take on negative values, $f_e(x)$ is assumed to be bounded from below as $f_e(x) > -1$ with probability 1. Then, a potential energy functional for the beam can be derived.

$$U_e[v(x,t)] = \int_0^{L_e} \left[\frac{1}{2}(EI)_e(x)\kappa^2(x,t) - P_e\left(1 - \sqrt{1 - v'(x,t)^2}\right) \right] dx \tag{6}$$

$$U_e[v(x,t)] \simeq \int_0^{L_e} \left[\frac{1}{2}(EI)_e(x)\left(v''(x,t)^2 + v''(x,t)^2 v'(x,t)^2\right) - \frac{1}{2}P_e\left(v'(x,t)^2 + \frac{1}{4}v'(x,t)^4\right) \right] dx \tag{7}$$

where P_e is the compressive normal force acting on the beam element in the statical referential state, Figure 1. In (7), the Taylor expansion of $U_e[v(x,t)]$ about the referential state has been truncated after the 4th order terms. $v(x,t)$ is determined from 4 local dof which are selected as the end displacements $v_1(t)$, $v_2(t)$ in the local y-direction and rotations $\theta_1(t)$, $\theta_2(t)$ of the end sections in the local z-direction. The random $v(x,t)$ is approximated as a linear combination of the indicated random nodal dof multiplied by deterministic cubic interpolation functions $\mathbf{N}(x)$, i.e.

$$v(x,t) = \mathbf{N}^T(x)\mathbf{v}^e(t) \ , \quad \mathbf{v}^e(t) = \begin{bmatrix} v_1(t) \\ \theta_1(t) \\ v_2(t) \\ \theta_2(t) \end{bmatrix} \ , \quad \mathbf{S} = \begin{bmatrix} 1 & 0 & -\frac{3}{L^2} & \frac{2}{L^3} \\ 0 & 1 & -\frac{2}{L_e} & \frac{1}{L^2} \\ 0 & 0 & \frac{3}{L^2} & -\frac{2}{L^3} \\ 0 & 0 & -\frac{1}{L_e} & \frac{1}{L^2} \end{bmatrix} \ , \quad \mathbf{p}(x) \doteq \begin{bmatrix} 1 \\ x \\ x^2 \\ x^3 \end{bmatrix} \tag{8}$$

$$\mathbf{N}(x) = \mathbf{S}\mathbf{p}(x)$$

It should be noted that cubic shape functions yield exact representation of $v(x,t)$ in the case of only nodally loaded deterministic linear beams with constant bending stiffness and zero normal force. The potential energy for the beam element can then be discretized inserting (8) into (7).

$$U_e(\mathbf{v}^e(t)) = \frac{1}{2}\left(K_{e,ij}v_i^e v_j^e + K_{e,ijkl}v_i^e v_j^e v_k^e v_l^e\right) \tag{9}$$

$$K_{e,ij} = \int_0^{L_e}(EI)_e(x)N_i''(x)N_j''(x)dx - P_e\int_0^{L_e}N_i'(x)N_j'(x)dx \tag{10}$$

$$K_{e,ijkl} = \int_0^{L_e}(EI)_e(x)N_i''(x)N_j''(x)N_k'(x)N_l'(x)dx - \frac{P_e}{4}\int_0^{L_e}N_i'(x)N_j'(x)N_k'(x)N_l'(x)dx \tag{11}$$

In (9) and below, summation convention has been applied over the dummy indices i,j,k,l which run from 1 to 4. Since the normal force in the referential state is assumed to be deterministic, the last terms in (10) and (11) are deterministic. As seen from (10) and (11), $K_{e,ij}$ is symmetric in indices i and j, whereas, $K_{e,ijkl}$ is symmetric in i and j, and, in k and l. Inserting (8) and (5) into (10) and (11) yields the stochastic second and 4th order stiffness tensors.

$$K_{e,ij}(\mathbf{X}^e) = K_{e,ij}^{(0)} + X_0^e\Delta K_{e,ij}^{(0)} + X_1^e\Delta K_{e,ij}^{(1)} + X_2^e\Delta K_{e,ij}^{(2)} \tag{12}$$

$$K_{e,ijkl}(\mathbf{X}^e) = K_{e,ijkl}^{(0)} + X_0^e\Delta K_{e,ijkl}^{(0)} + X_1^e\Delta K_{e,ijkl}^{(1)} + X_2^e\Delta K_{e,ijkl}^{(2)}$$
$$+ X_3^e\Delta K_{e,ijkl}^{(3)} + X_4^e\Delta K_{e,ijkl}^{(4)} + X_5^e\Delta K_{e,ijkl}^{(5)} + X_6^e\Delta K_{e,ijkl}^{(6)} \tag{13}$$

where $\mathbf{X}^{eT} = [X_0, ..., X_6]$ is a non-dimensional zero-mean vector of random variables defined as

$$X_i^e = \int_0^1 \eta^i f_e(\eta)\, d\eta \ , \qquad i = 0, 1, ..., 6 \ , \qquad \eta = \frac{x}{L_e} \tag{14}$$

Given the probabilistic character of $f_e(x)$, joint statistical moments of arbitrary order among X_i can be obtained from (14). These random variables which are representing the random field $f_e(\eta)$ are called "weighted integrals", [15]. They were introduced for linear frame analysis and the explicit expressions for the linear coefficient matrices $\Delta \mathbf{K}_e^{(0)}$, $\Delta \mathbf{K}_e^{(1)}$ and $\Delta \mathbf{K}_e^{(2)}$ without P_e contributions are given in [15]. $\mathbf{K}_0^e$ is the deterministic part or the expected value and $X_0^e\Delta \mathbf{K}_e^{(0)} + X_1^e\Delta \mathbf{K}_e^{(1)} + X_2^e\Delta \mathbf{K}_e^{(2)}$ is the stochastic part of the element stiffness matrix $\mathbf{K}_e$. These matrices are listed in compact index notation below.

$$K_{e,ij}^{(0)} = \overline{(EI)}_e \sum_{m=3}^{4} \sum_{n=3}^{4} \frac{(m-1)(m-2)(n-1)(n-2)}{m+n-5} S_{im} S_{jn} L_e^{m+n-5}$$

$$-P_e \sum_{m=2}^{4} \sum_{n=2}^{4} \frac{(m-1)(n-1)}{m+n-3} S_{im} S_{jn} L_e^{m+n-3} \tag{15}$$

$$\Delta K_{e,ij}^{(0)} = 4\overline{(EI)}_e L_e S_{i3} S_{j3} \quad , \quad \Delta K_{e,ij}^{(1)} = 12\overline{(EI)}_e L_e^2 (S_{i3} S_{j4} + S_{i4} S_{j3}) \quad , \quad \Delta K_{e,ij}^{(2)} = 36\overline{(EI)}_e L_e^3 S_{i4} S_{j4} \tag{16}$$

where S_{ij} indicates the components of the coefficient matrix S given in (8). As the expansion in (13) is due to the accounted nonlinearity and the weighted integral idea has been applied, it is termed "the nonlinear weighted integrals 4th order stiffness tensor" where $K_{e,ijkl}^{(0)}$ is the deterministic part or the expected value and $\sum_{m=0}^{6} X_i^e \Delta K_{e,ijkl}^{(m)}$ is the zero-mean stochastic part of the 4th order element stiffness tensor. The 4th order tensors can also be compactly written as

$$K_{e,ijkl}^{(0)} = \overline{(EI)}_e \sum_{m=3}^{4} \sum_{n=3}^{4} \sum_{r=2}^{4} \sum_{s=2}^{4} \frac{(m-1)(m-2)(n-1)(n-2)(r-1)(s-1)}{m+n+r+s-9} S_{im} S_{jn} S_{kr} S_{ls} L_e^{m+n+r+s-9}$$

$$-\frac{P_e}{4} \sum_{m=2}^{4} \sum_{n=2}^{4} \sum_{r=2}^{4} \sum_{s=2}^{4} \frac{(m-1)(n-1)(r-1)(s-1)}{m+n+r+s-7} S_{im} S_{jn} S_{kr} S_{ls} L_e^{m+n+r+s-7} \tag{17}$$

$$\Delta K_{e,ijkl}^{(0)} = 4\overline{(EI)}_e L_e S_{i3} S_{j3} S_{k2} S_{l2} \tag{18}$$

$$\Delta K_{e,ijkl}^{(1)} = \overline{(EI)}_e L_e^2 \left[12(S_{i4} S_{j3} S_{k2} S_{l2} + S_{i3} S_{j4} S_{k2} S_{l2}) + 8(S_{i3} S_{j3} S_{k3} S_{l2} + S_{i3} S_{j3} S_{k2} S_{l3}) \right] \tag{19}$$

$$\Delta K_{e,ijkl}^{(2)} = \overline{(EI)}_e L_e^3 \left[36 S_{i4} S_{j4} S_{k2} S_{l2} + 12(S_{i3} S_{j3} S_{k4} S_{l2} + S_{i3} S_{j3} S_{k2} S_{l4}) \right.$$

$$\left. +24(S_{i4} S_{j3} S_{k3} S_{l2} + S_{i4} S_{j3} S_{k2} S_{l3} + S_{i3} S_{j4} S_{k3} S_{l2} + S_{i3} S_{j4} S_{k2} S_{l3}) + 16 S_{i3} S_{j3} S_{k3} S_{l3} \right] \tag{20}$$

$$\Delta K_{e,ijkl}^{(3)} = \overline{(EI)}_e L_e^4 \left[24(S_{i3} S_{j3} S_{k4} S_{l3} + S_{i3} S_{j3} S_{k3} S_{l4}) + 48(S_{i3} S_{j4} S_{k3} S_{l3} + S_{i4} S_{j3} S_{k3} S_{l3}) \right.$$

$$\left. +36(S_{i3} S_{j4} S_{k2} S_{l4} + S_{i3} S_{j4} S_{k4} S_{l2} + S_{i4} S_{j3} S_{k2} S_{l4} + S_{i4} S_{j3} S_{k4} S_{l2}) + 72(S_{i4} S_{j4} S_{k3} S_{l2} + S_{i4} S_{j4} S_{k2} S_{l3}) \right] \tag{21}$$

$$\Delta K_{e,ijkl}^{(4)} = \overline{(EI)}_e L_e^5 \left[72(S_{i3} S_{j4} S_{k4} S_{l3} + S_{i3} S_{j4} S_{k3} S_{l4} + S_{i4} S_{j3} S_{k4} S_{l3} + S_{i4} S_{j3} S_{k3} S_{l4}) \right.$$

$$\left. +108(S_{i4} S_{j4} S_{k2} S_{l4} + S_{i4} S_{j4} S_{k4} S_{l2}) + 36 S_{i3} S_{j3} S_{k4} S_{l4} + 144 S_{i4} S_{j4} S_{k3} S_{l3} \right] \tag{22}$$

$$\Delta K_{e,ijkl}^{(5)} = \overline{(EI)}_e L_e^6 \left[108(S_{i3} S_{j4} S_{k4} S_{l4} + S_{i4} S_{j3} S_{k4} S_{l4}) + 216(S_{i4} S_{j4} S_{k3} S_{l4} + S_{i4} S_{j4} S_{k4} S_{l3}) \right] \tag{23}$$

$$\Delta K_{e,ijkl}^{(6)} = 324\overline{(EI)}_e L_e^7 S_{i4} S_{j4} S_{k4} S_{l4} \tag{24}$$

The element linear consistent deterministic mass tensor is given in (25).

$$M_{e,ij} = \sum_{m=1}^{4} \sum_{n=1}^{4} \frac{\mu_e}{m+n-1} S_{im} S_{jn} L_e^{m+n-1} \tag{25} \qquad v_i^e = A_{iI}^e V_I \tag{26}$$

In the global coordinates, each system node is allowed to 3 dof which are selected as the displacements in the X and Y directions and the rotation about the Z direction. The random local dof v^e are related to the corresponding random global dof V from the linear deterministic transformation as given in (26). In (26) and below, summation convention has been applied over the dummy indices I, J, K, L which run from 1 to d, where d signifies the number of global dof. The transformation tensor A_{iI} is merely a function of the angle Γ_e depicted in Figure 1 and the location of the beam element in the frame. $U(V(X,t), X)$ is made up of the sum of the potential energies of all N_e many beam elements of the frame. Upon inserting (26) into (9), we obtain

$$U(V(X,t), X) = \sum_{e=1}^{N_e} U_e(v^e) = \frac{1}{2} \left(K_{IJ}(X) V_I V_J + K_{IJKL}(X) V_I V_J V_K V_L \right) \tag{27}$$

$$K_{IJ}(X) = \sum_{e=1}^{N_e} K_{e,ij}(X^e) A_{iI}^e A_{jJ}^e \quad , \quad K_{IJKL}(X) = \sum_{e=1}^{N_e} K_{e,ijkl}(X^e) A_{iI}^e A_{jJ}^e A_{kK}^e A_{lL}^e \tag{28}$$

$K_{IJ}(\mathbf{X})$ signifies the components of the linear random stiffness matrix indicated in equation (1). It should be noted that the system random vector $\mathbf{X}$ includes all $7 \times N_e$ many weighted integral random variables besides random parameters specifying the system damping. Both $K_{IJ}(\mathbf{X})$ and $K_{IJKL}(\mathbf{X})$ are linear functions of the $7 \times N_e$ weighted integral random variables. However, $K_{IJ}(\mathbf{X})$ is only a linear function of first 3 basic variables of each element.

$$K_{IJ}(\mathbf{X}) = K_{IJ}^{(0)} + \sum_{m=1}^{N_e} \sum_{n=1}^{3} \Delta K_{IJ}^{(7(m-1)+n)} X_{7(m-1)+n} \quad , \quad K_{IJKL}(\mathbf{X}) = K_{IJKL}^{(0)} + \sum_{m=1}^{7N_e} \Delta K_{IJKL}^{(m)} X_m \quad (29)$$

where $K_{IJ}^{(0)}$, $K_{IJKL}^{(0)}$, $\Delta K_{IJ}^{(m)}$, $\Delta K_{IJKL}^{(m)}$, $m = 1, ..., 7N_e$ are deterministic tensor components obtained from $K_{e,ij}^{(0)}$, $K_{e,ijkl}^{(0)}$, $\Delta K_{e,ij}^{(m)}$ and $\Delta K_{e,ijkl}^{(m)}$ by the transformations of (28). Using the symmetry properties of K_{IJKL}, the nonlinear stochastic function $H_I(\mathbf{V}(\mathbf{X}, t), \mathbf{X})$ is determined from the stationarity condition of the potential energy in the dynamic equilibrium state as :

$$H_I(\mathbf{V}(\mathbf{X}, t), \mathbf{X}) = \frac{1}{2} \frac{\partial}{\partial V_I} \left(K_{JKLM}(\mathbf{X}) V_J V_K V_L V_M \right) = \left(K_{IJKL}(\mathbf{X}) + K_{JKIL}(\mathbf{X}) \right) V_J V_K V_L \quad (30)$$

3. Solution of The Nonlinear Stochastic Differential Equations

The approximate stationary solution of the nonlinear second order sode with random coefficients is based on a system reduction to a modal subspace and expansion by the deterministic linear eigenmodes. Consider the following deterministic linear eigenvalue problem where $\mathbf{K}_0$ with components $K_{IJ}^{(0)}$ is defined as the deterministic part of equation (29).

$$(\mathbf{K}_0 - \omega_{j,0}^2 \mathbf{M}) \mathbf{\Phi}_0^{(j)} = 0 \ , \qquad j = 1, ..., d \qquad (31)$$

$\omega_{0,j}$ is the jth circular undamped eigenfrequency and $\mathbf{\Phi}_0^{(j)}$ is the jth eigenvector of the linear deterministic problem, i.e. the mean structural system. Since $\mathbf{\Phi}_0^{(j)}$ form an independent basis in R^d, the following expansion is exact even for the nonlinear random system.

$$\mathbf{V}(\mathbf{X}, t) = \sum_{\alpha=1}^{d} y_\alpha(\mathbf{X}, t) \mathbf{\Phi}_0^{(\alpha)} \qquad (32)$$

The modal coordinates $y_\alpha(\mathbf{X}, t)$ are random processes dependent on the random loading vector $\mathbf{F}(t)$ and the basic random variables $\mathbf{X}$. Next, a system reduction is performed, i.e. the series expansion of (32) is truncated at $d_1 \leq d$ terms. d_1 is chosen such that the truncation yields negligible loss of accuracy in the response quantity interested. System reduction scheme has been applied to linear random vibration of random structures earlier in [11] and very high rate of convergence has been reported after a few terms for linear elastic frames. Consider the deterministic $d \times d_1$ truncated modal matrix $\mathbf{B} = [\mathbf{\Phi}_0^{(1)}, ..., \mathbf{\Phi}_0^{(d_1)}]$ and let $B_{I\alpha}$ be the components of the Ith row and α^{th} column. Then $V_I(\mathbf{X}, t)$ becomes

$$V_I(\mathbf{X}, t) \simeq B_{I\alpha} y_\alpha(\mathbf{X}, t) \qquad (33)$$

In (33) and below, summation convention is applied over dummy indices $\alpha, \beta, \gamma, \delta$ which range from 1 to d_1. Expressing the potential energy of (27) in the related modal coordinates by means of (33) and applying a stationarity condition similar to (30) yield the equations of motion in the modal coordinates.

$$m\ddot{y}(\mathbf{X}, t) + \mathbf{c}(\mathbf{X})\dot{y}(\mathbf{X}, t) + \mathbf{k}(\mathbf{X})y(\mathbf{X}, t) + \mathbf{h}(y(\mathbf{X}, t), \mathbf{X}) = \mathbf{f}_0 f(t) \qquad (34)$$

$$\mathbf{m} = \mathbf{B}^T \mathbf{M} \mathbf{B} \quad , \quad \mathbf{c}(\mathbf{X}) = \mathbf{B}^T \mathbf{C}(\mathbf{X}) \mathbf{B} \quad , \quad \mathbf{f}_0 = \mathbf{B}^T \mathbf{F}_0 \qquad (35)$$

$$k_{\alpha\beta}(\mathbf{X}) = K_{IJ}(\mathbf{X}) B_{I\alpha} B_{J\beta} = k_{\alpha\beta}^{(0)} + \sum_{m=1}^{N_e} \sum_{n=1}^{3} \Delta k_{\alpha\beta}^{(7(m-1)+n)} X_{7(m-1)+n} \qquad (36)$$

$$h_\alpha(y(\mathbf{X}, t), \mathbf{X}) = \left(k_{\alpha\beta\gamma\delta}(\mathbf{X}) + k_{\beta\gamma\alpha\delta}(\mathbf{X}) \right) y_\beta y_\gamma y_\delta \qquad (37)$$

$$k_{\alpha\beta\gamma\delta}(\mathbf{X}) = K_{IJKL}(\mathbf{X}) B_{I\alpha} B_{J\beta} B_{K\gamma} B_{L\delta} = k_{\alpha\beta\gamma\delta}^{(0)} + \sum_{m=1}^{7N_e} \Delta k_{\alpha\beta\gamma\delta}^{(m)} X_m \qquad (38)$$

Above, m, $c(X)$, $k(X)$ $h(y(X,t),X)$ are the diagonal deterministic modal mass matrix, the stochastic diagonal modal damping matrix, the stochastic modal stiffness matrix and the stochastic modal stiffness tensor. $k_{\alpha\beta}^{(0)}$, $k_{\alpha\beta\gamma\delta}^{(0)}$, $\Delta k_{\alpha\beta}^{(m)}$ and $\Delta k_{\alpha\beta\gamma\delta}^{(m)}$ are obtained from $K_{IJ}^{(0)}$, $K_{IJKL}^{(0)}$, $\Delta K_{IJ}^{(m)}$ and $\Delta K_{IJKL}^{(m)}$ by the transformations indicated in (36) and (38).

The mean value linear modal stiffness matrix $k^{(0)}$ with components $k_{\alpha\beta}^{(0)}$ is diagonal because of the chosen basis, i.e. $k_{\alpha\beta}^{(0)} = 0$, $\alpha \neq \beta$. If the eigenvectors Φ_0 are mass orthonormalized, m becomes the identity matrix and the diagonal elements of the proportional damping matrix reads

$$c_{ii}(X) = 2\zeta_i\omega_{i,0}, \qquad \zeta_i = \zeta_{i,0}(1 + X_{7N_e+i}), \qquad i = 1, 2, ..., d_1 \tag{39}$$

For simplicity, ζ_i are assigned as mutually independent random variables stochastically independent of the weighted integrals. Then the mean-zero random structural vector X is made up of $7 \times N_e$ weighted integral random variables and d_1 random damping ratios. The stationary unconditional covariances $\kappa_{V_I V_J}$ of the zero-mean stationary responses $V_I(t)$ and $V_J(t)$ can be derived from (33).

$$\kappa_{V_I V_J} = E[\kappa_{V_I V_J}(X)] = B_{I\alpha}B_{J\beta}E[\kappa_{y_\alpha y_\beta}(X)] \tag{40}$$

$\kappa_{y_\alpha y_\beta}(x)$ is the stationary conditional covariance of the modal coordinates on condition of the structure $X = x$ and $E[\kappa_{y_\alpha y_\beta}(X)]$ signifies the unconditional covariance according to the total probability theorem. Use of Wiener-Khintchine relations next provides

$$E[\kappa_{y_\alpha y_\beta}(X)] = 2\int_0^\infty E\big[Re[S_{y_\alpha y_\beta}(\omega, X)]\big]d\omega \tag{41}$$

where $S_{y_\alpha y_\beta}(\omega, x)$ signifies the components of the cross spectral density matrix of the nodal response vector process on condition of $X = x$. In (41), the symmetry property $Re[S_{y_\alpha y_\beta}(\omega, x)] = Re[S_{y_\alpha y_\beta}(-\omega, x)]$ for any realized structure has been applied. In order to determine the conditional cross spectral density function, an equivalent statistical linearization of the nonlinear restoring forces in (37) is applied on condition of $X = x$. This implies the following approximation.

$$k_{\alpha\beta}(x)y_\beta(x,t) + h_\alpha(y(x,t),x) \simeq k_{eq,\alpha\beta}(x)y_\beta(x,t) \tag{42}$$

The response $y(x,t)$ of the equivalent linear structure then becomes Gaussian. The least mean square estimate of $k_{eq,\alpha\beta}(x)$ for the conditional structure are, e.g. see [17],

$$k_{eq,\alpha\beta}(x) = k_{\alpha\beta}(x) + E\Big[\frac{\partial}{\partial y_\beta}h_\alpha(y(x,t),x)\Big] = k_{eq,\alpha\beta}^{(0)}(x) + \sum_{m=1}^{7N_e+d_1} \Delta k_{eq,\alpha\beta}^{(m)}(x)x_m \tag{43}$$

$$k_{eq,\alpha\beta}^{(0)}(x) = k_{\alpha\beta}^{(0)} + (k_{\alpha\beta\gamma\delta}^{(0)} + 2k_{\alpha\gamma\beta\delta}^{(0)} + 2k_{\beta\gamma\alpha\delta}^{(0)} + k_{\gamma\delta\alpha\beta}^{(0)})\kappa_{y_\alpha y_\beta}(x) \tag{44}$$

$$\Delta k_{eq,\alpha\beta}^{(m)}(x) = \Delta k_{\alpha\beta}^{(m)} + (\Delta k_{\alpha\beta\gamma\delta}^{(m)} + 2\Delta k_{\alpha\gamma\beta\delta}^{(m)} + 2\Delta k_{\beta\gamma\alpha\delta}^{(m)} + \Delta k_{\gamma\delta\alpha\beta}^{(m)})\kappa_{y_\alpha y_\beta}(x) \quad , m = 1, ...7N_e \tag{45}$$

In (44) and (45), the linear expansions (36) and (38) have been applied for $k_{\alpha\beta}(x)$ and $k_{\alpha\beta\gamma\delta}(x)$. x_m in (43) signifies the mth component of the realized vector x. As seen from (44) and (45), the expansion coefficients of the equivalent linear stiffness matrix are no longer constants, but depend on the conditional covariance matrix $\kappa_{y_\alpha y_\beta}(x)$ of the realized structure x.

The conditional cross-spectral density can be written as,

$$S_{y_\alpha y_\beta}(\omega, X) = Y_\alpha^*(\omega, X)Y_\beta(\omega, X)S_{ff}(\omega) \tag{46}$$

$$Y_\alpha(\omega, X) = H_{\alpha\beta}(\omega, X)f_{0,\beta} \quad , \quad H(\omega, X) = [(i\omega)^2 m + i\omega c(X) + k_{eq}(X)]^{-1} \tag{47}$$

In order to calculate (41), the following Taylor's expansion to the second order from the mean structure $E[X] = 0$ for the frequency response vector $Y(\omega, X)$ is applied.

$$Y(\omega, X) \simeq Y(\omega, 0) + \sum_{m=1}^{7N_e+d_1} \frac{\partial}{\partial x_m}(Y(\omega, 0))X_m + \frac{1}{2}\sum_{m=1}^{7N_e+d_1}\sum_{n=1}^{7N_e+d_1} \frac{\partial^2}{\partial x_m \partial x_n}(Y(\omega, 0))X_m X_n \tag{48}$$

$$Y(\omega, 0) = H(\omega, 0)f_0 \tag{49}$$

and partial derivatives from the mean structure becomes

$$\frac{\partial}{\partial x_m}Y(\omega, 0) = -H(\omega, 0)\frac{\partial}{\partial x_m}(k_{eq}(0) + i\omega c(0))Y(\omega, 0) \tag{50}$$

$$\frac{\partial^2}{\partial x_m \partial x_n} \mathbf{Y}(\omega, 0) = -\mathbf{H}(\omega, 0)\Big(\frac{\partial}{\partial x_m}\big(\mathbf{k}_{eq}(0) + i\omega\mathbf{c}(0)\big)\frac{\partial}{\partial x_n}\mathbf{Y}(\omega, 0) +$$

$$\frac{\partial}{\partial x_n}\big(\mathbf{k}_{eq}(0) + i\omega\mathbf{c}(0)\big)\frac{\partial}{\partial x_m}\mathbf{Y}(\omega, 0) + \Big(\frac{\partial^2}{\partial x_m \partial x_n}\mathbf{k}_{eq}(0)\Big)\mathbf{Y}(\omega, 0)\Big) \tag{51}$$

In (51), $\frac{\partial^2}{\partial x_m x_n}\mathbf{c}(0) = 0$ because $\mathbf{c}(\mathbf{X})$ is a linear function of $\mathbf{X}$, see (39). Due to page limitations, the analytically closed form of the partial deivatives in (50) and (51) are not listed in this paper. These are given in [12, 13]. The statistical moments $E[y_\alpha(t, 0)y_\beta(t, 0)]$ and $\frac{\partial}{\partial x_n} E[y_\alpha(t, 0)y_\beta(t, 0)]$ are needed to evaluate unconditional statistical moments. These moments are related to the mean structure $\mathbf{X} = 0$ and are determined iteratively from nonlinear analysis of the mean structure. Upon inserting (48), (46) into (41) and retaining all the terms up to second order in X_m yields

$$E[\kappa_{Y_\alpha Y_\beta}(\mathbf{X})] \simeq 2\int_0^\infty Re\Big[Y_\alpha^*(\omega, 0)Y_\beta(\omega, 0) + \sum_{m=1}^{7N_e+d_1}\sum_{n=1}^{7N_e+d_1}\Big(\frac{\partial}{\partial x_m}Y_\alpha^*(\omega, 0)\frac{\partial}{\partial x_n}Y_\beta(\omega, 0) +$$

$$\frac{1}{2}Y_\alpha^*(\omega, 0)\frac{\partial^2}{\partial x_m \partial x_n}Y_\beta(\omega, 0) + \frac{1}{2}\frac{\partial^2}{\partial x_m \partial x_n}Y_\alpha^*(\omega, 0)Y_\beta(\omega, 0)\Big)\kappa_{X_m X_n}\Big]S_{ff}(\omega)d\omega \tag{52}$$

where $\kappa_{X_\alpha X_\beta}$ signifies the components of the covariance matrix of the basic variables. The quadrature of (52) is computed numerically.

4. Numerical Example

Two 3-storey 1-bay frames, a linearly and a nonlinearly behaving one, with 9 elements ($N_e = 9$) and 9 dof ($d = 9$) subject to horizontal ground excitation modeled as white noise excitation with auto-spectral density $S_{ff}(\omega) = S_0$ is solved using the proposed formulation to illustrate the validity range of the second order perturbed random vibration analysis in terms of variability of the random damping ratios and both the variability and correlation structure of the random bending rigidity field. Then, $\mathbf{F}_0 = \mathbf{MA}$ with $\mathbf{A}^T = [1, 0, 0, 1, 0, 0, 1, 0, 0]^T$. The system parameters for the first frame are given below Figure 2. The first frame considered is pretty stiff and for this frame the axial load contributions to the stiffness are neglected, i.e. $P_e = 0$ for all elements, thus, the frame behaves linear. The standard deviation of the top storey displacement (V_7) and the top storey angular deformation (V_9) of the deterministic mean structure are $\sigma_7 = 0.043358762\ m$ and $\sigma_9 = 0.00050311157\ rad$ for $d_1 = 1$, and, $\sigma_7 = 0.043432072\ m$ and $\sigma_9 = 0.00056084531\ rad$ for $d_1 = 3$. Hence, the first mode totally governs the response for V_7 and convergence is observed after three modes for V_9 The autocorrelation function of the zero-mean nondimensional random field $f_e(x)$ of each element is assumed to be in the following form.

$$R_{ff}(x_1, x_2) = v_e^2 \exp\Big(-\big(\frac{x_1 - x_2}{b_e}\big)^2\Big) \tag{53}$$

where v_e is the standard deviation of the random field $f_e(x)$ and the coefficient of variation of the random bending stiffness field of element e. b_e is a nondimensional correlation parameter such that $b_e \to \infty$ and $b_e \to 0$ denote fully correlated and perfectly uncorrelated random fields, respectively. Then, the covariance structure of the weighted integrals, which is a monotonously increasing function of b_e and v_e, can be calculated easily using (14). For simplicity, the random fields of beam elements are assumed to be independent of each other and both b_e and v_e are taken as the same for each element . The modal damping ratios for each mode ζ_i are considered as independent uniformly distributed random variables with $U(0.02(1 - \sqrt{3}v_c), 0.02(1 + \sqrt{3}v_c))$ where v_c is the coefficient of variation of the random damping ratios.

The proposed second order perturbation method is a good approximation if $\mathbf{C}$, $\mathbf{K}$ and $\mathbf{H}(\mathbf{V})$ of equation (1) have small variabilities, i.e small v_c, v_e and small b_e. In Figures 3 and 4 the validity range of v_c, Figures 5 and 6 the validity range of v_e, Figures 7 and 8 the validity range of b_e are examined for V_7 and V_9. Convergence in solving equation (52) iteratively is assumed to be achieved when the sum of the relative errors of the modal variances are below 2.5×10^{-3}.

In all the studied cases, it is observed that the unconditional variance of the response of the random structure $\kappa_{V_I V_I}$ is greater than the variance of the deterministic structure with mean realization of the random variables $\kappa_{V_I V_I}(x = 0)$ for all I. In the following figures ordinate show the nondimensional quantity $\sqrt{\frac{\kappa_{V_I V_I} - \kappa_{V_I V_I}(x=0)}{\kappa_{V_I V_I}(x=0)}}$ for $I = 7, 9$ which is the coefficient of variation of the unconditional variance $\kappa_{V_I V_I}$ with respect to the conditional variance $\kappa_{V_I V_I}(x = 0)$ conditioned on $x = 0$. This quantity is denoted with v_I $I = 7, 9$ in the figures for short-hand notation. The notation of the plots 3-14 follows Table 1.

  ━━━━ perturbation results considering only the first mode ($d_1 = 1$)
  .... simulation results considering only the first mode ($d_1 = 1$)
  − − −− perturbation results considering three modes ($d_1 = 3$)
  −. − .− simulation results considering three modes ($d_1 = 3$)

Table 1. Explanation of the notation of Figures 3-14.

Therefore, ━━━━ and .... are to be compared in Figures 3-14, so are − − −− and −. − .−. The exact results are estimated via Monte Carlo simulations. 1000 realizations of the structure, the uniformly distributed independent random damping ratios and the Gaussian distributed random stiffness matrix, are generated and equation (33) is solved for each of these realizations using equivalent statistical linearization with Gaussian closure to obtain the stationary covariances of the response of that particular realization. Hence, the simulations are on condition that equivalent statistical linearization is a good approximation. This is shown to be a good approximation when the nonlinearities are small, [13, 17]. Then, the total probability theorem is calculated by averaging the results. For the simulations of the local and global stiffness tensors, there are 63 random variables; 7 random variables for each element. The covariance structure of these random variables are obtained using (53) in (14).

From the figures, it is concluded that the unconditional variances have large sensitivity on v_c, mean sensitivity on v_e and large sensitivity on b_e for small b_e and modest sensitivity on b_e for large b_e. Perturbation solutions seem to be satisfactory for variabilities up to 0.3 for v_c and 0.35 for v_e. Convergence in v_I $I = 7, 9$ in terms of number of modes retained in the expansion of (42) is observed in Figures 3 and 4 after the first mode. This implies that the first random modal damping ratio is very important for structures where the first mode totally governs the response.

Next, the same frame is excited with white noise of 50 times stronger intensity, i.e. $S_0 = 1 m^2/s^3$, and an axial force of $P_e = 1MN$ is introduced for all elements to observe nonlinear behaviour. Then, the standard deviation of V_7 and V_9 of the deterministic mean structure are $\sigma_7 = 0.3274519\,m$ and $\sigma_9 = 0.0038146174\,rad$ for $d_1 = 1$, and, $\sigma_7 = 0.3293790\,m$ and $\sigma_9 = 0.0042605987\,rad$ for $d_1 = 3$. Figures 9-14 studies the validity ranges similar to Figures 3-8. The second and third modes have more contribution to the results in this nonlinear example compared to the previous linear example. This shows that modal expansion is performing poorer for nonlinear structure compared to linear structure. Figures 9 and 10 are more or less the same with Figures 3 and 4 since the variability is only in the linear random modal damping ratios. A comparison of figures 11-14 with 5-8 show that v_I for $I = 7, 9$ increases with an increase in the nonlinear behaviour. An increase in the nonlinear behaviour is tantamount to an increase in the stiffness tensors $\Delta K_{IJKL}^{(m)}$ $m = 0, 1, ..., 6$ which means an increase in the randomness of the structural system. Figure 15 is to illustrate the axial load, P_e, affect. It should be noted that each data of the coefficient of variation of the unconditional variance $\kappa_{V_7 V_7}$ with respect to $\kappa_{V_7 V_7}(x = 0)$ in Figure 15 is obtained for different $\kappa_{V_7 V_7}(x = 0)$. The P_e and $\sigma_7(x = 0)$ pairs are $(0, 0.3065926\,m)$, $(1\,MN, 0.3274519\,m)$, $(3\,MN, 0.3813871\,m)$, $(5\,MN, 0.4625571\,m)$ and $(7\,MN, 0.6028258\,m)$.

5. Conclusion

A SFEM is derived for the stochastic response analysis of geometrically nonlinear structures with random stiffness and random damping properties subject to random stationary excitation. Nonlinear weighted integrals are introduced. The proposed second order perturbed random vibration analysis has been examined in detail with examples considering the variability of the random damping ratios and both the variability and correlation structure of the random fields. It is concluded that that the unconditional variance has comparably large sensitivity on the variability of the random damping ratio, mean sensitivity on the variability of the random bending rigidity field and modest sensitivity on the correlation length of the random bending rigidity field. The second order perturbation method proposed seem to be satisfactory for variabilities up to 25-30 percent.

6. References

1. Vanmarcke, E., Shinozuka, M., Nakagiri, S., Schueller, G. and Grigoriu, M. (1986) Random Fields and Stochastic Finite-Elements, *Structural Safety*, **3**, 143-166.

2. Benaroya, H. and Rehak, M. (1988) Finite Element Methods in Probabilistic Structural Analysis : A Selective Review, *Appl. Mech. Rev.*, **40**, 201-213.

3. Der Kiureghian, A., Li, C.C. and Zhang, Y. (1991) Recent Developments in Stochastic Finite Elements, *Lecture Notes in Engineering IFIP 76, Proc. Fourth IFIG WG 7.5 Conference*, Germany, ed. R. Rackwitz & P. Thoft-Christensen, Springer-Verlag.

4. Brenner C. (1991) Stochastic Finite Elements (Literature Review), *Internal Working Report No. 35-91, Institute of Engrg. Mech., University of Innsbruck, Austria.*

5. Ghanem, R. and Spanos, P.D. (1991) *Stochastic Finite Elements : A Spectral Approach*, Springer-Verlag, NewYork.

6. Kleiber, M. and Hien, T.D. (1992) *The Stochastic Finite Element Method*, Wiley, Chichester.

7. Hisada, T. & Nakagiri, S., Role of the Stochastic Finite Element Method in Structural Safety and Reliability, *ICOSSAR'85, 4th Int. Con. Str. Safety and Reliability I* ed. I. Konishi, A.H-S. Ang and M. Shinozuka, 1985, pp. 385-394.

8. Liu, W.K., Belytschko, T. and Mani, A. (1986) Probabilistic Finite Elements for Nonlinear Structural Dynamics, *Com. Meth. Appl. Mech. Engrg.*, **56**, 61-81.

9. Chang, C.C., and Yang, H.T.Y. (1991) Random Vibration of Flexible, Uncertain Beam Element, *J. Engrg. Mech.*, ASCE, **117**, 2369-2351.

10. Jensen, H. and Iwan, W.D. (1992) Response of Systems with Uncertain Parameters to Stochastic Excitation, *J. Engng. Mech.*, ASCE, **118**, 1012-1025.

11. Köylüoğlu, H.U. & Nielsen, S.R.K. (1993) Stochastic Dynamics of Linear Structures with Random Stiffness Properties and Random Damping subject to Random Loading, *Structural Dynamics, Eurodyn'93, Vol.2* ed. Moan, T. et al. 1993, pp. 705-711.

12. Köylüoğlu, H.U., Nielsen, S.R.K. and Çakmak, A.Ş. (1994) Stochastic Dynamics of Geometrically Nonlinear Structures subject to Stationary Random Excitation, accepted for publication in *Journal of Sound and Vibration*, provisionally scheduled for February 1996.

13. Köylüoğlu, H.U. (1995) *Stochastic Response and Reliability Analyses of Structures with Random Properties subject to Random Stationary Excitation*, Ph.D. thesis, Princeton University, NJ, U.S.A.

14. Köylüoğlu, H.U., Nielsen, S.R.K. and Çakmak, A.Ş. (1994) Solution of random structural system subject to nonstationary excitation : Transforming the equation with random coefficients to one with deterministic coefficients and random initial conditions, *Soil Dynamics and Earthquake Engineering*, **14**, 219-228.

15. Deodatis, G. (1991) Weighted Integral Method. I : Stochastic Stiffness Matrix, *J. Engrg. Mech.*, ASCE, **117**, 1851-1864.

16. El Nashie, M.S., *Stress, Stability and Chaos in Structural Engineering: An Energy Approach*, Mc Graw Hill Book Company, London, 1990.

17. Atalik, T.S. and Utku, S. (1976) Stochastic Linearization of Multi-Degree of Freedom Non-linear Systems, *Earthquake Engineering and Structural Dynamics*, **4**, 411-420.

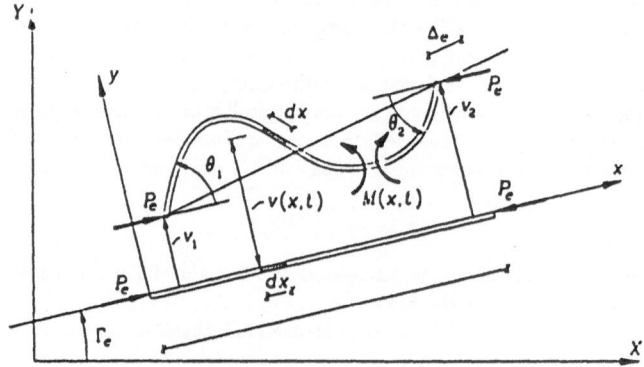

Figure 1. Beam element with 4 dof.

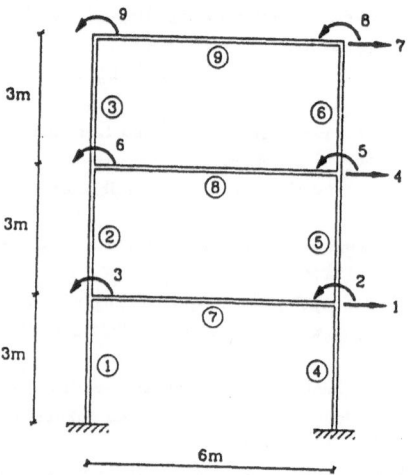

Figure 2. 3-storey 1-bay frame. All elements $E = 3.5 \times 10^{10}\ N/m^2$, Columns $I = 4.0 \times 10^{-4}\ m^4$, $\mu = 300\ kg/m$, Beams $I = 3.0 \times 10^{-3}\ m^4$, $\mu = 2800\ kg/m$ (for beams 7 and 8), $\mu = 1900\ kg/m$ (for beam 9), $S_0 = 0.02 m^2/s^3$. $\omega_{1,0} = 11.000\ rad/s$, $\omega_{2,0} = 31.789\ rad/s$, $\omega_{3,0} = 47.353\ rad/s$.

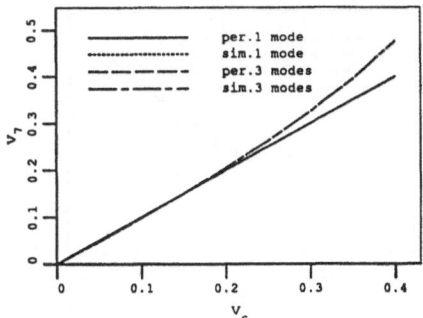

Figure 3. The coefficient of variation of the variance of the top storey displacement v_7 versus v_c. $v_e = 0$.

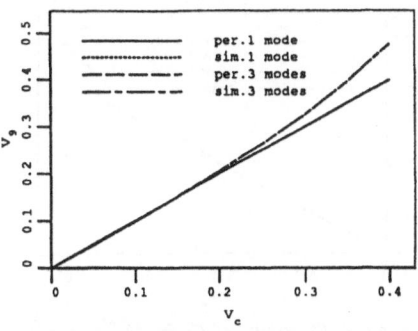

Figure 4. The coefficient of variation of the variance of the angular deformation v_9 versus v_c. $v_e = 0$.

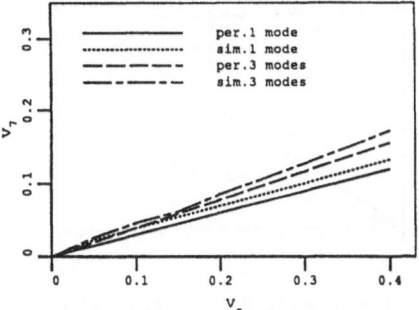

Figure 5. v_7 versus v_e. $b_e = 0.5$ and $v_c = 0$.

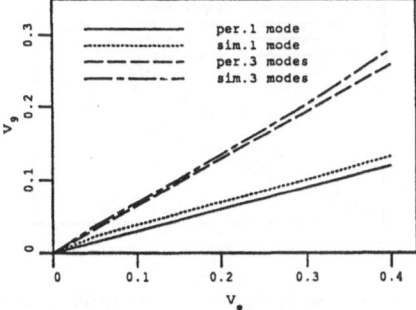

Figure 6. v_9 versus v_e. $b_e = 0.5$ and $v_c = 0$.

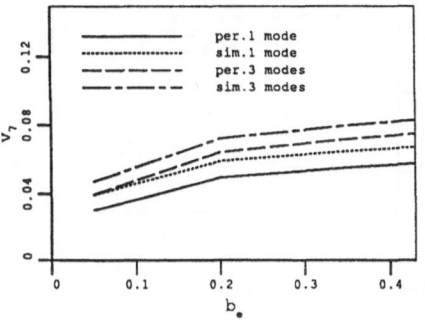

Figure 7. v_7 versus b_e. $v_e = 0.2$ and $v_c = 0$.

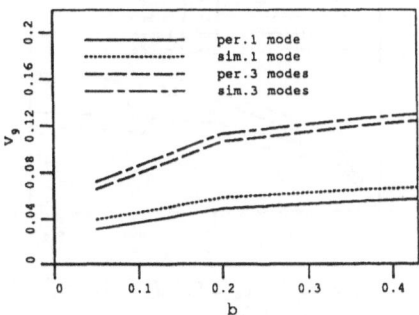

Figure 8. v_9 versus b_e. $v_e = 0.2$ and $v_c = 0$.

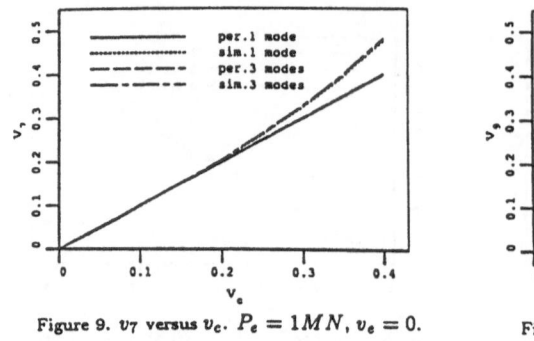

Figure 9. v_7 versus v_c. $P_e = 1MN$, $v_e = 0$.

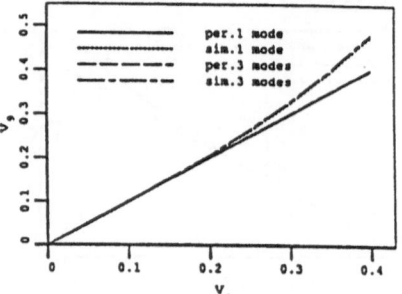

Figure 10. v_9 versus v_c. $P_e = 1MN$, $v_e = 0$.

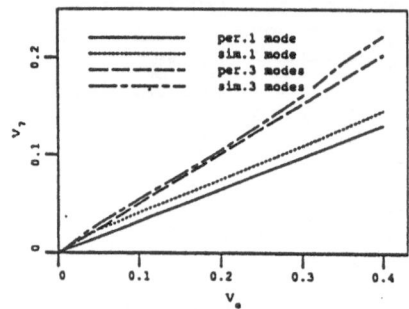

Figure 11. v_7 versus v_e. $b_e = 0.5$, $P_e = 1MN$, $v_c = 0$.

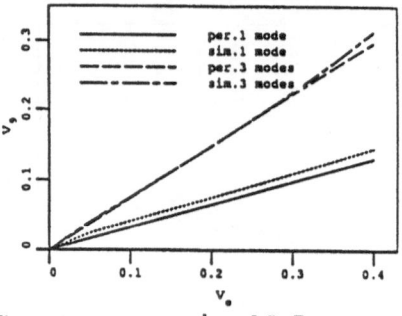

Figure 12. v_9 versus v_e. $b_e = 0.5$, $P_e = 1MN$, $v_c = 0$.

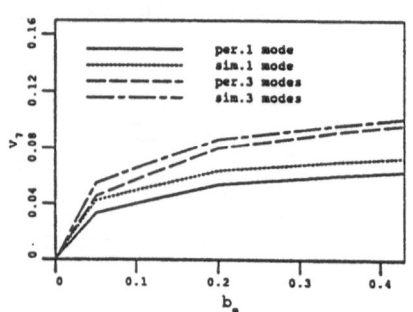

Figure 13. v_7 versus b_e. $v_e = 0.2$, $P_e = 1MN$, $v_c = 0$.

Figure 14. v_9 versus b_e. $v_e = 0.2$, $P_e = 1MN$, $v_c = 0$.

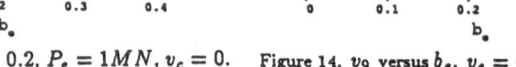

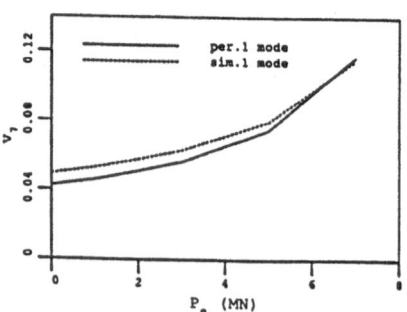

Figure 15. v_7 versus P_e in MN. $v_e = 0.2$, $b_e = 0.5$, and $v_c = 0$.

WIND FIELD COHERENCE AND DYNAMIC WIND FORCES

S. KRENK
Division of Mechanics
Lund Institute of Technology
Lund University
Box 118, S-221 00 Lund
Sweden

1. Introduction

Dynamic response of structures to turbulent wind depends on the temporal and spatial fluctuations of the wind. In principle the statistical properties could be combined into a random field model of the turbulent wind. However, in practice it turns out to be very difficult to combine the height variation of the mean wind with a fully satisfactory stochastic description of the turbulent fluctuations. Several models have been proposed, but none of them combine completeness and simplicity, and we shall restrict the attention to some basic characteristics. The combination of spatial and temporal fluctuations is made by Taylor's hypothesis of convected frozen turbulence. This amounts to convecting a time-invariant velocity field downstream with the mean velocity. Within this approximation the problem is reduced to a stochastic spatial wind velocity field. The classical account of the theory for homogeneous isotropic turbulence is Batchelor (1953). Several extensions to anisotropic turbulence have been proposed. The most promising is probably that of Mann (1994), in which the effect of shear due to the vertical gradient of the mean wind is accounted for as a perturbation in the Navier-Stokes equations. This model accounts for differences in the three velocity components and is easily adapted to simulation, Mann & Krenk (1993).

In practice engineering response analysis usually is based on the turbulence frequency spectrum $S_{11}(\omega)$ for the along-wind velocity component u_1 at a single point combined with a function representing the loss of correlation with distance. This factor is usually taken as an exponential in the

A. Naess and S. Krenk (eds.), IUTAM Symposium on Advances in Nonlinear Stochastic Mechanics, 269–278.
© *1996 Kluwer Academic Publishers.*

form

$$\psi(\omega r/U) = \exp\left(-c\frac{\omega r}{U}\right) \tag{1}$$

where r is the distance, U is the mean wind speed, and c is a non-dimensional attenuation parameter, Davenport (1977). This format - also adopted in the new Eurocode (1994) - has the advantage of simplicity, but also embodies two inconsistencies. The use of a function that is positive for any separation leads to a finite integral over a plane transverse to the wind in conflict with the definition of u_1 as a zero mean fluctuation component. In addition the use of the parameter $\omega r/U$ implies full correlation of the low frequency components, irrespective of distance.

In the following the theory of convected turbulence is used to identify a simple modified format for the spatial coherence function $\psi(r,\omega)$, where mean value consistency is obtained by including a suitable derivative, and correlation of the low frequency components is limited by replacing the wave number ω/U by a modified wavenumber κ_1 with finite value at zero frequency.

2. Mathematics of Homogeneous Turbulence

Some basic results from the theory of homogeneous isotropic turbulence are summarized as background for a derivation of the cross-spectra in convected turbulence

2.1. SPATIAL DESCRIPTION

A homogeneous turbulent velocity field (u_1, u_2, u_3) is described by its covariance function

$$R_{ij}(\mathbf{x}) = E[u_i(\mathbf{x_0})u_j(\mathbf{x_0}+\mathbf{x})] \tag{2}$$

For isotropic fields the covariance function is fully described by the radial and transverse correlation functions

$$R_{11}(x_1) = \sigma_u^2 f(x_1) \quad , \quad R_{22}(x_1) = \sigma_u^2 g(x_1) \tag{3}$$

Invariance arguments then give the general form

$$R_{ij}(\mathbf{x}) = \sigma_u^2 \left([f(x)-g(x)]\frac{x_i x_j}{x^2} - g(x)\delta_{ij}\right) \tag{4}$$

For an incompressible field $\partial R_{ij}/\partial x_j = 0$, and substitution of R_{ij} from (4) leads to the following expression for $g(x)$ in terms of $f(x)$.

$$g(x) = \frac{1}{2x}\frac{d}{dx}\left(x^2 f(x)\right) = \left(1 + \frac{x}{2}\frac{d}{dx}\right)f(x) \tag{5}$$

It is seen from the first form that the area integral of the transverse correlation $g(x)$ is zero. In fact this follows from ergodicity when the fluctuations (u_1, u_2, u_3) have zero mean. It will be seen that a similar condition holds for the individual frequency components.

The spectral wavenumber representation of the radial correlation function is

$$\sigma_u^2 f(x) = \int_{-\infty}^{\infty} F(k)\, e^{ikx}\, dk \tag{6}$$

This representation leads to the following spectral wavenumber representation of the covariance function

$$R_{ij}(\mathbf{x}) = \int_{-\infty}^{\infty}\int_{-\infty}^{\infty}\int_{-\infty}^{\infty} \frac{E(k)}{4\pi k^4}\left(k^2\,\delta_{ij} - k_i\,k_j\right) e^{i\,\mathbf{k}\cdot\mathbf{x}}\, dk_1 dk_2 dk_3 \tag{7}$$

where the energy density, determined by

$$E(k) = k^3 \frac{d}{dk}\left(\frac{1}{k}\frac{dF}{dk}\right) \tag{8}$$

represents the energy of the three velocity components for wavenumbers with magnitude in the interval $[k, k + dk]$.

2.2. CROSS-SPECTRAL DENSITIES IN CONVECTED TURBULENCE

The turbulent velocity field is now translated with the mean velocity $\mathbf{U}$ according to Taylor's hypothesis. The covariance function is then a function of the convected coordinates $\mathbf{x} - \mathbf{U}\tau$, where $\mathbf{x}$ is the coordinate differnce, and τ is the time difference. The frequency specrum can then be written as

$$S_{ij}(\mathbf{x},\omega) = \frac{1}{2\pi}\int_{-\infty}^{\infty} R_{ij}(\mathbf{x} - \mathbf{U}\tau)\, e^{i\tau\omega}\, d\tau \tag{9}$$

Substitution of the covariance function $R_{ij}(\mathbf{x} - \mathbf{U}\tau)$ from (7) gives after interchange of the order of integration

$$S_{ij}(\mathbf{x},\omega) = \frac{1}{U}\iiint \frac{E(k)}{4\pi k^4}\left(k^2\,\delta_{ij} - k_i\,k_j\right)\delta(\mathbf{k}\cdot\mathbf{U} - \omega)e^{i\,\mathbf{k}\cdot\mathbf{x}}\, dk_1 dk_2 dk_3 \tag{10}$$

When the mean wind $\mathbf{U}$ is in the direction of the x_1-axis the delta function implies that

$$k_1 = \frac{\omega}{U} \tag{11}$$

and the spectral densities are given by the two-dimensional integrals

$$S_{ij}(\mathbf{x},\omega) = \frac{1}{U}\iint \frac{E(k)}{4\pi k^4}\left(k^2\,\delta_{ij} - k_i\,k_j\right)e^{i\,\mathbf{k}\cdot\mathbf{x}}\, dk_2 dk_3 \tag{12}$$

The factors inside the parenthesis can be generated by differentiation of the exponential leading to the following form of the spectral density functions

$$S_{ij}(\mathbf{x},\omega) = \left(\frac{\partial}{\partial x_i}\frac{\partial}{\partial x_j} - \delta_{ij}\frac{\partial}{\partial x_m}\frac{\partial}{\partial x_m}\right)\frac{1}{U}\iint \frac{E(k)}{4\pi\,k^4}e^{i\,\mathbf{k}\cdot\mathbf{x}}\,dk_2 dk_3 \quad (13)$$

The integrand is axisymmetric in the $k_2 k_3$-plane and the two-dimensional exponential transform can therefore be replaced with a one-dimensional zero-order Bessel transform. The polar wave number component in the $k_2 k_3$-plane is denoted κ, and the polar distance in the $x_2 x_3$-plane is denoted r. The integral in (13) is then expressed as

$$A(x_1,r,\omega) = \frac{1}{U}\iint \frac{E(k)}{4\pi\,k^4}e^{i\,\mathbf{k}\cdot\mathbf{x}}\,dk_2 dk_3 = \frac{e^{i\,k_1 x_1}}{2U}\int_0^\infty \frac{E(k)}{k^4}J_0(\kappa r)\,\kappa\,d\kappa$$
$$(14)$$

by use of the fundamental integral representation of the zero order Bessel function $J_0(\ )$, see e.g. Abramowitz & Stegun (1964). The along-wind coordinate x_1 appears only explicitly in the exponential function. This is a consequence of the convected turbulence hypothesis, and limits the accuracy of the model for along-wind separation.

When Greek subscripts α,β are used to denote the x_2 and x_3 components, the cross-spectral densities from (13) take the form

$$S_{11} = -2\left(1 + \frac{r}{2}\frac{d}{dr}\right)\left(\frac{1}{r}\frac{dA}{dr}\right) \quad (15)$$

$$S_{1\alpha} = i\,k_1 x_\alpha\left(\frac{1}{r}\frac{dA}{dr}\right) \quad (16)$$

$$S_{\alpha\beta} = \delta_{\alpha\beta}k_1^2 A - \delta_{\alpha\beta}\left(\frac{1}{r}\frac{dA}{dr}\right) - \left(\delta_{\alpha\beta} - \frac{x_\alpha x_\beta}{r^2}\right)r\frac{d}{dr}\left(\frac{1}{r}\frac{dA}{dr}\right) \quad (17)$$

Thus, in convected isotropic turbulence all spectral density components are given explicitly in terms of the potential A and the two derivatives $B = r^{-1}dA/dr$ and $C = r\,dB/dr$.

The spectral component S_{11} is of particular importance in connection with wind buffeting calculations. This component has the general form (15). This form is identical to (5) for the transverse covariance function $g(r)$ when $f(r)$ is replaced by $-2r^{-1}dA/dr$. This form implies that the area integral of each frequency component vanishes. This is to be expected, but is violated by the commonly used exponential format (1).

3. A Family of Spectral Density Functions

A simple and mathematically tractable family of spectral density functions containing a length-scale l and a shape parameter γ is defined by the following one-dimensional spectral density.

$$F(k) = \frac{F_0}{[1 + (lk)^2]^\gamma} \tag{18}$$

The energy density function $E(k)$ follows from (7).

$$E(k) = \frac{4\gamma(\gamma + 1)F_0 (lk)^4}{[1 + (lk)^2]^{\gamma+2}} \tag{19}$$

The maximum of the energy density corresponds to $k_{max} = \sqrt{2/\gamma}\, l^{-1}$.

3.1. COVARIANCE FUNCTIONS

The corresponding one-dimensional correlation function $f(r)$ follows from the integral – Abramowitz & Stegun (1964) 9.6.25 –

$$\sigma_u^2 f(r) = 2F_0 \int_0^\infty \frac{\cos(kr)}{[1 + (lk)^2]^\gamma}\, dk = \frac{\sqrt{\pi}}{\Gamma(\gamma)} \frac{2F_0}{l} \left(\frac{r}{2l}\right)^{\gamma - \frac{1}{2}} K_{\gamma - \frac{1}{2}}\left(\frac{r}{l}\right) \tag{20}$$

$K_\nu(\ )$ is the modified Bessel function of the second kind of order ν. The variance σ_u^2 follows from the limit $r = 0$.

$$\sigma_u^2 = \sqrt{\pi} \frac{\Gamma(\gamma - \frac{1}{2})}{\Gamma(\gamma)} \frac{F_0}{l} \tag{21}$$

Boundedness corresponds to integrability of $f(r)$, i.e. $\gamma > \frac{1}{2}$.

A particularly simple form is obtained, when the order of the modified Bessel function is $\frac{1}{2}$, $K_{1/2}(z) = \sqrt{\pi/2z}\, e^{-z}$. Thus, for $\gamma = 1$ the variance is $\sigma_u^2 = \pi F_0/l$ and the one-dimensional radial correlation functions take the elementary form

$$f(r) = e^{-r/l} \quad , \quad g(r) = \left(1 - \frac{r}{2l}\right) e^{-r/l} \tag{22}$$

In this case $g(r)$ changes sign at $r = 2l$.

Theoretically the asymptotic behaviour for large values of the wavenumber is determined by energy transport and dissipation, leading to $E(k) \propto -\frac{5}{3}$, see e.g. Batchelor (1953). This corresponds to $\gamma = \frac{5}{6}$, giving the well known von Karman spectrum. Results for the cross-spectra are available for this particular spectrum, see e.g. Harris (1970) and Kristensen & Jensen (1979). Here we derive the cross-spectra for the general spectral family (18) and identify a simple modified exponential format in the limit $\gamma \simeq \frac{1}{2}$.

3.2. CROSS-SPECTRA

All cross-spectra are derived from the potential $A(x_1, r, \omega)$ by differentiation. The potential follows by substitution of the energy density function (19).

$$A(x_1, r, \omega) = 2\gamma(\gamma+1) F_0 l^4 \frac{e^{i k_1 x_1}}{U} \int_0^\infty \frac{J_0(\kappa r)}{[1 + (k_1 l)^2 + (\kappa l)^2]^{\gamma+2}} \kappa \, d\kappa \quad (23)$$

The constant in the denominator of the integrand is conveniently expressed in terms of a modified wavenumber κ_1, defined by

$$\kappa_1 = \sqrt{k_1^2 + l^{-2}} = \sqrt{\left(\frac{\omega}{U}\right)^2 + \left(\frac{1}{l}\right)^2} \quad (24)$$

The integral can be expressed in terms of a modified Bessel function of the second kind - see e.g. Abramowitz & Stegun (1964) 11.4.44, or Lebedev (1972) p.133 for a proof of the integral formula. The result is

$$A(x_1, r, \omega) = \frac{F(k_1)}{U} \frac{2e^{i k_1 x_1}}{\Gamma(\gamma)\kappa_1^2} \left(\frac{\kappa_1 r}{2}\right)^{\gamma+1} K_{\gamma+1}(\kappa_1 r) \quad (25)$$

The format of (15)–(17) is particularly convenient in the present case, because the derivatives $B = r^{-1} dA/dr$ and $C = r \, dB/dr$ follow by application of the differentiation rule $z^{-1} d(z^{\gamma+1} K_{\gamma+1}(z))/dz = -z^\gamma K_\gamma(z)$, whereby

$$B(x_1, r, \omega) = \frac{1}{r} \frac{dA}{dr} = -\frac{F(k_1)}{U} \frac{e^{i k_1 x_1}}{\Gamma(\gamma)} \left(\frac{\kappa_1 r}{2}\right)^\gamma K_\gamma(\kappa_1 r) \quad (26)$$

and

$$C(x_1, r, \omega) = r \frac{dB}{dr} = \frac{F(k_1)}{U} \frac{2e^{i k_1 x_1}}{\Gamma(\gamma)} \left(\frac{\kappa_1 r}{2}\right)^{\gamma+1} K_{\gamma-1}(\kappa_1 r) \quad (27)$$

The functions A, B and C all contain the frequency spectrum of the along-wind fluctuation component

$$S_{11}(\mathbf{0}, \omega) = \frac{F(k_1)}{U} = \frac{F_0}{U} \left[1 + \left(\frac{l\omega}{U}\right)^2\right]^{-\gamma} \quad (28)$$

as a factor.

3.3. ALONG-WIND COHERENCE

The along-wind spectrum S_{11} only depends on the transverse separation through the parameter $\kappa_1 r$, and this leads to some particularly simple results. The along-wind spectrum is factored in the form

$$S_{11}(\mathbf{x}, \omega) = S_{11}(\mathbf{0}, \omega) e^{i k_1 x_1} \psi_{11}(\kappa_1 r) \quad (29)$$

The factor $\psi_{11}(\kappa_1 r)$ is the coherence function. The general expression is

$$\psi_{11}(\kappa_1 r) \ = \ \frac{2}{\Gamma(\gamma)}\left(1 + \frac{r}{2}\frac{d}{dr}\right)\left[\left(\frac{\kappa_1 r}{2}\right)^\gamma K_\gamma(\kappa_1 r)\right]$$

$$= \ \frac{2}{\Gamma(\gamma)}\left[\left(\frac{\kappa_1 r}{2}\right)^\gamma K_\gamma(\kappa_1 r) - \left(\frac{\kappa_1 r}{2}\right)^{\gamma+1}K_{1-\gamma}(\kappa_1 r)\right] \quad (30)$$

For $\gamma = \frac{5}{6}$ the latter form corresponds to that given by Kristensen & Jensen (1979).

The coherence function $\psi_{11}(\kappa_1 r)$ is regular in the limit $\gamma = \frac{1}{2}$, and here takes the particularly simple form

$$\psi_{11}(\kappa_1 r) \ = \ \left(1 + \frac{r}{2}\frac{d}{dr}\right)e^{-\kappa_1 r} \ = \ \left(1 - \tfrac{1}{2}\kappa_1 r\right)e^{-\kappa_1 r} \quad (31)$$

This formula for the along-wind coherence for transverse separation r is quite similar to the formula (5) for the correlation of the transverse component for radial separation r. However, the special exponential case occurs for different values of the spectral shape parameter γ, and the reference length of the spectral component is the modified wavenumber κ_1 from (24).

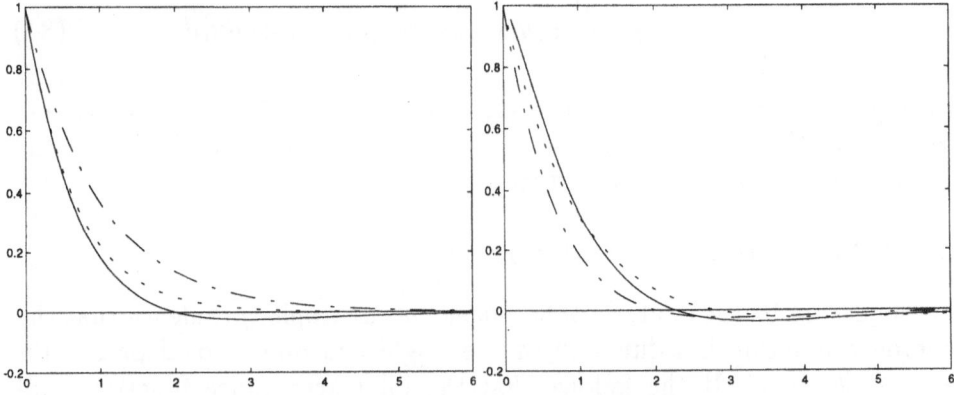

Figure 1.   Coherence functions: a) — $\psi_e(\kappa r)$, $-\cdot-$ $e^{-\kappa r}$, $\cdots$ $e^{-1.5\kappa r}$, b) — $\psi_\gamma(\kappa r)$, $-\cdot-$ $\psi_e(\kappa r)$, $\cdots$ $\psi_e(-0.7\kappa r)$.

Figure 1a shows the modified exponential coherence function ψ_e from (31) and the exponential approximations $e^{-\kappa r}$ and $e^{-1.5\kappa r}$. The latter of these, corresponding to common tangent at zero, gives a good approximation in the interval $0 < \kappa_1 r < 0.5$, but deteriorates for $\kappa_1 r > 0.5$. Figure 1b shows the coherence function ψ_γ from (30) with $\gamma = \frac{5}{6}$ and its approximation by the modified exponential ψ_e with argument $\kappa_1 r$ or $0.7\kappa_1 r$, corresponding to equal integrals.

4. The Joint Acceptance Function

It is commonly accepted to calculate fluctuating wind forces on slender structures from the undisturbed wind velocity field. The wind pressure is $p = \frac{1}{2}\rho(U + u_1)^2 \simeq \frac{1}{2}\rho U^2 + \rho U u_1$, where the last term is the fluctuating part. The general format is briefly illustrated below with reference to a slender horizontal structural member.

4.1. GENERAL FORMULATION

Let $\xi(y)$ be the mode-shape of a slender structural member of length a and height b. The corresponding generalized force from the fluctuating wind pressure then is

$$q(t) = \int_0^a bC\rho U\xi(y)u_1(y,t)\,dy \qquad (32)$$

The frequency spectrum of $q(t)$ is obtained by Fourier transformation of the covariance function in the form

$$S_q(\omega) = (abC\rho U)^2 J(\omega)^2 S_{11}(\omega) \qquad (33)$$

where C is the pressure coefficient, and

$$J(\omega)^2 = \frac{1}{a^2} \int_0^a \int_0^a \xi(y_1)\,\xi(y_2)\,\psi_{11}(|y_1 - y_2|)\,dy_1 dy_2 \qquad (34)$$

is the *joint acceptance function*, that combines the effect of mode-shape, size, and in the general case also variations in the mean wind U and the pressure coefficient C over the structure.

4.2. MODIFIED EXPONENTIAL FORMAT

It is suggested by the analysis of homogeneous isotropic turbulence that the coherence function is a function of the single non dimensional parameter $\kappa_1 r$, i.e. $\psi_{11}(\kappa_1 r)$. It the follows that the joint acceptance function is of the form $J(\kappa_1 a)^2$, where size and frequency are combined in the single parameter $\kappa_1 a$. This parameter takes the finite value a/l at zero frequency, corresponding to coherence less than unity for finite separation. In the traditional exponential format the non-dimensional parameter is $c\omega a/U$, with full correlation for $\omega = 0$.

The computation of the joint acceptance function is straightforward for exponential coherence, and many special cases are available – see e.g. Davenport (1977) and Dyrbye & Hansen (1988). If the joint acceptance function for exponential coherence is denoted $J_e(\ )$, the similar joint acceptance function corresponding to the limiting form (31) can be obtained directly as

$$J(\kappa_1 a)^2 = \left(1 + \frac{\kappa_1}{2}\frac{d}{d\kappa_1}\right) J_e(\kappa_1 a)^2 \tag{35}$$

For large size or frequency $J_e^2 \propto (\kappa_1 a)^{-1}$, and the modification introduced by the operator in (35) simply introduces a factor $\frac{1}{2}$.

TABLE 1.   Joint admittance functions

	$\xi = 1$	$\xi = 1 - 2y/a$
$J_e(\kappa)^2$	$\dfrac{2}{\kappa^2}(e^{-\kappa} - 1 + \kappa)$	$\dfrac{2}{\kappa^4}[4 - \kappa^2 + \frac{1}{3}\kappa^3 - (2+\kappa)^2 e^{-\kappa}]$
$J(\kappa)^2$	$\dfrac{1}{\kappa}(1 - e^{-\kappa})$	$\dfrac{2}{\kappa^4}[(2 + \kappa + \frac{1}{2}\kappa^2)(2+\kappa)e^{-\kappa} - 4 + \frac{1}{6}\kappa^3]$

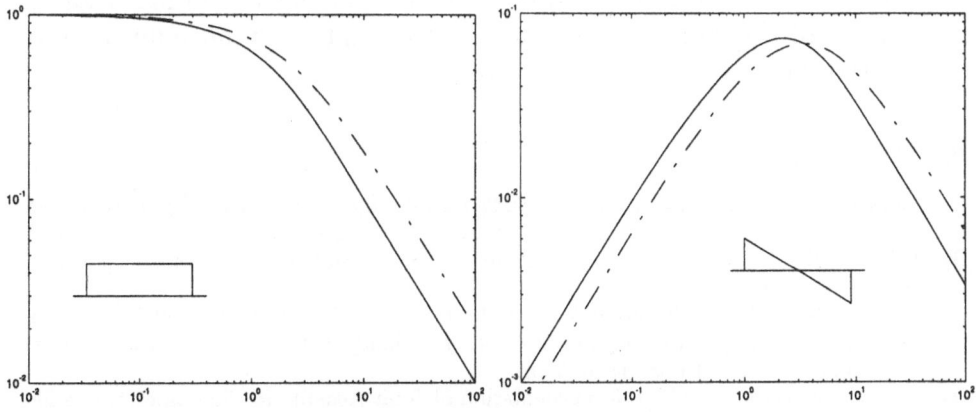

Figure 2.   Joint acceptance functions: —— $J(\kappa_1 a)^2$, — · — $J_e(\kappa_1 a)^2$.
a) $\xi = 1$, b) $\xi = 1 - 2y/a$.

Figure 2 shows the joint coherence function according to the exponential and the modified exponential format for two mode shapes, constant and linear with zero mean. The analytical expressions are given in Table 1. In both cases the high-frequency value is reduced by a factor of 0.5. For the constant mode shape the low-frequency behaviour is unaffected, while for the zero mean mode shape the maximum value is slightly higher and shifted to the left, and the low-frequency values are increased by the factor 1.5. As illustrated in Fig. 1a calibration of the exponential format against the modified exponential format for small to moderate separation would lead to the argument $1.5\kappa_1 a$. This correeponnds to a shift to the right of the exponential format curves in Fig. 2, that would lead to coincidence of the low-frequency behaviour in the two formats, while the high-frequency values would still be reduced by a factor of 0.75 in the modified exponential format.

5. Conclusions

A modified exponential format (31) has been proposed for the along-wind coherence component. It is expressed in terms of the modified wavenumber κ_1 (24) accounting for the length-scale of the wind turbulence spectrum and the lack of full coherence at finite separation. The joint acceptance function can be evaluated from its exponential counterpart by the simple homogeneous differential operator (35). The length scale of the modified exponential format is estimated in the interval 0.8–$1.0\kappa_1$. For small separation this corresponds to an exponential format with 1.2–$1.5\kappa_1$. In the traditional frequency format this gives an attenuation parameter of 7.5–9.5, corresponding well to experimental evidence, Simiu & Scanlan (1986). When compared with the exponential coherence function format, calibrated at small to moderate separation, the modified exponential format leads to a reduction of the high-frequency part of the joint acceptance function for slender structures with a factor of 0.75.

References

Abramowitz, M. and Stegun, I. (eds.) (1964) *Handbook of Mathematical Functions*. National Bureau of Standards, Applied Mathematics Series.

Batchelor, G.K. (1953) *The Theory of Homogeneous Turbulence*. Cambridge University Press, Cambridge.

Davenport, A.G. (1977) The prediction of the response of structures to gusty wind, in *Safety of Structures under Dynamic Loading*, Holand, Kavlie, Moe and Sigbjörnsson (eds.), pp. 257–284. Tapir, Trondheim.

Dyrbye, C. and Hansen, S.O. (1988) Calculation of Joint Acceptance Function for line-like structures, *Journal of Wind Engineering and Industrial Aerodynamics*, **31**, pp. 351–353.

Eurocode (1994) *Eurocode 1: Basis of Design and Actions on Structures, Part 2-4: Wind Actions*, CEN, European Committee for Standardization, Brussels.

Harris, R.I. (1970) The nature of the wind, in *Seminar on Modern Design of Wind-Sensitive Structures, 1970*, pp. 29–55. Construction Industry Research and Information Association, CIRIA, London.

Kristensen, L. and Jensen, N.O. (1979) Lateral coherence in isotropic turbulence and in the natural wind, *Boundary-Layer Meteorology*, **17**, pp. 353–373.

Lebedev, N.N. (1972) *Special Functions and their Applications*. Dover Publications, New York.

Mann, J. (1994) The spatial structure of neutral atmospheric surface-layer turbulence, *Journal of Fluid Mechanics*, **273**, pp. 141–168.

Mann, J. and Krenk, S. (1994) Fourier simulation of a non-isotropic wind field model, in *Structural Safety and Reliability*, Schueller, Shinozuka and Yao (eds.), pp. 1669–1674. Balkema, Rotterdam.

Simiu, E and Scanlan, R.H. (1986) *Wind Effects on Structures, Second Edition*. Wiley, New York.

WAVE PROPAGATION THROUGH RANDOMLY DISORDERED NEAR-PERIODIC STRUCTURES

R.S. LANGLEY
Department of Aeronautics and Astronautics
University of Southampton
Southampton SO17 1BJ, UK

Abstract

Various aspects of the statistics of wave transmission through a disordered one-dimensional waveguide are considered. An expression is derived for the statistical moments of the resistance of an N-1 bay disordered system, and it is shown that the natural logarithm of the resistance is Gaussian for large N. It is then shown that a weakly disordered system obeys one-parameter scaling, and a simple expression is derived for the localization factor produced by weak disorder. The present work extends results which have appeared recently in the solid state physics literature to the case of structural dynamic systems.

1. Introduction

Many engineering structures are nominally periodic, in the sense that the structure is composed of a number of identical units which are connected in a regular pattern. A periodically stiffened cylindrical shell for example forms the basic component of an aircraft fuselage or a submarine hull. The dynamic behaviour of such structures is of considerable practical interest, and much work has been performed on the wave bearing characteristics of "perfect" periodic structures (for example [1]). However, in recent years it has been recognised that manufacturing imperfections and other uncertainties will prevent the realization of a perfect periodic structure, and moreover it has been found that the dynamic behaviour of such structures can be extremely sensitive to the presence of small irregularities. Such irregularities tend to "localize" the structural vibration to the vicinity of an excitation source, and the degree to which this occurs is commonly measured by a quantity known as the "localization factor".

Previous work has considered the application of a variety of analytical and numerical techniques to the study of vibration localization in (for example) coupled oscillators [2-4], beams [5-8], and cyclic structures such as antennas [9] and turbine discs [10]. The analytical techniques adopted range from perturbation methods [2] to the study of

A. Naess and S. Krenk (eds.), IUTAM Symposium on Advances in Nonlinear Stochastic Mechanics, 279–288.
© *1996 Kluwer Academic Publishers.*

Lyapunov exponents [8]. In the present work a general waveguide model of a one-dimensional disordered periodic structure is adopted, which means that the analysis is applicable to a wide range of physical systems. The analysis is based on a number of recent advances in solid state physics which have yet to be applied to structural dynamic systems: in particular, it is shown that the statistical moments of the "resistance" (the inverse of the power transmission coefficient) of a disordered waveguide may be computed exactly by using a method developed by Pendry and his co-workers [11-14]. The method is reformulated here and extended to a general study of weakly disordered systems. It shown that weak disorder leads to a phenomenon known as one-parameter scaling, and simple expressions are derived for the localization factor of a weakly disordered structural dynamic system.

2. Transmission Statistics of a Disordered Waveguide

In this section the statistical properties of the transmission coefficient T_N of an N-1 bay disordered waveguide are investigated. Initially, the wave transmission properties of an ordered waveguide are reviewed, and the effect of disorder on the statistical moments of the resistance (defined as $|T_N|^{-2}$) is then considered. The statistical distribution of the quantity $x = \ln|T_N|^{-2}$ is then investigated, and this leads to an expression for the localization factor of the system for the case of weak disorder.

2.1 THE ORDERED WAVEGUIDE

A schematic of a periodic waveguide is shown in Figure 1: the system consists of a series of waveguides of length L which are coupled through identical joints. The joints

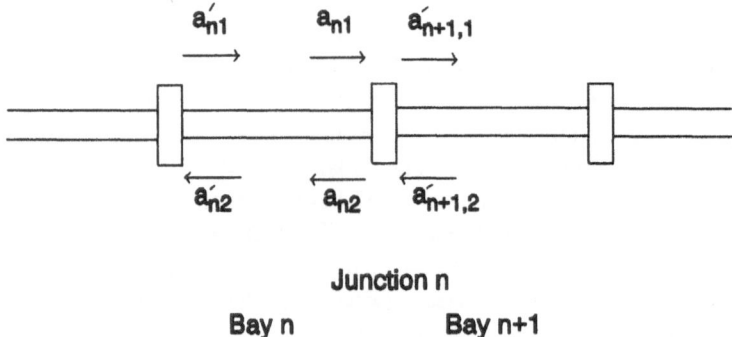

Figure 1 Schematic of a one-dimensional waveguide

are characterised here by the transmission coefficient T and the reflection coefficients for right and left going waves, R^+ and R^-. If the joints are conservative, then these coefficients satisfy the conditions [15]

$$T=te^{i\phi_t}, \quad R^+=re^{i(\phi_t+\phi_r+\phi)}=R^-e^{2\phi_r}, \quad t^2+r^2=1, \tag{1-3}$$

where t and r are the amplitudes of the coefficients, ϕ_t and ϕ_r are the phase angles, and $\phi=\pm\pi/2$. It is assumed that the waveguide carries only one type of propagating wave, which is taken to have wavenumber k (complex in the presence of damping) at frequency ω: throughout the following analysis simple harmonic motion is assumed, and the time dependency $\exp(i\omega t)$ is suppressed for ease of notation. The motion in the nth bay consists of a right going wave with complex amplitude a_{n1} and a left going wave with complex amplitude a_{n2}: as shown in Figure 1, these amplitudes are defined at the right hand end of the bay - the corresponding amplitudes at the left hand end are denoted by a_{n1}' and a_{n2}'. It follows from these definitions that

$$a_{n1}=e^{-ikL}a_{n1}', \quad a_{n2}=e^{ikL}a_{n2}' \tag{4,5}$$

$$a_{n+1,1}'=Ta_{n1}+R^-a_{n+1,2}', \quad a_{n2}=Ta_{n+1,2}'+R^+a_{n1}, \tag{6,7}$$

Equations (1-7) lead to the relation

$$a_{n+1}=\begin{bmatrix} e^{-ikL}/T^* & -e^{-ikL}(R^+/T)^* \\ -e^{ikL}(R^+/T) & e^{ikL}/T \end{bmatrix} a_n=Ta_n, \tag{8}$$

where $a_n=(a_{n1}\ a_{n2})^T$ and the transfer matrix T is defined accordingly. The nature of wave motion in a periodic system is normally investigated by using Bloch's Theorem [16], which in the present context states that $a_{n+1}=\lambda a_n$ where $\lambda=\exp(i\epsilon+\delta)$ is a frequency dependent constant, and ϵ and δ are known respectively as the phase and attenuation constants. In the absence of damping, the system typically displays "pass band" and "stop band" behaviour: a pass band is a frequency interval over which $\delta=0$, while a stop band has $\delta\neq0$. Physically, wave motion can propagate within a pass band, while all motion is subject to spatial decay within a stop band. A uniqueness convention is normally applied to the phase constant ϵ, so that ϵ ranges from 0 to π (or from π to 0) over a pass band.

It follows from equation (8) that λ is an eigenvalue of the transfer matrix, and by employing equations (1-3) the characteristic equation can be written in the form

$$\lambda^2-(T_{11}+T_{22})\lambda+1=0, \tag{9}$$

where T_{ij} is the ijth entry of T. The two roots yielded by equation (9) have the form $\lambda_1=1/\lambda_2$ - one root is associated with wave motion which propagates to the right, while the other relates to left propagating wave motion. The eigenvectors associated with these roots can be written in the form $(1\ \alpha_1)$ and $(\alpha_2\ 1)$, where the detailed form of α_i can be deduced from equation (8); by associating α_1 with the right propagating wave it follows that $|\alpha_i| \leq 1$.

The wave transmission properties of an N-1 bay (N joint) periodic system which is embedded in an otherwise homogeneous waveguide can be deduced from the foregoing analysis. An incident wave a_{01} at the left hand side of the structure will produce a reflected wave a_{02} and a transmitted wave a_{N1}. The wave motion at either side of the structure can be represented as a linear combination of the two "periodic waves" $(1\ \alpha_1)$ and $(\alpha_2\ 1)$; it can further be noted that these waves are modified by the factors λ_1^N and λ_2^N respectively in moving from the left to the right of the embedded system. By this method, the transmitted wave amplitude a_{N1} can be related to the incident wave amplitude a_{01} to yield the transmission coefficient T_N of the embedded system in the form

$$T_N = a_{N1}/a_{01} = \lambda_1^N(1 - \alpha_1\alpha_2)/[1 - (\lambda_1/\lambda_2)^N\alpha_1\alpha_2]. \tag{10}$$

Within a pass band the eigenvalues λ_i are such that $(\lambda_1/\lambda_2)^N = \exp(2iN\epsilon_1)$; given that ϵ_1 covers a range of π over a pass band, it then follows from equation (10) that for every pass band there will be $N+1$ frequencies at which $\exp(2iN\epsilon_1) = 1$. At each of these frequencies except $\epsilon_1 = 0$ and $\epsilon_1 = \pi$ (where $|\alpha_1| = 1$) equation (10) yields $|T_N| = 1$, which means that there are N-1 frequencies of perfect transmission within each pass band. In contrast, within a stop band the attenuation constant is non-zero so that λ_1^N will be small, and hence a low value of the transmission coefficient can be expected.

It has been shown that the transmission coefficient of an ordered waveguide is given by equation (10). In the following sections the statistical properties of T_N are investigated for the case of a randomly disordered waveguide.

2.2 THE STATISTICAL MOMENTS OF $|T_N|^{-2}$

In the solid state physics literature, the quantity $|T_N|^{-2}$ is usually referred to as the resistance. As shown by Pendry and his co-workers [11-14], the statistical moments of the resistance of a homogeneously disordered system can be derived from the statistical properties of the transfer matrix of an individual bay. Initially it can be noted that for a disordered waveguide with N joints, the Mth statistical moment of the resistance can be written in the form

$$E[|T_N|^{-2M}] = E[|a_{N2}/a_{02}|^{2M}_{a_{01}=0}], \tag{11}$$

where $E[]$ represents the expected value, and the notation is such that the ratio which appears on the right is to be evaluated under the condition $a_{01} = 0$ (thus meaning that the waveguide is subjected to an incident wave a_{N2} which produces a transmitted wave a_{02}). An expression for this ratio can be derived by considering initially the transmission properties of a single bay of the system - in particular, it follows from equation (8) that any Mth order product of the two wave amplitudes $a_{n+1,1}$ and $a_{n+1,2}$ can be written in the form

$$a_{n+1,j_1} a_{n+1,j_2} \cdots a_{n+1,j_M} = \sum_{k_1=1}^{2} \sum_{k_2=1}^{2} \cdots \sum_{k_M=1}^{2} T_{j_1 k_1} T_{j_2 k_2} \cdots T_{j_M k_M} a_{nk_1} a_{nk_2} \cdots a_{nk_M}, \tag{12}$$

where each $j_r = 1$ or 2, and $T_{j_r k_r}$ is the $j_r k_r$ entry of the transfer matrix. Equation (12) can be written in the abbreviated form

$$u_{n+1,j} = \sum_{k=0}^{M} A_{jk} u_{nk}, \tag{13}$$

where $u_{n+1,j} = a_{n+1,j1} a_{n+1,j2} \cdots a_{n+1,jM}$ with j representing the number of twos which are contained in the set $\{j_1 j_2 \ldots j_M\}$; there will be $M+1$ distinct products of this type, with j ranging from a minimum of 0 to a maximum of M. With this notation, it can be shown that the term A_{jk} which appears on the right of equation (13) has the form

$$A_{jk} = \sum_{p=0}^{\min(j,k)} {}^{j}C_p \, {}^{M-j}C_{k-p} T_{22}^{p} T_{21}^{j-p} T_{12}^{k-p} T_{11}^{M-j-k+p}. \tag{14}$$

Now the ratio which appears in equation (11) involves the modulus squared of an Mth order product of wave amplitudes. Further progress can be made towards the development of an explicit expression for this ratio by defining two vectors v_{n+1} and v_n as follows

$$v_{n+1,\alpha} = u_{n+1,j} u_{n+1,r}^{*}, \quad \alpha = j(M+1) + r + 1, \tag{15,16}$$

$$v_{n\beta} = u_{nk} u_{ns}^{*}, \quad \beta = k(M+1) + s + 1. \tag{17,18}$$

Given that j, k, r, and s all range from 0 to M, it follows that α and β will range from 1 to $(M+1)^2$. It can be deduced from equations (13) and (14) that v_{n+1} and v_n are related as follows

$$v_{n+1} = C v_n, \tag{19}$$

$$C_{\alpha\beta} = \sum_{p=0}^{\min(j,k)} \sum_{q=0}^{\min(r,s)} {}^{j}C_p \, {}^{M-j}C_{k-p} \, {}^{r}C_q \, {}^{M-r}C_{s-q} T_{11}^{M-j-k+p} \cdots$$

$$\cdots T_{11}^{*(M-r-s+q)} T_{12}^{k-p} T_{12}^{*(s-q)} T_{21}^{j-p} T_{21}^{*(r-q)} T_{22}^{p} T_{22}^{*q}. \tag{20}$$

It can be noted that the indexing system represented by equations (16) and (18) is also applicable to equation (20). Equation (19) is concerned with vibration transmission over a single bay of the disordered waveguide - it is clear that C plays the role of a transfer matrix, in that it relates wave amplitude products of order $2M$ at the left hand side of the bay to those at the right hand side. The corresponding result for a waveguide containing N joints will take the form

$$v_N = \left\{ \prod_{n=1}^{N} C_n \right\} v_0,$$ (21)

where C_n is the transfer matrix for bay n. By noting that $v_{N\alpha} = |a_{N2}|^{2M}$ for $\alpha = (M+1)^2$, it can be deduced from equation (11) that the final diagonal entry of the matrix which appears on the right of equation (21) is equal to $|T_N|^{-2M}$. It thus follows that

$$E[|T_N|^{-2M}] = w^T E[C]^N w,$$ (22)

where w is a vector of dimension $(M+1)^2$ whose entries are all zero apart from the final component, which is unity, and $E[C]$ is the expected value of each of the transfer matrices C_n. Equation (22) can be re-expressed in the form

$$E[|T_N|^{-2M}] = \sum_{j=1}^{(M+1)^2} \mu_j^N e_j e_j',$$ (23)

where μ_j is the jth eigenvalue of $E[C]$ and e_j and e_j' are the final entries of the associated left and right hand eigenvectors respectively. If N is sufficiently large then the summation on the right of equation (23) will be dominated by the eigenvalue of largest modulus, $\mu_{max}(M)$ say. In this case it follows that

$$E[|T_N|^{-2M}] \approx [\mu_{max}(M)]^N e_{max} e_{max}', \quad \ln E[|T_N|^{-2M}] \approx N \ln[\mu_{max}(M)].$$ (24,25)

Equation (25) will be used in the following section to investigate the statistics of the variable $x = \ln|T_N|^{-2}$; this encompasses a discussion of the "localization factor" which is commonly used as a measure of disorder.

2.3 THE STATISTICAL DISTRIBUTION OF $\ln|T_N|^{-2}$

The localization factor γ of a disordered system is defined as [2]

$$\gamma = \lim_{N \to \infty} (1/2N) E[\ln|T_N|^{-2}].$$ (26)

This factor is intended to give an indication of the degree of localization which can be expected in a typical system - the numerical value of the factor is strongly influenced by the statistical distribution of the variable $x = \ln|T_N|^{-2}$. Certain aspects of the statistics of x can be deduced from the analysis contained in the previous section; initially it can be noted that the cumulants of the random variable x are defined such that [17]

$$\ln M_x(\theta) = \sum_{n=1}^{\infty} (i\theta)^n c_n/n!, \qquad (27)$$

where $M_x(\theta) = E[\exp(i\theta x)]$ is the characteristic function of x, and c_n is the nth cumulant. It follows directly from equation (27) that

$$\ln E[\,|T_N|^{-2M}] = \sum_{n=1}^{\infty} M^n c_n/n!. \qquad (28)$$

Now if the approximation represented by equation (25) is valid then it follows from equation (28) that c_n is proportional to N for all n; it then follows that the ratio $c_n/c_2^{n/2}$ is proportional to $N^{1-n/2}$ for $n > 2$. By considering an Edgeworth or Gram-Charlier series expansion of the probability density function of x [18], it can be shown that the behaviour of the cumulants implies that x becomes Gaussian for large N. This result is well known in the solid state physics literature but perhaps less well known in structural dynamics. It is important to add that although x may be approximately Gaussian, there is no guarantee that either $|T_N|^{-2}$ or $|T_N|^2$ will be approximately log-normal: each of these variables is strongly affected by the behaviour of the tails of the distribution of x, which may remain non-Gaussian even if the main body of the distribution is reliably Gaussian. There is evidence to suggest that $|T_N|^{-2}$ might become log-normal for large N for weakly disordered systems [14].

The conclusions reached in the present section are based on equation (25), which is valid for large N. As detailed in the following section, further analytical results regarding the statistics of x can be derived for the special case of a weakly disordered undamped system.

2.4 WEAK DISORDER

Before considering the transmission statistics of a weakly disordered system, it can be noted that the transmission coefficient T_N is dependent on the basis which is used to describe the wave motion in the system. In sections 2.1 and 2.2 the wave components a_n were used, and it was deduced that an ordered system exhibits $N-1$ frequencies per pass band at which $|T_N| = 1$. A change of basis to a new set of coordinates $b_n = Sa_n$ say, for some matrix S, will generally change the value of T_N. Such a change of basis will modify the eigenvectors e_j and e_j' which appear in equations (23) and (24), but the eigenvalues μ_j of the matrix $E[C]$ will be unchanged, and thus equation (25) and the analysis contained in section 2.3 will be unaffected. It is analytically convenient to adopt a "periodic" wave basis when considering the statistics of a weakly disordered system, so that the transfer matrix of a single bay, G say, has the form

$$G = S^{-1}TS, \quad S = \begin{pmatrix} 1 & \alpha_2 \\ \alpha_1 & 1 \end{pmatrix}. \qquad (29,30)$$

Here T is the randomly disordered transfer matrix in the original wave coordinates, and the columns of S are the eigenvectors of the *ordered* (deterministic) transfer matrix T, as described in section 2.1. For an ordered system G is diagonal with $G_{11}=\lambda_1$ and $G_{22}=\lambda_2=1/\lambda_1$.

The results contained in sections 2.1-2.3 are equally applicable to the periodic wave basis providing all occurrences of the transfer matrix entries T_{ij} are replaced by G_{ij}. The influence of weak disorder on the behaviour of the system can now be investigated by considering the effect of a perturbation in the transfer matrix G on the matrix $E[C]$. Initially it can be noted that in the absence of disorder the matrix $E[C]$ is diagonal, and furthermore for an *undamped* system there are $M+1$ unit diagonals which arise from the terms with $j=r=k=s$ in equation (20). The effect of disorder on the largest eigenvalue of $E[C]$, $\mu_{max}(M)$, can be investigated by considering this $(M+1)\times(M+1)$ sub-block (C' say) alone [19]: it can be shown that in the presence of weak disorder the sub-block is tri-diagonal with the form

$$C'_{j,j-1}=j^2\Delta, \tag{31}$$

$$C'_{ij}=1+M\Delta+2j(M-j)\Delta, \tag{32}$$

$$C'_{j,j+1}=(M-j)^2\Delta, \tag{33}$$

where Δ is a small parameter which is defined such that

$$E[|G_{11}|^2]=E[|G_{22}|^2]=1+\Delta, \quad E[|G_{12}|^2]=E[|G_{21}|^2]=\Delta. \tag{34,35}$$

Now equations (31-33) imply that each row of the matrix C' has the following sum

$$\mu=C'_{j,j-1}+C'_{ij}+C'_{j,j+1}=1+M(M+1)\Delta, \tag{36}$$

and it thus follows that μ is an eigenvalue of C', with the associated eigenvector having each entry equal to unity. Furthermore, μ is the largest eigenvalue of C', since the row norm $\|C'\|_\infty$ is an upper bound on the largest eigenvalue and in this case $\|C'\|_\infty=\mu$. It can thus be deduced from equations (25), (28), and (36) that

$$c_1=E[\ln|T_N|^{-2}]=N\Delta, \quad c_2=var[\ln|T_N|^{-2}]=2N\Delta, \quad c_n=0 \ (n>2). \tag{37-39}$$

It follows that to the present level of approximation $x=\ln|T_N|^{-2}$ is Gaussian, and furthermore the variance and the mean are related through the condition $c_2=2c_1$: in the solid state physics literature this is known as *one-parameter scaling* - the present derivation has shown that this condition holds for large N and weak disorder, *regardless of the details of the specific model under consideration*. It should be noted that equations (31-39) relate to an undamped system - it has recently been shown that one-parameter scaling no longer applies in the presence of damping [20]. Furthermore, equation (24) requires the condition $N\Delta \gg 1$ to be met: in the parlance of solid state physics, this implies that the system length is much greater than the localization length.

Finally in this section it can be noted that equations (26) and (37) lead to the following expression for the localization factor of an undamped system

$$\gamma = \Delta/2. \tag{40}$$

This result, together with equation (34), gives a simple method of computing the localization factor of a weakly disordered undamped system. For a waveguide in which the bay lengths are subject to disorder it can readily be shown that

$$\gamma_{mid} = (r/t)^2 (k\sigma_L)^2/2, \tag{41}$$

where γ_{mid} is the localization factor at the mid pass band frequency, and σ_L is the standard deviation of the random bay length. Alternatively, for a waveguide in which the amplitude of the junction transmission coefficient is subject to disorder it can be shown that

$$\gamma_{mid} = \sigma_t^2/2r^2t^2, \tag{42}$$

where σ_t is the standard deviation of the transmission coefficient. Equations (41) and (42) can be used to assess rapidly whether a particular system is prone to localization by disorder.

3. Conclusions

Various aspects of the statistics of wave transmission through a disordered waveguide have been considered. It has been shown that the statistical moments of the resistance can be computed from equation (23); as shown in section 2.3, it follows from this result that for large N the variable $x = \ln|T_N|^{-2}$ is Gaussian. For weak disorder, one-parameter scaling occurs, in the sense that the variance of x is equal to twice the mean value. In this case, the localization factor is given in general by equation (40) and two specific examples of this result are given by equations (41) and (42). The present work has extended results which have appeared recently in the solid state physics literature to the case of structural dynamic systems.

4. References

1. Mead, D.J.: A general theory of harmonic wave propagation in linear periodic systems with multiple coupling, *Journal of Sound and Vibration* 27 (1973), 235-260.
2. Pierre, C.: Weak and strong vibration localization in disordered structures: a statistical investigation,

   Journal of Sound and Vibration **139** (1990), 111-132.

3. Castanier, M.P., and Pierre, C.: Individual and interactive mechanisms for localization and dissipation in a mono-coupled near-periodic structure, *Journal of Sound and Vibration* **168** (1994), 479-505.

4. Hodges, C.H., and Woodhouse, J.: Confinement of vibration by one-dimensional disorder, I: theory of ensemble averaging, *Journal of Sound and Vibration* **130** (1989), 237-251.

5. Cai, C.Q., and Lin, Y.K.: Statistical distribution of frequency response in disordered periodic structures, *American Institute of Aeronautics and Astronautics Journal* **30** (1992), 1400-1407.

6. Bouzit, D., and Pierre, C.: Vibration confinement phenomena in disordered, mono-coupled, multi-span beams, *Journal of Vibration and Acoustics* **114** (1992), 521-530.

7. Kissel, G.I.: Localization factor for multichannel disordered systems, *Physical Review* A **44** (1991), 1008-1014.

8. Ariaratnam, S.T., and Xie, W.-C.: Wave localization in randomly disordered nearly periodic long continuous beams, *Journal of Sound and Vibration* **181** (1995), 7-22.

9. Bendiksen, O.O.: Mode localization phenomena in large space structures, *American Institute of Aeronautics and Astronautics Journal* **25** (1987), 1241-1248.

10. Valero, N.A., and Bendiksen, O.O.: Vibration characteristics of mistuned shrouded blade assemblies, *American Society of Mechanical Engineers Journal of Engineering for Gas Turbines and Power* **108** (1986), 293-299.

11. Pendry, J.B.: The evolution of waves in disordered media, *Journal of Physics C: Solid State Physics* **15** (1982), 3493-3511.

12. Kirkman, P.D., and Pendry, J.B.: The statistics of one-dimensional resistances, *Journal of Physics C: Solid State Physics* **17** (1984), 4327-4344.

13. Slevin, K.M., and Pendry, J.B.: Statistics and scaling in one-dimensional disordered systems, *Journal of Physics: Condensed Matter* **2** (1990), 2821-2832.

14. Slevin, K.M., and Pendry, J.B.: Log-normal distribution as a description of fluctuations in one-dimensional disordered systems, *Physical Review B* **41** (1990), 10240-10242.

15. Mace, B.R.: Reciprocity, conservation of energy and some properties of reflection and transmission coefficients, *Journal of Sound and Vibration* **155** (1992), 375-381.

16. Brillouin, L.: *Wave Propagation in Periodic Structures*, Dover, New York, 1946.

17. Lin, Y.K.: *Probabilistic Theory of Structural Dynamics*, McGraw-Hill, New York, 1967.

18. Ibrahim, R.A.: *Parametric Random Vibration*, Research Studies Press, Letchworth, 1985.

19. Langley, R.S.: One-parameter scaling in weakly disordered one-dimensional systems, *Journal of Physics: Condensed Matter* **6** (1994), 8259-8268.

20. Langley, R.S.: The statistics of wave transmission through disordered one-dimensional waveguides, accepted for publication in *Journal of Sound and Vibration* (1996).

HARMONIC RESPONSE ANALYSIS OF STOCHASTIC RODS USING SPATIAL STOCHASTIC AVERAGING

C S MANOHAR and B R SHASHIREKHA
Department of Civil Engineering
Indian Institute of Science
Bangalore 560 012 India.

1. Introduction

The vibration analysis of engineering structures possessing randomly varying elastic, mass and damping properties constitutes a challenging class of problems in the area of stochastic structural mechanics, see references [1-4], for a survey of the state of art. These problems are of fundamental importance in the safety assessments of engineering structures, especially, when the frequency range of excitation encompasses several modes of vibration. In an ongoing program of research aimed at understanding the dynamical behavior of stochastic continuous systems, we have studied the free and forced vibration characteristics of random rod and beam elements [5-8]. Some of the major conclusions emerging from these studies are:

- Approaches based on Markov process theory are useful in characterizing eigensolutions of axially vibrating stochastic rods and, in particular, the stochastic averaging theorem of Stratonovich provides a powerful means for generating acceptable free vibration solutions for a class of problems [6,8].

- The complex frequency response curves for stochastic rods and beams tend to become statistically stationary for large values of driving frequencies. A statistical overlap factor can be formed that, given the statistical properties of the system physical properties, allows the response statistics to be characterized. A similar behavior at high frequencies in the spatial domain is also observed in regions away from boundaries [5,7].

- The frequency response functions of stochastic systems is pronouncedly non-gaussian and, in some cases, the probability density function (pdf) of the frequency response functions has a very long upper tail. This limits the usefulness of the response mean and standard deviation as descriptors of the system behavior. Lognormal and gamma distributions fit digitally simulated data on response frequency functions well [5].

The aim of this paper is to develop an analytical method for evaluating the dynamic response of second order continuous systems such as axially vibrating rods

289

A. Naess and S. Krenk (eds.), IUTAM Symposium on Advances in Nonlinear Stochastic Mechanics, 289–298.
© 1996 *Kluwer Academic Publishers.*

and shear beams with randomly varying mass and stiffness properties. The standard way of performing this analysis would be to do the free vibration analysis first, which, then is followed by eigenfunction expansion procedure, leading to the evaluation of dynamic response. In the context of stochastically defined systems, however, this approach has two drawbacks: firstly, it requires the determination of the joint statistics of natural frequencies and mode shapes, which, by no means, is an easy task and, secondly, the series expansion introduces a large number of random variables, which is, at least, twice as many as the number of modes retained in the modal expansion which, in turn, increases the size of integration on joint probability distributions while evaluating the response statistics. Besides, when the stochastic mode shapes are not determined exactly, which, most often, will be the case, the question of orthogonality of mode shapes requires careful interpretation. An alternative approach, which proves to be advantageous, would be to evaluate the dynamic stiffness matrix directly by solving the governing field equations. This, as will be shown in this paper, not only eliminates the necessity of determining the random eigensolutions, but also, restricts the number of random variables entering the formulations. An important step in this analysis consists of solving the governing field equation which, for axially vibrating stochastic rods, is a second order ordinary differential equation with stochastic coefficients. Exact solutions to this problem are not possible except under highly specialized circumstances [7] and, consequently, approximations become necessary. In the present study, it is proposed to employ the stochastic averaging principle to approximately analyze the field equation. The application of this principle requires that there be a clear cut separation between the characteristic length of the stochastic variations and that of the system. Numerical results on the response statistics of a harmonically driven, discretely damped stochastic shear beam as a function of driving frequency are presented. Satisfactory comparison with a limited amount of digital simulation results is also demonstrated.

2. Damped stochastic rods

The field equation for the vibration of an inhomogeneous, viscously damped rod element can be written as

$$\frac{\partial}{\partial x}[AE(x)\frac{\partial Y}{\partial x} + C_1(x)\frac{\partial^2 Y}{\partial x \partial t}] = \rho(x)\frac{\partial^2 Y}{\partial t^2} + C_2(x)\frac{\partial Y}{\partial t} - F(x,t) \qquad (1)$$

The stiffness $AE(x)$, mass per unit length $\rho(x)$, viscous damping coefficients $C_1(x)$ and $C_2(x)$ are obtained by randomly perturbing the respective constant mean values as follows:

$$AE(x) = AE_0[1 + \delta g(x)], \quad \rho(x) = \rho_0[1 + \epsilon f(x)],$$

$$C_1(x) = C_{10}[1 + \nu_1 q_1(x)], \quad C_2(x) = C_{20}[1 + \nu_2 q_2(x)]. \qquad (2)$$

Here, $g(x)$, $f(x)$, $q_1(x)$ and $q_2(x)$ are jointly stationary, meansquare bounded, zero mean random processes. The parameters ϵ, δ, ν_1 and ν_2 are deterministic quantities indicating the strengths of stochastic variations and are taken to be small compared to unity. The external excitation $F(x,t)$ can, in general, be a random field.

3. Dynamic stiffness matrix

The dynamic stiffness coefficient $D_{ij}(\omega)$ is defined as the harmonic force of frequency ω at nodal co-ordinate i due to a harmonic displacement of unit amplitude and of the same frequency at nodal co-ordinate j, all other co-ordinates being fixed [10]. For harmonic nodal excitations, the solution of equation (1) is sought in the form $Y(x,t) = y(x)\exp(i\omega t)$; $i = \sqrt{-1}$. This separates the time and space variables and leads to the ordinary differential equation

$$\frac{d}{dx}[\{1+\delta g(x)\}\frac{dy}{dx}+i\beta_1\{1+\nu_1 q_1(x)\}\frac{dy}{dx}]+\lambda^2[1+\epsilon f(x)]y-i\beta_2[1+\nu_2 q_2(x)]y = 0; \quad (3)$$

where

$$\lambda^2 = \frac{\rho_0\omega^2}{AE_0}; \quad \beta_1 = \frac{C_{10}\omega}{AE_0}; \quad \beta_2 = \frac{C_{20}\omega}{AE_0}. \quad (4)$$

In order to derive the elements of the dynamic stiffness matrix, the above equation needs to be solved for the following two sets of boundary conditions

$$y(0) = \delta_1; \quad y(L) = \delta_2 \quad (5)$$

and

$$\frac{dy}{dx}(0) = \frac{-P_1}{AE(0)}; \quad \frac{dy}{dx}(L) = \frac{P_2}{AE(L)}. \quad (6)$$

Equation (3) together with the above boundary conditions constitute a pair of complex stochastic boundary value problems. No exact solutions are currently available for this type of problems and, consequently, one has to take recourse to either Monte Carlo simulation procedures or to adopt approximate analytical methods.

4. Stochastic averaging analysis for undamped rods

When the coefficient processes $f(x)$, $g(x)$, $q_1(x)$ and $q_2(x)$ are modeled as broad band random processes, an approximate solution to equations (3) can be obtained by using the Stratonovich-Khasminiskii averaging theorem [9]. This method consists of eliminating minor rapid variations in the response and derive simplified equations for the dominant slowly varying components. Furthermore, the broad band stochastic variations are replaced by equivalent delta-correlated random processes. This method is extensively used in random vibration studies and its application to problems of system stochasticity is of recent origin [6,8]. We begin by focusing attention on undamped field equation and, therefore, the equation for further study reduces to

$$\frac{d}{dx}[\{1+\delta g(x)\}\frac{dy}{dx}] + \lambda^2[1+\epsilon f(x)]y = 0. \tag{7}$$

To proceed further, it is found advantageous to represent stiffness variations as

$$[1+\delta g(x)] = [1+\gamma h(x)]^{-1}. \tag{8}$$

Furthermore, the solution of equation (7) is sought in the form

$$y(x) = \exp[a(x)]\sin[\lambda x + \theta(x)]; \quad [1+\delta g(x)]\frac{dy}{dx} = \exp[a(x)]\lambda\cos[\lambda x + \theta(x)]. \tag{9}$$

Here, $\exp[a(x)]$ and $\theta(x)$ are, respectively, the amplitude and phase functions which can be shown to be governed by

$$\frac{da}{dx} = 0.5\lambda\sin2(\theta+\lambda x)[\gamma h(x) - \epsilon f(x)]; \quad \frac{d\theta}{dx} = \lambda\gamma h(x)\cos^2(\theta+\lambda x) + \epsilon\lambda f(x)\sin^2(\theta+\lambda x). \tag{10}$$

These equations are exactly equivalent to equation (7) and are in a form suitable for applying the averaging theorem. This consists of a combination of a spatial averaging and ensemble averaging and leads to a two-dimensional Markovian approximation to the solution vector $\{a(x), \theta(x)\}$ which can be shown to be governed by

$$\frac{da}{dx} = m_1 + \sigma_1 W_1(x); \quad \frac{d\theta}{dx} = m_2 + \sigma_2 W_2(x) \tag{11}$$

Here $W_1(x)$ and $W_2(x)$ are independent gaussian white noise processes with unit strength. The drift and diffusion coefficients appearing in the above equations can be shown to be given by

$$m_1 = 0.25\lambda^2 \int_{-\infty}^{0} [\gamma^2 < h(x)h(x+\tau) > + \epsilon^2 < f(x)f(x+\tau) > - \epsilon\gamma < h(x)f(x+\tau) >$$
$$- \epsilon\gamma < f(x)h(x+\tau) >]\cos2\lambda\tau d\tau \tag{12}$$

$$m_2 = 0.25\lambda^2 \int_{-\infty}^{\infty} < [\gamma h(x) - \epsilon f(x)][\gamma h(x+\tau) - \epsilon f(x+\tau)] > \sin2\lambda\tau \, d\tau \tag{13}$$

$$\sigma^2_1 = 0.125\lambda^2 \int_{-\infty}^{\infty} < [\gamma h(x) - \epsilon f(x)][\gamma h(x+\tau) - \epsilon f(x+\tau)] > \cos2\lambda\tau d\tau \tag{14}$$

$$\sigma^2_2 = 0.25\lambda^2 \int_{-\infty}^{\infty} < [\gamma h(x) - \epsilon f(x)][\gamma h(x+\tau) - \epsilon f(x+\tau)] > d\tau +$$
$$0.125\lambda^2 \int_{-\infty}^{\infty} < [\gamma h(x) - \epsilon f(x)][\gamma h(x+\tau) - \epsilon f(x+\tau)] > \cos2\lambda\tau \, d\tau. \tag{15}$$

Here $< \cdot >$ denotes the mathematical expectation operator. Furthermore, it may be deduced that $a(x)$ and $\theta(x)$, in this approximation, are stochastically independent and gaussian distributed. Substitution of solution of equation (11) into (9) leads to

$$y(x) = AF_1(x)\sin F_2(x) + BF_1(x)\cos F_2(x) \tag{16}$$

$$[1 + \delta g(x)]\frac{dy}{dx} = A\lambda F_1(x)\cos F_2(x) - \lambda B F_1(x)\sin F_2(x) \tag{17}$$

$$A = \exp(a_0)\cos\theta_0; \quad B = \exp(a_0)\sin\theta_0 \tag{18}$$

$$F_1(x) = \exp[m_1 x + \sigma_1 G_1(x)]; \quad F_2(x) = \lambda x + m_2 x + \sigma_2 G_2(x) \tag{19}$$

$$G_1(x) = \int_0^x W_1(s) \, ds; \quad G_2(x) = \int_0^x W_2(s) \, ds; \tag{20}$$

$$a_0 = a(0); \quad \theta_0 = \theta(0). \tag{21}$$

It may be noted that the field solution given above is nongaussian in nature. The arbitrary constants A and B, or equivalently, a_0 and θ_0, need to be determined based on the boundary conditions. Accordingly, the amplitudes of harmonic nodal displacements and forces can be shown to be related as follows:

$$\left\{ \begin{array}{c} P_1 \\ P_2 \end{array} \right\} = AE_0\lambda \left[\begin{array}{cc} \cot F_2(L) & -F_1(L)\cosec F_2(L) \\ -F_1(L)\cosec F_2(L) & \cot F_2(L) \end{array} \right] \left\{ \begin{array}{c} \delta_1 \\ \delta_2 \end{array} \right\} \tag{22}$$

Notice that as ϵ and δ tend to zero, the above result reduces to the exact solutions valid for uniform rods [10].

5. Models for damping

Inclusion of damping terms into the elements of dynamics stiffness matrix for homogeneous systems is straightforward. Thus, for a homogeneous damped rod, that is, with $\delta = 0$, $\nu_1 = 0$, $\nu_2 = 0$ and $\epsilon = 0$, the solution of equations (3) can be written as

$$y(x) = A_0 \cos \bar{\lambda}x + B_0 \sin \bar{\lambda}x; \quad \bar{\lambda} = \lambda\sqrt{\frac{1 - i\beta_2}{1 + i\beta_1}}. \tag{23}$$

Consequently, using the boundary conditions listed in equations (5,6), the elements of dynamics stiffness coefficients can be shown to be given by $D_{11}(\omega) = D_{22}(\omega) = AE_0\bar{\lambda}\cot\bar{\lambda}L$ and $D_{12}(\omega) = D_{21}(\omega) = -AE_0\bar{\lambda}\cosec\bar{\lambda}L$. Notice that these elements are now complex in nature. On the other hand, when the rod has stochastically varying properties, the averaging approach outlined in the previous section becomes more complicated when the effects of damping need to be included into the definition of $a(x)$ and $\theta(x)$ in equation (9). This is essentially because of complex nature of these functions arising due to the damping and the consequent increase in the dimension of the problem from two to four. This difficulty can be circumvented if an alternative *discrete* damping representation is employed instead of the distributed damping used in equation (1). This is shown schematically in figure 1 in which a discrete viscous damper is added in parallel to the undamped rod element. One way to model this damper would be to derive the contribution of damping terms to the dynamic stiffness of a damped uniform rod and add this to the undamped stochastic dynamic stiffness matrix. It can be easily shown that this contribution is given by

$$C = AE_0 \begin{bmatrix} (\bar{\lambda}\cot\bar{\lambda}L - \lambda\cot\lambda L) & -(\bar{\lambda}\mathrm{cosec}\bar{\lambda}L - \lambda\mathrm{cosec}\lambda L) \\ -(\bar{\lambda}\mathrm{cosec}\bar{\lambda}L - \lambda\mathrm{cosec}\lambda L) & (\bar{\lambda}\cot\bar{\lambda}L - \lambda\cot\lambda L) \end{bmatrix} \qquad (24)$$

Notice that the elements of this matrix are complex in nature and are trigonometric functions of driving frequency ω.

6. Digital simulations

The accuracy of the averaging results can be assessed by comparing the analytical results with digital simulations of statistics of sample solutions. For this purpose, the governing boundary value problem is first converted into a pair of equivalent initial value problems and these, in turn, are integrated using the Runge-Kutta method. Thus, two independent solutions, denoted by $y_1(x)$ and $y_2(x)$, are obtained by solving equation (7) under two different initial conditions, namely, $[y(0), \frac{dy}{dx}(0)] = (1,0)$ and $(0,1)$ respectively. The expressions for the dynamic stiffness coefficients can be derived by taking the general solution in the form

$$y(x) = a_1 y_1(x) + a_2 y_2(x) \qquad (25)$$

and are shown to be given by

$$D_{11}(\omega) = \frac{AE(0)y_1(L)}{y_2(L)}; \quad D_{12}(\omega) = \frac{-AE(0)}{y_2(L)};$$

$$D_{21}(\omega) = AE(L)[y_1'(L) - \frac{y_2'(L)y_1(L)}{y_2(L)}]; \quad D_{22}(\omega) = \frac{AE(L)y_2'(L)}{y_2(L)}. \qquad (26)$$

It must be noted that $D_{12}(\omega) = D_{21}(\omega)$ which can be used as a check to validate numerical work.

7. Numerical example and discussions

For the purpose of illustration, we consider the problem of seismic wave amplification through a stochastic soil layer modeled as a shear beam with randomly varying stiffness and mass properties. The soil layer is taken to be fixed at the base and free at the top and is assumed to have a depth of 30.5 m, shear modulus $G_0= 1.03 \times 10^9$ N/m^2, mass density $\rho_0=2005$ Kg/m^3; these data have been taken from reference [11]. The stiffness and mass along the beam length are perturbed by independent, stationary random gaussian processes with autocovariances given, respectively, by

$$R_{ff}(\tau) = \exp(-\alpha|\tau|) \quad \text{and} \quad R_{gg}(\tau) = \exp(-\xi|\tau|). \qquad (27)$$

In the numerical work it is assumed that $\epsilon = 0.05$, $\gamma = 0.03$, $\alpha = 20$ /m and $\xi =20$ /m. The wave amplification factor can be defined as the ratio of the amplitude of a harmonic displacement at the top of the soil layer due to a unit harmonic displacement at the base [11]. Using the undamped stochastic stiffness matrix of equation (22) together with the damping matrix of equation (24), it can easily be shown that the amplification factor is given by

$$\delta_2(\omega) = \sec F_2(L)\{\frac{G_0\lambda F_1(L) - C_{12}\sin F_2(L)}{G_0\lambda - C_{11}\tan F_2(L)}\} \tag{28}$$

It follows that $\delta_2(\omega)$ is a zero mean, nongaussian, nonstationary, complex random process evolving in ω. Also notice that $\delta_2(\omega)$ is a function of only two, independent, zero mean Gaussian random variables, namely, $G_1(L)$ and $G_2(L)$, see equations (19,20), which implies that, for calculating statistics of this function, not necessarily the first order statistics, one has to perform, at the most, a two dimensional integration on a pair of one dimensional Gaussian probability distributions. This, in the opinion of the authors, is a major simplification of the problem. Furthermore, for deterministic undamped system, that is, as $\epsilon \to 0$ and $\delta \to 0$, the amplification factor given above agrees with the result $\delta_2(\omega) = \sec\lambda L$ valid for deterministic systems [11]. The statistics of magnitude and phase of the amplification factor obtained using the averaging solution for a soil layer with strain rate dependent viscous damping with $C_{10} = 10^{-3}AE_0$ Ns are compared in figure 2-4 with a 100 samples Monte Carlo simulation results. From these figures, it can be observed that the passage of driving frequency through the system resonant frequencies induces nonstationarity into the random process $\delta_2(\omega)$. The analytical results capture this feature fairly accurately as evidenced by satisfactory comparisons which are observed to exist between simulation and analytical solutions. In the neighborhood of system resonance frequencies, the standard deviations are seen to be higher and the comparisons between theory and simulations are fairly good. The presence of damping ensures that the mean transfer functions tend to become statistically stationary for large values of ω. However, it must be noted that, for higher values of driving frequency, the characteristic length of the system becomes smaller as compared with correlation lengths of the mass and stiffness variations, and, consequently, the accuracy of the averaging approximation deteriorates with increases in ω. Conversely, as $\omega \to 0$, the averaging results on the mean amplification factor approach the known static limit of unity; the standard deviations, however, approach a nonzero limit instead of going to zero. This, clearly, is a limitation of the present analysis.

The formulation of dynamic stiffness coefficients also enables stationary random vibration analysis in the frequency domain. Here, one needs to consider variations in the response variables across the ensembles of both the systems and excitations. Thus, one can talk about sample realization of the response power spectral density functions and their ensemble across the realizations of the vibrating systems. This is illustrated in figures 6 in which the analytically determined statistics of power spectral density functions of the displacement of the top of the soil layer is compared with digital simulation results. It may be noted that this psd function is given by

$S_u(\omega) = |\delta_2(\omega)|^2 S_b(\omega)$ where $S_b(\omega)$ is the the psd of the support motion at the base, which is taken to be a bandlimited white noise with a frequency range of 0-100 rad/s and unit variance. Consequently, the trends of the mean and standard deviation variations in ω are similar to those seen in figure 2. Extension of this analysis to compute variability in response spectra is currently being carried out by us.

8. Conclusions

The paper outlines the development of a new technique for deriving statistical properties of dynamic stiffness coefficients of a rod with randomly varying elastic and mass properties. A distinguishing feature of this approach is that it enables the determination of forced vibration characteristics without having to first find the random eigensolutions. The analysis is based on the application of stochastic averaging principle to the solution of the governing field equation and it takes into account the mean, power spectral density and cross spectral density functions of the mass and stiffness processes. Satisfactory agreement is found to exist between the analytical results and a limited amount of digital simulations.

References

1. Ibrahim, R.A., Structural dynamics with parameter uncertainties, App. Mech. Rev., 40(3) (1987) 309-328.
2. Benaroya, H. and Rehak, M., Finite element methods in probabilistic structural analysis: A selective review, App. Mech. Rev., 41(5) (1988) 201-213.
3. Ghanem, R.G. and Spanos, P.D., *Stochastic finite elements: a spectral approach*, Springer Verlag, Berlin 1991.
4. Benaroya, H., Random eigenvalues, algebraic methods and structural dynamic models, Applied mathematics and computation, 52 (1992) 37-66.
5. Manohar, C.S. and Keane, A.J., Statistics of energy flows in spring coupled one-dimensional systems, Phil. Trans. of Roy. Soc. of Lond., A-346 (1994) 525-542.
6. Manohar, C.S. and Iyengar, R.N., Free vibration analysis of stochastic strings, J. of Sound and Vib., 176(1) (1994) 35-48.
7. Manohar, C.S. and Keane, A.J., Axial vibration of a stochastic rod, J. of Sound and Vib., 165(2) (1993) 341-359.
8. Manohar, C.S. and Iyengar, R.N., Probability distribution of eigenvalues of systems governed by stochastic wave equation, Prob. Engg Mech., 8(1) (1993) 57-64.
9. Roberts, J.B. and Spanos, P.D., Stochastic averaging: An approximate method of solving random vibration problems, Int. J. of Nonlinear Mech., 21 (1986) 111-134.
10. Paz, M., *Structural dynamics*, CBS publishers, New Delhi.
11. Hansen, R.J. (Ed), *Seismic Design for nuclear power plants*, The MIT press, Cambridge, 1970.

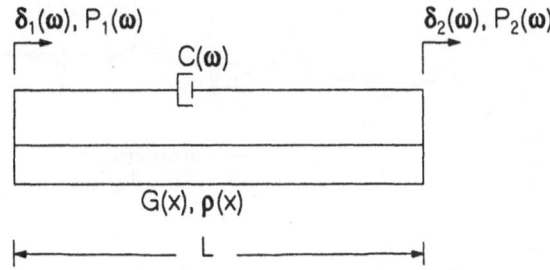

Figure 1 Discretely damped stochastic rod element

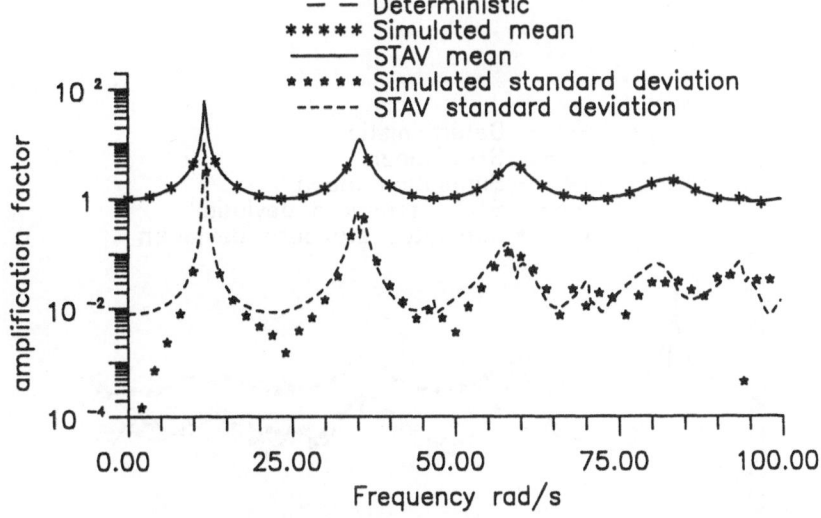

Figure 2 Amplitude of amplification factor

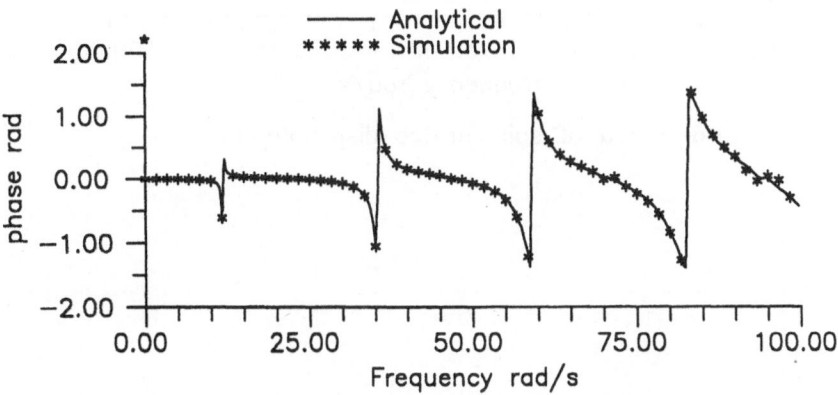

Figure 3 Mean of the phase of amplification factor.

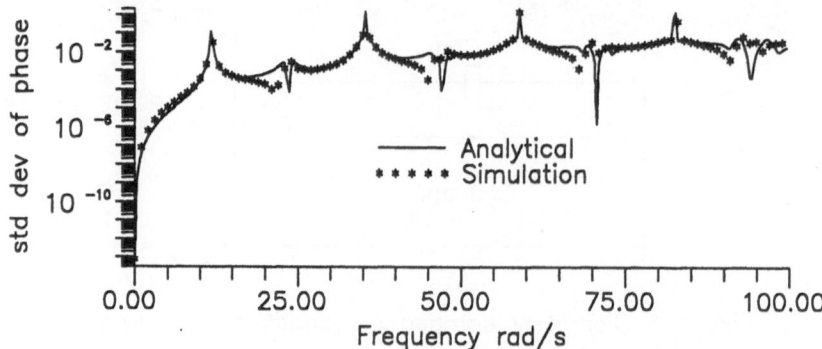

Figure 4 Standard deviation of phase of amplification factor.

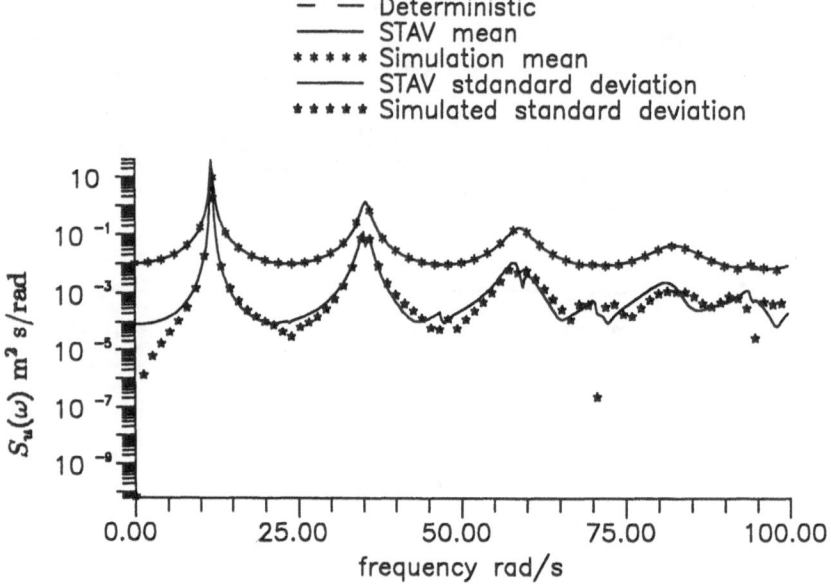

Figure 5 psd of soil surface displacement.

SHIP CAPSIZING IN RANDOM SEA WAVES AND THE

MATHEMATICAL PENDULUM

N. K. MOSHCHUK*, R. A. IBRAHIM*,
R. Z. KHASMINSKII** AND P. L. CHOW**

Wayne State University, Detroit, MI 48202, U.S.A.

* Department of Mechanical Engineering
** Department of Mathematics

1. Introduction

The ship capsizing is treated as a first-passage problem by using the method of asymptotic expansion solutions of the Pontryagin's differential equation [1]. The analysis includes first- and second-order asymptotic approximations for the mean exit time based on perturbation analysis of diffusion processes [2-4]. Related work of first-passage problem has been considered by others [5-9]. The ship governing equation of motion is related to a great extent to the motion of the mathematical pendulum in the rotational motion regime. In this case the analysis is extended to include an approximate solution of the Fokker-Planck equation for stationary probability density. Conditions for stochastic bifurcation in probability are obtained.

2. Statement of the Problem

The uncoupled roll motion of a ship in irregular seas may be modeled by the nonlinear equation of motion

$$\frac{d^2\phi}{dt^2} + a_1\frac{d\phi}{dt} + a_2\frac{d\phi}{dt}\left|\frac{d\phi}{dt}\right| + \Gamma(\phi) = x(t) \tag{1}$$

where ϕ is the ship roll angle, $x(t)$ is the wave excitation moment, a_1 is the linear damping coefficient, a_2 is the non-linear Morison type damping coefficient, $\Gamma(\phi)$ is the restoring moment which is nonlinear function in ϕ. Restoring moment vanishes at the capsizing angle ϕ_s. The following representation of $\Gamma(\phi)$ is proposed in the present analysis

$$\Gamma(\phi) = \omega_n^2 \sin(\phi\frac{\pi}{\phi_s}) + \bar{\gamma}(\phi\frac{\pi}{\phi_s}) \tag{2}$$

299

A. Naess and S. Krenk (eds.), IUTAM Symposium on Advances in Nonlinear Stochastic Mechanics, 299–309.
© *1996 Kluwer Academic Publishers.*

where ω_n is the ship roll natural frequency and the function $\tilde{\gamma}$ accounts for the difference between the exact function $r(\phi)$ and $\omega_n^2 \sin(\phi\pi/\phi_s)$.

Introducing the non-dimensional parameters

$$\tau = \omega_n \sqrt{\frac{\pi}{\phi_s}} t, q = \phi \frac{\pi}{\phi_s}, \quad \frac{a_1}{\omega_n \sqrt{\frac{\pi}{\phi_s}}} = \varepsilon\zeta, \quad a_2\frac{\pi}{\phi_s} = \varepsilon\eta, \quad \frac{\tilde{\gamma}(q)}{\omega_n^2} = \varepsilon\gamma(q) \qquad (3)$$

and assuming that $X(\tau)/\omega_n^2 = \sqrt{\varepsilon}W(\tau) + \varepsilon\varkappa$, where $0 \le \varepsilon \ll 1$ is a small parameter, ζ, η and $\varkappa$ are some constants of order one and $W(\tau)$ is white noise of intensity v, equation (1) can be written in the form

$$\begin{aligned} q' &= p \\ p' &= -\sin q - \varepsilon\zeta p - \varepsilon\eta p|p| + \varepsilon\gamma(q) + \varepsilon\varkappa + \sqrt{\varepsilon}\,W(\tau) \end{aligned} \qquad (4)$$

where a prime denotes differentiation with respect to τ.

Setting $\varepsilon = 0$ in equations (4) gives the well known free oscillation of a simple pendulum

$$q' = p, \qquad p' = -\sin q \qquad (4a)$$

with the Hamiltonian

$$H = \frac{p^2}{2} + 2\sin^2\frac{q}{2} \qquad (5)$$

Each level of energy H corresponds to a periodic closed orbit in the phase space {q,p} Level H=2 corresponds to the unstable position of equilibrium $q = \pm\pi$ and is characterized by two different homoclinic orbits. Periodic motion is only restricted in the domain $D = \{(q,p) \mid H \le H_c\}$; where $H_c = 2 - \hbar$ ($\hbar > 0$ is sufficiently small) is the critical energy level above which capsizing takes place and the trajectories will be structurally unstable. Introducing the transformation $\sin q/2 = k \sin\psi$ and $p = 2k \cos\psi$, the general solution of the unperturbed system when H<2 can be written in the form

$$\tau = \int_0^\psi \frac{d\overline{\psi}}{\sqrt{1 - k^2\sin^2\overline{\psi}}} = F(\psi,k) \qquad (6)$$

where $\sin\psi = \pm\sqrt{2H - p^2}/\sqrt{2H}$, $F(\psi,k)$ is known as the incomplete elliptic integral of the first kind, $k = \sqrt{H/2}$. The upper limit ψ is referred to as the amplitude of the function $F(\psi,k)$ and is written as $\psi = am\tau$.

The ship roll capsizing can be treated as a first-passage problem. It is convenient to write equations (4) in the standard form of the Ito stochastic differential equation

$$dZ = f(Z,\tau)d\tau + b(Z,\tau)dB(\tau), \qquad Z = [Z_1 \dots Z_n]^T \qquad (7)$$

in some region D, with the boundary of D: $\partial D = \Gamma$. $\mathbf{B}(\tau)$ is a vector of Brownian motion processes. In this case the expected exit time $U(\mathbf{z})$ from the region D satisfies the following Pontryagin's partial differential equation

$$LU(\mathbf{z}) = -1 \tag{8}$$

where L is the generator of the Ito process which governs many response functions (such as the expected value of the exit time from region D)

$$L(\cdot) = \mathbf{f}(\mathbf{Z},\tau)\frac{\partial(\cdot)}{\partial \mathbf{Z}} + \frac{1}{2}\mathrm{Tr}\{[\frac{\partial}{\partial \mathbf{Z}}\frac{\partial^\mathrm{T}}{\partial \mathbf{Z}}(\cdot)][\mathbf{b}(\mathbf{Z},\tau)\mathbf{v}\mathbf{b}^\mathrm{T}(\mathbf{Z},\tau)]\} \tag{9}$$

and $\mathbf{v}$ is the intensity matrix of the vector $d\mathbf{B}(\tau)$.
The Pontryagin equation (8) of system (4) takes the form

$$p\frac{\partial U}{\partial q} - \sin q\frac{\partial U}{\partial p} + \varepsilon\{[-\zeta p - \eta p|p| + \gamma(q) + \bar{x}]\frac{\partial U}{\partial p} + \frac{v}{2}\frac{\partial^2 U}{\partial p^2}\} = -1 \tag{10}$$

with the boundary conditions $U|_{H(q,p)=H_c} = 0$, $|U(0,0)| < \infty$
The generator L can be decomposed into two operators $L = L_1 + \varepsilon L_2$ where

$$L_1 = p\frac{\partial}{\partial q} - \sin q\frac{\partial}{\partial p} , \quad \text{and} \quad L_2 = [-\zeta p - \eta p|p| + \gamma(q) + \bar{x}]\frac{\partial}{\partial p} + \frac{v}{2}\frac{\partial^2}{\partial p^2} \tag{11}$$

Note that the first operator L_1 represents derivative in the direction of motion of the unperturbed system and the second operator L_2 is due to perturbations. The solution of the partial differential equation (8) is represented as the sum of two functions

$$U(q,p) \approx U_n(q,p) + G_n(q,p) \tag{12}$$

where the function U_n approximates function U inside the domain D and G_n represents the boundary layer function.
We need to write the state equations (4) in terms of H, and θ coordinates through the following transformation:

$$\sin\frac{q}{2} = k\,\mathrm{sn}\,\theta, \quad \cos\frac{q}{2} = \mathrm{dn}\,\theta, \quad p = 2k\,\mathrm{cn}\,\theta \tag{13}$$

where $\mathrm{sn}\,\theta$, $\mathrm{cn}\,\theta$ and $\mathrm{dn}\,\theta$ are Jacobian elliptic functions. It is important to note that (H,θ) and $(H,\theta+4K(k))$ correspond to the same position of the system so $\theta = \theta(\mathrm{mod}\,4K(k))$. The unperturbed system in the new coordinates is described by the simple equations: $H' = 0$; $\theta' = 1$.
The ship roll equations of motion in terms of the new coordinates H and θ instead of q and p can now be obtained from (1) where q and p must be replaced according to (13) by using the Ito formula. These equations have the form of the Ito stochastic equations

$$H' = \varepsilon f_1(H,\theta) + \sqrt{\varepsilon}\, b_1(H,\theta)\, W(\tau)$$

$$\theta' = 1 + \varepsilon f_2(H,\theta) + \sqrt{\varepsilon}\, b_2(H,\theta)\, W(\tau) \qquad (14)$$

where the drift f_i and diffusion b_i coefficients are given in reference [10].

3. First-Order Approximation

In terms of the new coordinates H, θ relations (11) take the form

$$L = L_1 + \varepsilon L_2 \;,\;\; L_1 = \frac{\partial}{\partial \theta} \;,\; L_2 = f_1\frac{\partial}{\partial H} + f_2\frac{\partial}{\partial \theta} + \sigma_1\frac{\partial^2}{\partial H^2} + \sigma_2\frac{\partial^2}{\partial \theta^2} + \sigma_3\frac{\partial^2}{\partial H\partial \theta} \qquad (12a)$$

where $\sigma_1 = \frac{v}{2}b_1^2$, $\sigma_2 = \frac{v}{2}b_2^2$, $\sigma_3 = vb_1b_2$ and boundary conditions $U|_{H=H_c} = 0$, $|U(0,0)| < \infty$. The function U_n in the new coordinates H,θ can be written in the form

$$U_n(H,\theta) = \frac{1}{\varepsilon}u_0(H,\theta) + u_1(H,\theta) + \varepsilon u_2(H,\theta) + \ldots + \varepsilon^{n-1}u_n(H,\theta) \;,\;\; n=0,1,\ldots \qquad (15)$$

while the boundary layer part G_n is expressed in the form (n=1,2,...)

$$G_n(H,\theta) = g_1(\frac{H-H_c}{\sqrt{\varepsilon}},\theta) + \sqrt{\varepsilon}g_2(\frac{H-H_c}{\sqrt{\varepsilon}},\theta) + \varepsilon g_3(\frac{H-H_c}{\sqrt{\varepsilon}},\theta) + \ldots + \varepsilon^{n-1/2}g_{2n}(\frac{H-H_c}{\sqrt{\varepsilon}},\theta) \qquad (16)$$

Later all functions g_i will be selected such that $g_i(\frac{H-H_c}{\sqrt{\varepsilon}},\theta) \to 0$, as $\frac{H-H_c}{\sqrt{\varepsilon}} \to -\infty$. Substituting (15) into equation (8) yields the following set of equations

$$L_1u_0 = 0, \;\; L_1u_1 + L_2u_0 = -1\,, \;\; L_1u_2 + L_2u_1 = 0 \quad \text{and so on} \qquad (17)$$

The first equation (15) in the new coordinates has the form $L_1u_0 \equiv \partial u_0(H)/\partial\theta = 0$. From this equation follows that $u_0 = u_0(H)$. The second equation is integrated with respect to θ from θ_0 to $\theta_0 + 4K(k)$ and taking into account that $u_1(H,\theta_0) = u_1(H,\theta_0 + 4K(k))$ we obtain the following equation for the unknown function $u_0(H)$

$$\beta(H)\frac{d^2u_0}{dH^2} + \alpha(H)\frac{du_0}{dH} = -1 \qquad (18)$$

where

$$\beta(H) = \langle f_1(H,\theta) \rangle = -2\zeta[H - 2 + \frac{2E(k)}{K(k)}] - \eta\frac{2}{K(k)}[(H-1)\cos^{-1}(1-H) + \sqrt{H(2-H)}] + \frac{v}{2} \qquad (18a)$$

$$\alpha(H) = \langle \sigma_1(H,\theta) \rangle = v[H - 2 + \frac{2E(k)}{K(k)}]\;, \;\; \langle \bullet \rangle = \frac{1}{4K(k)}\int_0^{4K(k)}(\bullet)d\theta$$

Equation (18) subject to the boundary conditions: $u_0(H_c) = 0$, $|u_0(0)| < \infty$ has the following solution

$$u_0(H) = \int_H^{H_c} R(h)\, dh \tag{19}$$

where

$$R(h) = \frac{1}{2\nu} \frac{\exp\left(\frac{2\zeta}{\nu}h + \frac{2\eta}{\nu}\Phi(h)\right)}{E(\sqrt{h/2}) - (1 - h/2)K(\sqrt{h/2})} \int_0^h K(\sqrt{y/2}) \exp\left(-\frac{2\zeta}{\nu}h - \frac{2\eta}{\nu}\Phi(y)\right) dy \tag{19a}$$

$$\Phi(H) = \int_0^H \bar{\Phi}(h)\, dh \ , \quad \bar{\Phi}(h) = \frac{\langle p^2|p|\rangle}{\langle p^2\rangle} = \frac{(h-1)\cos^{-1}(1-h) + \sqrt{h(2-h)}}{K(\sqrt{h/2})(h-2) + 2E(\sqrt{h/2})}$$

The function $R(h)$ has singularity at $h=0$ and $h=2$, but $\lim_{h \to 0} R(h) = 2/\nu$ and $\lim_{h \to 2} R(h)$ also exists and is finite.

4 Second-Order Approximation

The second-order approximation is carried out by considering the second equation of (17) and substituting the known function $u_0(H)$ into the right hand side of this equation. However, the function $u_1(H,\theta)$ cannot be determined uniquely from the equation

$$L_1 u_1 = \frac{\partial u_1}{\partial \theta} = -(L_2 u_0 + 1) \tag{20}$$

The dependence of this solution on H can be obtained by considering the third equation of (17). Accordingly, the function $u_1(H,\theta)$ may be sought in the form

$$u_1(H,\theta) = u_{11}(H,\theta) + u_{12}(H) \tag{21}$$

where u_{12} is some function of H, which will be determined later and

$$u_{11}(H,\theta) = -\int_0^\theta (L_2 u_0 + 1) = -\int_0^\theta \left[1 + f_1(H,\bar{\theta})\frac{du_0(H)}{dH} + \sigma_1(H,\bar{\theta})\frac{d^2 u_0(H)}{dH^2}\right] d\bar{\theta} \tag{22}$$

Taking into account (18) and performing integration this expression becomes

$$u_{11}(H,\theta) = R(H)\, F_1(H,\theta) + \frac{dR(H)}{dH}\, F_2(H,\theta) \tag{22a}$$

where $R(H) = du_0/dH$ and functions $F_1(H,\theta)$, $F_2(H,\theta)$ are periodic with respect to θ and evaluated in reference [10].

Substituting $u_1(H,\theta) = u_{11}(H,\theta) + u_{12}(H)$ into the last equation of (17), taking into account that $\langle L_1 u_2 \rangle = 0$, gives

$$\langle L_2 u_{12}(H) \rangle + \langle L_2 u_{11}(H,\theta) \rangle = 0 \tag{23}$$

Condition (23) leads the following differential equation for $u_{12}(H)$

$$\beta(H)\frac{d^2 u_{12}}{dH^2} + \alpha(H)\frac{du_{12}}{dH} = -\langle L_2 u_{11}(H,\theta) \rangle \tag{24}$$

with the boundary conditions: $u_{12}(H_c) = u_{12}^* = \text{const}$, $\left| u_{12}(0) \right| < \infty$.

The constant value of u_{12}^* will be determined later. The solution of the equation (24) is given by expressions (19) and (19a) where the integrand in relation (19a) must be multiplied by $\langle L_2 u_{11} \rangle$.

The contribution of the boundary layer part (16) is obtained by introducing the boundary layer function $G_n(H,\theta)$ into equation (8). In order to solve this equation for G_n, the coefficients of the operator L_2 given by relations (11) and (12a) are expanded into Taylor series near $H=H_c$. Now the operator $L=L_1+\varepsilon L_2$ can be expressed in asymptotic powers in $\sqrt{\varepsilon}$ instead in ε, i.e.

$$L = \Lambda_0 + \sqrt{\varepsilon}\Lambda_1 + \dots \tag{25}$$

where $\Lambda_0 = \dfrac{\partial}{\partial\theta} + \sigma_1(H_c,\theta)\dfrac{\partial^2}{\partial\xi^2}$, $\Lambda_1 = f_1(H_c,\theta)\dfrac{\partial}{\partial\xi} + \sigma_1^{(1)}(\theta)\xi\dfrac{\partial^2}{\partial\xi^2} + \sigma_3(H_c,\theta)\dfrac{\partial^2}{\partial\xi\partial\theta}$,

$$\sigma_1^{(1)}(\theta) = \frac{\partial\sigma_1}{\partial H}\bigg|_{H=H_c}$$

Therefore the differential equation $LG_n=0$ for the boundary layer part takes the form

$$LG_n(H,\theta) = (\Lambda_0 + \sqrt{\varepsilon}\Lambda_1 + \dots)[g_1(\xi,\theta) + \sqrt{\varepsilon}g_2(\xi,\theta) + \dots] =$$

$$\Lambda_0 g_1(\xi,\theta) + \sqrt{\varepsilon}[\Lambda_0 g_2(\xi,\theta) + \Lambda_1 g_1(\xi,\theta)] + \dots = 0 \tag{26}$$

Setting all coefficients of like powers of $\sqrt{\varepsilon}$ equal to zero gives

$$\Lambda_0 g_1 = 0 \ , \ \Lambda_0 g_2 = -\Lambda_1 g_1 \ , \ \dots \tag{27}$$

The functions g_i (i=1,2,...) should satisfy the following boundary conditions

$$g_1(\xi,\theta)|_{\xi=0} = -u_1(H_c,\theta) \ , \ g_2(\xi,\theta)|_{\xi=0} = 0 \ ,\dots \tag{28}$$

$$g_i(\xi,\theta) \to 0 \quad \text{when} \quad \xi \to -\infty \quad (i=1,2,..) \tag{28a}$$

It is possible to show that each of the following problems

$$\Lambda_0 g_1(\xi,\theta) \equiv \frac{\partial g_1(\xi,\theta)}{\partial \theta} + \sigma_1(H_c,\theta)\frac{\partial^2 g_1(\xi,\theta)}{\partial \xi^2} = 0 \;,\; g_1(0,\theta) = -u_1(H_c,\theta) \tag{29a}$$

$$\Lambda_0 g_2(\xi,\theta) \equiv \frac{\partial g_2(\xi,\theta)}{\partial \theta} + \sigma_1(H_c,\theta)\frac{\partial^2 g_2(\xi,\theta)}{\partial \xi^2} = -\Lambda_1 g_1(\xi,\theta) \;,\; g_2(0,\theta)=0 \tag{29b}$$

has a unique solution satisfying condition (28a). This is done by changing the coordinate θ (mod $4K(k)$) to α (mod 2π), where

$$\alpha = c_1 \int_0^\theta \sigma_1(H_c,s)ds \;,\; c_1 = 2\pi / \int_0^{4K(k)} \sigma_1(H_c,s)ds \tag{30}$$

and using a Fourier expansion (with respect to α) of the unknown functions g_1 and g_2. The final expressions of g_1 and g_2 are given in [10]. Condition (28a) gives the following expression for u_{12}^*

$$u_{12}^* = -\frac{1}{2\pi}\int_0^{2\pi} u_{11}(H_c,\alpha)d\alpha = -\frac{\gamma}{3}[4 - H_c \frac{E(k_c)}{E(k_c) - (1 - k^2)K(k_c)}] \tag{31}$$

This completes the procedure for evaluating the first- and second-order approximation terms. Summing up these terms gives

$$U \approx U_1 + G_1 = \frac{1}{\varepsilon}u_0(H) + u_{11}(H,\theta) + u_{12}(H) + g_1(\frac{H-H_c}{\sqrt{\varepsilon}},\theta) + \sqrt{\varepsilon}g_2(\frac{H-H_c}{\sqrt{\varepsilon}},\theta) \tag{32}$$

It can be rigorously proved that for the exact solution $u(H,q)$ of equation (8) the following estimate is valid uniformly in D

$$u(H,\theta) - [\frac{1}{\varepsilon}u_0(H) + u_{11}(H,\theta) + u_{12}(H) + g_1(\frac{H-H_c}{\sqrt{\varepsilon}},\theta) + \sqrt{\varepsilon}g_2(\frac{H-H_c}{\sqrt{\varepsilon}},\theta)] = O(\varepsilon) \tag{33}$$

Equation (32) establishes the mean exit time u estimated up to second-order. It includes the contribution of the boundary layer which was dropped in the first-order approximation. This solution does not involve singularity. In order to appreciate the contribution of second-order approximation the mean exit time is plotted in Figure 1 for different values of the critical value of energy level H_c. For $H_c=2$, the first-order approximation gives similar results with resolvable singularities. However, the second-order approximation involves singularity, only at $H_c=2$, which is difficult to eliminate.

5. Mathematical Pendulum in a Random Medium

The analysis of the previous sections considers the asymptotic expansion of the Prontryagin's equation for estimating the mean exit time as a first passage problem of ship roll motion. It is possible to treat the ship motion as a mathematical pendulum and to examine the stochastic bifurcation of its motion in probability. This will be achieved by considering equations (4) on the cylinder $\{q(\bmod 2\pi), -\infty < p < \infty\}$. The main goal of this section is to obtain the first-approximation for the stationary probability density function $f(q,p)$. Taking the time derivative of H gives

$$H' = \varepsilon[-\zeta p^2 - \eta p^2|p| + \gamma(q)p + \bar{x}p + \frac{y}{2}] + \sqrt{\varepsilon}pW(\tau) \qquad (34)$$

where the Ito correction term is included. Taking averaging of the drift and diffusion coefficients gives

$$<H'> = \varepsilon[-\zeta\langle p^2\rangle - \eta\langle p^2|p|\rangle + \langle\gamma(q)p\rangle + \bar{x}\langle p\rangle + \frac{y}{2}] + \sqrt{\varepsilon\langle p^2\rangle}\,W(\tau) \qquad (35)$$

where $< >$ denotes averaging over one cycle with respect to the unperturbed motion. It is known that the phase space of the unperturbed system splits into the oscillatory domain (H<2) and rotational domains (H>2) with either p>0 or p<0. So averaging is different in oscillatory and rotational domains. Evaluating the averages in (33) gives

$$\langle p^2\rangle = 2(H-2) + \frac{4E(\sqrt{H/2})}{K(\sqrt{H/2})}\ \text{for}\ 0 < H < 2 \ \text{or}\ \langle p^2\rangle = 2H\frac{E(\sqrt{2/H})}{K(\sqrt{2/H})}\ \text{for}\ H > 2$$

$$\langle p\rangle = 0,\ \text{for}\ 0 < H < 2,\ \text{or}\ \langle p\rangle = \sqrt{H/2}\ \frac{\pi}{K(\sqrt{2/H})}\ \text{sgn}\,(p),\ \text{for}\ H > 2$$

$$\qquad (36)$$

$$\langle p^2|p|\rangle = \frac{2}{K(\sqrt{H/2})}\ [(H-1)\cos^{-1}(1-H) + \sqrt{2H-H^2}]\ ,\ \text{for}\ 0 < H < 2$$

$$\langle p^2|p|\rangle = \sqrt{2H}\ \frac{\pi}{K(\sqrt{2/H})}\ (H-1),\ \ \text{for}\ H > 2\ ,\ \text{and}\ \langle\gamma(q)p\rangle = 0$$

The stationary solution of the Fokker-Planck equation of system (35) is

$$f(H) = c\mu(H)\exp\,[-\frac{2\zeta}{v}H - \frac{2\eta}{v}\int_0^H \frac{\langle p^2|p|\rangle}{\langle p^2\rangle}dH + \frac{2\bar{x}}{v}\int_0^H \frac{\langle p\rangle}{\langle p^2\rangle}dH] \qquad (37)$$

where c is the normalization constant and

$$\mu(H) = \frac{1}{2}K(\sqrt{\frac{H}{2}}),\ \text{for}\ H < 2,\ \text{or}\ \mu(H) = \sqrt{\frac{2}{H}}\,K(\sqrt{\frac{2}{H}}),\ \text{for}\ H > 2$$

The stationary distribution for the angle θ is the uniform distribution.
The function $f(H)$ goes to infinity when H approaches 2. It is convenient to use probability density function in terms of the original variables q and p by using the

transformation $f(q,p) = f(H,\theta)|\partial(H,\theta)/\partial(q,p)|$ which can be determined by substituting the explicit expressions for the H and θ in terms of q and p. This process results in the following expression for the stationary probability density function

$$f(q,p) = C_N exp\left[-\frac{2\zeta}{v}H - \frac{2\eta}{v}\int_0^H \frac{\langle p^2|p|\rangle}{\langle p^2 \rangle}\, dh + \chi(H)\right] \tag{38}$$

where $\chi(H) = 0$, for $0 < H < 2$ or $\chi(H) = \{\frac{\bar{x}\pi}{v\sqrt{2}}\int_2^H \frac{dh}{\sqrt{h}E(\sqrt{2/h})}\}sgn\, p$, for $H > 2$ $\qquad$ (38a)

and H should be replaced by its expression in terms of q and p.

The function f(q,p) depends on the energy level H and for H>2 it also depends on the sign of p (whether the pendulum rotates clockwise or anti-clockwise). It follows from (38) and (38a) that for sufficiently low values of $|\bar{x}|$ there is only one maximum at q=p=0. However if $|\bar{x}|$ exceeds a certain critical value, then there is another maximum of f. Figure (2) shows three dimensional plot of f(q,p) for p>0, $\zeta=\eta=0.5$ and $v=1$, and excitation mean value $\bar{x} =4$. The figure reveals two hills one is located at the equilibrium (q,p)=(0.0, 0.0) while the other at (q,p)$\neq$(0.0, 0.0). The other peak will be on the opposite side if p<0. Figure (3) shows the probability density function as a function of H for two values of $\bar{x} =1.5$ and 4. For $\bar{x} =1.5$ the pdf has only one peak at H=0, and exhibits singularity at H=2. For $\bar{x} =4$, the response pdf has the same singularity but possesses two peaks, one at H=2, while the other is at $H \approx 3.5$.

In order to obtain the bifurcation in probability we need to determine the condition for df/dH=0 for H>2. This condition gives

$$\frac{\pi\eta}{2\zeta}k + \frac{\bar{x}\pi}{4\zeta}k - \frac{\pi\eta}{\zeta}\frac{1}{k} = E(k) \tag{39}$$

where $k = \sqrt{2/H}$ is the modulus, 0<k<1. The behavior of the function E(k) is well known. It gradually decreases from $\pi/2$ at k=0 to 1 at k=1. The left side of (39) has negative value at k=0. Therefore equation (39) has a solution if and only if the left hand side at k=1 has a value greater than 1, (see Figure 4), i.e.

$$\frac{\bar{x}\pi}{4\zeta} - \frac{\pi\eta}{2\zeta} > 1 \qquad \text{or} \qquad \bar{x}_c = \frac{4\zeta}{\pi} + 2\eta \tag{40}$$

$x \geq \bar{x}_c$ is the condition for the existence of another peak in the rotational region. If p<0 then this condition becomes $-\bar{x} > (4\zeta/\pi) + 2\eta$. Figure 4 shows the dependence of E(k) and the left-hand side of (39) on the parameter k for different values of $\bar{x}$. The points of intersection establish the points of bifurcation for each excitation mean value $\bar{x}$.

6. Conclusions

The first-passage problem of ship roll capsizing is treated within the framework of the first-order solution of asymptotic expansion technique due to [3]. The solution is obtained in a closed form and does not involve singularity. In order to appreciate the contribution of the second-order approximation together with the boundary layer effect the mean exit time is plotted in Figure 1 for different values of the critical value of the ship energy level H_c. For $H_c=2$, the first-order approximation gives similar results with resolvable singularities. However, for the second-order approximation, the value $H_c=2$ results in singularities which requires separate investigation. The equation of motion of the ship can also describe the motion of the mathematical pendulum at energy level exceeding the homoclinic orbits, i.e. the rotational motion of the pendulum. During this motion regime the analysis is extended to estimate the stationary probability density function. For excitation mean level which exceeds a certain critical value, the probability density experienced bifurcation manifested by two peaks, one at the equilibrium position while the other at different position.

Acknowledgment

This research is supported by grant from the Office of Naval Research under a grant number N000149310936. Dr. Julia Abrahams is the Scientific Officer.

References

1. Pontryagin, L.S., Andronov, A. A., and Vitt, A. A.: On statistical consideration of dynamical systems, *J. Exper. and Theor. Physics* **3** (1933), 165.
2. Feller, W.: Diffusion processes in one-dimension, *Trans. Amer. Math. Society* **77**(1954), 1.
3. Khasminskii, R.: On diffusion pocesses with small parameter, *Izv. USSR Academy of Sciences. Math.* **27**(1963), 1281.
4. Matkowsky, B. J. and Schuss, Z.: A singular perturbation approach to the computation of the mean first-passage time to a nonlinear filter, *SIAM Journal of Applied Mathematics* **42**(1982), 174.
5. Kozin, K.: First passage time: some results, Proceedings Int. Workshop on *Stochastic Structural Mechanics*, Insbruck, 28 (1983).
6. Roberts, J.B.: A stochastic theory for nonlinear ship rolling in irregular seas, *Journal of Ship Research* **26**(1982), 229.
7. Roberts, J. B.: First-passage probabilities for randomly excited systems: diffusion methods. *Prob. Engn. Mech.* **1**(1986), 66.
8. Cai, G. Q. and Lin, Y. K.: On statistics of first-passage failure, *ASME Journal of Applied Mechanics* **61** (1994) 93.
9. Cai, G. Q, Yu, J. S., and Lin, Y. K.: Ship rolling in random sea, in *Stochastic Dynamics and Reliability of Nonlinear Ocean Systems*, ASME WAM, GE Vol. 77, New York (1994).
10. Moshchuk, N. K., Ibrahim, R. A., Khasminskii, R. Z., and Chow, P.: Asymptotic expansion of ship capsizing in random sea waves. Part I and II. *Int. J. of Non-linear Mechanics*, in press (1995).

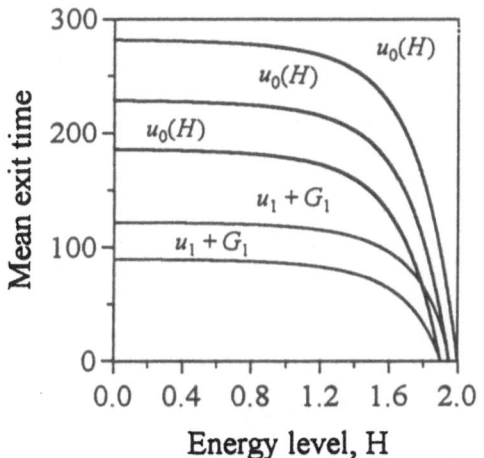

Figure 1. First-order mean exit time —— $u_0(H)$ and components of second-order averaging —— u_1+G_1

Figure 2. Probability density function characteristics in the rotational regime.

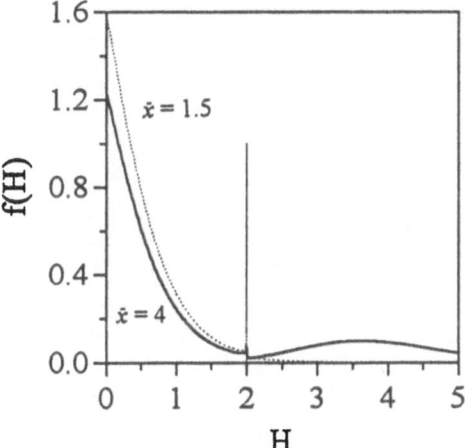

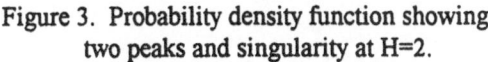

Figure 3. Probability density function showing two peaks and singularity at H=2.

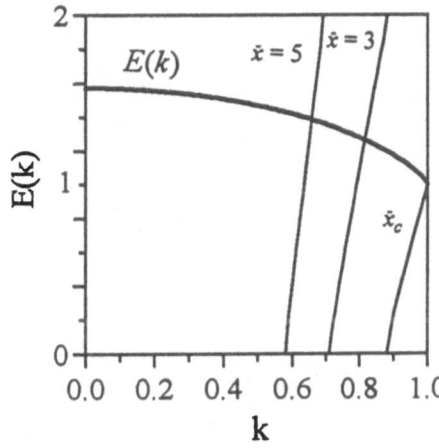

Figure 4. Regions of single and two peaks of the response probability density function.

RANDOM VIBRATION OF SHIP HULLS

A NAESS
Faculty of Civil Engineering
The Norwegian Institute of Technology
Rich. Birkelands v. 1a
N-7034 Trondheim, Norway

1. Introduction

Measurements of the bending moment induced stresses in a ship hull when the ship is moving in a sea-way, have revealed a marked asymmetry, cf. Juncher Jensen and Terndrup Pedersen (1979, 1981). It has been observed that the longitudinal peak stresses in the bottom plates of the hull are generally larger in tension (sagging) than in compression (hogging). This phenomenon will not be captured by a linear theory of wave forces and response. To have available a practical prediction tool it is therefore necessary to set up a nonlinear theory for these vibration responses. This work was initiated by Juncher Jensen and Terndrup Pedersen (1979, 1981), who developed a quadratic strip theory which makes it possible to calculate the vertical vibration response of the ship hull taking due account up to second order of nonlinearities in the exciting waves, the nonvertical ship sides and nonlinear hydrodynamic forces. The flexibilty of the ship is also taken into account by modelling the ship hull as a nonprismatic Timoshenko beam with variable mass and stiffness.

On the basis of this theory it is possible to express the vibration response as a sum of two components, viz. a linear and a quadratic transformation of the sea surface elevation process. More specifically, the response process can be written as a Volterra series truncated after the quadratic term.

Using a Gram-Charlier type expansion of the pertinent probability density functions (PDFs), Juncher Jensen and Terndrup Pedersen (1979, 1981) showed that the quadratic theory predicts the observed response statistics fairly well. The present paper will describe the results of applying an alternative stochastic analysis method for quadratic systems, cf. Naess (1985, 1990) and Naess and Johnsen (1992). The advantage of this alternative method is that the relevant PDFs can be calculated with high accuracy.

2. The Ship Hull Vibration Response

In accordance with the theory developed by Juncher Jensen and Terndrup Pedersen (1979, 1981), the wave induced bending moment in a ship hull subjected to a random

A. Naess and S. Krenk (eds.), IUTAM Symposium on Advances in Nonlinear Stochastic Mechanics, 311–320.
© 1996 *Kluwer Academic Publishers.*

sea way can be written as a sum of a linear and a quadratic component. Specifically, let $M(t;x)$ denote the bending moment response process at the local position x along the hull, see Fig. 1. Then

$$M(t;x) = M^{(1)}(t;x) + M^{(2)}(t;x) \tag{1}$$

where the superscript 1 signifies the linear component while superscript 2 refers to the quadratic part. In this report, the two components will be expressed as

$$M^{(1)}(t;x) = \sum_{\substack{i=-n}}^{n}{}_{o}\, q_i B_i e^{i\omega_i t} \tag{2}$$

where

$$q_i = \hat{M}_1(\sigma_i; V, \varphi, x)\, [\,\tfrac{1}{2} S_X(|\sigma_i|)\Delta\sigma\,]^{1/2} \tag{3}$$

and

$$M^{(2)}(t;x) = \sum_{\substack{i=-n}}^{n}{}_{o} \sum_{\substack{j=-n}}^{n}{}_{o}\, Q_{ij} B_i B_j^{*} e^{i(\omega_i - \omega_j)t} \tag{4}$$

where

$$Q_{ij} = \frac{1}{2}[S_X(|\sigma_i|) S_X(|\sigma_j|)]^{1/2}\,\Delta\sigma\; \hat{M}_2(\sigma_i, -\sigma_j; V, \varphi, x) \tag{5}$$

The index zero on a summation sign signifies that the summation index omits zero. $0 \le \sigma_1 < \sigma_2 < \ldots < \sigma_n$ is an equidistant discretization of the positive frequency axis of the specified sea state, $\Delta\sigma = \sigma_{i+1} - \sigma_i$ and $\sigma_{-i} = -\sigma_i$. The assumption of an equidistant discretization is adopted for simplicity and is not necessary. The formulas are easily adapted to cover the situation of nonequidistant discretization. V denotes the speed

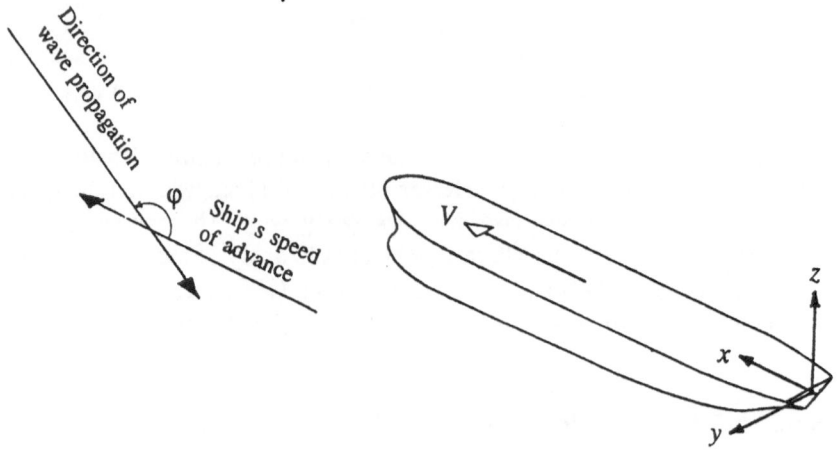

Figure 1 Definition of local coordinate system

of the ship, φ denotes the angle of advance relative to the long-crested sea way, cf. Fig. 1. The resulting encounter frequency ω corresponding to a wave frequency σ is assumed to be given by the relation

$$\omega = \omega(\sigma) = sign(\sigma)|\sigma - \frac{\sigma|\sigma|}{g}V\cos\varphi| \qquad (6)$$

where g denotes the gravitational acceleration. This formula ensures that the encounter frequency always has the same sign as the wave frequency, which is convenient for the mathematical treatment where both positive and negative frequencies are needed. $\{B_i\}$ is a set of independent, complex Gaussian N(0,1)-variables with independent, identically distributed real and imaginary parts. These variables are assumed to satisfy the relation $B_{-i} = B_i^*$, where * signifies complex conjugation. The use of the letter i to denote the imaginary unit $\sqrt{-1}$ and i to denote an index in expressions like $e^{i\omega_i t}$ should cause no confusion.

The expressions for the linear transfer function (LTF) $\hat{M}_1(\sigma_i) = \hat{M}_1(\sigma_i;V,\varphi,x)$ and the quadratic transfer function (QTF) $\hat{M}_2(\sigma_i,-\sigma_j) = \hat{M}_2(\sigma_i,-\sigma_j;V,\varphi,x)$ have been calculated by Juncher Jensen and Terndrup Pedersen (1979).

The aggregate of terms Q_{ij} indexed by $i,j = -n,...,-1,1,...,n$ is represented by a $2n \times 2n$ matrix $Q = (Q_{ij})$, which is Hermitian, that is $Q = Q^H = (Q^*)^T$. By invoking the representation theorem proved by Naess (1987, 1990) and expressing it in discretized form, it follows that the eigenvalue problem that has to be solved to obtain the desired representation in the present context can be written as

$$Qw = \mu w \qquad (7)$$

It is assumed that Q is nonsingular. Let μ_α and $w_\alpha = (w_{\alpha,-n},..., w_{\alpha,-1}, w_{\alpha,1},..., w_{\alpha,n})^T$, $\alpha = -n,...,-1, 1,..., n$, denote the eigenvalues and orthonormal eigenvectors of Q. Without loss of generality, it can also be assumed that $w_{\alpha,-i} = w_{\alpha,i}^*$.

The following decomposition of Q (see Naess, 1990) is obtained.

$$Q_{ij} = \sum_{\alpha=-n}^{n} {}_o\ \mu_\alpha w_\alpha(\sigma_i)w_\alpha(\sigma_j)^* \qquad (8)$$

where $w_\alpha(\sigma_i) = w_{\alpha,i}$.

By substituting for Q_{ij} given by equation (8) into equation (4), it is obtained that

$$M^{(2)}(t;x) = \sum_{i=-n}^{n} {}_o\ \mu_\alpha W_\alpha(t)^2 \qquad (9)$$

Here

$$W_\alpha(t) = \sum_{i=-n}^{n} {}_o\ w_\alpha(\sigma_i)B_i e^{i\omega_i t} \qquad (10)$$

which is a zero-mean, stationary and real Gaussian process of unit variance. For t fixed, the $W_\alpha = W_\alpha(t)$, $\alpha = -n,...,-1,1,...,n$ constitute a set of independent variables. Note that the (one-sided) spectral density $S_\alpha(\omega)$ of the $W_\alpha(t)$ process, expressed in terms of the encounter frequency ω, is given as follows

$$S_\alpha(\omega_i) = 2 \sum_{\substack{j \\ \omega(\sigma_j)=\omega_i}} |w_\alpha(\sigma_j)|^2 \left|\frac{d\omega_i}{d\sigma_j}\right|^{-1}, \qquad 0<\omega_1<\omega_2<...<\omega_m \tag{11}$$

To obtain the corresponding representation for the first-order response $M^{(1)}(t;x)$, the assumed completeness of the orthonormal set $\{w_\alpha\}$ is invoked. This implies that the following expansion is valid

$$q_i = \sum_{\substack{\alpha=-n}}^{n} c_\alpha\, w_\alpha(\sigma_i), \qquad i = -n,...,-1,1,...,n \tag{12}$$

The orthonormality of the eigenvectors then leads to the expressions ($q_{-i} = q_i^*$, $w_\alpha(\sigma_{-i}) = w_\alpha(\sigma_i)^*$)

$$c_\alpha = \sum_{\substack{i=-n}}^{n} q_i\, w_\alpha(\sigma_i)^* \tag{13}$$

By substituting from equation (12) into equation (2), it is seen that $M^{(1)}(t;x)$ can be represented as follows

$$M^{(1)}(t;x) = \sum_{\substack{\alpha=-n}}^{n} c_\alpha\, W_\alpha(t) \tag{14}$$

To get a first idea of how much the probability distribution of the response deviates from a Gaussian law, one may calculate the coefficients of skewness and excess. To this end we shall write down the expression for the cumulants k_j, $j=1,2,...$, of the total response $M(t;x)$. Specifically, it is obtained that (Naess, 1987)

$$k_j = \sum_{\substack{\alpha=-n}}^{n} 2^{j-1}(j-1)!\,\mu_\alpha^j\{1 - (1-\delta_{j1})j(c_\alpha/2\mu_\alpha)^2\} \tag{15}$$

3. The Probability Density Function of Vibration Response

Having achieved the desired representation of the response process $M(t,x)$, our goal is to compute the probability density function (PDF) $f_{TOT}(\cdot)$ of the random variable $M = M(t,x)$ (t and x fixed). This will be done by first calculating the characteristic function $g(\cdot)$ of M.

The characteristic function is the Fourier transform of the PDF, that is

$$g(\theta) = E[e^{i\theta M}] = \int_{-\infty}^{\infty} f_{TOT}(m)\, e^{i\theta m}\, dm \tag{16}$$

From equations (1), (9) and (14) it is found that

$$g(\theta) = \prod_{\substack{\alpha=-n}}^{n} \left[\frac{1}{\sqrt{1-2i\mu_\alpha\theta}} \exp\left\{ -\frac{c_\alpha^2\theta^2}{2(1-2i\mu_\alpha\theta)} \right\} \right] \tag{17}$$

The PDF of the random variable M can be obtained from the Fourier inversion of the characteristic function, and generally this has to be done numerically. By making the variable substitution $w = i\theta$, the integral we want to evaluate is

$$f_{TOT}(m) = \frac{1}{2\pi i} \int_{-i\infty}^{+i\infty} \exp\{-wm + \sum_{\alpha=-n}^{n} [-\frac{1}{2}\ln(1-2\mu_\alpha w) +$$

$$+ \frac{c_\alpha^2 w^2}{2(1-2\mu_\alpha w)}]\} dw \qquad (18)$$

Rice (1980) proposed a method for numerical evaluation of a similar integral using the so-called saddle point method. The idea behind the saddle point method is that in the integration of an analytic function, one is free to choose the path of integration between two points as long as it stays within the domain of analyticity. This method turns out to be ideally suited to calculate the integral in equation (18). It is described in detail by Naess and Johnsen (1992), who show that the method can be used to calculate the PDFs with high accuracy.

4. Extreme Value Prediction

The estimation of the combined extreme bending moment can be done in several alternative ways. The focus of our work here is on severe to extreme sea states. For such conditions, it has been shown that the linear part of the vibration response largely dominates the quadratic response, cf. Juncher Jensen and Dogliani (1993). Therefore, asymptotic methods for the estimation of combined extremes based on the quadratic response should not be used.

A possible approach is to use the method described in Naess (1993). This is based on the separate estimation of the extreme linear and quadratic responses combined with information on statistical coupling of large response values derived from the appropriate PDF's. The advantage of this method is the possibility to incorporate the effect of bandwidth on the extreme values of the quadratic response component, which includes resonant vibrations. However, this effect is of importance mostly in cases where the resonant part of the quadratic response component, which in our case corresponds to the sum-frequency response, dominates the quadratic response, and is of approximately equal or larger magnitude as compared to the linear response component. As already stated, in the example cases to be studied here, the linear response dominates in a certain sense, and in addition, it turns out that the sum-frequency response does not in fact dominate the total quadratic response. Hence one may expect that the bandwidth effect will not be of significance for our calculations.

Another approach to the problem of estimating the combined extreme response that incorporates the coupling between the linear and the quadratic responses in a similar manner as the previous method will be adopted here. By making appropriate simplifying assumptions, the chosen approach has a more direct implementation.

Let $\hat{M}(T) = \max\{M(t); 0 \le t \le T\}$ denote the extreme value of the bending moment response during the time period T. The extreme value distribution $F_{\hat{M}(T)}(m)$ of the bending moment is then calculated by the formula

$$F_{\hat{M}(T)}(m) = \exp\{ - v_{TOT}^+(m)\,T\} \qquad (19)$$

where $v_{TOT}^+(m)$ denotes the mean level upcrossing rate of the combined response process $M(t)$.

To simplify the calculation of $v_{TOT}^+(m)$, we introduce the basic assumption that $M = M(t;x)$ and $\dot{M} = \dot{M}(t;x)$ are statistically independent, which means that $f_{M\dot{M}}(\cdot,\cdot) = f_M(\cdot)f_{\dot{M}}(\cdot)$. This property is satisfied exactly in the case of Gaussian responses. Since the combined response is largely Gaussian, the adopted assumption is expectedly satisfied to good approximation. By invoking Rice's formula for the mean m-upcrossing rate $v_{TOT}^+(m)$ of the total response M, it is found that

$$v_{TOT}^+(m) = \int_0^\infty sf_{M\dot{M}}(m,s)\,ds = \int_0^\infty sf_{\dot{M}}(s)\,ds \cdot f_M(m) \qquad (20)$$

Hence, to calculate the mean level upcrossing rate $v_{TOT}^+(m)$, it is seen that the PDF of $\dot{M} = \dot{M}(t;x) = \dot{M}^{(1)}(t;x) + \dot{M}^{(2)}(t;x)$ is required. This can be calculated in precisely the same manner as described in the previous section. What is needed is the expressions corresponding to equations (2) - (5) for the present case. It is recognized that the following relations obtain

$$\dot{M}^{(1)}(t;x) = \sum_{i=-n}^{n} \dot{q}_i B_i e^{i\omega_i t} \qquad (21)$$

where $\dot{q}_i = i\omega_i q_i$, and

$$\dot{M}^{(2)}(t;x) = \sum_{i=-n}^{n} \sum_{j=-n}^{n} \dot{Q}_{ij} B_i B_j^* e^{i(\omega_i - \omega_j)t} \qquad (22)$$

where $\dot{Q}_{ij} = i(\omega_i - \omega_j)Q_{ij}$.

Solving equation (7) with Q replaced by $\dot{Q} = (\dot{Q}_{ij})$, the representations of $\dot{M}^{(1)}(t;x)$ and $\dot{M}^{(2)}(t;x)$ can then be calculated and the PDF of $\dot{M} = \dot{M}(t;x)$ determined as described in section 5. The integral appearing in equation (20) can then be calculated numerically.

Instead of calculating the integral of equation (20), one may observe that equation (20) leads to the result

$$v_{TOT}^+(m) = \frac{v_{TOT}^+(m_{ref})}{f_M(m_{ref})} \cdot f_M(m) \qquad (23)$$

where m_{ref} denotes some suitable reference level, typically the mean response level or similar. Since $f_M(m)$ is calculated with high accuracy, and $v_{TOT}^+(m_{ref})$ can be estimated by various techniques, equation (19) together with equation (23) provides a convenient way to estimate the extreme value distribution.

For the sake of completeness, we shall also briefly discuss the estimation of the long-term extreme value distribution. Assume that the vector $S = \{H_s, T_z, V, \varphi\}$ specifies uniquely the short-term condition in the sense that the ship hull vibration

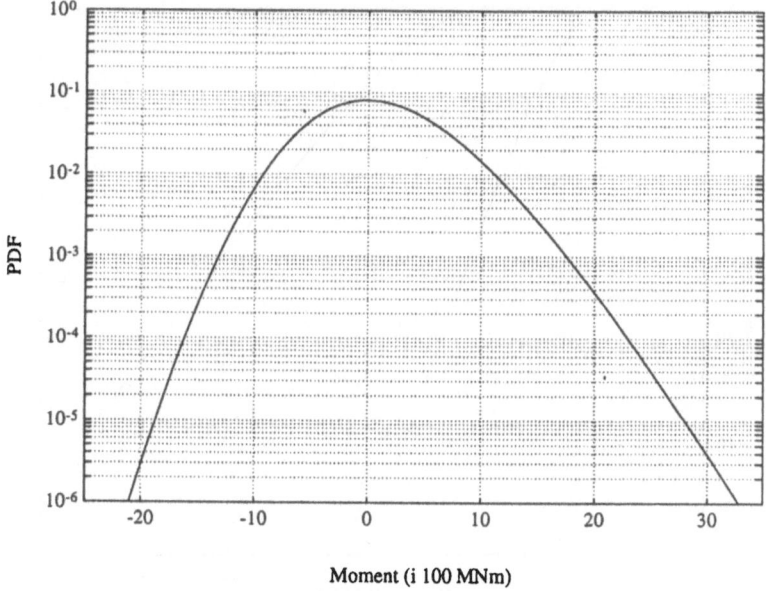

Figure 2 Logarithmic plot of the PDF of the bending moment amidship for case 1

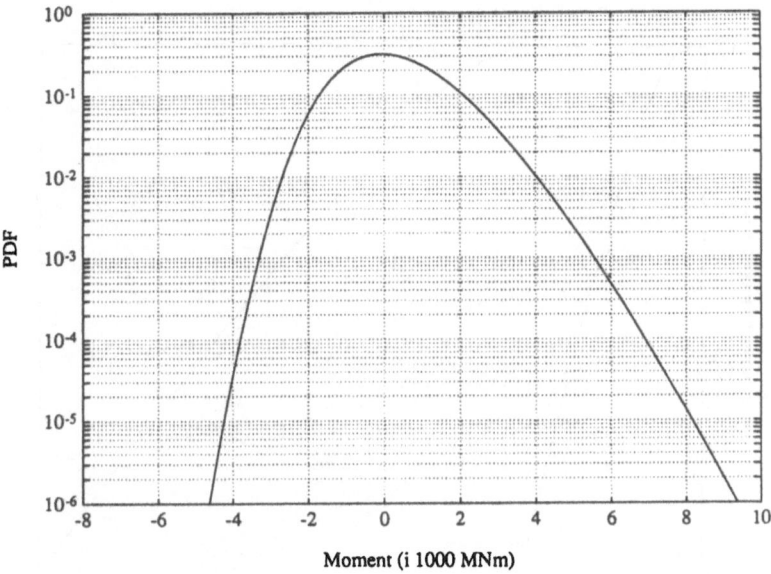

Figure 3 Logarithmic plot of the PDF of the bending moment amidship for case 2

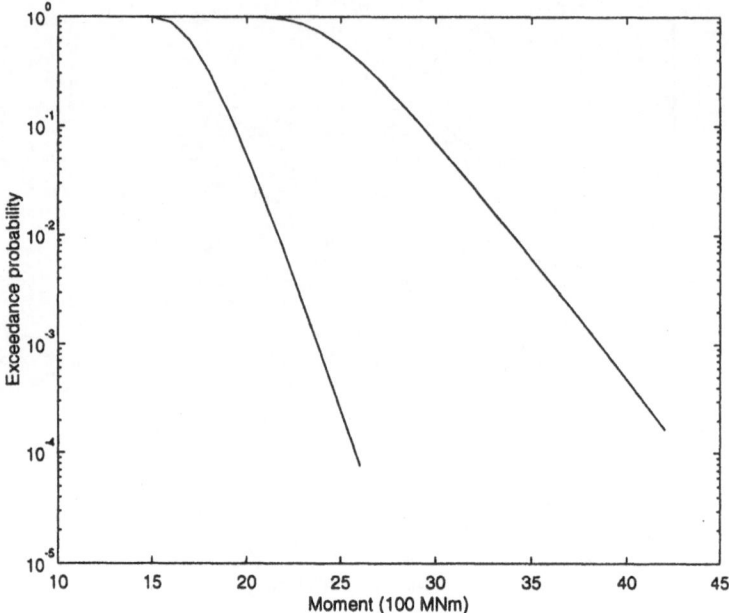

Figure 4 Exceedance probabilities for the bending moment amidship for a time period $T = 10,000$ s for
case 1

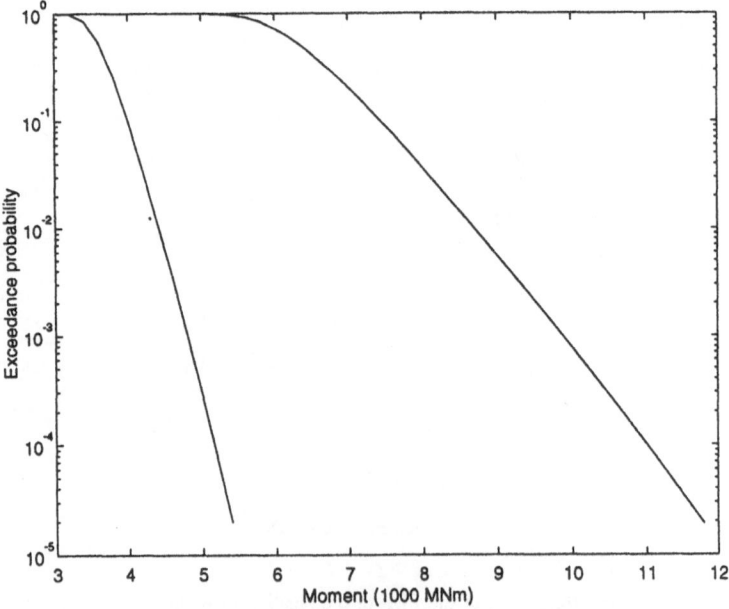

Figure 5 Exceedance probabilities for the bending moment amidship for a time period $T = 10,000$ s for
case 2

response can be considered stationary for each specific value of S. To underline this, the short-term mean level upcrossing rate may be written as $v_{TOT}^+(m|s)$. Let $f_S(s)$ denote the long-term averaging function of the conditions S, that is, $Tf_S(s)$ expresses the average fraction of a long time interval of length T that the condition vector S assumes the value s. As shown by Naess (1984), the long-term extreme value distribution can then be expressed as

$$F_{\hat{M}(T)}(m) = \exp\{ - \bar{v}_{TOT}^+(m) T \} \tag{24}$$

where

$$\bar{v}_{TOT}^+(m) = \int_s v_{TOT}^+(m|s) f_S(s)\, ds \tag{25}$$

5. Numerical Examples

The numerical results presented in this paper pertains to the Flokstra ship as detailed in Juncher Jensen and Dogliani (1993). The linear and quadratic transfer functions used in the numerical examples have been calculated according to the theory described by Jensen and Terndrup Pedersen (1979, 1981).

The numerical case studies presented concern two short-term wave conditions specified in terms of a significant wave height H_s and mean zero-crossing wave period T_z. These relate to a random sea way modelled by an ITTC wave spectrum. For case 1: $H_s = 6$ m, $T_z = 8.7$ s and forward speed $V = 12.6$ m/s. For case 2: $H_s = 15$ m, $T_z = 12$ s and forward speed $V = 3$ m/s. In both cases the ship is assumed rigid and to move in countering seas, i.e. $\varphi = 180$ degrees.

The numerical calculation of the probability density functions (PDFs) of the combined first- and second-order bending moment amidship has been carried out by using the method described in section 3. The results are given in Figures 2 and 3, which shows logarthmic plots of the PDFs. The conspicuous feature of these figures is the distinct asymmetry of the PDF, reflecting the observed fact that sag moments are generally larger than hog moments. It is seen that the asymmetry increases with the severity of the sea state. On the basis of calculated cumulants, it is found that the values of the skewness coefficient $\gamma_1 = k_3/k_2^{3/2}$ for the two cases studied are given as follows. For case 1: $\gamma_1 = k_3/k_2^{3/2} = 0.33$; while for case 2: $\gamma_1 = k_3/k_2^{3/2} = 0.50$, confirming the increasing asymmetry.

The exceedance probabilities $Q_{\hat{M}(T)}(m) = 1 - F_{\hat{M}(T)}(m)$ for a time period of 10,000 seconds have been calculated for the two example cases, and the results have been plotted in Figures 4 and 5, displaying very clearly the asymmetry between large sag and hog moments.

6. Conclusions

A quadratic theory for calculating response statistics of wave induced ship hull vibrations based on the Kac-Siegert expansion method has been described. The method may accommodate hull flexibility. The analysis partly applies a numerical procedure for calculating the PDF of combined linear and quadratic vibration responses of the ship hull in random seas. The method, which is based on the so-called method of steepest descent, is quite simple and it does not require extensive computer capacity. In fact, it can be implemented on any standard PC.

A main advantage of the described method for calculating the PDF is that no approximations are made except those inherent in the numerical calculations by discretization. However, the errors incurred by discretization can be controlled.

7. Acknowledgements

The author is greatly indebted to Professor J. Juncher Jensen and Dr. M. Dogliani for many helpful discussions and comments. The data for the linear and quadratic transfer functions used in this paper have been provided by Professor Juncher Jensen. The work presented here has been carried out as part of the BRITE/EURAM Project 4554 'Reliability Methods for Ship Structural Design' under EU contract BREU-CT-91-0501. The financial support from this project is gratefully acknowledged.

8. References

Flokstra, C. (1974), "Comparison of Ship Motion Theories with Experiments for a Container Ship", *International Shipbuilding Progress*, Vol. 21, June 1974.

Juncher Jensen, J. and Terndrup Pedersen, P. (1979), "Wave-Induced Bending Moments in Ships - a Quadratic Theory", *RINA, Supplementary Papers*, Vol. 121, pp. 151-165.

Juncher Jensen, J. and Terndrup Pedersen, P. (1981), "Bending Moments and Shear Forces in Ships Sailing in Irregular Seas", *Journal of Ship Research*, Vol. 24, No.4, pp. 243-251.

Juncher Jensen, J. and Dogliani, M. (1993), "Wave-induced Ship Hull Vibrations in Stochastic Seaways", submitted to *Journal of Marine Structures*.

Naess, A. (1984), "On the long-term statistics of extremes". *Applied Ocean Research*, Vol. 6, No. 4, pp. 227-228.

Naess, A. (1985), "Statistical Analysis of Second-Order Response of Marine Structures", *Journal of Ship Research*, Vol. 29, No. 4, pp. 270-284.

Naess, A. (1987), "Response Statistics of Non-Linear, Second-Order Transformations to Gaussian Loads", *Journal of Sound and Vibration*, Vol. 115, No. 1, pp. 103-127

Naess, A. (1990), "Statistical Analysis of Nonlinear, Second-Order Forces and Motions of Offshore Structures in Short-Crested Random Seas", *Probabilistic Engineering Mechanics*, Vol. 5, No. 4, pp. 192-204.

Naess, A. (1993), "Statistics of Combined Linear and Quadratic Springing Response of a TLP in Random Waves", *Proc. 12th International Conference on Offshore Mechanics and Arctic Engineering*, Glasgow, UK, June 1993.

Naess, A and Johnsen. J.M. (1992), "An Efficient Numerical Method for Calculating the Statistical Distribution of Combined First-Order and Wave-Drift Response", *Journal of Offshore Mechanics and Arctic Engineering*, Transactions of the ASME, Vol. 114, No. 3, 1992.

Rice, S.O. (1980), "Distribution of Quadratic Forms in Normal Random Variables - Evaluation by Numerical Integration", *Siam Journal Sci Stat Comput*, Vol. 1, No. 4, pp. 438-448.

STOCHASTIC RESPONSE OF A SYSTEM WITH SPACE IMPERFECTIONS UNDER A MOVING LOAD

J. NÁPRSTEK and L. FRÝBA
Institute of Theoretical and Applied Mechanics,
Prosecká 74, 190 00 Prague 9, Czech Republic

1. Introduction

Dynamic behaviour of railway track subjected to moving vehicles is affected by random deviations of its parameters from their nominal values. The most important factor of stochastic character is the rigidity of the substructure which is correlated with the random track irregularities. The first approximation of the load which is moving along the longitudinal axis of a beam (rail) may be represented by a lumped mass or by a system whose transfer functions are known.

The deterministic problem (dynamics of a beam on elastic foundation) was tackled by several authors [1], [2], [3] while the stochastic character was represented in [4] using the analytical method and later on in [5] using the finite element method. The last paper is based on the stochastically orthogonal decompositions (for a detailed description see [6]), together with the method of perturbations. It means, that the zero correlation of input and output in neighboring nodes is introduced. This assumption may be accepted only near the point of application of the load where the stochastic component of the response is low. The papers [7], [8] and [9] tried to take into account the correlations mentioned above.

The integral spectral decomposition method has been used for an analytical solution in [7], [9] while in the present paper, the effect of partially correlated imperfections of the substructure and of track irregularities is studied using FEM. Both input processes are assumed as centered, homogeneous, ergodic and Gaussian.

2. Basic Relations for a Beam Under Moving Load

The rail is idealized by an infinitely long beam on Winkler foundation with Voigt viscous damping. A concentrated (lumped) mass m is moving with a constant speed c along the beam. The problem is governed by the equation

$$\mathbf{L}^*\{v(x',t')\} + r_\epsilon(x')v(x',t') = \left[-mg + m\frac{\mathrm{d}^2}{\mathrm{d}t'^2}\left(v(x',t') + \varphi(x')\right)\right]\delta(x'-ct') \quad (1)$$

$\mathbf{L}^*\{\cdot\}$ − differential operator describing the behaviour of the basic system without imperfections (fixed coordinates):

$$\mathbf{L}^*\{\cdot\} \equiv \mathrm{EJ}\frac{\partial^4}{\partial x'^4} + \mu\frac{\partial^2}{\partial t'^2} + 2\mu\omega_b\frac{\partial}{\partial t'} + r_0 \quad (2)$$

A. Naess and S. Krenk (eds.), IUTAM Symposium on Advances in Nonlinear Stochastic Mechanics, 321–330.
© 1996 *Kluwer Academic Publishers.*

Boundary conditions

$$\lim_{x' \to \pm\infty} v(x', t') = 0 \quad ; \quad \lim_{x' \to \pm\infty} \frac{\partial v(x', t')}{\partial x'} = 0 \tag{3}$$

x', t', x, t − fixed and moving coordinates, $x = x' - ct', t = t'$;

ω_b − constant coefficient of damping

E, J, μ − constant modulus of elasticity, moment of inertia, mass per unit length, respectively

$\delta(\cdot)$ − Dirac function

$r(x')$ − rigidity of Winkler foundation

$$r(x') = r_0 + r_\varepsilon(x') \; ; \; |r_\varepsilon(x')| \ll r_0 \tag{4}$$

r_0 − constant mean value of the foundation rigidity

$r_\varepsilon(x'), \varphi(x')$ − fluctuating component of the rigidity or the track irregularities, respectively (centered, homogeneous, ergodic, Gaussian random processes).

Taking into account only a linear approximation of the effect of random processes $r_\varepsilon(x'), \varphi(x')$, the response $v(x, t)$ may be written in the form

$$v(x', t') = v_0(x', t') + v_\varepsilon(x', t') \tag{5}$$

$v_0(x', t')$ − mean value of the response; deterministic function

$v_\varepsilon(x', t')$ − stochastic component of the response due to imperfection of the foundations and of the track (random pseudo–Gaussian process).

The equation (1) is transformed into the moving coordinate system bound together with the moving load. With respect to (2), (4), (5) and analysing the total differential on the right hand side, the equation (1) transforms into

$$\mathbf{L}\{v_0(x, t) + v_\varepsilon(x, t)\} + r_\varepsilon(x + ct)(v_0(x, t) + v_\varepsilon(x, t)) =$$

$$= \left[-mg + m\left(\ddot{v}_0(x, t) + \ddot{v}_\varepsilon(x, t) + c^2\varphi''(x + ct)\right)\right]\delta(x) \tag{6}$$

where $\mathbf{L}\{\cdot\}$ is the following differential operator:

$$\mathbf{L}^*\{\cdot\} \to \mathbf{L}\{\cdot\} \equiv EJ\frac{\partial^4}{\partial x^4} + \mu\frac{\partial^2}{\partial t^2} + 2\mu\omega_b\frac{\partial}{\partial t} + r_0 + \mu c^2\frac{\partial^2}{\partial x^2} - 2\mu c\left(\frac{\partial^2}{\partial x \partial t} + \omega_b\frac{\partial}{\partial x}\right)$$

with corresponding boundary conditions, see (3).

3. Stochastic Finite Elements for Non-Selfadjoint Operator

The finite element method is the wide spread method for the calculation of the response of structures. Its classic formulation for linear problems is based on the principle of minimization of quadratic functional in the sense of Ritz method, see

[10]. The operator $\mathbf{L}\{\cdot\}$ in (6) is non–selfadjoint making this approach inapplicable. However the Galerkin method can be used. The solution itself should be understood as a formal process of the solution of the stochastic differential equation (6), and not as a classical solution of a variational problem of the linear theory of elasticity. Indeed, the operator $\mathbf{L}\{\cdot\}$ is a differential operator and it may be written in the form

$$\mathbf{L}\{\cdot\} = \mathbf{L}_s\{\cdot\} + \mathbf{L}_n\{\cdot\} \tag{7}$$

$\mathbf{L}_s\{\cdot\}$ – selfadjoint differential operator of the order $2k = 4$ which becomes a positive definite one if the length of the beam is restricted to a finite length (in moving coordinates)

$\mathbf{L}_n\{\cdot\}$ – non-selfadjoint differential operator of the order at most $2k - 1$, in our case of the order 1.

Restricting the beam to a finite length, the coefficients become integrable with quadrate and it may be proved, using some theorems in [11], that $\mathbf{L}_n\{\cdot\}$ is a totally continuous operator. As $\mathbf{L}_s\{\cdot\}$ is positively definite and $\mathbf{L}_n\{\cdot\}$ totally continuous and of lower order than $\mathbf{L}_s\{\cdot\}$, the Galerkin method applied to the equation (6) converges to the unique solution under the assumption that the base system is complete and that it satisfies the conditions required by the Galerkin method. These conditions are fulfilled if every base function satisfies all boundary conditions and if they are continuous together with their derivatives with respect to x up to the order $k - 1$, in our case up to order 1. Then a uniform convergence can be expected for the seeked function together with the derivatives up to the order $k - 1$ and the convergence in the mean for the derivatives of the order k. In our case, the deflection and rotation of the beam will converge uniformly.

The length of the beam is restricted to both sides from the load up to the distance where a response can hardly be expected. The beam can be divided now into $n\,(x > 0)$ and $m\,(x < 0)$ elements of constant length d. The deflection $v(x,t)$ is written in the form

$$v(x,t) \approx \sum_{j=-m+1}^{n-1} \mathbf{n}_j^t(x)\mathbf{u}_j(t) \tag{8}$$

$\mathbf{u}_j(t) = |v_j(t), \zeta_j(t)|$ – vector which components are deflection and rotation of the j-th node, respectively

$\mathbf{n}_j(x) = |n_{vj}(x), n_{\zeta j}(x)|$ – base functions

$$n_{vj}(x), n_{\zeta j}(x) \begin{cases} = 0 & ; \quad -md \leq x \leq (j-1)d \\ \neq 0 & ; \quad (j-1)\bar{d} < x < (j+1)d \\ = 0 & ; \quad (j+1)d \leq x \leq nd \end{cases} \tag{9}$$

The stochastic process $r_\varepsilon(x')$ is modelled by a gradual function which is constant within each element and equal to the mean value at boundary points:

$$r_\varepsilon(x + ct) \approx \frac{1}{2} \sum_{j=-m+1}^{n-1} \Delta_j(x)r_j(ct) \; ; \quad r_j(ct) = r_\varepsilon(x_j + ct) \tag{10}$$

Putting (8), (10) into (6) and demanding the orthogonality with every base function for $-m + 1 < i < n - 1$ successively, the following relation appears:

$$\sum_{l=0}^{1}\sum_{j=i-1}^{i}\left[\mathbf{n}_i(x)(EJ\mathbf{n}_{j+l}^{t\,\prime\prime\prime}(x)+\mu c^2\mathbf{n}_{j+l}^{t\,\prime}(x))-\mathbf{n}_i'(x)EJ\mathbf{n}_{j+l}^{t\,\prime\prime}(x))\right]\Big|_{x_{i-1+l}}^{x_{i+l}}\mathbf{u}_{j+l}(t)+$$

$$+\left[\mu\mathbf{a}_{i,j}^{j+l}\cdot\ddot{\mathbf{u}}_{j+l}(t)+2\mu(\omega_b\mathbf{a}_{i,j}^{j+l}-c\mathbf{b}_{i,j}^{j+l})\cdot\dot{\mathbf{u}}_{j+l}(t)+\right.$$

$$\left.+(EJ\mathbf{d}_{i,j}^{j+l}-\mu c^2\mathbf{g}_{i,j}^{j+l}+r_0\mathbf{a}_{i,j}^{j+l}-2\mu c\mathbf{b}_{i,j}^{j+l})\cdot\mathbf{u}_{j+l}(t)+r_j^s(ct)\mathbf{a}_{i,j}^{j+l}\mathbf{u}_{j+l}(t)\right]=$$

$$=(-mg+mc^2\chi_0(ct))\sum_{j=i-1}^{i}\mathbf{h}_j+\sum_{l=0}^{1}\sum_{j=i-1}^{i}m\hat{\mathbf{a}}_{i,j}^{j+l}\cdot\ddot{\mathbf{u}}_{j+l}(t)\qquad(11)$$

where the following symbols for (2×2) matrices have been used:

$$\mathbf{a}_{i,j}^{j+l}=\int_{x_j}^{x_{j+1}}\mathbf{n}_i(x)\mathbf{n}_{j+l}^t(x)dx\quad;\quad\mathbf{b}_{i,j}^{l}=\int_{x_j}^{x_{j+1}}\mathbf{n}_i(x)\mathbf{n}_{j+l}^{t\,\prime}(x)dx$$

$$\mathbf{d}_{i,j}^{j+l}=\int_{x_j}^{x_{j+1}}\mathbf{n}_i''(x)\mathbf{n}_{j+l}^{t\,\prime\prime}(x)dx\quad;\quad\mathbf{g}_{i,j}^{l}=\int_{x_j}^{x_{j+1}}\mathbf{n}_i'(x)\mathbf{n}_{j+1}^{t\,\prime}(x)dx$$

$$\mathbf{h}_i=\int_{x_j}^{x_{j+1}}\mathbf{n}_i(x)\delta(x)dx=\left|\begin{matrix}h\\0\end{matrix}\right|\text{ for }i=0;\quad\left|\begin{matrix}0\\0\end{matrix}\right|\text{ for }i\neq0;\text{ where the first}$$

element represents displacement, the second one rotation, usually $h=1$

$$\hat{\mathbf{a}}_{i,j}^{j+l}=\int_{x_j}^{x_{j+1}}\mathbf{n}_i(x)\mathbf{n}_{j+l}^t(x)\delta(x)dx\quad;\quad\chi_0(x')=\varphi''(x')$$

$$\Bigg\}\qquad(12)$$

The discretized system may be described by the system of differential equations

$$(\mu\mathbf{A}+mg\mathbf{P}\mathbf{P}^t)\ddot{\mathbf{U}}(t)+2\mu(\omega_b\mathbf{A}-c\mathbf{B})\dot{\mathbf{U}}(t)+(EJ\cdot\mathbf{D}-\mu c^2\mathbf{G}+r_0\mathbf{A}-2\mu c\mathbf{B})\mathbf{U}(t)+$$

$$+\sum_{i=-m+1}^{n}(r_i(ct)\mathbf{A}_i)\cdot\mathbf{U}(t)=-\mathbf{P}\cdot mg+\mathbf{P}\cdot mc^2\chi(ct)\qquad(13)$$

$\mathbf{U}^t(t)=|\mathbf{u}_{-m+1}^t(t),\ \ldots\ ,\mathbf{u}_0^t(t),\ \ldots\ ,\mathbf{u}_{n-1}^t(t)|$ – vector of node displacement, length $2(m+n-1)$

$\mathbf{P}$ – column vector containing zeros with the exception of the element corresponding to $v_0(t)$ which contains 1 ; this vector has the same length as $\mathbf{U}(t)$.

$\mathbf{A},\mathbf{B},\mathbf{D},\mathbf{G}$ – square matrices with dimensions $2(m+n-1)\times2(m+n-1)$. They are formed of matrices $\mathbf{A}_i^0,\mathbf{B}_i^0,\mathbf{D}_i^0,\mathbf{G}_i^0$, dimension (4×4), as described above. The matrices $\mathbf{A}_{-m+1}^0,\ldots\mathbf{G}_{-m+1}^0$ or $\mathbf{A}_n^0,\ldots\mathbf{G}_n^0$ apply only to their right bottom or left upper quarter (2×2), i.e. the submatrices $\mathbf{a}_{-m+1,-m+1}^{-m+1}$ resp. $\mathbf{a}_{m-1,n-1}^n$

$\mathbf{A}_i$ – square matrices of the same dimensions as $\mathbf{A}$. There is a matrix $\mathbf{A}_i^0$ lying on the main diagonal and having its element $(1,1)$ on the element $(2(i-1),2(i-1))$ of the matrix $\mathbf{A}_i$; the remaining elements of $\mathbf{A}_i$ are zero.

$$\mathbf{A}_i^0=\left|\begin{matrix}\mathbf{a}_{i-1,i-1}^i&\mathbf{a}_{i-1,i}^i\\\mathbf{a}_{i,i-1}^i&\mathbf{a}_{i,i}^i\end{matrix}\right|\quad;\text{ and in a similar way }\mathbf{B}_i^0,\mathbf{D}_i^0,\mathbf{G}_i^0,\qquad(14)$$

If the local coordinates $s = x - x_{i-1}$; $s\epsilon(0, d)$ are introduced within the i-th element and it is denoted (compare (8)):

$$N_i^t(s) = |n_{i-1}^t(x), n_i^t(x)| = |\psi_{1i}(s), \psi_{2i}(s), \psi_{3i}(s), \psi_{4i}(s)| \qquad (15)$$

the vector $N_i(s)$ or functions $\psi_{1i}(s), \ldots, \psi_{4i}(s)$ may be assumed the shape functions of the i-th element. As the lengths of all elements are the same, the index i in $A_i^0, \ldots, G_i^0$ can be omitted. Their submatrices are then denoted $a_{1,1} = a_{i-1,i-1}^i$, $a_{2,1} = a_{i,i-1}^i$ etc.

The functions $\psi_1(s), \ldots, \psi_4(s)$ can be selected in the form of cubic polynomials in a way that the conditions (9) layed down on the base system are satisfied, see [5],[10], with the exception of points x_{-m}, x_n, where the functions $n_{-m+1}(x), n_{n-1}(x)$ satisfy only the main boundary conditions:

$$\psi_1(s) = \left(1 - \frac{s}{d}\right)^2 \left(1 + 2\frac{s}{d}\right) \quad ; \psi_2(s) = s\left(1 - \frac{s}{d}\right)^2$$
$$\psi_3(s) = \left(\frac{s}{d}\right)^2 \left(3 - 2\frac{s}{d}\right) \quad ; \psi_4(s) = \frac{s^2}{d}\left(\frac{s}{d} - 1\right)^2 \qquad (16)$$

and from here the matrices $A^0, \ldots, G^0$ are calculated

$$A^0 = |A_{kl}| = \left|\int_0^d \psi_k(s)\psi_l(s)\mathrm{d}s\right| = \frac{d}{420}\begin{vmatrix} 156 & 22d & 54 & -13d \\ 22d & 4d^2 & 13d & -3d^2 \\ 54 & 13d & 156 & -22d \\ -13d & -3d^2 & -22d & 4d^2 \end{vmatrix} \qquad (17)$$

$$B^0 = |B_{kl}| = \left|\int_0^d \psi_k(s)\psi_l'(s)\mathrm{d}s\right| = \frac{1}{60}\begin{vmatrix} -30 & 6d & 30 & -6d \\ -6d & 0 & 6d & -d^2 \\ -30 & -6d & 30 & 6d \\ 6d & d^2 & -6d & 0 \end{vmatrix} \qquad (18)$$

$$D^0 = |D_{kl}| = \left|\int_0^d \psi_k''(s)\psi_l''(s)\mathrm{d}s\right| = \frac{1}{d^3}\begin{vmatrix} 12 & 6d & -12 & 6d \\ 6d & 4d^2 & -6d & 2d \\ -12 & -6d & 12 & -6d \\ 6d & 2d^2 & -6d & 4d^2 \end{vmatrix} \qquad (19)$$

$$G^0 = |G_{kl}| = \left|\int_0^d \psi_k'(s)\psi_l'(s)\mathrm{d}s\right| = \frac{1}{30d}\begin{vmatrix} 36 & 3d & -36 & 3d \\ 3d & 4d^2 & -3d & -d^2 \\ -36 & -3d & 36 & -3d \\ 3d & -d^2 & -3d & 4d^2 \end{vmatrix} \qquad (20)$$

4. Mean Values and Dispersals of the Response

The vector of node displacements with $2(m + n - 1)$ elements introduced in (8) and (13) contains both deterministic and stochastic part of the response

$$U^t(t) = U_0^t + U_\epsilon^t(t) = |u_{-m+1,0}^t + u_{-m+1,\epsilon}^t, \quad \cdots \quad , u_{n-1,0}^t + u_{n-1,\epsilon}^t| \qquad (21)$$

It can be shown that the mathematical mean value of the response U_0 almost does not depend on time, see [9]. The total independence on time occurs on an infinitely long beam. This state is achieved at FEM if the numbers md and nd are sufficiently high. The other assumption is the homogeneity of random processes $r(x'), \chi(x')$.

The stochastic part of the response $\mathbf{U}_\epsilon(t)$ is written in the form

$$\mathbf{U}_\epsilon(t) = \sum_{j=-m+1}^{n-1} \int_{-\infty}^{\infty} \mathbf{U}_{jr}(\omega) e^{i\omega t} dV_j(\omega) \ ; \ \ \omega = c \cdot \alpha \tag{22}$$

$dV_j(\omega)$ — spectral differential of the deflection in the nodes

$\mathbf{U}_{jr}(\omega)$ — unknown vectors

The imperfections $r_i(ct)$ and track irregularities $\chi(t)$ are expressed in a similar way

$$r_i(ct) = \int_{-\infty}^{\infty} e^{i\omega t} dR_i(\omega) \ ; \ \ \chi(t) = \int_{-\infty}^{\infty} e^{i\omega t} d\Psi_0(\omega) \ ; \ \ d\Psi_j(\omega) = 0 \ \ j \neq 0 \tag{23}$$

where $dR_i(\omega)$, $d\Psi_0(\omega)$ are corresponding spectral differentials depending also on the speed of the load movement.

After changing the order of summation and integration and putting (21), (22), (23) into (13) arises:

$$\mathbf{M} \cdot \mathbf{U}_0 - \sum_{j=-m+1}^{n} \sum_{i=-m+1}^{n} \int_{-\infty}^{\infty} \mathbf{A}_i \mathbf{Q}^{-1}(\omega) \mathbf{A}_j S_{ij}^{rr}(\omega) d\omega \mathbf{U}_0 =$$

$$= -\mathbf{P}mg - \sum_{j=-m+1}^{n} \sum_{i=-m+1}^{n-1} \int_{-\infty}^{\infty} \mathbf{Q}^{-1}(\omega) \mathbf{P} \cdot mc^2 S_{ij}^{r\chi}(\omega) d\omega \cdot \delta(j) \tag{24}$$

where we have denoted

$$\begin{array}{ll} \mathbf{H} = \mu\mathbf{A} + mg\mathbf{P}\mathbf{P}^t & \mathbf{K} = 2\mu(\omega_b\mathbf{A} - c\mathbf{B}) \\ \mathbf{M} = EJ \cdot \mathbf{D} - \mu c^2\mathbf{G} + r_0\mathbf{A} - 2\mu c\mathbf{B} \end{array} \tag{25}$$

$$\mathbf{Q}(\omega) = (-\omega^2\mathbf{H} + 2i\omega\mathbf{K} + \mathbf{M}) \tag{26}$$

The equation (24) determines the vector of the mathematical mean value $\mathbf{U}_0$ of the response and demonstrates the way of dependancy of $\mathbf{U}_0$ upon foundation imperfections and track irregularities. Eq.(24) implies that $\mathbf{U}_0$ is influenced by irregularities of track only by means of a statistic connection with foundation imperfections.

If $c \to 0$ the interaction of $\mathbf{U}_0$ and of track irregularities vanishes and only the static member $\mathbf{P}mg$ remains on the right hand side of (24). In this case, the integral on the left hand side (24) disappears and the equation takes the form

$$\left(\mathbf{M} - \sum_{j=-m+1}^{n} \sum_{i=-m+1}^{n} \mathbf{A}_i \mathbf{M}^{-1}(\omega) \mathbf{A}_j (\sigma_{ij}^{rr})^2 \right) \mathbf{U}_0 = -\mathbf{P}mg \tag{27}$$

$(\sigma_{ij}^{rr})^2$ – covariance of imperfections between the points i, j. It is valid with respect to (10)

$$(\sigma_{ij}^{rr})^2 = (\sigma^{rr}(id - jd))^2 \qquad (28)$$

which is the autocorrelation function of imperfections in fixed coordinates.

The equation (27) describes the static effect of a load on the beam resting on foundation with homogeneous imperfections according (28).

The homogeneous processes of foundation and track imperfection may be described by relation

$$S_{ij}^{rr}(\omega) = S^{rr}\left(\frac{\omega}{c}\right)\exp\left(i\frac{(i-j)d}{c}\omega\right) \quad ; \quad S_{i0}^{rx}(\omega) = S^{rx}\left(\frac{\omega}{c}\right)\exp\left(i\frac{id\omega}{c}\right) \qquad (29)$$

$S^{rr}(\omega)$ – spatial spectral density of foundation imperfections ($\omega = c \cdot \alpha$)

$S^{rx}(\omega)$ – spatial cross spectral density of foundation and track imperfections.

The numeric solution of the integro-algebraic equation (24) is possible. It is, of course, laborious although the integration interval $(\omega_{min}, \omega_{max})$ can be very limited due to quickly decreasing $S^{rr}(\omega), S^{rx}(\omega)$ for $|\omega| > 0$, and although many terms of the sum for $i \neq j$ can be neglected. The spectral densities $S^{rr}(\omega), S^{rx}(\omega)$ (29) can be usually expressed in the form of very simple rational fraction. The matrix $\mathbf{Q}^{-1}(\omega)$ may take the form

$$\mathbf{Q}^{-1}(\omega) = \sum_{k=-2m+2}^{2n}\left(\mathbf{W}_k\frac{1}{\omega - i\omega_k} + \overline{\mathbf{W}_k}\frac{1}{\omega + i\omega_k}\right) \qquad (30)$$

$$\mathbf{W}_k = \left|\frac{\Delta_{ji}(\omega_k)}{\Delta'(\omega_k)}\right| ; \qquad \overline{\mathbf{W}_l} = \left|\frac{\Delta_{ji}(\overline{\omega_k})}{\Delta'(\overline{\omega_k})}\right| \qquad (31)$$

$\Delta(\omega), \Delta'(\omega)$ – the determinant of the matrix $\mathbf{Q}(\omega)$ and its derivative in the point ω ;

$\Delta_{ji}(\omega)$ – subdeterminant of the matrix $\mathbf{Q}(p)$ corresponding to its (j, i)-th element in the point ω ;

$\omega_k, \overline{\omega_k}$ – complex conjugate roots of the equation $\Delta(p) = 0$.

The definition of variance gives with respect to (22):

$$\Gamma = \mathbf{E}\{\mathbf{U}_\epsilon(t)\overline{\mathbf{U}_\epsilon(t)}\} = \sum_{i=-m+1}^{n-1}\int_{-\infty}^{\infty} \mathbf{U}_i(\omega)\overline{\mathbf{U}_j^!(\omega)} \cdot S_{ij}(\omega)d\omega \qquad (32)$$

Γ – matrix$(2(n+m-1) \times 2(n+m-1))$ of dispersals and cross dispersals of the response at individual nodes of the system.

For practical application of (32) it is suitable to break down $S_{ij}^{vv}(\omega)$ going back to (22); every spectral differential of the deflection is composed of foundation and of track imperfection components. Therefore, every vector $\mathbf{U}_j(\omega)$ possesses two components

$$\mathbf{U}_\epsilon(t) = \sum_{i=-m+1}^{n-1} \int_{-\infty}^{\infty} \left(\mathbf{U}_{jr}(\omega)\mathrm{d}R_j(\omega) + \mathbf{U}_{j\chi}(\omega)\mathrm{d}\Psi_j(\omega)\delta(j)\right)\mathrm{d}\omega \qquad (33)$$

and applied in (32) gives:

$$\Gamma = \sum_{i=-m+1}^{n-1} \int_{-\infty}^{\infty} \left(\mathbf{U}_{ir}(\omega)\overline{\mathbf{U}_{jr}^t(\omega)} \cdot S_{ij}^{rr}(\omega) + \mathbf{U}_{ir}(\omega)\overline{\mathbf{U}_{j\chi}^t(\omega)} \cdot S_{ij}^{r\chi}(\omega)\delta(j) + \right.$$

$$\left. + \mathbf{U}_{i\chi}(\omega)\overline{\mathbf{U}_{jr}^t(\omega)} \cdot S_{ij}^{\chi r}(\omega)\delta(i) + + \mathbf{U}_{i\chi}(\omega)\overline{\mathbf{U}_{j\chi}^t(\omega)} \cdot S_{ij}^{\chi\chi}(\omega)\delta(i)\delta(j)\right)\mathrm{d}\omega \qquad (34)$$

In order to enumerate (34) it is necessary to know all vectors $\mathbf{U}_j(\omega)$ and $\mathbf{U}_{0\chi}(\omega)$. With respect to (33) and taking into account that $\mathbf{U}_0$ is already known, we can put (21), (22), (23) into (13) and change the order of summation and integration. Multiplying the result by spectral differential $d\overline{R}_k(\omega)$ and applying the mean value operator on the result the following algebraic system can be derived:

$$\mathbf{Q}(\omega) \sum_{i=-m+1}^{n-1} \left(\mathbf{U}_{jr}(\omega)S_{kj}^{rr}(\omega) + \mathbf{U}_{j\chi}(\omega)S_{kj}^{r\chi}(\omega) \cdot \delta(j)\right) =$$

$$= \sum_{i=-m+1}^{n-1} \mathbf{A}_j \mathbf{U}_0 S_{kj}^{rr}(\omega) - \sum_{i=-m+1}^{n-1} \mathbf{P}mc^2 S_{kj}^{r\chi}(\omega)\delta(j) \qquad (35)$$

The system (35) possesses a hyper-element structure where the hyper-element size corresponds to the size of the matrix $\mathbf{Q}(\omega)$. It has $(m+n-1)$ hyper-rows (index k) and $(m+n)$ hyper-columns. One hyper-row is added to the (35) which arises by a similar way as in the previous case using instead $d\overline{R}_k(\omega)$ the spectral differential $d\Psi_0(\omega)$:

$$\mathbf{Q}(\omega) \sum_{i=-m+1}^{n-1} \left(\mathbf{U}_{jr}(\omega)S_{kj}^{\chi r}(\omega)\delta(k) + \mathbf{U}_{j\chi}(\omega)S_{kj}^{\chi\chi}(\omega)\delta(k)\delta(j)\right) =$$

$$= \sum_{i=-m+1}^{n-1} \mathbf{A}_j \mathbf{U}_0 S_{kj}^{\chi r}(\omega)\delta(k) - \sum_{i=-m+1}^{n-1} \mathbf{P}mc^2 S_{kj}^{\chi\chi}(\omega)\delta(k)\delta(j) \qquad (36)$$

Thus, the system is complete. If the imperfections are quite independent in every element and independent on track irregularities the system (35), (36) falls into independent system along the hyperdiagonal. The vectors $\mathbf{U}_{jr}(\omega)$ and $\mathbf{U}_{0\chi}(\omega)$ would be independent and they would take a simple form

$$\mathbf{U}_{jr}^0(\omega) = \mathbf{Q}^{-1}(\omega)\mathbf{A}_j \mathbf{U}_0 \; ; \qquad \mathbf{U}_{0\chi}^0(\omega) = -\mathbf{Q}^{-1}(\omega)\mathbf{P}mc^2 \qquad (37)$$

Also the result (35) would by very simple

$$\Gamma = \sum_{i=-m+1}^{n-1} \int_{-\infty}^{\infty} \mathbf{Q}^{-1}(\omega)\left(\mathbf{A}_i \mathbf{U}_0 \overline{\mathbf{U}_0^t \mathbf{A}_i^t} S_{ii}^{rr}(\omega) + \mathbf{PP}^t m^2 c^4 S_{ii}^{\chi\chi}(\omega)\delta(i)\right)\overline{\mathbf{Q}^{-1}(\omega)}\mathrm{d}\omega$$

$$(38)$$

In usual case, the integral may be directly calculated using (30) and the residual theorem. The simple case $c \to 0$ can be written immediately

$$\Gamma^0 = \sum_{i=-m+1}^{n-1} \mathbf{M}^{-1}\mathbf{A}_i\mathbf{U}_0\overline{\mathbf{U}_0^t}\mathbf{A}_i^t\mathbf{M}^{-1}\left(\sigma_{ii}^{rr}\right)^2 \tag{39}$$

The assumption going to the degeneration of the system (35), (36) into a hyper-diagonal structure is too presumptuous. It would be possible for high ω because $S^{rr}(\omega)$, $S^{\chi\chi}(\omega)$ are quickly decreasing, see (29). However, the spectral densities could have sharp local peaks (e.g.white noise filtered through the 2nd order filter) and the system (35), (36) is recomendable. This system is very large because it contains $2(m+n-1)(m+n)$ unknowns. Every hyperelement of the matrix is the matrix $\mathbf{Q}(\omega)$ multiplied by the coefficient $S_{kj}^{rr}(\omega)$ or $S_{kj}^{r\chi}(\omega)$ according to the position of the hyperelement. Thus, the hyperelements of the inverse matrix may be written according to Kronecker algebra in the form

$$\mathbf{Q}(\omega)S_{kj}^{\alpha\beta}(\omega) \longrightarrow \mathbf{Q}^{-1}(\omega)F_{kj}^{\alpha\beta}(\omega) \; ; \qquad \alpha,\beta = r,\chi \tag{40}$$

where $F_{kj}^{\alpha\beta}(\omega)$ is the corresponding element of the inverse matrix formed from the elements $S_{kj}^{\alpha\beta}(\omega)$.

The matrix set up of elements $S_{kj}^{\alpha\beta}(\omega)$ may be written with respect to (29)

$$\left| \begin{array}{ccc} \mathbf{E}^{rr}(\omega)\cdot S^{rr}(\omega) & ; & e^{r\chi}(\omega)\cdot S^{r\chi}(\omega) \\ \overline{e^{r\chi t}(\omega)\cdot S^{r\chi}(\omega)} & ; & 1\cdot S^{\chi\chi}(\omega) \end{array} \right| \tag{41}$$

$\mathbf{E}^{rr}(\omega)$ — square matrix of "correlation coefficients" with $(m+n-1)\times(m+n-1)$ elements, there are 1 on the main diagonal and $\exp(|i-j|\cdot d\cdot w_c)$ on every super or sub diagonal

$e^{r\chi}(\omega)$ — column vector with $(m+n-1)$ elements, see (29)

If the inverse matrix to (41) is multiplied by right hand sides of (35), (36) a complete algorithm for the solution $\mathbf{U}_{jr}(\omega)$ and $\mathbf{U}_{0\chi}(\omega)$ arises. The inverse matrix $\mathbf{E}^{rr}(\omega)$ depends only on the length of the element d and it may be easy calculated. The inverse matrix $\mathbf{Q}^{-1}(\omega)$ is provided by (30).

The structure of (41) and of right hand sides (35), (36) shows that the vectors $\mathbf{U}_{jr}(\omega)$ and $\mathbf{U}_{0\chi}(\omega)$ are dependent on the external load. This fact makes impossible to define them directly as characteristics of the basic system without perturbations. However this dependency is very weak being given only by mutual spectral density $S^{r\chi}(\omega)$ influencing only unimportant terms while the principal ones remain independent from this processes. This fact can be simply shown by putting $S^{r\chi}(\omega) = 0$.

The structure of (41) and (35), (36) enables to study the other special cases: one or both input random processes are zero, special mutual relation between foundation and track imperfections etc.

Although the case (37) is not too much realistic it enables to show the physical meaning of $\mathbf{U}_{jr}(\omega)$ and $\mathbf{U}_{0\chi}(\omega)$. The vectors $\mathbf{U}_{jr}^0(\omega)$ represent the deflection of an ideal beam subjected at the j—th element by an unit perturbation of foundation matrix multiplied by the "importance factor" given by the mean value of the response at point j . The norm $\mathbf{U}_{jr}^0(\omega)$ diminishes with increasing $|j|$. The vector $\mathbf{U}_{0\chi}^0(\omega)$ represents the deflection due to single harmonic force acting at point $j = 0$.

5. Conclusion

The railway track and its substructure is idealized by an infinitely long beam subjected to a moving lumped mass. The speed of the movement is held constant. The beam rests on Winkler foundation the stiffness of which is a random function of the length coordinate. The surface of rails possesses random track irregularities.

Stochastic finite element method in version for a non–selfadjoint problem has been presented and a uniqueness and convergence of such solution has been shown.

At least a diffuse model of imperfections should be assumed. Only in this case, it is possible to keep the law of energy equilibrium for infinitely long systems.

The difference between the classic solution without imperfections and the mathematical mean value of the response is growing with increasing distance from the point of load application and in a certain distance the energy transits from the deterministic into the stochastic component of the response.

The deflection of rails is charged by greater dispersion than their stresses. The track imperfections wear out the rail vehicles more than the rails. The overall dynamic and fatigue effects of vehicles on track with imperfections are greater than for ideal case without imperfections.

Acknowledgement - *The supports of the Grant Agency of the Czech Republic under grant No.103/96/0017 and of the Grant Agency of the Academy of Sciences of the Czech Republic, grant No.271406 are gratefully acknowledged.*

6. References

1. Kenney, J.T.: Steady – State Vibrations of Beam on Elastic Foundation for Moving Load, *Journal of Appl. Mech.*, 21, No 4 (1954), pp. 359–364.

2. Frýba, L.: *Vibration of Solids and Structures under Moving Loads*, Noordoff International Publishing, Groningen, 1972.

3. Filippov, A.P.: *Vibrations of Deformable Systems* (in Russian), Mashinostroienie, Moskva, 1970.

4. Frýba, L.: Random Vibration of a Beam on Elastic Foundation under Moving Force, in I. Elishakoff and R.H. Lyon (eds.), *Random Vibration - Status and Recent Developments*, Elsevier, Amsterdam (1986), pp. 127–147.

5. Frýba, L., Nakagiri, S., Yoshikawa, N.: Stochastic Finite Elements for a Beam on a Random Foundation with Uncertain Damping under a Moving Force, *Journal of Sound and Vibration*, 163, No 1 (1993), pp. 31–45.

6. Pugachev, V.S., Sinitsyn, I.N.: *Stochastic Differential Systems — Analysis and Filtering*, J.Wiley, Chichester, 1987.

7. Náprstek, J.: Dispersion of Longitudinal Waves Propagating in a continuum with a Randomly Perturbated Parameters, in S.Prakash (editor), *3rd Int. Conf. on Recent Adv. in Geo. Earthquake Engineering and Soil Dynamics*, Univ. of Missouri Rolla, (1995), pp. 705–708.

8. Náprstek, J., Frýba, L.: Interaction of a Long Beam on Stochastic Foundation with a Moving Random Load, in N.S.Ferguson, H.F.Wolfe and C.Mei (eds.), *5th Int. Conf. on Recent Advances in Structural Dynamics*, Inst. of Sound and Vibr. Res., Southampton, vol II, (1994), pp. 714–723.

9. Náprstek, J., Frýba, L.: Stochastic Modelling of Track and its Substructure, in K.Knothe, S.L.Grassie and J.A.Elkins (eds.), *Int. Jour. of Vehicle System Dynamics*, Suppl. 24, 1995, pp. 297–310.

10. Zienkiewcz, O.C., Taylor, R.L.: *The Finite Element Method*, vol. 1, 2, McGraw–Hill, London, New York, 1989.

11. Michlin S.G.: *Variational Methods in Mathematical Physics* (in Russian), Gostechizdat, Moskva, 1957.

MOMENT EQUATIONS FOR NON-LINEAR SYSTEMS UNDER RENEWAL-DRIVEN RANDOM IMPULSES WITH GAMMA-DISTRIBUTED INTERARRIVAL TIMES

S.R.K. NIELSEN
Department of Building Technology and Structural Engineering
Aalborg University, Sohngaardsholmsvej 57, 9000 Aalborg, Denmark

R. IWANKIEWICZ
Institute of Materials Science and Applied Mechanics
Technical University of Wrocław, Wybrzeże Wyspiańskiego 27, 50
370 Wrocław, Poland

AND

P.S. SKJÆRBÆK
Department of Building Technology and Structural Engineering
Aalborg University, Sohngaardsholmsvej 57, 9000 Aalborg, Denmark

Abstract

The moment equations technique is devised for non-linear dynamic systems subjected to random trains of impulses driven by an ordinary renewal point process with gamma-distributed integer parameter interarrival times (Erlang process). Since the renewal point process has not independent increments the state vector of the system, consisting of the generalized displacements and velocities, is not a Markov process. Based on the fact that for this class of renewal processes the renewal events are every kth Poisson events (k - being the integer parameter of the gamma distribution) the renewal impulse process is recast in such a way as to express it in terms of the stationary Poisson counting process. This results in the introduction of additional state variables, for which the stochastic equations are also formulated. The resulting state vector augmented by the additional variables is now a Markov vector process.

Next, the equations for the joint central moments of the state variables are obtained based on the generalized Itô's differential rule valid for Poisson driven processes. As the example problem the Duffing oscillator is considered subjected to the renewal impulse processes with $k = 2$, $k = 3$ and $k = 4$. The cumulant neglect closure is used to truncate the equations for moments at fourth order moments. The computed response mean values and variances are verified against the results of Monte Carlo simulations.

A. Naess and S. Krenk (eds.), IUTAM Symposium on Advances in Nonlinear Stochastic Mechanics, 331–340.
© *1996 Kluwer Academic Publishers.*

1   Introduction

In some problems of engineering the excitation, or its discontinuous component, may be characterized as a train of pulses arriving at random times.

In the simplest approach such excitations may be modelled as Poisson-distributed trains of impulses corresponding to exponential distributed inter-arrival times. The dynamic response of non-linear systems with polynomial non-linearities under such excitations has previously been considered by the authors combined with an ordinary cumulant-neglect closure technique [1], or with a modified cumulant-neglect closure technique [2].

In the paper [3] the technique of moment equations has been extended from a Poisson to a renewal impulse process, with gamma distributed, with $k = 2$, interarrival times of impulses (an Erlang process with $k = 2$). The aim of the present paper is further extension of the moment equations technique to the whole class of renewal driving processes, with gamma distributed, with an arbitrary integer parameter k, interarrival times. However the approach presented herein is completely different and new.

A suitable transformation is proposed which allows to recast the original renewal driven system into an equivalent augmented dynamic system driven by a compound Poisson process. Different cosine and sine transformations of the Poisson counting process constitute the additional state variables. At the expense of augmenting the state vector the original non-Markov problem is then converted to a Markov one, for which effective solution procedures are available.

2   Governing integro-differential equations for renewal driven non-Markov problems

Let us confine the considerations to a class of Erlang renewal processes, i.e. the ones for which the interarrival times $X_n = t_n - t_{n-1}$ are independent, gamma distributed random variables, i.e. $G(k-1, \nu)$, with the probability density function given by

$$g(t) = \nu^k t^{k-1} \exp(-\nu t)/(k-1)!, \quad t > 0 \tag{1}$$

where $k = 1, 2, 3, \ldots$. The case $k = 1$ corresponds to a negative exponential density function, hence it is the case of a Poisson process. Since the gamma distributed random variable with parameters $k-1, \nu$ has the same distribution as the sum of k independent, negative exponential distributed random variables, with parameter ν, the events driven by an Erlang process with parameter k can be regarded as every kth Poisson events, taken out of a stationary Poisson process with the mean arrival rate ν, cf. e.g. [4].

The idea is to recast the renewal-driven impulse process, or the excitation term of the dynamic system in such a way as to obtain a non-zero impulse magnitude for every $k, 2k, 3k, \ldots$ i.e. every kth Poisson event and zero magnitudes for all other Poisson events. The Poisson counting process $\{N(t), t \in [0, \infty[\}$ is defined as the number of events in the time interval $[0, t[$, hence the additional assumption is made: $\Pr\{N(0) = 0\} = 1$.

The governing stochastic equations can then be written

$$\frac{d}{dt}\mathbf{Y}(t) = \mathbf{c}\left(\mathbf{Y}(t), t\right) + \mathbf{d}(\mathbf{Y}(t), t) \sum_{i=1}^{N(t)} \rho(N\left(t_i\right)) P_i \delta\left(t - t_i\right) \tag{2}$$

where $\mathbf{c}\left(\mathbf{Y}(t), t\right)$ is the drift vector, $\mathbf{d}(\mathbf{Y}(t), t)$ is an analogue of the diffusion vector in the white noise driven problems and $\rho\left(N(t_i)\right)$ is the required transformation of the Poisson counting process $N(t_i)$, such that $\rho\left(N(t_i)\right) = 1$ for every kth Poisson event and $\rho\left(N(t_i)\right) = 0$ for all other Poisson events. $N(t_i)$ is the number of past Poisson events, not including the one which occurs at the time t_i.

The stochastic integro-differential equations governing the system state vector can alternatively be written as

$$d\mathbf{Y}(t) = \mathbf{c}\left(\mathbf{Y}(t), t\right) dt + \mathbf{d}(\mathbf{Y}(t), t) \rho\left(N(t)\right) \int_{\mathcal{P}} p M(dt, t, dp, p) \tag{3}$$

where $M(dt, t, dp, p)$ is a Poisson random measure [5], which gives the random number of impulses in the time interval $[t, t + dt[$ with the random magnitudes from the interval $[p, p + dp[$. The expectation of this measure is, in the case of the stationary Poisson process

$$E[M(dt, t, dp, p)] = \nu f_P(p) dt dp \tag{4}$$

where ν is the constant mean arrival rate of impulses and $f_P(p)$ is the probability density function of the random impulse magnitude P.

3 Converting the non-Markov problem to a Markov one

The transformation satisfying the required property is found to be

$$\rho\left(N(t)\right) = \frac{1}{k} \sum_{j=0}^{k-1} \exp\left(i 2\pi \frac{j(N(t) + 1)}{k}\right) = \frac{1}{k} \sum_{j=0}^{k-1} U_j \tag{5}$$

For $N(t) = 0, 1, 2, \ldots, k-1$ the first term at the right-hand side of (5) is the Discrete Fourier Transform of the k-dimensional sequence $\{0, 0, \ldots, 0, 1\}$. From the periodicity properties it follows that

$$\rho\left(N(t)\right) = \begin{cases} 1 & , \quad \text{for } N(t) = k-1, \; 2k-1, \; 3k-1, \ldots \\ 0 & , \quad \text{else} \end{cases} \tag{6}$$

which means that $\rho\left(N(t)\right) = 1$ as every kth Poissonian impulse arrives.

Seeing that $U_j = U_{k-j}^*$, where $*$ denotes the complex conjugate, the right-hand side of (5) can be evaluated as

$$\rho\left(N(t)\right) = \begin{cases} \frac{1}{k}\left(1 + 2\sum_{j=1}^{k_0-1} C_j\right) & , \quad \text{for } k \text{ odd} \\ \frac{1}{k}\left(1 + 2\sum_{j=1}^{k_0-1} C_j + C_{k_0}\right) & , \quad \text{for } k \text{ even} \end{cases} \tag{7}$$

where

$$k_0 = \left[\frac{k+1}{2}\right] \qquad (8)$$

[...] being the integer part and

$$C_j = \Re(U_j) = \cos\left(2\pi\frac{j(N(t)+1)}{k}\right), \quad j = 1, 2, \ldots, k_0 - 1 \qquad (9)$$

$$S_j = \Im(U_j) = \sin\left(2\pi\frac{j(N(t)+1)}{k}\right), \quad j = 1, 2, \ldots, k_0 - 1 \qquad (10)$$

$$C_{k_0} = \exp\left(i\pi(N(t)+1)\right) = (-1)^{N(t)+1} = -\cos\left(\pi N(t)\right), \quad \text{for } k \text{ even} \quad (11)$$

These transformations of a Poisson counting process $N(t)$ will be regarded as additional, auxiliary, state variables. The stochastic equations for these variables are obtained from

$$dU_j(t) = U_j(t + dt) - U_j(t) =$$

$$\exp\left(i2\pi\frac{i(N(t+dt)+1)}{k}\right) - \exp\left(i2\pi\frac{i(N(t)+1)}{k}\right) =$$

$$\exp\left(i2\pi\frac{i(N(t)+dN(t)+1)}{k}\right) - \exp\left(i2\pi\frac{i(N(t)+1)}{k}\right) = \qquad (12)$$

$$U_j(t)\left(\exp\left(i2\pi\frac{i}{k}dN(t)\right) - 1\right) =$$

$$U_j(t)\left(\exp\left(i2\pi\frac{i}{k}\right) - 1\right)dN(t), \quad j = 1, 2, \ldots, k_0$$

The equivalence of the last two statements of (12) follows from the fact that the right-hand sides give the same result for both $dN(t) = 0$ and $dN(t) = 1$. Specifically, the equations for the real and imaginary parts become

$$dC_j = \left(C_j\left(\cos\left(2\pi\frac{j}{k}\right) - 1\right) - S_j\sin\left(2\pi\frac{j}{k}\right)\right)dN(t), \quad j = 1, 2, \ldots, k_0 - 1 \qquad (13)$$

$$dS_j = \left(C_j\sin\left(2\pi\frac{j}{k}\right) + S_j\left(\cos\left(2\pi\frac{j}{k}k\right) - 1\right)\right)dN(t), \quad j = 1, 2, \ldots, k_0 - 1 \qquad (14)$$

$$dC_{k_0} = -2C_{k_0}dN(t), \quad \text{for } k \text{ even} \qquad (15)$$

It is seen that the additional state variables C_j, S_j and for k even also C_{k_0} have been introduced. The state vector augmented by these new variables is governed by the stochastic equations

$$d\mathbf{Z}(t) = \mathbf{a}\left(\mathbf{Z}(t), t\right)dt + \int_P \mathbf{b}\left(\mathbf{Z}(t), t, p\right)M(dt, t, dp, p) \qquad (16)$$

where for k even

$$\mathbf{Z}(t) = \begin{bmatrix} Y(t) \\ C_1 \\ S_1 \\ C_2 \\ S_2 \\ \cdot \\ \cdot \\ \cdot \\ C_{k_0-1} \\ S_{k_0-1} \\ C_{k_0} \end{bmatrix}, \quad \mathbf{a}(\mathbf{Z}(t),t) = \begin{bmatrix} c(Y(t),t) \\ 0 \\ 0 \\ 0 \\ 0 \\ \cdot \\ \cdot \\ \cdot \\ 0 \\ 0 \\ 0 \end{bmatrix} \tag{17}$$

$$\mathbf{b}(\mathbf{Z}(t),t,p) = \begin{bmatrix} \frac{1}{k}\left(1 + 2\sum_{j=1}^{k_0-1} C_j + C_{k_0}\right)d(Y(t),t)p \\ C_1\left(\cos\left(2\pi\frac{1}{k}\right) - 1\right) - S_1\sin\left(2\pi\frac{1}{k}\right) \\ C_1\sin\left(2\pi\frac{1}{k}\right) + S_1\left(\cos\left(2\pi\frac{1}{k}\right) - 1\right) \\ C_2\left(\cos\left(2\pi\frac{2}{k}\right) - 1\right) + S_2\sin\left(2\pi\frac{2}{k}\right) \\ C_2\sin\left(2\pi\frac{2}{k}\right) + S_2\left(\cos\left(2\pi\frac{2}{k}\right) - 1\right) \\ \cdot \\ \cdot \\ \cdot \\ C_{k_0-1}\left(\cos\left(2\pi\frac{k_0-1}{k}\right) - 1\right) - S_{k_0-1}\sin\left(2\pi\frac{k_0-1}{k}\right) \\ C_{k_0-1}\sin\left(2\pi\frac{k_0-1}{k}\right) + S_{k_0-1}\left(\cos\left(2\pi\frac{k_0-1}{k}\right) - 1\right) \\ -2C_{k_0} \end{bmatrix} \tag{18}$$

and for k odd

$$\mathbf{Z}(t) = \begin{bmatrix} Y(t) \\ C_1 \\ S_1 \\ C_2 \\ S_2 \\ \cdot \\ \cdot \\ \cdot \\ C_{k_0-1} \\ S_{k_0-1} \end{bmatrix}, \quad \mathbf{a}(\mathbf{Z}(t),t) = \begin{bmatrix} c(Y(t),t) \\ 0 \\ 0 \\ 0 \\ 0 \\ \cdot \\ \cdot \\ \cdot \\ 0 \\ 0 \end{bmatrix} \tag{19}$$

$$
\mathbf{b}\big(\mathbf{Z}(t),t,p\big) =
\begin{bmatrix}
\frac{1}{k}\left(1 + 2\sum_{j=1}^{k_0-1} C_j\right) \mathbf{d}(\mathbf{Y}(t),t)p \\
C_1\left(\cos\left(2\pi\frac{1}{k}\right) - 1\right) - S_1\sin\left(2\pi\frac{1}{k}\right) \\
C_1\sin\left(2\pi\frac{1}{k}\right) + S_1\left(\cos\left(2\pi\frac{1}{k}\right) - 1\right) \\
C_2\left(\cos\left(2\pi\frac{2}{k}\right) - 1\right) + S_2\sin\left(2\pi\frac{2}{k}\right) \\
C_2\sin\left(2\pi\frac{2}{k}\right) + S_2\left(\cos\left(2\pi\frac{2}{k}\right) - 1\right) \\
\vdots \\
\vdots \\
C_{k_0-1}\left(\cos\left(2\pi\frac{k_0-1}{k}\right) - 1\right) - S_{k_0-1}\sin\left(2\pi\frac{k_0-1}{k}\right) \\
C_{k_0-1}\sin\left(2\pi\frac{k_0-1}{k}\right) + S_{k_0-1}\left(\cos\left(2\pi\frac{k_0-1}{k}\right) - 1\right)
\end{bmatrix}
\tag{20}
$$

The state vector $\mathbf{Z}(t)$, augmented by additional, auxiliary state variables, as governed by equation (16) is driven by a Poisson process, and hence it is a Markov vector process.

4 Differential equations for moments

Equations for the mean values $\mu_i(t) = E[Z_i(t)]$ are obtained by direct averaging of the governing stochastic equations

$$
\dot{\mu}_i(t) = E[a_i(\mathbf{Z}(t),t)] + \nu \int_{\mathcal{P}} E\left[b_i\big(\mathbf{Z}(t),t,p\big)\right] f_P(p)dp
\tag{21}
$$

where $P(t)$ is the random magnitude of the impulse arriving in $[t, t+dt[$.

The following form of the differential rule will prevail [5]

$$
df\left(\mathbf{Z}^0(t)\right) = \mathcal{K}_{\mathbf{z},t}^T\left[f\left(\mathbf{Z}^0(t)\right)\right]dt = \sum_i a_i^0\left(\mathbf{Z}^0(t),t\right)\frac{\partial f\left(\mathbf{Z}^0(t)\right)}{\partial Z_i^0}dt
$$
$$
+ \int_{\mathcal{P}}\left[f\left(\mathbf{Z}^0(t) + \mathbf{b}^0\big(\mathbf{Z}^0(t),p\big)\right) - f\left(\mathbf{Z}^0(t)\right)\right]M(dt,t,dp,p)
\tag{22}
$$

where
$$
dZ_i^0(t) = Z_i(t) - \mu_i(t)
\tag{23}
$$
$$
a_i^0\left(\mathbf{Z}^0(t),t\right) = a_i\left(\mathbf{Z}^0(t) + \mu(t),t\right) - E[a_i\left(\mathbf{Z}^0(t) + \mu(t),t\right)] - \nu E\left[b_i\big(\mathbf{Z}(t),t,P(t)\big)\right]
\tag{24}
$$
$$
b_i^0\left(\mathbf{Z}^0(t),t,p\right) = b_i\left(\mathbf{Z}^0(t) + \mu(t),t,p\right) = b_i\big(\mathbf{Z}(t),t,p\big)
\tag{25}
$$

Using $f\left(\mathbf{Z}^0(t)\right) = Z_{i_1}^0(t)\ldots Z_{i_N}^0(t)$ in (22) and next taking the expection, the differential equation for the joint central moment $\kappa_{i,\ldots i_N}(t) = E\left[Z_{i_1}^0(t)\cdots Z_{i_N}^0(t)\right]$ is obtained. For $N = 2,3,4$ is obtained.

$$\dot{\kappa}_{ij}(t) = 2 \left\{ E \left[Z_i^0 \left(a_j^0(\mathbf{Z}^0(t)) + \nu b_j^0(\mathbf{Z}^0(t), P(t)) \right) \right] \right\}_s$$
$$+ \nu E \left[b_i^0(\mathbf{Z}^0(t), P(t)) b_j^0(\mathbf{Z}^0(t), P(t)) \right] \tag{26}$$

$$\dot{\kappa}_{ijk}(t) = 3 \left\{ E \left[Z_i^0 Z_j^0 \left(a_k^0(\mathbf{Z}^0(t)) + \nu b_k^0(\mathbf{Z}^0(t), P(t)) \right) \right] \right\}_s$$
$$+ 3\nu \left\{ E \left[Z_i^0 b_j^0(\mathbf{Z}^0(t), P(t)) b_k^0(\mathbf{Z}^0(t), P(t)) \right] \right\}_s \tag{27}$$
$$+ \nu E \left[b_i^0(\mathbf{Z}^0(t), P(t)) b_j^0(\mathbf{Z}^0(t), P(t)) b_k^0(\mathbf{Z}^0(t), P(t)) \right]$$

$$\dot{\kappa}_{ijkl}(t) = 4 \left\{ E \left[Z_i^0 Z_j^0 Z_k^0 \left(a_l^0(\mathbf{Z}^0(t)) + \nu b_l^0(\mathbf{Z}^0(t), P(t)) \right) \right] \right\}_s$$
$$+ 6\nu \left\{ E \left[Z_i^0 Z_j^0 b_k^0(\mathbf{Z}^0(t), P(t)) b_l^0(\mathbf{Z}^0(t), P(t)) \right] \right\}_s$$
$$+ 4\nu \left\{ E \left[Z_i^0 b_j^0(\mathbf{Z}^0(t), P(t)) b_k^0(\mathbf{Z}(t), P(t)) b_l^0(\mathbf{Z}^0(t), P(t)) \right] \right\}_s \tag{28}$$
$$+ \nu E \left[b_i^0(\mathbf{Z}^0(t), P(t)) b_j^0(\mathbf{Z}^0(t), P(t)) b_k^0(\mathbf{Z}^0(t), P(t)) b_l^0(\mathbf{Z}^0(t), P(t)) \right]$$

where $\{\ldots\}_s$ denotes the symmetrizing operation, i.e. the arithmetic mean of all the terms similar to the one in the brackets, obtained by all possible permutations of the indices, the number of these terms being equal to the factor in front of the brackets.

The initial conditions associated with the equations for moments are non-zero, since some of the auxiliary state variables C_j, S_j and C_{k_0} as defined by equations (9)–(11), and hence their mean values assume non-zero initial values.

5 Example: Duffing oscillator to renewal impulse process, with k = 2, k = 3 and k = 4

Consider a Duffing oscillator, where $\mathbf{Y}(t)$, $\mathbf{c}(\mathbf{Y}(t), t)$ and $\mathbf{d}(\mathbf{Y}(t), t)$ of equation (2) are given by

$$\mathbf{Y}(t) = \begin{bmatrix} Y(t) \\ \dot{Y}(t) \end{bmatrix}, \mathbf{c}(\mathbf{Y}(t), t) = \begin{bmatrix} \dot{Y}(t) \\ -2\zeta\omega_0\dot{Y}(t) - \omega_0^2 Y(t) - \varepsilon\omega_0^2 Y^3(t) \end{bmatrix}, \mathbf{d} = \begin{bmatrix} 0 \\ 1 \end{bmatrix} \tag{29}$$

ζ is the damping ratio, ω_0 is the circular eigenfrequency of the corresponding linear oscillator and ε is the non-linearity parameter.

The mean value and the variance of the stationary response of a linear oscillator to Poissonian impulses with the mean arrival rate $\frac{\nu}{k}$ is, given by

$$E[Y] = \mu_1 = \frac{\nu}{k} \frac{E[P]}{\omega_0^2} \quad , \quad \sigma_Y^2 = \kappa_{11} = \frac{\nu}{k} \frac{E[P^2]}{4\zeta\omega_0^3} \tag{30}$$

In the example the fraction $\frac{\nu}{k}$ is kept constant. The linear responses of the comparable Poisson driven system is then the same. Moreover the data for the random variable P is assumed in such a way that σ_Y^2 as given by (30) has a unit value.

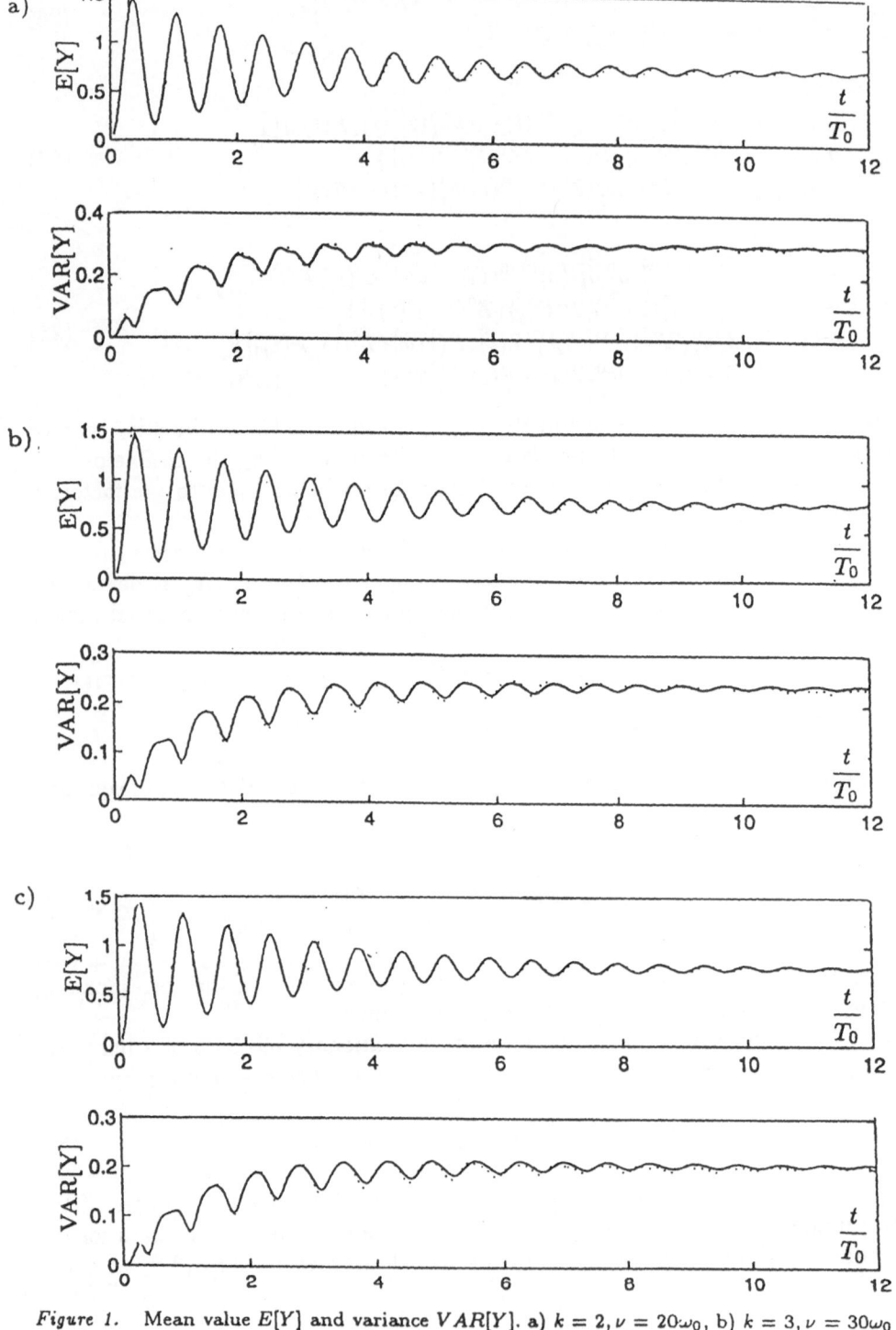

Figure 1. Mean value $E[Y]$ and variance $VAR[Y]$. a) $k = 2, \nu = 20\omega_0$, b) $k = 3, \nu = 30\omega_0$ and c) $k = 4, \nu = 40\omega_0$. ——: Analytical results. $\cdots$: Simulated results.

The data assumed for the Duffing oscillator is: $\omega_0 = 1s^{-1}, \zeta = 0.05, \varepsilon = 0.5$.

With this value of the parameter ε the non-linearity should be regarded as quite strong, since the mean value and the variance of the response of the Duffing oscillator to the Poisson train of impulses are then substantially different from the statistics of the response of a linear oscillator, cf. [1], [2]. The values of ν for the cases $k = 2$, $k = 3$ and $k = 4$ are assumed, respectively as $\nu = 20\omega_0$, $\nu = 30\omega_0$ and $\nu = 40\omega_0$ The impulses magnitudes are assumed to be Rayleigh-distributed random variables, $P \sim R(\sigma_P^2)$, with $\sigma_P = 0.1s^{-1}$.

The equations for moments up to and including the fourth order moments have been derived. Since the drift terms $a_i^0(\mathbf{Z}^0(t))$ are the cubic forms in the state variables, the fifth order moments appear in the equations for third order moments, see (27), and fifth- and sixth order moments appear in the equations for fourth order moments, see (28). These redundant moments have been evaluated with the help of the modified cumulant neglect closure scheme devised by the authors for Poisson driven pulse problems [2].

To verify the approximate analytical results, the Monte Carlo simulations have been performed. The simulated results were obtained based on averaging over an ensemble of 32000 independent response sample curves, each obtained by numerical integration of governing equation of motion (2).

Both the differential equations for moments and the governing equation of motion (in the simulation technique) have been solved with the help of 4th order Runge-Kutta numerical integration technique. In order to obtain the excitation sample functions, the interarrival times of impulses are generated with the help of negative exponential distributed random variables. In each case of a renewal process the interarrival times are constructed as a sum of k auxiliary negative exponential distributed variates. Next the impulses magnitudes generated from a Rayleigh distribution are assigned to the impulses arrival times. The response sample curves are obtained by numerical integration of the homogeneous governing equation of motion (2) between the impulses arrival times, whereas at each arrival time the velocity is increased by a jump, which gives the updated initial condition for the next interarrival time interval.

The results for the mean value $\mu_1(t) = E[Y]$ and variance $\kappa_{11}(t) = \sigma_Y^2(t) = VAR[Y]$ evaluated for the cases $k = 2$, $k = 3$ and $k = 4$ are shown in Figs. 1a-c, respectively, versus relative time t/T_0, where T_0 is the eigenperiod of the corresponding linear oscillator. Agreement of the approximate analytical results with the simulated ones is very good.

The mean responses of the Duffing oscillator are in all three cases practically the same. It should be noted that the mean response of a Duffing oscillator to a renewal impulses with $k = 2$ is only slightly different (larger) than the response to a comparative Poisson impulse process, cf.[3].

The variance of the response decreases as the parameter k increases.

6 Concluding remarks

The moment equations technique has been devised for non-linear dynamic systems subjected to random trains of impulses driven by an ordinary renewal point process with gamma-distributed, with integer parameter k, interarrival times. The renewal impulse process is recast by expressing it in terms of a stationary

Poisson counting process, which results in the introduction of additional state variables, for which the governing stochastic equations are also formulated. The augmented state vector is a Poisson driven process, hence at the expense of augmenting the state vector, the non-Markov problem is converted to a Markov one. Equations for the joint central moments of the response are derived based on the generalized Itô's differential rule for Poisson driven processes.

The approximate analytical results obtained for a Duffing oscillator problem are verified against the Monte Carlo simulations. The developed moment equations technique, combined with a modified cumulant neglect closure technique, appears to be valid and effective.

Even though the considered trains of impulses are rather dense, i.e. the mean arrival rates are relatively high, the response (variance) behaviour differs very much from the Poisson one (the latter being in the case of dense trains of impulses close to a Gaussian).

Acknowledgements

The present research was made within the project **Response and Reliability of Partially Damaged Reinforced Concrete Structures subject to Random Dynamic Loads** supported by the Danish Technical Research Council under the grant No. 16-5453-1.

References

1. R. Iwankiewicz and S.R.K. Nielsen, Dynamic response of non-linear systems to Poisson-distributed random impulses, J. Sound and Vibration, 3 (1992) 407-423.

2. R. Iwankiewicz, S.R.K. Nielsen and P. Thoft-Christensen, Dynamic response of non-linear systems to Poisson-distributed pulse trains: Markov approach, Structural Safety, 8 (1990) 223-238.

3. R. Iwankiewicz and S.R.K. Nielsen, Dynamic response of non-linear systems to renewal-driven random pulse trains, Int. J. Non-linear Mechanics, 29 (1994) 555-567.

4. S. Osaki, Applied Stochastic System Modelling, Springer-Verlag (1992).

5. D.L. Snyder, Random Point Processes, John Wiley, New York (1975).

MICROMECHANICALLY BASED CONSTITUTIVE LAWS AND RANDOM FIELDS IN SOLID MECHANICS: ELASTICITY, PLASTICITY, AND FRACTURE

MARTIN OSTOJA-STARZEWSKI
Department of Materials Science and Mechanics
Michigan State University, East Lansing, MI 48824, USA

1. Introduction

A stochastic formulation of constitutive laws in solid and structural mechanics, as is typically needed in the setting up of stochastic finite elements and differences [1, 2, 3], has to be based on micromechanical analyses. In this paper we give an account of our recent studies on the development and use of such laws in three areas: elasticity, plasticity and fracture/damage phenomena.

2. Elastic Microstructures

Let us consider an elastic *random microstructure* (or *random medium*) to be a family $B = \{B(\omega); \omega \in \Omega \}$ of deterministic media $B(\omega)$, where ω is an indicator of a given realization, and Ω is an underlying sample space. Next, we introduce a *window* $B_\delta(\omega)$ of scale

$$\delta = \frac{L}{d} \tag{1}$$

where L is the window (or sample) size and d is the size of a heterogeneity, Fig. 1.

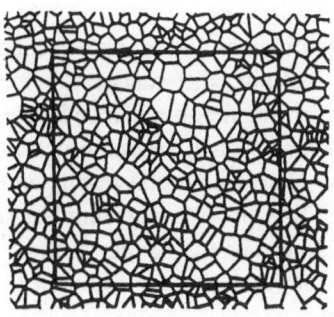

Figure 1. One realization $B(\omega)$ of a random medium and the window of scale δ.

A. Naess and S. Krenk (eds.), IUTAM Symposium on Advances in Nonlinear Stochastic Mechanics, 341–350.
© 1996 Kluwer Academic Publishers.

In order to define effective properties one has to consider δ values larger 1, and two types of boundary conditions:

a) essential (Dirichlet, "displacement-controlled")

$$u = \varepsilon \cdot \vec{x} \tag{2}$$

which yield a stiffness tensor $\mathbf{C}_\delta^e$ ('e' stands for essential boundary conditions), where u is the displacement, $\overline{\nabla u}$ is the spatial average strain, and $\vec{x}$ is a position vector;

b) natural (Neumann, "stress-controlled"):

$$\sigma = \vec{t} \cdot \vec{n} \tag{3}$$

which yield the tensor $\mathbf{C}_\delta^n = \left(\mathbf{S}_\delta^n\right)^{-1}$ ('n' stands for natural boundary conditions), where $\vec{\sigma}$ is stress, t is spatial average traction, and $\vec{n}$ is the outer unit normal to the window's boundary. In the above we employ ⟶ for a vector, boldface for a second-rank tensor, and an overbar for a spatial average over the window domain. Note that $\delta \to \infty$ is the conventional limit typically sought in the effective medium theories, while numerical simulations correspond to some finite δ.

It is important to observe that $\mathbf{C}_\delta^e$ is different from $\mathbf{C}_\delta^n$ as it provides an *upper estimate*, while the latter a *lower estimate* on the effective stiffness of the given specimen. In fact, it can be shown [4, 5] from the variational principles of minnimum potential energy and complementary energy that the effective macroscopic stiffness tensor $\mathbf{C}^{eff}$ is bounded by two tensors $\langle \mathbf{C}_\delta^e \rangle$ and $\langle \mathbf{S}_\delta^n \rangle$, where $\langle\ \rangle$ denotes the ensemble averaging, i.e., averaging over the space of all realizations Ω. Both bounds get ever closer as the scale δ increases, so that one has a hierarchy of δ-dependent bounds on $\mathbf{C}^{eff}$

$$\mathbf{C}^R \equiv \left(\mathbf{S}^R\right)^{-1} \equiv \langle \mathbf{S}_1^n \rangle^{-1} \leq \langle \mathbf{S}_{\delta'}^n \rangle^{-1} \leq \langle \mathbf{S}_\delta^n \rangle^{-1} \leq \mathbf{C}^{eff} \leq \langle \mathbf{C}_\delta^e \rangle \leq \langle \mathbf{C}_{\delta'}^e \rangle \leq \langle \mathbf{C}_1^e \rangle \equiv \mathbf{C}^V$$
$$\forall \delta' < \delta \tag{4}$$

In the following we assume that the microstructure has spatially homogeneous and isotropic statistics, which implies that $\mathbf{C}^{eff}$ is isotropic, i.e. $\mathbf{C}^{eff} = \delta C^{eff}$; δ is the Kronecker delta. In (4) $\mathbf{C}^V$ and $\mathbf{C}^R$ denote the (elementary) Voigt and Reuss bounds and correspond to taking a window on the smallest scale $\delta = 1$. In other words, effective response depends on the boundary conditions, and the influence of the latter disappears as the body becomes infinite. Here, for two second rank tensors $\mathbf{A}$ and $\mathbf{B}$, the order relation $\mathbf{B} \leq \mathbf{A}$ means $\vec{t} \cdot \mathbf{B} \cdot \vec{t} \leq \vec{t} \cdot \mathbf{A} \cdot \vec{t}$ for any vector $\vec{t} \neq 0$.

In order to solve the field equations of a two-phase composite we employ a very fine spring network (equivalently, a finite difference) scheme. Numerical examples of calculation of bounds (4) are given for the in-plane and out-of-plane elasticity problems in [6] and [7]; the latter reference also gives second-order as well as two-point statistics of both scale-dependent tensors. Here we present results on the upper and lower bounds for a random two-phase lattice. It is possibly the simplest setting in which to calculate the hierarchy of bounds for a large range of length scales. First, we note that it has a continuum counterpart in a random two-phase chessboard [8], which for a 50% volume fraction of both phases results in

$$C^{eff} = \sqrt{C_1 C_2} \tag{5}$$

where C_1 and C_2 are the properties of both phases. In Figure 2 we show, for the whole range of volume fraction, both bounds $\langle \underline{C}^e_\delta \rangle$ and $\langle \underline{S}^n_\delta \rangle$ for the contrast ratio $C_2/C_1 = 1000$. Window sizes are $\delta = 4$, 10 and 20; also plotted are the Voigt, Reuss and Hashin upper and lower bounds [9]: C^V, C^R, C^H_u, and C^H_l. As expected, one can observe from this figure a progressive improvement in bounds with increasing δ, which leads, already at $\delta = 4$, to estimates comaprable to, or better than, the Hashin bounds.

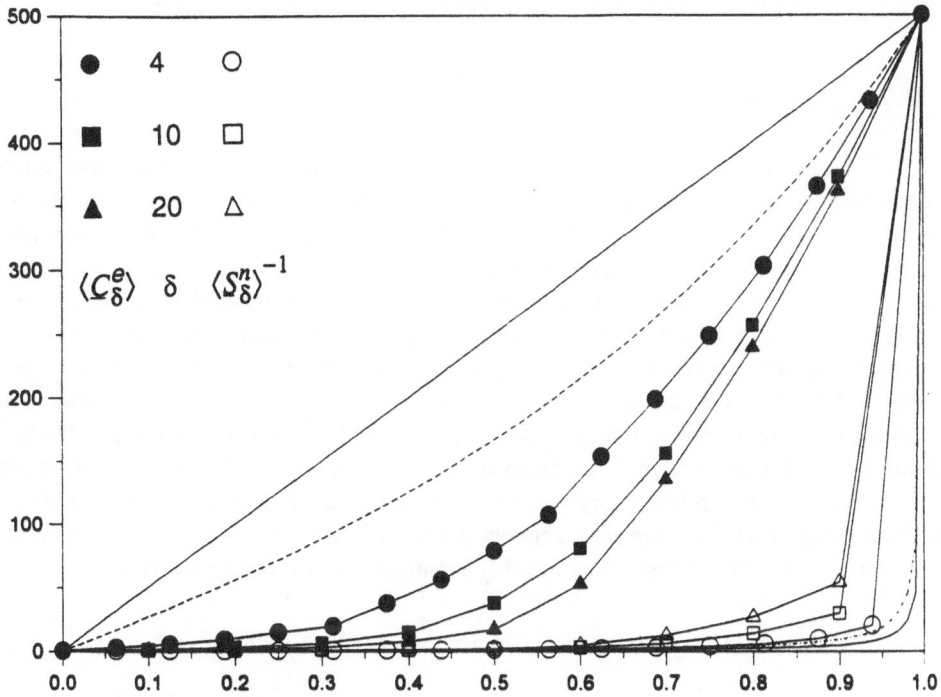

Figure 2. δ-dependent bounds, as well as the Voigt, Reuss, and Hashin bounds, for a random chessboard.

These plots were obtained from simulations involving one node of a lattice per one board of the chessboard. This is, of course, a crude approximation with respect to the singularity problem at the corners and, therefore, does not represent a piecewise continuum system for which the Hashin bounds are correct. Nevertheless, it can be seen that our results agree very well with those of formula (5).

We end this section with function fits of the finite size scaling of $\langle \underline{C}^e_\delta \rangle$ and $\langle \underline{S}^n_\delta \rangle^{-1}$ [10]. It has been found that the bounds scale as follows

$$\langle \underline{C}^e_\delta \rangle \sim \exp\left(-\delta^{-p}\right) \qquad \langle \underline{S}^n_\delta \rangle^{-1} \sim \exp\left(\delta^{-q}\right) \tag{6}$$

where p and q are themselves functions of the contrast α. The latter are actually of hyperbolic forms

$$p(\alpha) = 3.8\alpha^{0.14} \qquad q(\alpha) = 2.4\alpha^{0.59} \tag{7}$$

Note that the limits of $\alpha \rightarrow 1$ and $\alpha \rightarrow \infty$ are correctly recovered here. Based on some recent calculations of random disk composites and random polygon mosaics, we conjecture that these are universal relations for finite size scaling of both bounds in two-phase random systems. We end this section with an observation that the above outlined procedure provides a micromechanical basis for stochastic finite element methods.

3. Plastic Microstructures

It is well known that in case of rigid-plastic media, a boundary value problem may conveniently be solved by a finite difference net of characteristics. Let us consider a situation where, given any two points 1 and 2 that do not lie on the same characteristic, a new point N is to be found. While a homogeneous medium problem results in a unique point N, there is a diffused region of total probability 1, where the new point is located in case of a random plastic medium, Fig. 4a). Considering the initial points in the setting of a material microstructure (Fig. 4b) we see that the spacing of the finite difference net is not infinite with respect to the grain size, and hence, there is a difference in plastic responses of all quadrangle shaped windows corresponding to points 1, 2, and N. Clearly, the spacing of the finite difference net defines a scale δ of a mesocontinuum resolution with respect to the microscale. In fact, any given laboratory experiment is conducted on specimens that are finitely (δ times) larger that the grain size, whereby a scatter in response is observed. Thus, supposing we work with a Mohr-Coulomb (MC) material, whose yield condition is

$$\frac{1}{4}(\sigma_x - \sigma_y)^2 + \tau_{xy}^2 = \frac{(\sin\rho)^2}{4}(\sigma_x + \sigma_y + 2H)^2 \tag{8}$$

the angle of internal friction ρ and the ultimate resistance H are actually functions of δ, and so, we need to introduce a vector random field $\{\rho, H\}_\delta$ where

$$\rho_\delta(\underline{x}, \omega) = \langle \rho_\delta \rangle + \rho_\delta'(\underline{x}, \omega) \qquad \langle \rho_\delta'(\underline{x}, \omega) \rangle = 0$$
$$H_\delta(\underline{x}, \omega) = \langle H_\delta \rangle + H_\delta'(\underline{x}, \omega) \qquad \langle H_\delta'(\underline{x}, \omega) \rangle = 0 \tag{9}$$

The criterion (9) has two well known special cases:
i) Materials of the Tresca (or Huber-Mises) type

$$(\sigma_x - \sigma_y)^2 + 4\tau_{xy}^2 = 4k^2 \tag{10}$$

equivalent, in two dimensions, to plasticity of media governed by the τ_{max} criterion [1, 11], whereby

$$k_\delta(\underline{x}, \omega) = \langle k_\delta \rangle + k_\delta'(\underline{x}, \omega) \qquad \langle k_\delta'(\underline{x}, \omega) \rangle = 0 \tag{11}$$

ii) Cohesionless soils with H = 0 in (10), whereby only the ρ_δ field is required; a need to consider randomness in this type of yield condition has been pointed out in [12].

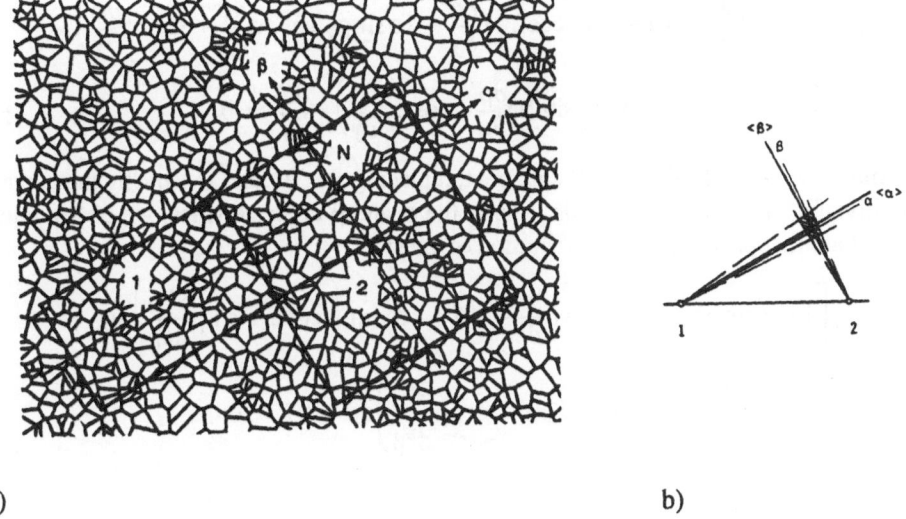

a)   b)

Figure 3. a) Windows and characteristics of a mesocontinuum approximation that are
involved in a finite difference solution, superposed onto the microstructure;
b) scatter in the characteristics emanating from points 1 and 2 towards point N.

In the following we assume:
- scale $\delta \geq r_c$ (a correlation length) so that the random field of material properties may be
treated as white-noise on that scale;
- k_δ is space-homogeneous, ergodic, and has a high signal-to-noise ratio.
For the sake of reference we will also need a *deterministic homogeneous (plastic) medium*
$B_{det} = B(k_{det})$ with $k_{det} = \langle k_\delta \rangle$. The *deterministic homogeneous (plastic) medium* is now
specified as $B_{det} = B(\rho_{det}, H_{det})$ with $\rho_{det} = \langle \rho_\delta \rangle$ and $H_{det} = \langle H_\delta \rangle$.

It follows that any boundary value problem in plasticity of random media is governed
by a stochastic, rather than a deterministic, quasi-linear hyperbolic system. Accordingly, a
finite difference method has to be developed to deal with spatial inhomogeneity of the yield
condition for any body $B(\omega)$, so that a repeated solution, in a Monte Carlo sense, for a number of realizations $\omega \in \Omega$ will result in a stochastic solution.

Following the calssical formulation of continuum plasticity of granular media [13, 14],
two variables σ and φ are now introduced as follows

$$\sigma_x = \sigma(1 + \sin\rho\cos2\varphi) - H, \quad \sigma_y = \sigma(1 - \sin\rho\cos2\varphi) - H, \quad \tau_{xy} = \sigma\sin\rho\sin2\varphi \quad (12)$$

By substituting equations (12) into the equilibrium equations $\sigma_{ij,j} = 0$, we get

$$[1 + \sin\rho\cos2\varphi]\frac{\partial\sigma}{\partial x} + \sin\rho\sin2\varphi\frac{\partial\sigma}{\partial y} - 2\sigma\sin\rho\left(\sin2\varphi\frac{\partial\varphi}{\partial x} - \cos2\varphi\frac{\partial\varphi}{\partial y}\right)$$

$$+ \sigma\cos\rho\left(\cos2\varphi\frac{\partial\rho}{\partial x} + \sin2\varphi\frac{\partial\rho}{\partial y}\right) - \frac{\partial H}{\partial x} = 0 \quad (13)$$

and

$$\sin\rho\sin2\varphi\frac{\partial\sigma}{\partial x} + (1 - \sin\rho\cos2\varphi)\frac{\partial\sigma}{\partial y} + 2\sigma\sin\rho\left(\cos2\varphi\frac{\partial\varphi}{\partial x} + \sin2\varphi\frac{\partial\varphi}{\partial y}\right)$$

$$+ \sigma\cos\rho\left(\sin2\varphi\frac{\partial\rho}{\partial x} - \cos2\varphi\frac{\partial\rho}{\partial y}\right) - \frac{\partial H}{\partial y} = 0 \qquad (14)$$

Setting $\varphi = -\varepsilon$ in (13) and (14) and, next, replacing $\partial\ /\partial x$ and $\partial\ /\partial y$ by the tangential dertivatives $\partial\ /\partial s_\alpha$ and $\partial\ /\partial s_\beta$ along the α and β characteristics, we obtain equations that are independent of the orientation of axes:

$$\left[1 + (\sin\rho)^2 - \sin\rho\cos\rho\frac{ds_\alpha}{ds_\beta}\right]d\sigma + 2\sigma\sin\rho\left(\cos\rho + \sin\rho\frac{ds_\alpha}{ds_\beta}\right)d\varphi$$

$$+ \sigma\cos\rho\left(\sin\rho - \cos\rho\frac{ds_\alpha}{ds_\beta}\right)d\rho - dH = 0 \qquad (15)$$

and

$$\left[-\sin\rho\cos\rho\frac{ds_\beta}{ds_\alpha} + (\cos\rho)^2\right]d\sigma + 2\sigma\sin\rho\left(\sin\rho\frac{ds_\beta}{ds_\alpha} - \cos\rho\right)d\varphi$$

$$- \sigma\cos\rho\left(\cos\rho\frac{ds_\beta}{ds_\alpha} + \sin\rho\right)d\rho - dH = 0 \qquad (16)$$

The corresponding directions of the two families of characteristics, α and β are $dy/dx = \tan(\varphi \mp \varepsilon)$. System (15-16) can now be solved for σ and φ at a new point N by using a finite difference approach. This is, in fact, coupled with the solution for the position x_N, y_N of the new point N according to

$$y_N - y_1 = \frac{1}{2}(x_N - x_1)[\tan(\varphi_1 + \pi/4) + \tan(\varphi_N + \pi/4)]$$

$$y_N - y_2 = \frac{1}{2}(x_N - x_2)[\tan(\varphi_2 - \pi/4) + \tan(\varphi_N - \pi/4)] \qquad (17)$$

In Fig. 4 a) and b) we demonstrate the difference, in a characteristic boundary value problem, between two media: one governed by the MC and another by the HM criterion. In both cases the α-characteristic (going to the right) is the same and specified by

$$y_\alpha = x - \frac{x^2}{10} \qquad (18)$$

whereby the β-characteristic follows according to the type of yield condition used. Moreover, the first case is specified by a 5% noise about the mean $\langle k_\delta \rangle = 0.6$, that is $k_\delta' \in [-0.025, 0.025]$ taken as s a uniform random variate. The second case is specified by 5% noises about the means $\langle \rho_\delta \rangle = 15°$ and $\langle H_\delta \rangle = 0.6$, that is $\rho_\delta' \in [-0.375, 0.375]$ and $H_\delta' \in [-0.015, 0.015]$ both of which are independent uniform random variates. The relatively stronger sensitivity of the MC-type material versus the Tresca-type material, accompanied by a departure from the orthogonality of the slip-line net, is a typical feature here.

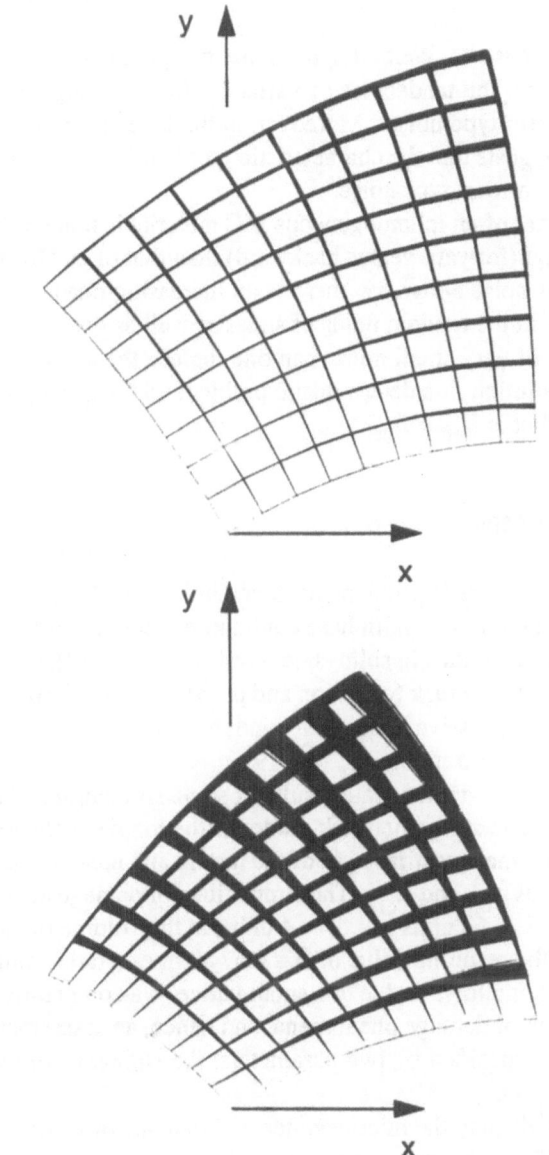

Figure 4. The characteristic boundary value problem for HM (a) and MC (b) material.

An extensive study, focusing especially on the sensitivity of the slip-line net geometry and the stress field, for uniform vis-à-vis Weibull noise in the yield parameters, was conducted in [15]. Follwoing are its principal conclusions:
i) The scatter in dependent field variables increases continuously with decreasing angle ρ and with the increase in the ultimate resistance to uniform three dimensional tension H.
ii) Parameter ρ has a stronger perturbing effect than H.

iii) The average of the inhomogeneous stress distribution is always smaller than that of the deterministic one, and the difference increases with the noise and inhomogeneity of the boundary data.

iv) Although in case of Weibull type noise the position and the field variables have smoother variations, the tendencies of variation do not change compared to those corresponding to uniform type noise. Moreover, in the latter case, the standard deviations are greater, which suggests that the characteristic problem is less sensitive to the Weibull type noise than to the uniform type noise.

v) The slip-line net of an inhomogeneous MC material is much more dependent upon the integration method (forward versus backward) than that of an HM material.

vi) For values of noise above 5% there is an increasing change in the average, and very amplified scatter, in the random fields of stress as well as position.

vi) Only in case of very small noise, can one replace the average solution of a stochastic problem by the solution of a deterministic problem with $\rho_{eff} = \rho_{det} = \langle \rho_\delta \rangle$ as well as $H_{eff} = H_{det} = \langle H_\delta \rangle$; that is $B_{eff} = B_{det}$.

4. Elastic-Brittle Microstructures

We now return to the elastic problem discussed in Section 2, and, specifically, to the spring network representation of a multiphase continuum medium. Following an approach advanced in the mideighties in solid state physics [e.g. 16, 17], one can employ such a spring network to study crack formation and propagation in elastic-brittle composites; this is done through a progressive removal of bonds according to the local fracture criterion and stress (or strain) concentrations, e.g. [18, 19].

In the following we demonstrate results of some simulations of damage in matrix-iclusion composites. The spring network model of the matrix-inclusion composite is characterized by four parameters: stiffnesses of the matrix and inclusion phases, C^m and C^i, and respective strengths ε_{cr}^m and ε_{cr}^i. The simulations of damage are carried out using a window of mesh size L = 64 units and d = 14 units, at the volume (area) fraction of inclusions f = 35%. While the resulting value of $\delta \cong 4.57$ defines a rather small scale of an effective continuum model, it allows - due to a small lattice - a comparatively rapid, and high resolution, simulation of damage phenomena, and hence, an assessment of the entire parameter space which is specified by two parameters: the *stiffness ratio* C^i/C^m and the *strain-to-failure ratio* $\varepsilon_{cr}^i/\varepsilon_{cr}^m$.

In Fig. 5a) we display the microgeometry of damage of a typical disordered composite $B_\delta(\omega)$, at $\delta \cong 4.57$, of the post-peak type with strain softening, specified by $C^i/C^m = 0.1$ and $\varepsilon_{cr}^i/\varepsilon_{cr}^m = 10.0$. Fig. 5b) shows the stress-strain curves of this and nineteen other such specimens, all of them plotted in coordinates normalized by σ_{cr}^m and ε_{cr}^m. Simulations of that type allow one to assess the statistics of key constitutive parameters such as σ_{max} and ε_{max}, Fig. 5c), for a range of the stiffness ratio, strength ratio, and volume fraction parameters. Furthermore, by covering a range of δ-scales and loading types, one can establish a complete statistical characterization of matrix-inclusion composites. This, in principle, forms the basis for a *stochastic damage mechanics*, which replaces the conventional, phenomenological, deterministic *continuum damage mechanics*.

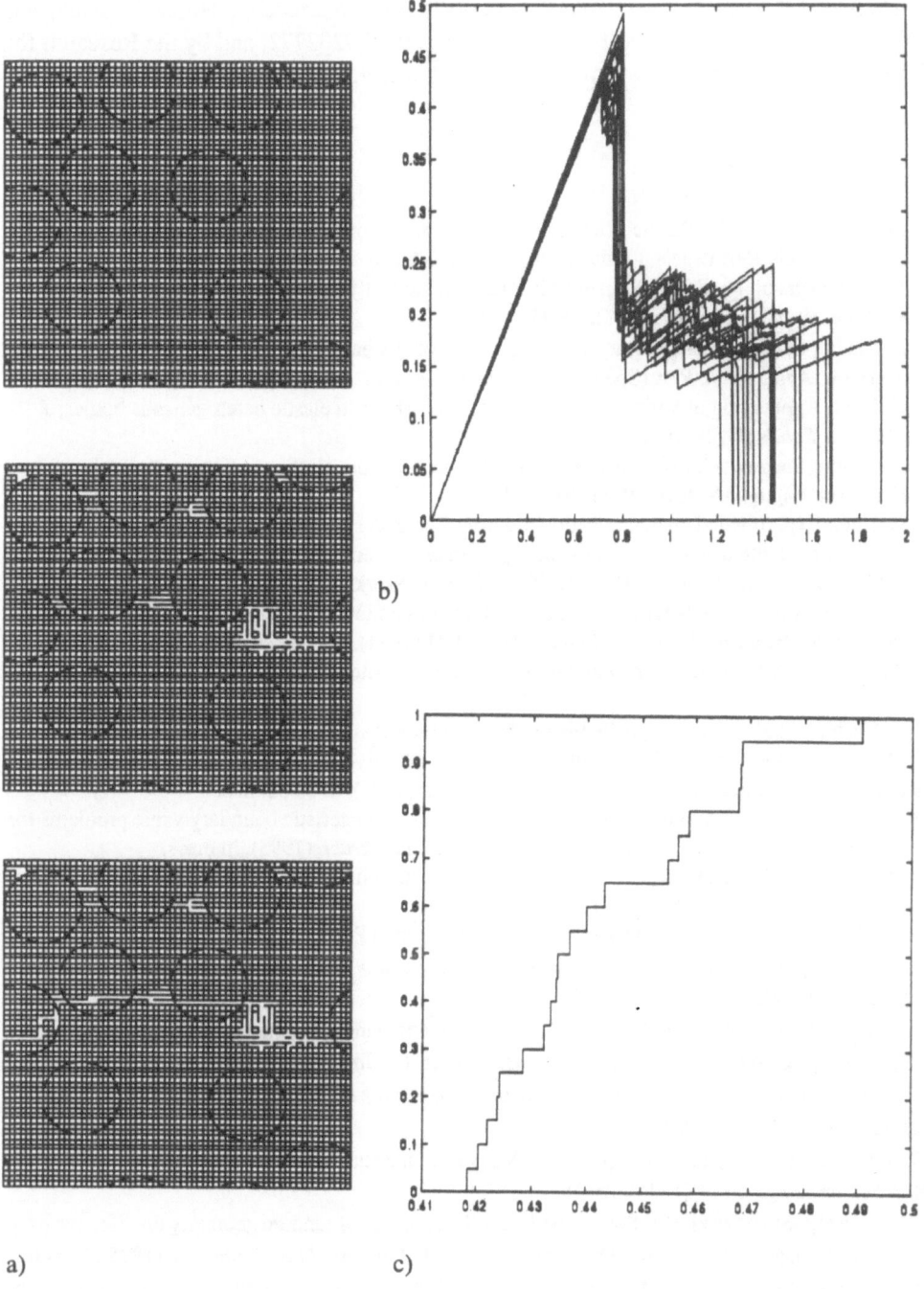

b)

c)

a)

Figure 5. a) Damage evolution in a random composite at $C^i/C^m = 0.1$ and $\varepsilon^i_{cr}/\varepsilon^m_{cr} = 10.0$; b) effective stress-strain curves of twenty specimens; c) probability distribution of σ_{max}.

Acknowledgment

This research is, in part, a result of collaboration with K. Alzebdeh, I. Horea, I. Jasiuk, and P.Y. Sheng. Support by the NSF under grant No. MSS 9202772, and by the Research for Excellence Fund from the State of Michigan, is gratefully acknowledged.

References

1. M. Ostoja-Starzewski: Plastic flow of random media: micromechanics, Markov property and slip-lines, *Appl. Mech. Rev.* (Special Issue: *Material Instabilities*) **45** (1992), S75-S82.
2. K. Alzebdeh and M. Ostoja-Starzewski: Micromechanically based stochastic finite elements, *Finite Elem. Anal. Design* **15** (1993), 35-41.
3. M. Ostoja-Starzewski: Micromechanics as a basis of stochastic finite elements and differences - an overview, *Appl. Mech. Rev.* (Special Issue: *Mechanics Pan-America 1993*) **46** (1993), S136-S147.
4. C. Huet: Application of variational concepts to size effects in elastic heterogeneous bodies, *J. Mech. Phys. Solids* **38** (1990), 813-841.
5. M. Ostoja-Starzewski: Micromechanics as a basis of random elastic continuum approximations, *Probabilistic Engng. Mech.* **8** (1993), 107-114.
6. C. Huet: Experimental characterization, micromechanical simulation and spatio-stochastic approach of concrete behaviours below the representative volume, *Probabilities and Materials - Tests, Models and Applications*, Proc. NATO Adv. Res. Workshop **E-269** (1994), 241-271.
7. M. Ostoja-Starzewski: Micromechanics as a basis of continuum random fields, *Appl. Mech. Rev.* (Special Issue: *Micromechanics of Random Media*) **47** (1994), S221-S230.
8. J.B. Keller: A theorem on the conductivity of a composite medium, *J. Math. Physics* **5** (1964), 548-549.
9. Z. Hashin: Analysis of composite materials - A survey, *J. Appl. Mech.* **50**, 481-505 (1983).
10. M. Ostoja-Starzewski and J. Schulte: Bounding of effective thermal conductivities of multiscale materials by essential and natural boundary conditions, (1995), to be published.
11. M. Ostoja-Starzewski and H. Ilies: The Cauchy and characteristic boundary value problems for weakly random rigid-perfectly plastic media, *Intl. J. Solids Struct.* (1995), in press.
12. G. Alpa and L. Gambarotta: Probabilistic failure criterion for cohesionless frictional materials, *J. Mech. Phys. Solids* **38** (1990), 491-503.
13. V.V. Sokolovskii: *Statics of Granular Media*, Pergamon Press, (1965).
14. W. Szczepinski: *Limit States and Kinematics of Granular Media* (in Polish), Polish Scientific Publlishers (1974).
15. M. Ostoja-Starzewski and I. Horea: Plasticity of random inhomogeneous media, ASME Summer Ann. Mtg., *Symp. on Fracture and Plastic Instabilities* (N. Ghoniem, Ed.), (1995), in press.
16. H.J. Herrmann and S. Roux (eds.): *Statistical Models for the Fracture of Disordered Systems*, Elsevier/North-Holland (1990).
17. P.M. Duxbury, S.G. Kim and P.L. Leath: Size effect and statistics of fracture in random materials, *Mat. Sci. and Eng. A* **176** (1994), 25-31.
18. M. Ostoja-Starzewski, P.Y. Sheng and I. Jasiuk: Influence of random geometry on effective properties and damage formation in 2-D composites, *ASME J. Engng. Mat. Techn.* **116** (1995), 384-391.
19. I. Jasiuk, P.Y. Sheng and M. Ostoja-Starzewski: Influence of random fiber arrangement on crack propagation in brittle-matrix composites, *Brittle-Matrix Composites* 4 (A.M. Brandt, V.C. Li and I.H. Marshall, Eds.), (1994), 200-208.

RESPONSE CORRELATIONS OF LINEAR SYSTEMS WITH WHITE NOISE LINEARLY PARAMETRIC INPUTS

M. DI PAOLA and G. FALSONE
Dipartimento di Ingegneria Strutturale & Geotecnica
Università di Palermo, Viale delle Scienze,
I-90128 Palermo, Italy

Abstract
Relationships between moments and correlations of the response of linear systems subjected to linearly parametric normal white noise inputs are here reported. They are obtained by extensively using the properties of the stochastic integral calculus.

1. Introduction

The fundamental goal in the stochastic analysis of linear and nonlinear systems subjected to randomly varying loads is finding the full probabilistic description of the stochastic response process once that the probabilistic characterization of the load processes is known. The characterization of any stochastic process can be fully given by the knowledge of its correlation functions, or, equivalently, of the means at multiple times [1-3], of order r (with $r = 1, 2, \cdots, \infty$), which are function of r variables (the instant times $t_1, t_2, \cdots, t_r$). In the case of normal processes, it is well known that only the first and the second order correlations are enough to characterize them. For non-normal processes, the knowledge of higher order correlations is required. But, even in the case of linear systems, it is rather difficult to find the differential equations governing the response correlations. It is simpler to find the differential equations governing the response moments or the cumulants, which are strictly connected each other and which are functions of a single time instant t, giving a partial description of the processes. However, for linear systems under both normal or non-normal delta-correlated inputs, it is possible to obtain the response correlations of any order from the cumulants of the corresponding order by means of simple relationships where the fundamental matrix [4] of the linear system appears [5].

When the system is nonlinear and is subjected to delta-correlated inputs, the differential equations governing the moments can be found by means of the Itô differential rule [6] (if the input is normal) or by means of its extension [3] (if the input is non-normal). But no analogous relationships between cumulants (or moments) and correlations have been found in literature for the response of nonlinear systems.

A. Naess and S. Krenk (eds.), IUTAM Symposium on Advances in Nonlinear Stochastic Mechanics, 351–360.

The aim of this paper is finding the relationships between the moments and the correlations for a very important class of systems: the linear systems subjected to white noise linearly parametric inputs. The importance of studying these systems lies on the fact that any nonlinear system can be optimally approximated by one of them [7]; hence, the procedure here presented could be used in order to obtain the response correlations of nonlinear systems subjected to normal white noise inputs. Moreover, the probabilistic characterization of the response of linear systems wiht linearly parametric white noise input can be exactly obtained because the differential equations governing the response moments do not create an infinite hierarchy. Hence the correlation functions are also obtained in a closed form. At last it is important to note that, even if the input is normal, the response is a non-normal process, so that the correlations of order greater than two are necessary in order to characterize it. In literature there are some papers devoted to the evaluation of the response correlations of linear systems subjected to parametric white noise inputs [8,9], but they are limited to the second order correlations and to the cases in which the Wong-Zakai correction terms do not arise. In the present work these limits are overcome.

In the first section of the paper, the solution of a linear differential equation, characterized by an external deterministic input and by a linearly parametric normal white noise input, is given, in the case in which the Itô differential calculus is used. Then, by starting from this solution, the first and the second order response correlations are obtained, by using the fundamental rules of the stochastic integration. At last, an alternative approach for obtaining the correlations is introduced; it allows us to write a first order differential equation (in which the variable is a single time instant) governing the behaviour of the response correlation function of any order; the solution of this equation gives the explicit expression of the correlation. This procedure gives the relationships between the response correlations of any order and the corresponding moments in a very simple recursive form in which the fundamental function related to the quasi-linear system appears. The correctness of this last procedure is proved by the fact that the expression for the first lesser order correlations are coincident with those obtained by the previous procedure (for this reason these procedures are both reported here). But the last one has the great advantage of being easily available for higher order correlations.

Here, for clarity's sake, all the formulation is presented for a first order SDOF system, but the extension to greater order and MDOF systems is straightforward. Moreover in the following we use the term correlation even if we are dealing with a moment at multiple times, because they are strictly connected by means of simple relationships [1,2].

2. Linear Stochastic Differential Equation with Parametric White Noise Input

Let us consider the following differential equation:

$$\dot{Z}(t) = [a(t) + g(t)W(t)]Z(t) + f(t) \tag{1}$$

where $a(t)$, $g(t)$ and $f(t)$ are deterministic time dependent functions, $W(t)$ is a white noise stochastic process, characterized by an intensity equal to q, $Z(t)$ is the response function and the dot over a variable means time derivative. As the process $W(t)$ is not mean-square integrable, it is well known that equation (1) has not a strong mathematical sense; in order to find the solution of this equation, it is opportune to consider the incremental equation connected to it. Following the two fundamental stochastic differential calculi, that is the Itô one and the Stratonovich one, this incremental equation assumes two different forms; the first one is now analysed; if the Itô differential calculus [6,10] is adopted, the incremental form connected to equation (1) is:

$$dZ(t) = \left[a(t) + \frac{1}{2}g^2(t)q \right] Z(t)dt + f(t)dt + g(t)Z(t)dB(t) \qquad (2)$$

where $B(t)$ is the Wiener process connected to the white noise $W(t)$ by means of the following formal relationship [11]:

$$W(t) = \frac{dB(t)}{dt} \qquad (3)$$

In equation (2) the term $1/2g^2(t)q$ in the drift coefficient represents the Wong-Zakai correction term [12], that is necessary because the presence of a parametric white noise input and because the Itô calculus is adopted.

With the aim of finding the solution of equation (2), let us now consider the homogeneous equation associated to it, that is:

$$d\bar{Z}(t) = \left\{ \left[a(t) + \frac{1}{2}g^2(t)q \right] dt + g(t)dB(t) \right\} \bar{Z}(t) \qquad (4)$$

and let us introduce the new variable:

$$Y(t) = \ln \bar{Z}(t) \qquad (5)$$

whose differential, by taking into account the Itô differential rule, is given by:

$$dY(t) = \frac{1}{\bar{Z}(t)}d\bar{Z}(t) - \frac{1}{2}\frac{1}{\bar{Z}^2(t)}[d\bar{Z}(t)]^2 = a(t)dt + g(t)dB(t) \qquad (6)$$

The solution of equation (6) is:

$$Y(t) = Y(t_0) + \int_{t_0}^{t} a(\rho)d\rho + \int_{t_0}^{t} g(\rho)dB(\rho) \qquad (7)$$

t_0 being the start time instant. Hence, the solution $\bar{Z}(t)$ of the homogeneous differential equation (4) is given by:

$$\bar{Z}(t) = \Phi(t, t_0)\bar{Z}(t_0); \quad \Phi(t, t_0) = \exp\left\{ \int_{t_0}^t a(\rho)d\rho + \int_{t_0}^t g(\rho)dB(\rho) \right\}$$
$$(8a, b)$$

Hence, $\Phi(t, t_0)$ can be considered as the fundamental function [4] related to equation (1).

In order to find the solution $Z(t)$ of equation (2), let us consider such a new variable that:

$$X(t) = \Phi^{-1}(t, t_0)Z(t) \qquad (9)$$

By using the Itô differential rule, we can evaluate $dX(t)$ as:

$$dX(t) = \Phi^{-1}(t, t_0)dZ(t) + Z(t)d\left[\Phi^{-1}(t, t_0)\right] + d\left[\Phi^{-1}(t, t_0)\right]dZ(t)$$
$$+ \frac{1}{2}Z(t)d^2\left[\Phi^{-1}(t, t_0)\right] \qquad (10)$$

which, by taking into account the expression of $\Phi(t, t_0)$ given into equation (8b) and the expression of $dZ(t)$ given into equation (2), gives at last the following relationship:

$$dX(t) = \Phi^{-1}(t, t_0)f(t)dt \qquad (11)$$

Hence $Z(t)$ is given by:

$$Z(t) = \Phi(t, t_0)Z_0 + \int_{t_0}^t \Phi(t, \tau)f(\tau)d\tau; \qquad Z_0 = Z(t_0) \qquad (12)$$

It is important to note that the evaluation of the fundamental function $\Phi(t, t_0)$ needs the evaluation of the integrals appearing into equation (8b); the first one of these integrals is an ordinary Cauchy-Riemann deterministic integral, while the second one is a stochastic integral which has to be considered in mean-square sense [2,10], that is:

$$\int_{t_0}^t g(\rho)dB(\rho) = l.i.m._{n\to\infty} \sum_{i=1}^n g(\tau_i)[B(t_{i+1}) - B(t_i)] \qquad (13)$$

where $t_1 \equiv t_0$ and $t_{n+1} \equiv t$; τ_i is any time instant included in the step $[t_i, t_{i+1}]$; at last $l.i.m.$ means limit in mean square. As $g(t)$ is a deterministic function, the value of this integral is just the same if Itô or Stratonovich type integral is used.

3. First Order Moments

By applying the mean operator $E[\cdot]$ to both the members of equation (12) and by assuming, for simplicity, $Z_0 = 0$, we obtain:

$$E[Z(t)] = \int_{t_0}^{t} E[\Phi(t,\tau)]f(\tau)d\tau \tag{14}$$

The mean $E[\Phi(t,\tau)]$ appearing in this expression is given by (see equation (8b)):

$$E[\Phi(t,\tau)] = \exp\left\{\int_{\tau}^{t} a(\rho)d\rho\right\} E\left[\exp\left\{\int_{\tau}^{t} g(\rho)dB(\rho)\right\}\right] \tag{15}$$

It is important to note that the stochastic process $\int_{t_0}^{t} g(\rho)dB(\rho)$ is a zero-mean normal process; hence, we can calculate the mean appearing into equation (15) by considering that, for any normal stochastic variable V, the following relationship holds:

$$E[\exp(V)] = \exp\left\{\frac{1}{2}E[V^2]\right\} \tag{16}$$

Equation (15) is then rewritten as follows:

$$E[\Phi(t,\tau)] = \exp\left\{\int_{\tau}^{t} \left[a(\rho) + \frac{1}{2}g^2(\rho)q\right] d\rho\right\} = \Theta(t,\tau) \tag{17}$$

where the following property of the stochastic integral has been considered [10]:

$$E\left[\int_{\tau}^{t} g(\rho)dB(\rho) \int_{\tau}^{t} g(\rho)dB(\rho)\right] = \int_{\tau}^{t} g^2(\rho)qd\rho \tag{18}$$

By replacing equation (17) into equation (14), we obtain:

$$E[Z(t)] = \int_{t_0}^{t} \Theta(t,\tau)f(\tau)d\tau \tag{19}$$

This equation allows us to evaluate the mean response for any time t. It is easy to verify that the second member of equation (19) is the solution of the first order differential equation governing the first order moment of $Z(t)$, which can be obtained by using the Itô differential rule and which has the following expression:

$$\dot{E}[Z(t)] = \left\{a(t) + \frac{1}{2}g^2(t)q\right\} E[Z(t)] + f(t) \tag{20}$$

It is important to note that equations (19) and (20) show that the function $\Theta(t, \tau)$, introduced into equation (17), is the fundamental function connected to the differential equation governing the first order moment.

4. Second Order Correlation

By taking into account equation (12) where $Z_0 = 0$ is assumed, the second order correlation function of $Z(t)$ can be expressed as follows (even if it is a moment at multiple times, we use the term correlation because they have a similar significance and are connected by means of simple relationships [1,2]):

$$E[Z(t_1)Z(t_2)] = \int_{t_0}^{t_1} \int_{t_0}^{t_2} E\left[\Phi(t_1, \tau_1)\Phi(t_2, \tau_2)\right] f(\tau_1)f(\tau_2)d\tau_1 d\tau_2 \quad (21)$$

where, due to the equation (8b), $E\left[\Phi(t_1, \tau_1)\Phi(t_2, \tau_2)\right]$ is given by:

$$
\begin{aligned}
E\left[\Phi(t_1, \tau_1)\Phi(t_2, \tau_2)\right] &= \exp\left\{\int_{\tau_1}^{t_1} a(\rho)d\rho + \int_{\tau_2}^{t_2} a(\rho)d\rho\right\} \\
&\times E\left[\exp\left\{\int_{\tau_1}^{t_1} g(\rho)dB(\rho) + \int_{\tau_2}^{t_2} g(\rho)dB(\rho)\right\}\right]
\end{aligned}
\quad (22)
$$

As the process $\int_{\tau_1}^{t_1} g(\rho)dB(\rho) + \int_{\tau_2}^{t_2} g(\rho)dB(\rho)$ is zero-mean and normal, we can apply the property given into equation (16), in such a way that equation (22) can be rewritten as follows:

$$
\begin{aligned}
E\left[\Phi(t_1, \tau_1)\Phi(t_2, \tau_2)\right] &= \exp\left\{\int_{\tau_1}^{t_1} a(\rho)d\rho + \int_{\tau_2}^{t_2} a(\rho)d\rho\right\} \\
&\times \exp\left\{\frac{1}{2}\int_{\tau_1}^{t_1} g^2(\rho)qd\rho + \frac{1}{2}\int_{\tau_2}^{t_2} g^2(\rho)qd\rho\right\} \\
&\times \exp\left\{E\left[\int_{\tau_1}^{t_1} g(\rho)dB(\rho) \int_{\tau_2}^{t_2} g(\rho)dB(\rho)\right]\right\}
\end{aligned}
\quad (23)
$$

where the property given into equation (18) has been considered. Equation (23), by taking into account equation (17) and the properties of the stochastic integrals, gives:

$$E\left[\Phi(t_1, \tau_1)\Phi(t_2, \tau_2)\right] = \Theta(t_1, \tau_1)\Theta(t_2, \tau_2)\exp\left\{\int_{max(\tau_1, \tau_2)}^{min(t_1, t_2)} g^2(\rho)qd\rho\right\}$$

$$(24)$$

By replacing equation (24) into equation (21) and by assuming $t_1 < t_2$, we obtain:

$$E[Z(t_1)Z(t_2)] =$$

$$\int_{t_0}^{\tau_2} \int_{t_0}^{t_2} \Theta(t_1,\tau_1)\Theta(t_2,\tau_2)\exp\left\{\int_{\tau_2}^{t_1} g^2(\rho)q d\rho\right\} f(\tau_1)f(\tau_2)d\tau_1 d\tau_2$$

$$+ \int_{t_0}^{t_1} \int_{t_0}^{\tau_1} \Theta(t_1,\tau_1)\Theta(t_2,\tau_2)\exp\left\{\int_{\tau_1}^{t_1} g^2(\rho)q d\rho\right\} f(\tau_1)f(\tau_2)d\tau_1 d\tau_2 \qquad (25)$$

$$+ \int_{t_0}^{t_1} \int_{t_0}^{t_2} \Theta(t_1,\tau_1)\Theta(t_2,\tau_2)f(\tau_1)f(\tau_2)d\tau_1 d\tau_2$$

If equation (19) is taken into account and if we note that the first two integrals of equation (25) are coincident, equation (25) can be simplified as follows:

$$E[Z(t_1)Z(t_2)]$$

$$= 2\int_{t_0}^{t_1} \Theta(t_1,\tau_2)\Theta(t_2,\tau_2)\exp\left\{\int_{\tau_2}^{t_1} g^2(\rho)q d\rho\right\} E[Z(\tau_2)]f(\tau_2)d\tau_2 \qquad (26)$$

$$+ E[Z(t_1)]\{E[Z(t_2)] - \Theta(t_2,t_1)E[Z(t_1)]\}$$

This equation gives the expression of the second order correlation of the response $Z(t)$ of the system governed by equation (1).

If we are interested to the evaluation of the second order moment $E[Z^2(t_1)]$, we have to particularize equation (26) for $t_2 = t_1$, obtaining:

$$E[Z^2(t_1)] = 2\int_{t_0}^{t_1} \Theta(t_1,\tau)^2 \exp\left\{\int_{\tau}^{t_1} g^2(\rho)q d\rho\right\} E[Z(\tau)]f(\tau)d\tau \qquad (27)$$

It is important to note that the last expression is the solution of the first order differential equation governing the second order moment, that can be written by using the Itô differential rule and has the following form:

$$\dot{E}[Z^2(t_1)] = \{2a(t_1) + 2g^2(t_1)q\} E[Z^2(t_1)] + 2E[Z(t_1)]f(t_1) \qquad (28)$$

By comparing equations (26) and (27), it is easy to find the following relationship between the second order correlation and moment of $Z(t)$:

$$E[Z(t_1)Z(t_2)] = \Theta(t_2,t_1)E[Z^2(t_1)]$$
$$+ E[Z(t_1)]\{E[Z(t_2)] - \Theta(t_2,t_1)E[Z(t_1)]\} \qquad (29)$$

Hence, if we know the value of the second order moment of $Z(t)$ at a certain instant t_1, we are able to evaluate the correlation function $E[Z(t_1)Z(t_2)]$, with $t_1 < t_2$, by means of the last equation. This relationship can be considered as an extension of that one reported in [5] which holds for linear systems response.

4.1. ALTERNATIVE APPROACH

An alternative approach able to simply give the second correlation function of $Z(t)$ is here presented.

By taking into account equation (2), we can write:

$$
\begin{aligned}
E[Z(t_1)dZ(t_2)] &= a(t_2)E[Z(t_1)Z(t_2)]dt_2 + f(t_2)E[Z(t_1)]dt_2 \\
&+ \frac{1}{2}g^2(t_2)qE[Z(t_1)Z(t_2)]dt_2 + g(t_2)E[Z(t_1)Z(t_2)dB(t_2)]
\end{aligned}
\tag{30}
$$

If we assume that $t_1 < t_2$ and use the properties of the Itô calculus, following the which, if $t_1 < t_2$, the last mean appearing into equation (30) is zero (because $Z(t)$ is a non-anticipating function with respect to $B(t)$), we obtain the following differential equation:

$$
\frac{\partial}{\partial t_2}\{E[Z(t_1)Z(t_2)]\} = \left\{a(t_2) + \frac{1}{2}g^2(t_2)q\right\}E[Z(t_1)Z(t_2)] + f(t_2)E[Z(t_1)]
\tag{31}
$$

whose solution, by considering $t_2 = t_1$ as start time instant, is:

$$
E[Z(t_1)Z(t_2)] = \Theta(t_2,t_1)E[Z^2(t_1)] + E[Z(t_1)]\int_{t_1}^{t_2}\Theta(t_2,\tau)f(\tau)d\tau
\tag{32}
$$

If, now, equation (27) is taken into account, equation (32) is rewritten as follows:

$$
\begin{aligned}
E[Z(t_1)Z(t_2)] &= \\
2\Theta(t_2,t_1)&\int_{t_0}^{t_1}\Theta(t_1,\tau)\exp\left\{\int_{\tau}^{t_1}g^2(\rho)qd\rho\right\}E[Z(\tau)]f(\tau)d\tau \\
&+ E[Z(t_1)]\left\{\int_{t_0}^{t_2}\Theta(t_2,\tau)f(\tau)d\tau - \Theta(t_1,t_2)\int_{t_0}^{t_1}\Theta(t_1,\tau)f(\tau)d\tau\right\}
\end{aligned}
\tag{33}
$$

which, after some algebra, gives a relationship that is coincident with equation (29).

It is important to note the simplicity in obtaining the second order correlation of $Z(t)$ with respect to the other approach before presented; moreover, as this simplicity is holden even for higher order correlations, in the next section we will use only this last approach.

5. Higher Order Correlations

In this section the correlations of $Z(t)$ of higher order than two are evaluated by using the alternative approach presented in the previous section. At this purpouse, let us consider the third order correlation function $E[Z(t_1)Z(t_2)Z(t_3)]$, with $t_1 < t_2 < t_3$. By taking into account equation (2) and the properties of the Itô calculus, we can write:

$$E[Z(t_1)Z(t_2)dZ(t_3)] = \left\{ a(t_3) + \frac{1}{2}g^2(t_3)q \right\} E[Z(t_1)Z(t_2)Z(t_3)]dt_3 \qquad (34)$$
$$+ f(t_3)E[Z(t_1)Z(t_2)]dt_3$$

from which it is possible to obtain the following partial differential equation:

$$\frac{\partial}{\partial t_3} \{ E[Z(t_1)Z(t_2)Z(t_3)] \} = $$
$$\left\{ a(t_3) + \frac{1}{2}g^2(t_3)q \right\} E[Z(t_1)Z(t_2)Z(t_3)] + f(t_3)E[Z(t_1)Z(t_2)] \qquad (35)$$

whose solution is:

$$E[Z(t_1)Z(t_2)Z(t_3)] = \Theta(t_3, t_1)E[Z^2(t_1)Z(t_2)]$$
$$+ E[Z(t_1)Z(t_2)] \int_{t_1}^{t_3} \Theta(t_3, \tau)f(\tau)d\tau \qquad (36)$$

that, after some algebra, can be rewritten as:

$$E[Z(t_1)Z(t_2)Z(t_3)] = \Theta(t_3, t_1)E[Z^2(t_1)Z(t_2)]$$
$$+ E[Z(t_1)Z(t_2)] \{ E[Z(t_3)] - \Theta(t_3, t_1)E[Z(t_1)] \} \qquad (37)$$

This equation gives the third order correlation function once that $E[Z^2(t_1)Z(t_2)]$ and the correlations of lesser order are known; in order to obtain the correlation $E[Z^2(t_1)Z(t_2)]$ a similar procedure can be considered, obtaining the following relationship:

$$E[Z^2(t_1)Z(t_2)] = \Theta(t_2, t_1)E[Z^3(t_1)]$$
$$+ E[Z^2(t_1)] \{ E[Z(t_2)] - \Theta(t_2, t_1)E[Z(t_1)] \} \qquad (38)$$

Hence, by means of equation (37), in which equation (29) and (38) have been considered, it is possible to express the third correlation in terms of the moments.

In general, for the correlation of order n $E[Z(t_1)Z(t_2)\cdots Z(t_n)]$, with $t_1 < t_2 < \cdots < t_n$, by using the same approach considered before, it is possible to write the following relationship:

$$E[Z(t_1)Z(t_2)\cdots Z(t_n)] = \Theta(t_n,t_1)E[Z^2(t_1)Z(t_2)\cdots Z(t_{n-1})]$$
$$+ E[Z(t_1)Z(t_2)\cdots Z(t_{n-1})]\{E[Z(t_n)] - \Theta(t_n,t_1)E[Z(t_1)]\} \tag{39}$$

By recursively applying equation (39) it is possible to express the correlation function of every order of $Z(t)$ in terms of the corresponding moments. This is a great advantage in the identification of the response process because finding and solving the differential equations governing the moments is very simpler than finding and solving the equations governing the correlations.

6. Conclusions

The relationships between the response correlations of any order and the corresponding moments of a SDOF first order linear system, subjected to an external deterministic input and to a linearly parametric normal white noise, are given in this paper. These relationships are very important because allow us to obtain the response correlations once that the response moments are known; the last ones can be easily obtained by solving the corresponding first order differential equations, written by using the Itô differential rule.

References

1. Lin, Y.K. (1967) *Probabilistic Theory of Structural Dynamics*, McGraw-Hill, New York.
2. Stratonovich, R.L. (1963) *Topics in the Theory of Random Noise*, Gordon and Breach, New York.
3. Di Paola, M. and Falsone, G. (1993) Stochastic Dynamics of Nonlinear Systems Driven by Non-Normal Delta-Correlated Processes. *J.Appl.Mech., ASME*, **60**, 141-148.
4. Muller, P.C. and Schielen, W.O. (1985) *Linear Vibrations*, Martinus Nijhoff Publ., Dordrecht.
5. Falsone, G. (1994) Cumulants and Correlations for Linear Systems under Non-Stationary Delta-Correlated Processes. *Prob.Engng.Mech.*, **9**, 157-165.
6. Itô, K. (1951) Stochastic Differential Equations. *Mem.Am.Math.Soc.*, **4**.
7. Di Paola, M. (in press) Quasi-Linear Oscillators under Poisson Pulses. *Jorn. Appl. Mechs., ASME*.
8. Benaroya, H. and Rehak, M. (1989) Response and Stability of a Random Differential Equation: Part I - Moment Equation Method. *J. Appl. Mechs., ASME*, **56**, 192-195.
9. Hou, Z.K. and Iwan, W.D. (1991) Nonstationary Response of Linear Systems under Uncorrelated Parametric and External Excitations. *Prob. Engng. Mechs*, **6**, 74-81.
10. Gardiner, C.W. (1985) *Handbook of Stochastic Methods*. Springer-Verlag, Berlin.
11. Grigoriu, M. (1987) White Noise Processes. *J. Engng. Mech. Div., ASCE*, **119**, 757-765.
12. Wong, E. and Zakai, M. (1965) On the Relation between Ordinary and Stochastic Differential Calculus. *Int. J. Engng. Sci.*, **3**, 213-229.

PARAMETER ESTIMATION FOR RANDOMLY EXCITED NON-LINEAR SYSTEMS

A Method based on Moment Equations and Measured Response

J. B. ROBERTS, J.F. DUNNE AND A. DEBONOS

School of Engineering
University of Sussex
Falmer, Brighton, BN1 9QT, UK.

Abstract

The problem of estimating unknown parameters in a non-linear randomly excited dynamic system, when the excitation is unmeasurable, is considered. It is shown that, if the excitation is modelled stochastically as a Gaussian process, with a prescribed spectral form, it is possible to estimate the parameters from response data alone using either moment equations or a spectral input-output relationship. When applied to simulated data for a particular non-linear oscillator, as an example, it is found that the use of moment equations leads to a very good estimation of the stiffness parameters but is incapable of yielding estimates of the absolute level of damping. However the latter can be found accurately by applying a spectral relationship. Improvements in the accuracy of estimation for the damping parameters, and the input intensity, are achieved by using a theoretical expression for the distribution of the energy envelope of the response in combination with statistical linearisation.

1. Introduction

A number of techniques have been developed, over the past few decades, for predicting the response of non-linear systems to random excitation, using stochastic process theory (e.g. see Roberts and Spanos 1990). However, in practice, some or all of the parameters in the governing equations of motion are not known a-priori, with any degree of precision. This is especially true of parameters relating to damping. Techniques for estimating the parameters, from measured input and output data, are therefore of great importance and have been successfully applied in a number of engineering applications including, for example, ship roll motion (Gawthrop et al 1988, Kountzeris et al 1991).

A. Naess and S. Krenk (eds.), IUTAM Symposium on Advances in Nonlinear Stochastic Mechanics, 361–372.
© *1996 Kluwer Academic Publishers.*

There are, however, many situations where the input is of an environmental origin and it is impractical, or impossible, to measure sample functions of the excitation process.   Examples include structures responding to wind or seismic excitation, the monitoring of land vehicles traversing random surfaces and the motion of ships and offshore structures due to wave excitation. As shown recently by the authors (Roberts et al 1992,1994,1995), it is possible to estimate unknown parameters from measurements of the response process alone, provided that a stochastic model of the excitation process is assumed - e.g. a zero-mean Gaussian process with an assumed parametric form for the power spectrum.

In this paper an alternative approach is introduced, based on the use of moment equations.   It is shown that this method may be combined with the spectral method (Roberts et al 1995) and the energy envelope method (Roberts et al 1994), to yield estimates of all the unknown parameters, with improved accuracy.

2. Theoretical Relationships

A general state vector form of the equations of motion of a non-linear dynamic system is

$$\dot{x} = G(x, \lambda) + f(t) \quad \text{or} \quad g(x, \dot{x}, \lambda) = f(t) \tag{1}$$

where

$$g(x, \dot{x}, \lambda) = \dot{x} - G(x, \lambda) \tag{2}$$

x is the stochastic response process, f is a stationary stochastic excitation process, g and G are non-linear functions and λ is a m-vector of parameters.

It will be supposed that sample functions of the stationary response state variables can be measured directly, or obtained through processing of measured data, over an interval of time $0 \le t \le T$.   In contrast, f is assumed to be unmeasurable and known only in terms of a stochastic model.   Specifically, here f is taken to be a zero-mean Gaussian process with a known correlation function

$$R(s) = E\left\{f(t)f^T(t+s)\right\} \tag{3}$$

where $E\{.\}$ is the expectation operator.   The elements of the parameter vector λ are assumed to be unknown and must be estimated from experimental data. λ can be augmented to include parameters relating to the input correlation function (or power spectrum).

The estimation problem can be stated as follows: how can the sample functions of the response be processed to yield estimates of λ ?

2.1. MOMENT RELATIONSHIPS

If the excitation process is modelled as a stationary vector white noise: i.e.

$$R(s) = D\delta(s) \tag{4}$$

where D is a matrix of constants then it is possible to derive a set of moment differential equations (e.g. see Roberts and Spanos 1990). For the stationary response these can be derived from the relationship

$$H(k_1, k_2, \ldots, k_n) \equiv \sum_{j=1}^{n} E\left\{ G_i \frac{\partial h}{\partial x_j} \right\} + \frac{1}{2} \sum_{i,j=1}^{n} D_{ij} \frac{\partial^2 h}{\partial x_i \partial x_j} = 0 \qquad (5)$$

where

$$h(x) = x_1^{k_1}(t) x_2^{k_2}(t) \ldots x_n^{k_n}(t) \qquad (6)$$

By choosing various sets of integer values for k_1, k_2, etc., $\Omega_1, \Omega_2, \ldots, \Omega_N$ say, one can, from equation (5), generate N equations relating various moments of the response to the elements of D. As is well known, for a specified parameter vector, λ, this set can not be solved, in the non-linear case, without applying a closure technique (see Roberts and Spanos 1990). However, for the inverse problem of estimating λ, the closure problem does not exist. Estimates of all the moments can, at least in principle, be derived directly by applying averaging procedures to the measured response data since one can always generate more equations than unknowns. Thus λ can be estimated by minimising the least-square cost function

$$J = \sum_{\Omega_1, \Omega_2, \ldots, \Omega_N} H^2(k_1, k_2, \ldots, k_n) \qquad (N > m) \qquad (7)$$

If the excitation is non-white the above relationships are still applicable provided that a shaping filter is introduced, thus modelling the excitation as a filtered white noise. This has the effect, of course, of increasing the order of the combined system.

The computation is much simplified if the equation of motion is of the linear-in-the-parameters type

$$d(x, \dot{x}) + e(x, \dot{x})\lambda = f(t) \qquad (8)$$

Then the moments equations are also linear-in-the-parameters and of the form

$$A\lambda = B \qquad (9)$$

where A and B are constant matrices. Least-square minimisation then gives

$$\lambda = \left(A^T A\right)^{-1} A^T B \qquad (10)$$

2.2. CORRELATION AND SPECTRAL RELATIONSHIPS

As will be demonstrated through a specific example, in section 3, moments related at one time only, such as $E\{h(x)\}$, do not provide a suitable basis for estimating the absolute level of damping. This is because only the ratio between each damping parameter and an appropriate intensity matrix element, D, can be evaluated from the moments. To obtain the damping level explicitly it is necessary to consider the correlation functions. An input-output correlation relationship can be established by multiplying equation (1) with itself, at two different times, and then by taking expectations throughout; thus

$$V(s) = E\{g[x(t)\lambda]g^T[x(t+s)\lambda]\} = R(s) \tag{11}$$

It is usually more convenient to work with spectra, rather than correlation functions. The spectral matrix of the excitation can be defined by

$$S(\omega) = \frac{1}{2\pi} \int_{-\infty}^{\infty} R(s)e^{-i\omega s} ds \tag{12}$$

and the Fourier transform of V, $F(\omega)$, can be defined similarly. λ can now be estimated by minimising the cost function

$$J = \sum_{k}^{n_k} \sum_{i,j} \left| \hat{F}_{ij}(\omega_k) - S_{ij}(\omega_k) \right|^2 \tag{13}$$

where $\hat{F}(\omega)$ is an estimate of $F(\omega)$, obtained by applying appropriate averaging procedures to the measured data. Clearly the number of frequencies, n_k, chosen for the minimisation must exceed the number of parameters to be estimated, m. It is noted that parameters relating to $S(\omega)$ may also be included in λ.

If the equation of motion is of the linear-in-the-parameters form given by equation (8), then the task of forming averages can be decoupled from the specification of λ. This considerably reduces the computational burden involved in minimising the cost function (Roberts et al. 1995)

3. Application to a Single Degree of Freedom System

3.1. EQUATION OF MOTION

As an illustration, a non-linear oscillator with the following equation of motion will be considered:

$$\ddot{x} + a_1\dot{x} + n_1\dot{x}|\dot{x}| + a_2x + n_2x^3 = f(t) \tag{14}$$

$f(t)$ is assumed to be a stationary stochastic process with zero mean. If the bandwidth of this input process is significantly greater than that of the response, $f(t)$ may be approximated as an ideal white noise with a correlation function of the form of equation (4) where D is scalar. $D = 2\pi S_0$, where S_0 is the constant spectral level of the white noise process which approximates $f(t)$. Equation (14) can be used to model the non-linear rolling motion of a ship in irregular seas (Roberts 1982, Roberts and Dacunha 1985).

Later it is convenient to consider a partly linearised version of equation (14). This involves the replacement

$$a_1\dot{x} + n_1\dot{x}|\dot{x}| \rightarrow a_{1eq}\dot{x} \tag{15}$$

where a_{leq} is the equivalent linear damping parameters. Standard statistical linearisation provides a relationship between a_{leq}, a_1, n_1 and the standard deviation of the velocity, $\sigma_{\dot{x}}$ (Roberts and Spanos 1990).

The specific identification problem which will be addressed here is as follows: how can a sample function of the response, $x(t)$, over a period $0 \leq t \leq T$, be processed to yield estimates of the system parameters a_1, n_1, a_2, n_2 and D?

3.2. MOMENT EQUATIONS

Using equation (5) it can be shown that the moments

$$m_{rs} = E\left\{x^r \dot{x}^s\right\} \qquad m_{rst} = E\left\{x^r \dot{x}^s y^t\right\} \tag{16}$$

where $y = \dot{x}|\dot{x}|$ and $r,s = 1,2,...$ satisfy the following equations, up to sixth order.

second order

$$-2a_1 m_{02} - 2n_1 m_{011} + D = 0 \tag{17}$$
$$m_{02} - n_1 m_{101} - a_2 m_{20} - n_2 m_{40} = 0 \tag{18}$$

fourth order

$$-2a_1 m_{04} - 2n_1 m_{031} - 2a_2 m_{13} - 2n_2 m_{33} + 3D m_{02} = 0 \tag{19}$$
$$m_{04} - 3a_1 m_{13} - 3n_1 m_{121} - 3a_2 m_{22} - 3n_2 m_{42} = 0 \tag{20}$$
$$2m_{13} - 2a_1 m_{22} - 2n_1 m_{211} + D m_{20} = 0 \tag{21}$$
$$3m_{22} - n_1 m_{301} - a_2 m_{40} - n_2 m_{60} = 0 \tag{22}$$

sixth order

$$-2a_1 m_{06} - 2n_1 m_{051} - 2a_2 m_{15} - 2n_2 m_{35} + 5D m_{04} = 0 \tag{23}$$
$$m_{06} - 5a_1 m_{15} - 5n_1 m_{141} - 5a_2 m_{24} - 5n_2 m_{44} + 10D m_{13} = 0 \tag{24}$$
$$m_{15} - 2a_1 m_{24} - 2n_1 m_{231} - 2a_2 m_{33} - 2n_2 m_{53} + 3D m_{22} = 0 \tag{25}$$
$$m_{24} - a_1 m_{33} - n_1 m_{321} - a_2 m_{42} - n_2 m_{62} = 0 \tag{26}$$
$$4m_{33} - 2a_1 m_{42} - 2n_1 m_{411} + D m_{40} = 0 \tag{27}$$
$$5m_{42} - n_1 m_{501} - a_2 m_{60} - n_2 m_{80} = 0 \tag{28}$$

Higher order moment equations can be generated, in like fashion.

Of the various moments appearing in the above equations the following are found, by estimating them from simulated data, to be much smaller than the other moments, when the damping is light:

$$m_{13}, m_{101}, m_{14}, m_{121}, m_{33}, m_{321}, m_{15}, m_{141}, m_{53}, m_{35}$$

Moreover, they become progressively less significant as the damping is reduced. It is noted that, as opposed to the other moments, they will all be exactly zero if the joint density function, $w(x, \dot{x})$, of x and $\dot{x}$ is symmetrical about the $\dot{x}$ axis. In general this is not true but it can be shown to be asymptotically correct, as the damping approaches zero magnitude (or is linear), using stochastic averaging theory (see section 3.4).

Accordingly the moment equations can be approximated by omitting the ones listed above. The result can be recast in matrix format as follows:

$$a\eta = b, \qquad c\mu = d \tag{29}$$

where

$$\eta = [e_1, e_2] \,, \quad \mu = [a_2, n_2], \quad e_1 = a_1 / D, \quad e_2 = n_1 / D \tag{30}$$

$$
a = \begin{bmatrix} m_{02} & m_{011} \\ m_{04} & m_{031} \\ m_{22} & m_{211} \\ m_{06} & m_{051} \\ m_{24} & m_{231} \\ m_{42} & m_{411} \end{bmatrix}
\quad
b = \begin{bmatrix} 1 \\ 3m_{02} \\ m_{20} \\ 5m_{04} \\ 3m_{22} \\ m_{40} \end{bmatrix}
\quad
c = \begin{bmatrix} m_{20} & m_{40} \\ 3m_{22} & 3m_{42} \\ m_{40} & m_{60} \\ 5m_{24} & 5m_{44} \\ m_{42} & m_{62} \\ m_{60} & m_{80} \end{bmatrix}
\quad
d = \begin{bmatrix} m_{02} \\ m_{04} \\ 3m_{22} \\ m_{06} \\ m_{24} \\ 5m_{42} \end{bmatrix}
\tag{31}
$$

Thus there are now two independent sets of equations, one for the stiffness parameters and one for the parameters e_1 and e_2. This shows that the damping parameters can not be decoupled from D. Thus, it is not possible to find the absolute level of damping from the moment equations alone.

The parameters can be estimated by minimising a least square cost function, as given by equation (7), using the expression given in equation (10). There is a choice concerning the number of moment equations which one includes.

3.3. SPECTRAL RELATIONSHIP

A spectral relationship can be derived fairly readily from equation (14) by following the procedure outlined earlier, in section 2.2. Using standard spectral relationships for the derivatives of processes, the following equation is eventually obtained:

$$F(\omega) = F_\ell(\omega) + F_n(\omega) = S_o \tag{32}$$

where

$$F_\ell(\omega) = (\omega^4 + a_1^2 \omega^2 - 2a_2 \omega^2 + a_2^2) S_{xx} \tag{33}$$

$$
\begin{aligned}
F_n(\omega) &= n_1^2 S_{yy} + n_2^2 S_{zz} + 2n_1 n_2 S_{yzr} \\
&\quad + 2(a_2 - \omega^2)(n_1 S_{xyr} + n_2 S_{xzr}) + 2a_1 \omega(n_1 S_{xyi} + n_2 S_{xzi})
\end{aligned}
\tag{34}
$$

Here $y = \dot{x}|\dot{x}|$, as before, $z = x^3$ and $S_{uv}(\omega)$ are (cross) spectra defined by the Fourier relationship

$$S_{uv}(\omega) = \frac{1}{2\pi} \int_{-\infty}^{\infty} R_{uv}(s) e^{-i\omega s} ds \tag{35}$$

where

$$R_{uv}(s) = E\{u(t)v(t+s)\} \tag{36}$$

R_{uv} is a (cross) correlation function and $S_{uv}(\omega)$ is a complex function of frequency; the additional subscripts, r and i, appearing in equation (34) denote, respectively, the real and imaginary parts of the cross-spectrum. $F_\ell(\omega)$ and $F_n(\omega)$ represent, respectively, the linear and non-linear contributions to equation (32).

All the spectral quantities in equations (32) to (34) can be estimated by processing a response sample function, using standard spectral estimation techniques based on the FFT algorithm. Numerical differentiation of the response sample function, to obtain $\dot{x}$, may be necessary in some cases. To obtain estimates of the parameters one can then minimise a least-square cost function, J, as given by equation (12), using a standard non-linear optimisation algorithm. The accuracy of the results obtained is, generally, not sensitive to the precise choice of frequencies (Roberts et al. 1995).

3.4. SEPARATION OF THE DAMPING COMPONENTS

The moment scheme outlined above enables one to accurately estimate the stiffness parameters, a_2 and n_2 whereas the spectral method enables the equivalent linear damping parameter, a_{1eq}, and the excitation strength, D to be found accurately (but gives poorer estimation of a_2 and n_2 than the moment method (Debonos 1993)). However, the spectral method is not capable of separating out the individual contributions from the linear and non-linear damping components (Roberts et al 1995).

This separation can be achieved by using estimates of e_1 and e_2 but such estimates obtained by the moment method are not very accurate (Debonos 1993). More accurate estimation of e_1 and e_2 is possible using the probability distribution of the response. An approximate theoretical expression for the stationary probability density function, $w(E)$ of the energy envelope process, $E(t)$, can be found where $E(t)$ is defined as the sum of the kinetic and potential energies of the oscillator. This expression is accurate when the bandwidth of the excitation process is significantly greater than that of the response (Roberts and Spanos 1990). This situation occurs when the overall rate of energy dissipation is small; i.e., the energy lost through damping, during any one "cycle" in the response is a small fraction of the total energy in that cycle. The result can be expressed as follows:

$$log_e[w(E)/A(E)] = C - 2e_1E - 2e_2g(E) \qquad (37)$$

where the functions $A(E)$ and $g(E)$, which depend on the stiffness parameters a_2 and n_2, can be evaluated from expressions given by Roberts et al (1992): C is a constant.

Fitting equation (37) to estimates of $log_e[w(E)/A(E)]$, derived from the histogram of the estimated energy levels, and using the estimates of a_2 and n_2 obtained by the moment method described earlier, enables fairly accurate estimates of e_1 and e_2 to be generated.

Two possibilities now exist for finding the values of a_1 and n_1. Firstly, and most obviously, one can use the estimated value of D, obtained from the spectral method, directly. Alternatively, one can combine the estimate of a_{1eq} obtained from the spectral method with the estimates of e_1 and e_2 from the energy histogram, and employ the statistical linearisation relationship. This yields a non-linear algebraic equation for D, which is readily solved; hence a_1 and n_1 can be found (Roberts et al 1995). This latter method was found to be more accurate, in general, and is used to obtain the results given in section 3.

Using stochastic averaging it can be shown, as the damping approaches zero, that the joint density function $w(x,\dot{x})$ of x and $\dot{x}$ is given by

$$w(x,\dot{x}) = w(E)/T(E) \tag{38}$$

where $T(E)$ is the period of undamped free vibration, at energy level E and $w(E)$ may be found from equation (37). This density function is symmetric with respect to both axes. The asymptotic result given by equation (38) serves to explain why some of the moments occurring in section 3.2 are negligibly small when the damping is light.

3.5. SUMMARY OF ESTIMATION METHOD

The combined technique for estimating the parameters can now be summarised as follows:
1. moment method to estimate a_2 and n_2.
2. spectral method to estimate a_{1eq}
3. energy envelope method to estimate e_1 and e_2
4. statistical linearisation equations to find a_1, n_1 and D.

4. Validation through Simulation

To test the proposed method, using the example considered in the previous section, sample functions of $x(t)$ were generated by numerically integrating equation (14), using a Runge-Kutta algorithm. Two types of Gaussian excitation processes were used in the simulation studies - white noise and correlated excitation. The method of generating sample functions of these processes is described by Roberts et al (1994).

Four particular cases are studied here, as summarised in Table 1.

TABLE 1. Parameters values and inputs used in the four test cases.

Case	Input	a_1	n_1	a_2	n_2	$d=\sqrt{D}$	e_1	e_2
A	white	0.008	0.190	0.504	0.0	0.057	2.42	58.5
B	white	0.008	0.190	0.504	3.0	0.105	0.73	17.2
C	correlated	0.008	0.190	0.504	0.0	0.059	2.30	54.6
D	correlated	0.008	0.190	0.504	3.0	0.112	0.64	15.2

In each case the equivalent damping level, as computed from statistical linearisation, is 4.1% of critical.

The mean and statistical variability of the parameter estimates generated were obtained by processing 100 independent response histories, each of duration 1800 seconds, with a time step of 0.09 seconds.

In the case of correlated excitation the target power spectrum was based on the so-called Jonswap spectrum, transformed from wave elevation to wave moment using linear hydrodynamic theory for a particular hull, as described by Roberts and Dacunha (1985). It is shown in Fig.1 of Roberts et al (1994).

In all cases the bandwidth of the correlated excitation is significantly greater than that of the response, as required by the energy envelope method for separating the damping contributions.

In the case of correlated excitation the D values in Table 1 correspond to the level of the excitation spectrum at the frequency at which the response spectrum peaks. Thus they give the strengths of the equivalent white noises which can be compared with the D values returned by the estimation technique. To compute D the theoretical target excitation spectrum was used and the peak frequency was estimated using the equivalent linear stiffness.

4.1. MOMENT BASED ESTIMATION

The values of the various moments appearing in equations (29) to (31) were estimated by suitably time averaging each sample function of the response. Then estimates of a_2, n_2, e_1 and e_2 were obtained by minimising the equation error, in a least-square sense, using the result given by equation (10).

Table 2 shows the mean and standard deviation (the latter in brackets) obtained for each of the four cases, using all moment equations up to sixth order. It is observed that the stiffness estimation is very accurate, in all cases, with low statistical variability (compare with the true values, given in Table 1). As expected, the results for case D are the least accurate because of the approximations involved in treating the input as an equivalent white noise. In contrast, the statistical variability of the estimates of e_1 and e_2 is high, resulting in substantial deviations from the expected values (also given in Table 2). In fact there is a strong correlation between these estimates - i.e a "trade-off" between their numerical values. Similar results have been obtained at higher level of damping (Debonos 1993).

TABLE 2. Estimates of the linear and non-linear stiffness parameters, and e_1 and e_2, from the moment method.

Case	$\hat{a}_2$	$\hat{n}_2$	expected		estimated	
			e_1	e_2	$\hat{e}_1$	$\hat{e}_2$
A	0.501	-0.001	2.46	58.5	8.07	42.3
	(0.010)	(0.061)			(4.88)	(17.3)
B	0.500	2.998	0.73	17.2	2.36	13.7
	(0.018)	(0.105)			(2.26)	(5.2)
C	0.498	-0.003	2.30	54.6	2.95	59.3
	(0.003)	(0.040)			(6.23)	(25.7)
D	0.492	2.784	0.64	15.2	-2.00	25.8
	(0.016)	(0.073)			(2.51)	(6.3)

Numerical experiments have shown that the accuracy of the estimation procedure does not significantly reduce if only second and fourth order moment equations are employed.   Moreover, further simplification, using only four equations from this reduced set, leading to the estimators

$$\hat{a}_2 = \frac{m_{02}m_{60} - 3m_{22}m_{40}}{m_{20}m_{60} - m_{40}^2} \qquad \hat{n}_2 = \frac{m_{02} - \hat{a}_2 m_{20}}{m_{40}} \tag{39}$$

$$\hat{e}_1 = \frac{1 - n\hat{e}_2 m_{011}}{m_{02}} \qquad \hat{e}_2 = \frac{m_{04} - 3m_{02}^2}{m_{011}m_{04} - m_{031}m_{02}} \tag{40}$$

also gave very satisfactory results with no further loss of accuracy.   Equations (39) and (40) provide very simple estimators for the parameters.

4.2. SPECTRAL BASED ESTIMATION

The equivalent linear damping coefficient, a_{leq}, and the input strength, d, were estimated, independently of the moment equations by minimising the cost function defined by equation (13), in conjuction with equations (32) to (34) (setting $a_1 = a_{leq}$ and $a_2 = 0$.   Spectra were computed at 70 frequencies, in the vicinity of the peak in the response spectrum.

Table 3 shows the mean and standard deviation of the estimates of a_{leq} and D compared with their true values.   In all cases the estimated parameters agree reasonably well with the true values, also shown in the Table , when the level of statistical variability (and in the case of correlated excitation the effect of the white noise approximation) is taken into account.

TABLE 3.   Estimated values of the equivalent linear damping coefficient and the input intensity, from the spectral method.

Case	a_{leq}		$d = \sqrt{D}$	
	expected	estimated	expected	estimated
A	0.059	0.057 (0.008)	0.057	0.057 (0.002)
B	0.085	0.085 (0.023)	0.105	0.122 (0.004)
C	0.059	0.053 (0.014)	0.059	0.060 (0.004)
D	0.085	0.061 (0.044)	0.112	0.088 (0.015)

4.3. USE OF THE ENERGY ENVELOPE

Alternative, more accurate, estimates of e_1 and e_2 were found by fitting the histogram of the energy envelope to the theoretical expression given by equation (37).   Then, using the estimate of a_{leq} obtained from the spectral method, together with the statistical linearisation result linking a_1, n_1, a_{1eq} and d, final estimates of the damping parameters, and d, were generated. The results are shown in Table 4.

TABLE 4. Final estimates of the linear and non-linear damping parameters, and the input intensity.

Case	$\hat{a}_1$	$\hat{n}_1$	$\hat{d} = \sqrt{D}$
A	0.011 (0.015)	0.164 (0.062)	0.055 (0.004)
B	0.023 (0.031)	0.154 (0.084)	0.105 (0.016)
C	0.005 (0.012)	0.179 (0.048)	0.055 (0.005)
D	0.008 (0.027)	0.159 (0.114)	0.089 (0.025)

Figures 1(a) to (d) show for all four cases the variation of a non-linear damping function $Q(E)$, plotted against response amplitude, as computed from the estimated values of a_1 and n_1. This function is defined as:

$$Q(E) = \text{mean rate of energy dissipation} \times \frac{T(E)}{4\pi E} \qquad (41)$$

Full details concerning the evaluation of $Q(E)$ are given by Debonos (1993). It is noted that when both stiffness and damping are linear $Q(E) = \zeta$ where ζ is the usual non-dimensional critical damping factor. For comparison the solid line shows the true variation of $Q(E)$, obtained using the true values of a_1 and n_1. Here, and for the other three cases (not shown), a good agreement is obtained.

5. Conclusions

Based on the results for the specific example chosen for study, the following principal conclusions are drawn:

1. All the parameters can be estimated from response measurements alone
2. Moment equations lead to good estimates of the stiffness parameters.
3. Equivalent linear damping can be estimated accurately by using a spectral relationship.
4. Separation of linear and non-linear damping contributions is more difficult - it can be achieved by using the energy envelope distribution.
5. A white noise approximation is adequate, for the cases studied. The equivalent white noise intensity is correctly estimated.

Acknowledgements

The work described in this paper was jointly funded by the Marine Technology Directorate and by British Maritime Technology (BMT). The authors also gratefully acknowledge the assistance received from Dr. R. Standing and Dr. A. Morrall of BMT.

References

Debonos, A. (1993) *Estimation of Non-linear Ship Roll Parameters using Stochastic Identification Techniques*. D.Phil. Thesis, University of Sussex.

Gawthrop, P. J., A. Kountzeris, A. &. Roberts, J.B. (1988) Parametric Identification of Non-linear Ship Roll Motion from Forced Roll Data. *J. Ship Research,* **32,** 101-111.

Kountzeris, A., Roberts, J.B. & Gawthrop, P.J. (1991) Estimation of Ship Roll Parameters from Motion in Irregular Seas. *Trans. Royal Institution of Naval Architects,* **132,** 253-266.

Roberts, J.B. (1982) A Stochastic Theory for Non-linear Ship Rolling in Irregular Seas, *J. Ship Research,* **26,** 229-245.

Roberts, J.B. and Dacunha, N.M.C. (1985) The Roll Motion of a Ship in Random Beam Waves: Comparison between Theory and Experiment. *J. Ship Research,* **29,** 112-126

Roberts, J. B., Dunne, J.F. & Debonos, A. (1992) Estimation of Ship Roll Parameters in Random Waves. *ASME J. Offshore Mechanics and Arctic Engineering,* **114,** 114-121.

Roberts, J.B., Dunne J.F. & A. Debonos, A. (1994) Stochastic Estimation Methods for Non-linear Ship Roll Motion. *Probabilistic Engineering Mechanics* **9,** 83-93.

Roberts, J.B., Dunne, J.F. and A. Debonos, A. (1995) A Spectral Method for Estimation of Non-linear System Parameters from Measured Response. *Probabilistic Engineering Mechanics* (to be published).

Roberts, J. B. & Spanos, P.D. (1986) Stochastic Averaging: An Approximate Method for Solving Random Vibration Problems. *International Journal of Non-linear Mechanics,* **21,** 111-134.

Roberts, J. B. & Spanos, P.D. (1990) *Random Vibration and Statistical Linearization.* J. Wiley & Sons, Chichester, UK.

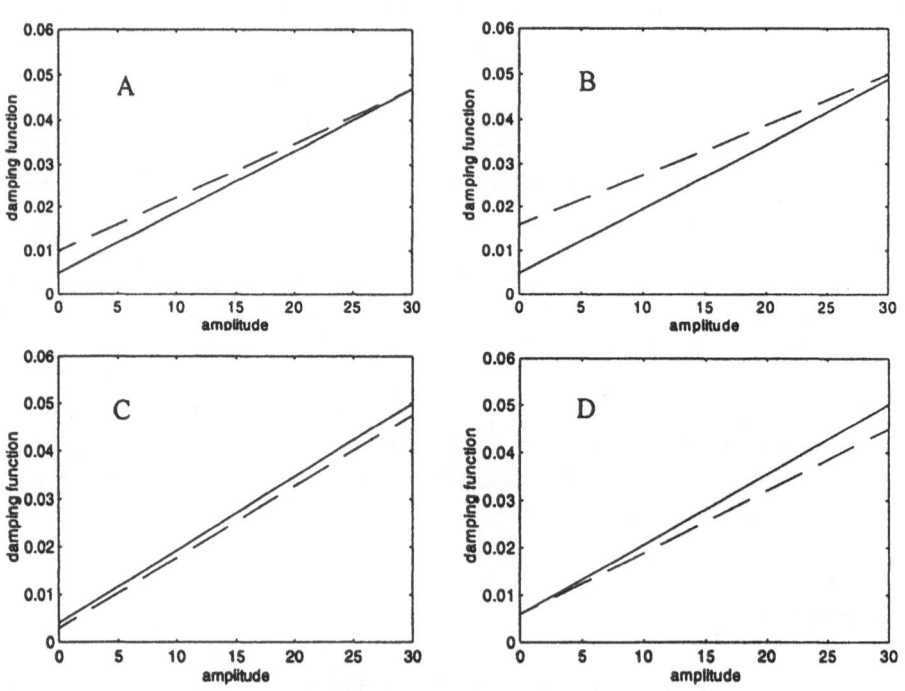

FIGURE 1. Variation of the non-linear damping function with displacement amplitude.
----- expected, - - - estimated.

AMPLITUDE BOUNDS OF STOCHASTIC
NONLINEAR MULTIBODY SYSTEMS

W. SCHIEHLEN, B. HU AND S. SCHAUB
Institute B of Mechanics
University of Stuttgart
Pfaffenwaldring 9
70550 Stuttgart, Germany

Abstract. For evaluation of the boundedness of stochastic nonlinear multibody systems, amplitude bounds with respect to the initial conditions are introduced in this paper. The amplitude bounds of stochastic nonlinear systems are analyzed based on the simulations using the Monte Carlo method. Simulations are available for systems with arbitrary finite dimensions and nonlinearities. By comparing the distribution of the random amplitude bounds, the boundedness of the system can be evaluated quantitatively in a statistical sense.

1. Introduction

Stability of dynamical systems has been studied extensively in recent decades. For the stability analysis, two sets play a crucial role. One set is composed of initial conditions and is denoted here as $\aleph_0$. The other set denoted by $\aleph$ is the range of all trajectories corresponding to the initial conditions in the set $\aleph_0$. It is clear that $\aleph \supset \aleph_0$. The sets $\aleph_0$ and $\aleph$ are usually chosen as hyberballs given by the Euclidean norm. However, for the computational analysis of responses, it is more suitable to choose them as hybercubes given by the uniform norm. The side lengths of the hypercubes $\aleph_0$ and $\aleph$ are denoted by δ and ε, respectively. The side length ε is a function of δ.

Some typical stability problems were summaried by Kushner (1967) as relations between the side lengths δ and ε. If $\lim_{\delta \to 0} \varepsilon = 0$, then, the system is stable in the sense of Ljapunov (Ljapunov, 1966). The Ljapunov

A. Naess and S. Krenk (eds.), IUTAM Symposium on Advances in Nonlinear Stochastic Mechanics, 373–382.
© *1996 Kluwer Academic Publishers.*

stability provides only a qualitative result. For a quantitative evaluation of the Ljapunov stability, the Ljapunov exponent and the degree of damping gives useful information (Benettin et al., 1980; Müller and Schiehlen, 1987). These quantities are based on the eigenvalues and characterize the rate of the convergence or divergence of the dynamical responses. However, for investigations concerned with safety and reliability, the rate of the convergence for a stable system is less important than the maximum deviation of the dynamical responses with respect to the equilibrium point. Therefore, to the authors' opinion, the relation between the maximum deviation of the trajectories ε and the maximum deviation of the initial conditions δ may be investigated, too.

Recently, a concept called stability measures was presented by Schiehlen (1993) to provide quantitative stability information on the global dynamical behavior of a deterministic nonlinear system. The stability measure is defined as δ/ε. It was shown that the stability number $\lim_{\delta \to 0} \delta/\varepsilon$ has a direct relation to the Ljapunov stabilty and characterizes the Ljapunov stabilty quantitatively (Hu et al, 1995). Inspired by these results, efforts have been made to extend this concept to stochastic systems. Since stability is defined in terms of boundedness and convergence, and since convergence of a sequence of random variables can be interpreted in different ways, different definitions for stochastic stability are possible (Kozin, 1972; Lin and Cai, 1993). Therefore, it is difficult to provide a quantitative measure for the stochastic stability. In this paper, however, a random amplitude bound with respect to the initial conditions is introduced, which provides some quantitative information about the boundedness of a stochastic system.

2. Definition of amplitude bounds

Dynamical equations describing stochastic nonlinear multibody systems are represented in the state space as

$$\dot{\mathbf{x}} = \mathbf{f}(\mathbf{x}, \xi_t, \mathbf{p}), \quad \mathbf{x}(t_0) = \mathbf{x}_0, \tag{1}$$

where $\mathbf{x}$ is the vector of state variables, $\mathbf{x}_0$ is an $n \times 1$ random initial state vector, ξ_t is an $l \times 1$ stochastic vector process including also random parameters, $\mathbf{p}$ is an $m \times 1$ deterministic parameter vector and $\mathbf{f}$ is an $n \times 1$ deterministic function depending on $\mathbf{x}$, ξ_t and $\mathbf{p}$.

For the classical stability analysis, it is assumed that the origin $\mathbf{x} = \mathbf{0}$ is an equilibrium position, i.e. $\mathbf{f}(\mathbf{0}, \xi_t, \mathbf{p}) = \mathbf{0}$. However, for the boundedness analysis, the origin $\mathbf{x} = \mathbf{0}$ can be simply regarded as a nominal position and the function $\mathbf{f}(\mathbf{0}, \xi_t, \mathbf{p})$ need not to be equal to zero. For the definition of the amplitude bound, the uniform norm or the absolute value norm for the sample solution $\mathbf{x}(t)$ is used. The time–variant norm of the $n \times 1$–state vector $\mathbf{x}(t) = [x_1(t) \;\cdots\; x_n(t)]^T$, $t \in [t_0, \infty)$ is defined as

$$\|\mathbf{x}(t)\| := \max_{1 \le i \le n} |x_i(t)| . \tag{2}$$

The time–interval norm reads as

$$\|\mathbf{x}(t)\|_T := \max_{t \in [t_0, t_0 + T]} \|\mathbf{x}(t)\| \tag{3}$$

where time T may approach infinity,

$$\|\mathbf{x}(t)\|_\infty := \max_{t \in [t_0, \infty)} \|\mathbf{x}(t)\| \tag{4}$$

The two sets $\aleph_0$ and $\aleph$ mentioned in the introduction section are chosen as hypercubes with side length δ and ε respectively, i.e.

$$\aleph_0 = \{\mathbf{x}_0 \mid \|\mathbf{x}_0\| \le \delta\}, \qquad \aleph = \{\mathbf{x} \mid \|\mathbf{x}\| \le \varepsilon\} . \tag{5}$$

It yields $\varepsilon \ge \delta$ and

$$\varepsilon = \varepsilon(\delta) = \max_{\|\mathbf{x}_0\| \le \delta} \|\mathbf{x}(t)\|_\infty . \tag{6}$$

A stochastic system is called almost surely stable or sample stable (in the sense of Ljapunov) if

$$P\{\lim_{\delta \to 0} \varepsilon(\delta) = 0\} = 1 . \tag{7}$$

Based on this definition, an amplitude bound with respect to an initial condition for a sample trajectory is defined as

$$B(\xi_t, \mathbf{p}, \mathbf{x}_0) := \frac{\|\mathbf{x}(t)\|_\infty}{\|\mathbf{x}_0\|} . \tag{8}$$

For the statistical analysis, its inverse

$$IB(\xi_t, \mathbf{p}, \mathbf{x}_0) := \frac{1}{B(\xi_t, \mathbf{p}, \mathbf{x}_0)} , \tag{9}$$

is preferable, since the amplitude bound can be infinite. It yields

$$B \in [1, +\infty) \quad \text{and} \quad IB \in [0, 1] . \tag{10}$$

In numerical analysis the integration interval is limited. The amplitude bound has to be replaced by

$$B_T(\xi_t, \mathbf{p}, \mathbf{x}_0) := \frac{1}{IB_T(\xi_t, \mathbf{p}, \mathbf{x}_0)} := \frac{\|\mathbf{x}(t)\|_T}{\|\mathbf{x}_0\|} \tag{11}$$

for finite time interval. Similarly, it remains $B_T \in [1, +\infty)$ and $IB_T \in [0, 1]$.

It should be noted that the inverse amplitude bound corresponds to the local stability measure for deterministic systems (Schiehlen, 1993). However, the amplitude bound and its inverse defined here are random variables. They are characterized by their probability distributions.

3. Amplitude bounds and stability

There exist some relations between the stochastic stability and the random amplitude. In the discussion of stochastic stability, the initial conditions are assumed to be deterministic. Since the Ljapunov stability concerns only about the dynamical behavior in the neighborhood of an equilibrium, one needs only to consider the random amplitude bounds for those initial conditions which belong to some small hypercubes $\aleph_0$ with side length δ. If there exists a small hypercube $\aleph_0(\delta)$ and a large real number M, such that if $\mathbf{x}_0 \in \aleph_0(\delta)$ then

$$P\{B \leq M\} = 1, \tag{12}$$

i.e. the amplitude bound is bounded in $\aleph_0(\delta)$ with probability 1, then the system is almost surely stable and stable in the mean. Since Equation (12) implies

$$P\{\|x(t)\|_\infty \leq M\|\mathbf{x}_0\|\} = P\{\varepsilon(\delta) \leq M\delta\} = 1 \tag{13}$$

and

$$E[\|\mathbf{x}(t)\|_\infty] \leq M\|\mathbf{x}_0\|, \tag{14}$$

it follows

$$P\{\lim_{\delta \to 0} \varepsilon(\delta) = 0\} = 1 \quad \text{and} \quad \lim_{\delta \to 0} E[\varepsilon(\delta)] = 0, \tag{15}$$

where $E[\cdot]$ means the mathematical expectation of a random variable.

Moreover, a stronger sufficient condition for the almost sure stability is, that the amplitude bound is globally bounded with probability 1, i.e. there exists a large number M such that $P\{B \leq M\} = 1$.

When a system is almost surely stable, the system is also stable in probability. On the other hand, if a system is instable in probability, this system is also not almost surely stable. A sufficient condition for the instability in probability is, that there exists a series of initial conditions $\{\mathbf{x}_{0_n}, n = 1, 2, \cdots\}$ satisfying

$$\lim_{n \to \infty} \|\mathbf{x}_{0_n}\| = 0 \quad \text{and} \quad \lim_{n \to \infty} P\{IB(\xi_t, \mathbf{p}, \mathbf{x}_{0_n}) = 0\} > 0, \tag{16}$$

which is also a sufficient condition for the instability in the mean.

In the engineering applications, the almost sure stability plays a more important role than the stability in probability and in the mean (Kozin, 1972).

4. Numerical approach

For the evaluation of the boundedness of a stochastic system, the probability distribution of the random amplitude bound B can be used as a measure. But in the numerical analysis, the inverse of the amplitude bound IB is preferred since it is a normalized random variable in $[0,1]$. For simple evaluations, the expectation $E[IB]$ can be used as a measure. The larger the expectation $E[IB]$ of a system is, the better is the boundedness of the system. For an accurate evaluation, a statistical inference has to be used where the probability distribution plays a crucial role.

For nonlinear systems, the analytical computation of the amplitude bound is in general impossible. In the numerical analysis of the amplitude bound, the initial conditions can be regarded as random variables, too. The randomness of the amplitude depends on the stochastic process ξ_t and the random initial conditions $\mathbf{x}_0$.

For numerical computation, the Monte Carlo simulation is used. Since the accuracy of the Monte Carlo simulation is not restricted by the number of equations of motion involved and the kind of the nonlinearity (Pradl-warter, 1994), it is especially suitable for nonlinear multibody systems. However, the Monto Carlo method requires much computer time. In order to compute the amplitude efficiently, some efforts are devoted to the simulation of stationary processes and the numerical integration. The simulation of stationary processes is based on the spectral representation. An algorithm developed by Shinozuka and Deodatis (1991) is adopted. Sample functions of a stochastic process are described using a cosine formula and are generated by the FFT technique. For the numerical integration, an algorithm from Shampine and Gordon (1975) is used. Since the sample functions of stochastic processes are a series of discrete values, they are interpolated with cubic polynoms. In order to reduce integration time, only the extrema of the dynamical responses are computed. Comparing all extrema, one can get a sample of the random amplitude. By repetition of this procedure independently fairly often, the frequency distribution of the amplitude bound can be obtained. With sufficient samples of the amplitude bound, the frequency distribution can be regarded as the probability distribution of the amplitude bound. Then, using this distribution, one can quantitatively evaluate the boundedness of a stochastic system.

5. Application

The Duffing oscillator represents a wide range of dynamical systems with nonlinear spring restoring forces. As an example, it is used to illustrate the application of the presented approach. The equation of motion reads as

$$\ddot{y} + c\dot{y} + y + \alpha y^3 = \xi_t, \tag{17}$$

where c is a damping coefficient, and α is a parameter describing the non-linearity. The excitation ξ_t is a stationary Gaussian process. We consider the boundedness of the stochastic response for a finite time interval. In this paper, it is assumed that the initial time is $t_0 = 0$ and the time interval is set to $T = 200$.

At first, it is assumed that $\xi_t = 0$, i.e. only free vibrations of the Duffing oscillator are discussed. The damping coefficient c is assumed to be a uniform random variable in $(0, 0.2)$. The initial conditions y_0 and $\dot{y}_0$ are chosen as standard normal random variables with a mean value equal to zero and a variance equal to one. For simplicity, the three random variables c, y_0 and $\dot{y}_0$ are assumed to be independent of each other. Figure 1 shows the distributions of the random amplitude bound for different values of α. It can be seen that the linear system ($\alpha = 0$) show the best boundedness. Moreover, it is of interest to note that the responses of the nonlinear system for $\alpha = -0.1$ are not globally bounded even though its equlibrium $y = 0, \dot{y} = 0$ is almost surely stable. The statistical results for different values of α are also shown in Figure 1. It can be seen that the maximum of the inverse of the amplitude bound remains constant under variation of the parameter α, while the minimum varies inhomogenously and shows a discontinuity at $\alpha = 0$. For $\alpha > 0$, all sample trajectories remain bounded. However, for $\alpha < 0$, some sample trajectories are unbounded. Therefore, it reveals a bifurcation for global boundedness at $\alpha = 0$.

Next, the damping coeffient is set to $c = 0.1$. Two stationary excitations ξ_t are considered. One excitation is a wide-band white noise in $0 - 50$ Hz, its spectral density reads as

$$S_w(\omega) = \begin{cases} 0.002 & \text{for} \quad \omega \le 100\pi \\ 0 & \text{else} \end{cases}. \tag{18}$$

The second excitation is a narrow-band stationary process with the central frequency at $\omega = 2$. It has a rational spectral density

$$S_n(\omega) = \frac{0.01749}{(-\omega^2 + 4)^2 + (0.4\omega)^2}. \tag{19}$$

For the linear system ($\alpha = 0$), the two excitations lead to a stationary response $x(t)$ with the same variance $\sigma_y^2 = 0.01$. The initial conditions y_0 and $\dot{y}_0$ are still independent standard normal random variables. Figure 2 shows 1000 samples of the random amplitude bound for the wide-band white noise excitation for different parameters α. Figure 3 shows 10000 samples of the random amplitude bound for the narrow-band excitation. The statistical analysis shows that for both excitations the linear system has the best boundedness.

6. Conclusion

Multibody systems result in highly nonlinear equations of motion typical for engineering dynamics and experience often random disturbances. In engineering applications only bounded motions are acceptable. In order to evaluate the boundedness of a stochastic system, the random amplitude bound is introduced in this paper. The random amplitude bound has some direct relations to the stochastic stability and can be simulated by the Monto Carlo method. By means of statistical inference, the boundedness of a stochastic system can be quantitatively evaluated and optimal values for parameters can be chosen so that the system is robust against the random disturbances.

References

Benettin, G.; Galgani, L. and Giorgolli, A. (1980), Ljapunov characteristic exponents for dynamical systems and for Hamiltonian systems: a method for computing all of them, *Meccanica* **15**, 9 – 30.

Hu, B., Schaub, S. and Schiehlen, W. (1995) Boundedness Evaluation of Structural systems with Randomly Distributed Initial Conditions, submitted to *J. of Nonlinear Mechanics*.

Kozin, F. (1972) Stability of the linear stochastic systems, In: Curtain, R.F. (ed.): *Stochastic Stability, Lecture notes in Mathematics No. 294* , Springer-Verlag, Berlin, pp. 186 – 229.

Kushner, H.J. (1967) *Stochastic Stability and Control*, Academic Press, New York and London.

Lin, Y.K. and Cai, G.Q. (1993) Stability in probability of some nonlinear stochastic systems. In: Schuëller, G.I., Shinozuka, M and Yao, J.T.P. (eds.), *Structural Safety and Reliability*, ICOSSAR'93, Vol. 1, A. A. Balkema Publishers, Rotterdam, pp. 165–172.

Ljapunov, A.M. (1966) *Stability of Motion*, Academic Press, London. (English translation of the in 1893 published original work in Russian)

Müller, P.C. and Schiehlen, W. (1985) *Linear Vibration*, Kluwer, Dordrecht.

Pradlwarter, H.J., Schuëller, G. I. and Melnik-Melnikov, P.G. (1994) Reliability of MDOF-systems. *Probabil. Engng. Mech.* **9**(4), 235–243.

Schiehlen, W. (1993) Nonlinear Oscillations in Multibody Systems –Modeling and Stability Assessment– , In: Kreuzer, E. and Schmidt, G. (eds): *1st European Nonlinear Oscillation Conference*, Academic-Verlag, Berlin, pp. 85 – 106.

Shampine, L.F. and Gordon, M.K. (1975) *Computer Solution of Ordinary Differential Equations*, Freeman, San Francisco.

Shinozuka, M and Deodatis, G. (1991) Simulation of stochastic processes by spectral representation. *Appl. Mech. Rev.* **44**(4), 191–203.

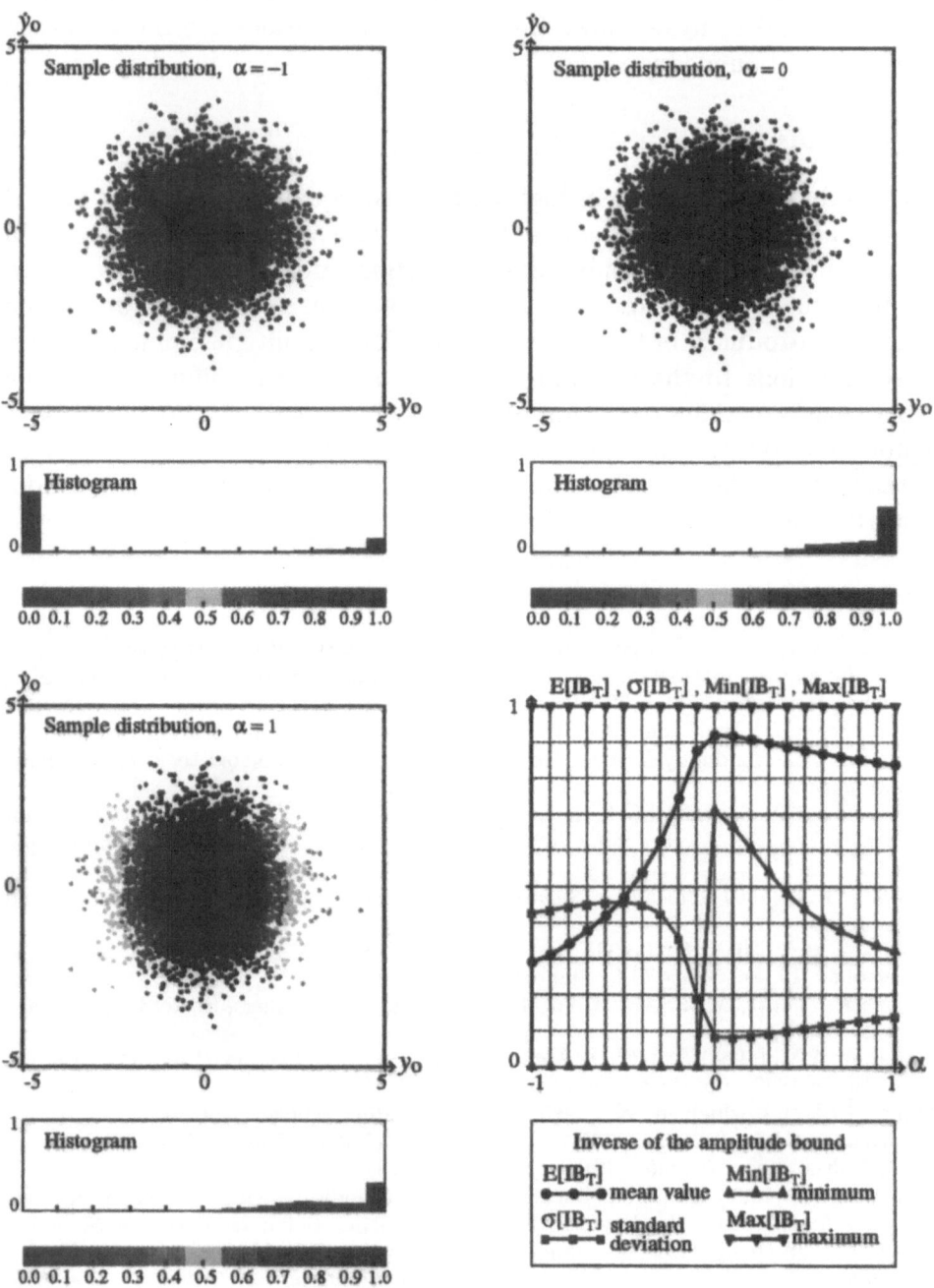

Figure 1. The distribution of IB_T and statistical quantities for different α with the uniform random damping coefficient c.

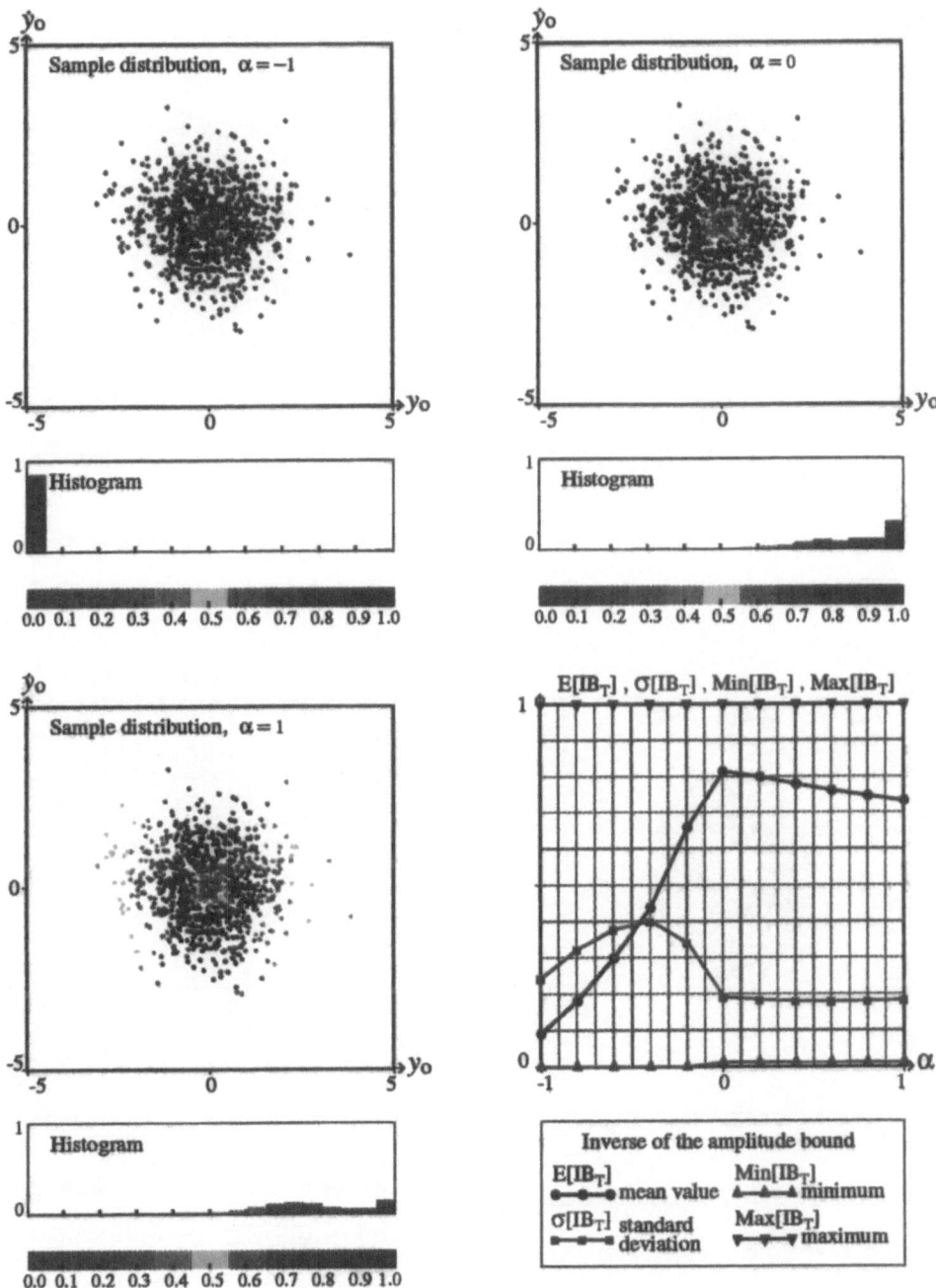

Figure 2. The distribution of IB_T and statistical quantities for different α with the wide–band white noise excitation.

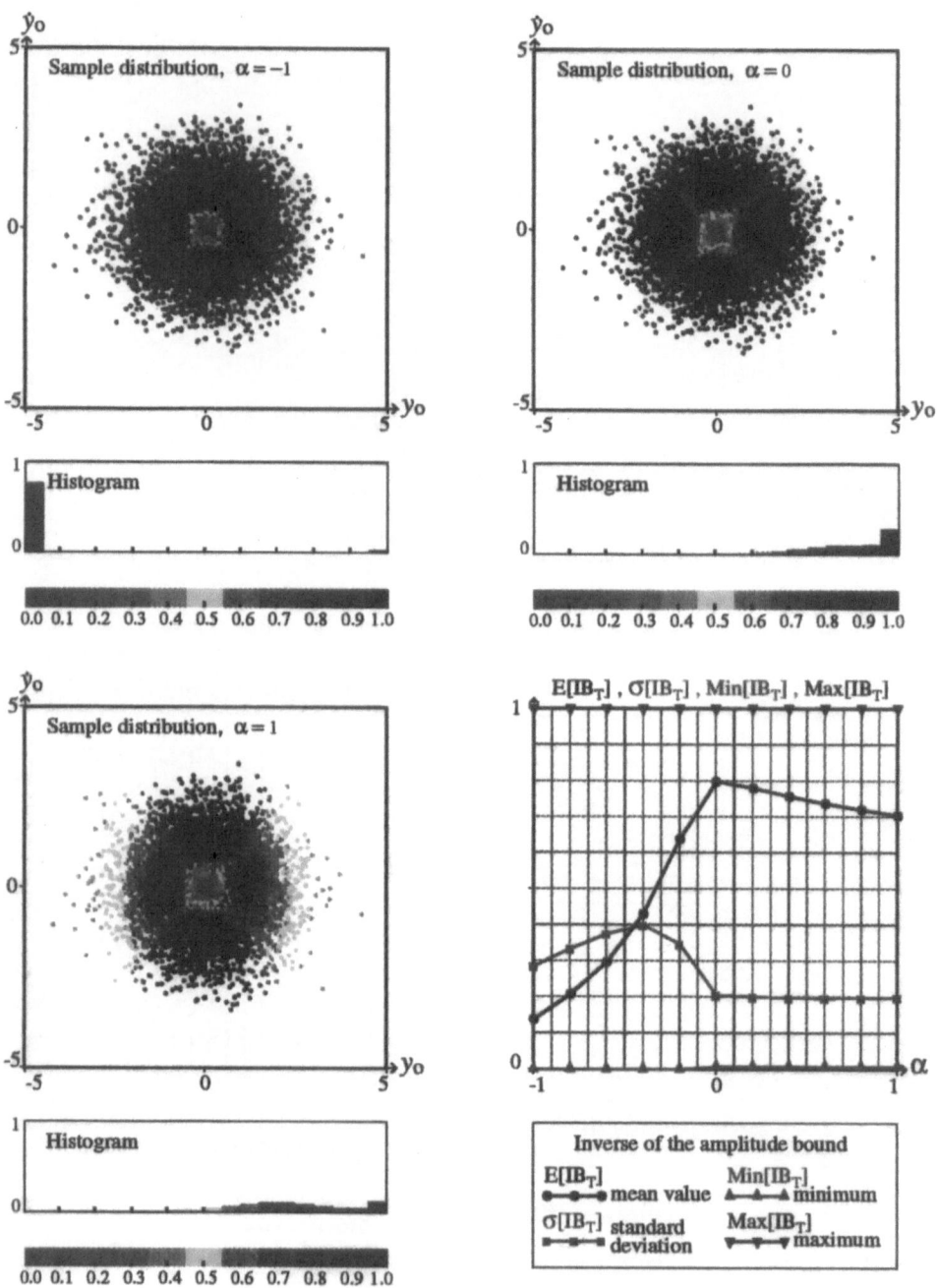

Figure 3. The distribution of IB_T and statistical quantities for different α with the narrow–band excitation.

ENGINEERING APPLICATIONS OF STOCHASTIC MECHANICS -

Achievements and Prospects

G.I. SCHUËLLER
Institute of Engineering Mechanics
University of Innsbruck
Innsbruck, Austria

Abstract: Various procedures of stochastic structural mechanics are discussed in view of their potential to be utilized for engineering applications. These procedures encompass direct and advanced Monte-Carlo simulation, response surface methodology, statistical equivalent linearization and hybrid methods. Subsequently, representative classes of areas of applications of these procedures are discussed i.e. stochastic structural analysis (systems reliability), stochastic fracture and fatigue analysis, stochastic Finite Elements and stochastic structural dynamics. Then the respective states of application of these methods in three selected fields of engineering i.e. aerospace, nuclear and offshore engineering are reviewed.

Finally some conclusions are drawn, particularly with respect to future prospects of stochastic mechanics in engineering practice.

1. Introduction

Almost 70 years went by since it was suggested first, that the uncertainties inherent in the analysis and design of structures and their components should be modeled by probability distributions [1]. Hence it is certainly timely to attempt to review the current status of stochastic methods in structural analysis and mechanics as applied in the engineering practice. Questions w.r.t. the requirements for acceptance of stochastic methods by the practicing engineering community are most relevant, also for the prospects, i.e. future developments of the field. Already *A.M. Freudenthal* - the founder of the field of Structural Reliability as we know it today - envisaged, that "probabilistic reasoning and the application of statistical methods must become an integral part of the procedures of structural design and analysis" [2]. In his recently held *Freudenthal*-Lecture at ICOSSAR'93, *J.N. Yang* stated in his conclusions [3]: "To date, the reliability method has not been widely used as an analysis or design tool in industry. "

Although this assessment is not too encouraging, one might be interested in analyzing the various stages of application. This implies to identify those methods which

A. Naess and S. Krenk (eds.), IUTAM Symposium on Advances in Nonlinear Stochastic Mechanics, 383–402.
© *1996 Kluwer Academic Publishers.*

(a)   have already surpassed the academic stage,
(b)   are used for a special type of analysis (as a "back up" for deterministic analyses in terms of sensitivity analyses)
(c)   are utilized as design tool for regular design (reliability oriented design)
(d)   are used for licensing procedures and codified design
(e)   are utilized for risk studies.

There appears to be not much disagreement, that, despite of the significant progress made so far, particularly over the last 30 years, a large number of procedures are still in the stage of development, i.e. have not yet passed the academic stage. Although, by and large, reliability oriented design has not yet entered the day-to-day design procedures, it is sometimes used in context with sensitivity analyses, following a deterministic design. Generally, licensing procedures are still carried out without utilizing the possibilities of rationally quantifying the uncertainties. However, just recently codified design procedures of civil engineering structures are going through a considerable change, i.e. by adapting semi-probabilistic concepts, where the (partial) safety factors are related to particular target failure probabilities (see e.g.[4]). Finally methods of stochastic structural mechanics and reliability are used - at least to a certain extent - for risk studies of technological systems.

As a discussion or review of engineering applications of Stochastic Mechanics in no way can be exhaustive, it will be concentrated here on a limited number of subject areas, such as Stochastic Structural Analysis (including systems reliability), Stochastic Fracture Mechanics and Fatigue, Stochastic Finite Elements and Stochastic Dynamics. In this context various procedures such as the response surface method (RSM), direct and advanced Monte Carlo Simulation (MCS), statistical equivalent linearization (EQL) as well as hybrid methods are discussed. Engineering applications of some of these procedures are limited to representative areas such as aerospace-, nuclear- and offshore structures.

2. Procedures

2.1 GENERAL REMARKS

The processing of the additional information characterizing the inherent uncertainties in the analysis and design process requires the development and application of procedures which are, by and large, still not yet common practice. Although it should be a declared goal to utilize the state-of-the-art in mechanical modeling, when applying stochastic methods in structural mechanics i.e. to use those mechanical models which are applied in deterministic design, this is quite frequently not the case, and hence costs credibility and may also contribute to the reluctance of the engineering community to utilize these methods at all. The processing of the uncertainties requires certainly increased efforts and, moreover, additional computational procedures. In the following some of the procedures which are believed to have the potential of being used in the engineering practice are discussed.

2.2 MONTE CARLO SIMULATION PROCEDURES

Monte Carlo Simulation (MCS) is certainly the most general approach among all available procedures. It is applicable to linear and nonlinear problems in the same way, hence no distinction between linear and nonlinear problems is required. And, most important, the accuracy of MCS is also independent of the dimensionality of the problem, where the accuracy of the procedure depends only on the sample size and the capability of the random number generator to produce pseudo statistically independent and equally distributed sets of random variables. These two outstanding properties make MCS especially suitable when dealing with large and strongly nonlinear structural systems, discretized by FE-models. In order to increase the efficiency of the direct MCS, so called variance reduction techniques have been developed or applied (see e.g. [5-8]). All these procedures increase the density of Monte Carlo realizations in the region of interest, i.e. in the region which contributes most to the total failure probability. These variance reduction methods are utilized for nonlinear static and dynamic reliability problems involving a limited set of important random variables which may be determined by prior sensitivity analyses. Realizations in the low probability tails are only obtainable for cases where the weight of the realizations are considerably decreased. Applying procedures based on MCS, independence among all realizations appears to be important. Now to each realization a weight is assigned and normalized such that they add up to one. These weights $w_n(t)$ play then the role of discrete probabilities and the cumulative distribution function CDF(x) is approximated by

$$CDF(x;t) \cong \sum_{n=1}^{nSim} I[X_n(t),x] \cdot w_n(t) \qquad (1)$$

where nSim denotes the sample size and $I[X_n,x]$ is an indicator function. This function assumes the value 1 in case each of the components $X_{nk} < x_k$, k = 1,2,..., nState, of the n-th state vector X_n is smaller than the components of the vector x. Otherwise, the indicator function assumes 0, i.e. $I[X_n,x] = 0$. In case direct MCS is applied, all weights are obviously equal to 1/nSim. As a consequence, no estimates can be obtained for the region x with CDF(x)< 1/nSim or CDF(x) > 1.0-1/nSim.

2.3 RESPONSE SURFACE METHODOLOGY (RSM)

The response surface is usually expressed by a simple continuous mathematical model, for example by a first or second order polynomial. This analytical expression is then fitted to the data by regression analysis. The RSM has been adapted to treat structural problems. One of the major advantages of the RSM applied to structural problems is also its generality. Any available experiment and analytical method can be utilized to establish an approximate relation between important parameters and the response surface. Hence, the treatment of highly nonlinear systems does not cause any substantial difficulties. Due to this feature the RSM is well suited for solving structural reliability problems. In early developments the response was established in terms of response properties, such as critical displacements, damage and other relevant response quantities. On the basis of the obtained response surface, the reliability expressed by the failure probability can be obtained (see e.g. [9]). In a further development, the response surface has been established in terms of the so called limit state function (see e.g.[10]). The limit state function divides the set of input variables in a so called safe domain (i.e. input sets leading to survival) and an unsafe domain. The procedure is well established for nonlinear static problems and also dynamic structural problems in case the stochastic process can be represented sufficiently well by a few random variables. Note that the procedure is also applicable for structural problems where the loading is represented by a stochastic process. The RSM is utilized to determine points on the limit state, i.e. set of input variables resulting in a transition from safe to unsafe states. In general, as stated above, the limit state function is the approximated by a second order polynomial,

$$\overline{g}(x_j) = a_0 + \sum_{i=1}^{r} b_i \cdot x_{ij} + \sum_{i=1}^{r} \sum_{k=1}^{i} c_{ik} \cdot x_{ij} \cdot x_{kj} \tag{2}$$

where the regression coefficients a_0, b_i and c_{ik} are determined by a weighted least square procedure. An overbar is used in eq. (2) in order to indicate that $\overline{g}(\mathbf{x})$ is an approximation of the true limit state function $g(\mathbf{x})$.

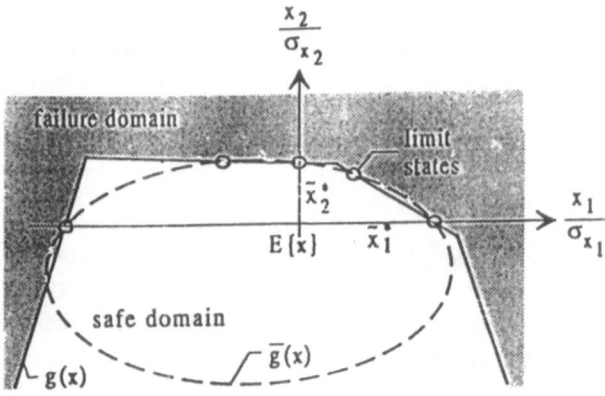

Figure 1: The limit state function and its response surface approximation

To determine a single point at the limit state function several nonlinear FE-analyses are required. Each FE-analysis leads to the outcome defined by failure or no failure respectively. Note, that this procedure has no restriction with respect to mechanical modeling, i.e. any mechanical model reflecting the current state-of-the-art can be utilized. The efficiency of the procedure, however, depends strongly on the number of random variables considered, i.e. the number of varying input parameters. In order to avoid an extensive numerical effort, the number of random variables should not exceed a certain size, preferably less than twenty. A sensitivity analysis may lead to the most significant random variables, i.e. those variables with the largest effect on the failure probability. The remaining variables are considered in an approximate manner, for example by Monte Carlo simulation, described above.

2.4 STATISTICAL EQUIVALENT LINEARIZATION

The procedure of Statistical Equivalent Linearization (EQL) applies to nonlinear systems under dynamic external excitation. However, it does not apply, to parametric excitation and for uncertain structural parameters. In the procedure of EQL the set of coupled nonlinear differential equations is replaced by a "statistically equivalent" linear set for which well developed and efficient procedures exist to determine the stochastic response (see e.g.[11]). One basic drawback of the method is that only second moment properties of the response quantities are obtained. Since the nonlinear response is known to be non-Gaussian, the first two moments are certainly insufficient for a complete characterization of the response. The method provides an overall description of the stochastic response, but in its conventional form it cannot be used for reliability analysis, because in the latter the probability density function at the tails must be utilized. The conventional EQL assumes a jointly normal distribution of the response. Generally, this assumption does not hold for nonlinear systems, even if the excitation is Gaussian. It would be possible to utilize other types of distributions (as shown e.g. in [28]), provided this type would be known. However, the type or shape of distribution needs to be assumed a priori and is only known in special cases. To overcome this deficiency of the method, a so called non-Gaussian statistical EQL has been suggested by utilizing a suitable nonlinear transformation between the normally distributed and the non-normally distributed state vector respectively (see e.g. [12]). Although the conventional procedure of statistical EQL has been developed already to a stage where it can be considered as user friendly (see e.g. [13]), there are still some difficulties to overcome to adapt it to the Finite Element (FE) method (see e.g.[14, 15]) which is used almost exclusively by the engineering practice.

As mentioned above, the procedure of EQL replaces a nonlinear system by a linear system by minimizing the mean square error between the two systems. For example, the equation of motion of a MDOF-system can be expressed in general form by

$$\mathbf{M\ddot{u} + C\dot{u} + Ku + Qq = f(t)} \tag{3}$$

where $\mathbf{M}$, $\mathbf{C}$ and $\mathbf{K}$ represent the mass, damping and stiffness matrices respectively; $\mathbf{f}(t)$ stands for the stochastic excitation vector; $\mathbf{u}, \dot{\mathbf{u}}, \ddot{\mathbf{u}}$ are the displacement and its time derivative vectors respectively; $\mathbf{q}$ is the nonlinear term of the restoring force. This leads to the following nonlinear differential equation

$$\dot{\mathbf{q}} = \mathbf{g(x)} \tag{4}$$

where $\mathbf{x}^T = \left\{ \mathbf{u}^T, \dot{\mathbf{u}}^T, \mathbf{q}^T \right\}$ This equation is now being linearized

$$\dot{\mathbf{q}} \approx \mathbf{v} = \mathbf{v}_0 + \mathbf{Ax} \tag{5}$$

where the matrix $\mathbf{A}$ contains the linearization coefficients. To obtain these coefficients a minimization criterion, in a general form (see e.g.[16])

$$\sum_i E\left\{ \left\| \mathbf{v}_j - g_j(\mathbf{x}) \right\|_2^{2mj} \right\} \quad \rightarrow \text{Min} \tag{6}$$

can be used. In context with the utilization of the FE method eq. (3) can be certainly formulated for each Finite Element *e*. Eq.(5) then reads

$$\dot{q}^{(e)} \approx v^{(e)} = v_o^{(e)} + A^{(e)} x^{(e)} \tag{5a}$$

For the solution of the equation above with respect to the applied criterion of eq.(6), analytical procedures are generally not available. Their development, though, would require enormous manpower. However, on the element level numerical and/or hybrid methods are straight forward and general, and hence applicable for all types of nonlinear elements. The required increased computational efforts would not be substantial as the major part goes into the solution of the linear stochastic differential equations (SDE). On the systems level, it should be noted, the current analytical solutions are still restricted to 1-D and 2-D elements. So far no bridge of statistical EQL to nonlinear deterministic FE-analysis, e.g. to available FE packages in nonlinear mechanics, is available.

2.5. HYBRID METHODS

The requirement as described above, which in fact would significantly improve the acceptance rate of the procedure of statistical EQL by the practicing engineer, leads directly into the development of a hybrid method, in which analytical and numerical procedures are combined. In this context samples of the state vector of the element $x^{(e)}$, as indicated in eq.(5a), are generated by MCS procedures. For each element the required expectation can then be approximated by the sample mean and subsequently the linear equation system for the unknown coefficients of A solved. This procedure would be applicable and sufficiently accurate for any type of nonlinear element. The interface to nonlinear elements must still be provided - preferably in close cooperation with FE- developers. For example nonlinear elements in commercially available FE-packages include rate-independent plasticity, nonlinear elasticity as well as rate-dependent plasticity. Note that general nonlinear FE's in the 2-D space have e.g. 8 or 24 DOF's etc., in the 3-D space 24 DOF's, etc..

3. Areas

3.1 GENERAL REMARKS

Some of the applications of the procedures, i.e. tools as described in section 2 to certain subject areas are now discussed in the following. For this purpose the problem areas of Stochastic Methods in Structural Analysis (e.g. Systems Reliability), Fracture and Fatigue, Finite Elements and Dynamics are selected.

3.2 STOCHASTIC STRUCTURAL ANALYSIS - SYSTEM RELIABILITY

Although, in principle, this subject area would encompass quite a large field, it is generally confined to the analysis of frame structures under static loading only. A number of methods have been determined so far - some of them are reviewed e.g. in [17] - however only three procedures have emerged to be more advantageous for practical application, i.e. the so called "Branch and Bound" method [18], the Response Surface Methodology (RSM) and Monte Carlo Simulation (MCS). While the first procedure is confined to static loads, the other two methods, as described above, are also applicable to structures under dynamic loading.

The "branch and bound" procedure is a so called "failure path" method, applicable to truss and frame structures. By this method - which is also based on the idealization of plastic hinge formation - the sequence of plastic hinges is determined by the probability of a cross section to satisfy the yield condition. Although the algorithm - if required - is capable of finding all failure modes, the modes of small probabilities maybe discarded by completely cutting off (bounding) failure paths of low probability of occurrence. The bounding criterion can be specified by the user. The algorithm, which starts the failure path at the cross section with the largest yield probability of all cross sections, modifies the structure by introducing a plastic hinge (and the corresponding plastic moment capacities as fictitious loads) at the respective cross section, and continues until a failure mechanism is found. Hence, each step requires an update of the stiffness matrix. Besides, the approach is limited to elasto-plastic behaviour. Moreover it is unable to take into account PΔ effects. In many cases the probability that instability occurs before the pre-determined number of hinges have developed may not be neglected. Moreover, when dealing with larger systems - due to the large number of mode combinations - it is not certain, that the algorithm finds the stochastically most relevant mode first. In view of the bounding procedure this may adversely affect the accuracy of the method.

At this stage it can be said, that the RSM and advanced MCS provide more accurate and reliable results than branch and bound. Although the RMS is most efficient, the computational effort increases considerably with the increase of the dimension. Hence future developments will have to concentrate on a combination between advanced MCS, such as directional sampling and the RSM, where adaptive algorithms have to be developed to increase the accuracy of the RS in the respective regions of interest. So far the procedures are applicable to problems of smaller size (e.g. three storey - two bay frames, etc.) only. In order to apply these procedures to problems of larger size as common to engineering practice, considerable additional developments are still required.

3.3 STOCHASTIC FRACTURE AND FATIGUE ANALYSIS

An extensive description of the status of the current developments of this field, particularly w.r.t. reliability estimation is presented in [3]. In modern developments this should be seen in context with quality assurance and maintenance. Hence reliability methods using S-N fatigue data (Wöhler diagram) are of less importance. Stochastic fatigue crack growth models have been discussed in detail e.g. in [19, 20]. The crack propagation rate can be described in general by the form

$$\frac{da}{dn} = f\left(a, S_{max}, S_{min}\right) \tag{7}$$

where a is the crack length, S_{max} and S_{min} are the maximum and minimum values of the far-field stress in the n-th load cycle respectively. For most practical cases, which are of linear type, the well known *Paris* law

$$\frac{da}{dn} = C(\Delta K)^m \tag{8}$$

describes the crack propagation rate da/dn sufficiently well. In the above equation ΔK represents the stress intensity range; the parameters C and m are being regarded as material properties. It should be noted that, as ΔK is derived from linear theory of fracture mechanics, it is only valid within the assumptions of this concept. In order to account for the stochastic characteristics of the crack propagation, the problem can be formulated in terms of a stochastic differential equation (SDE) of the type

$$\frac{da}{dn} = X(t)\, g(\Delta K,\, R) \tag{9}$$

where t denotes either time or number of cycles, X(t) a random process and R the stress ratio. The random variable approach to eq (9), where X(t) is replaced by a random variable X, is the simplest, and hence utilized quite frequently in engineering practice. In this context the model in which the evolution of the crack size a(t) is a discrete Markov Chain (see e.g. [21, 22]) should be also mentioned. Finally, again, MCS proves to be a useful and most practical tool for this purpose as well.

Aside from aerospace structures, where the crack initiation process plays a very important role, for welded structures the realistic modeling of the *initial crack distribution* is of significance. In this context information on the non detection probability of various non destructive crack detection devices has to be known (see e.g.[23, 24])

A failure analysis of structures prone to fatigue i.e. brittle failure must also take into account the possibility of ductile failure. These two failure conditions, i.e. failure modes can be combined by using the so-called "Two-Criteria" approach (Fig.2).

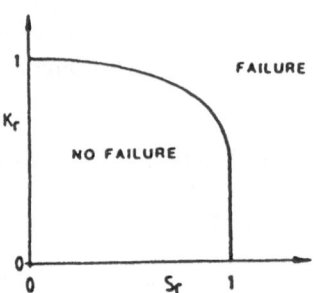

Figure 2: Schematic sketch of the "Two Criteria Approach"

Brittle failure is defined by $K_I \geq K_{IC}$, i.e. the stress intensity factor exceeds its critical value of the fracture mode I, whereas ductile failure occurs if the stress $\sigma >$ σ_{fail}, i.e. the stress exceeds a value, at which failure is assumed (e.g. yield stress, rupture stress, etc.). Note, that the mixed failure mode may also be treated by this approach. From Fig.2 it can be seen clearly that failure is expected to occur when the fracture checking point lies outside the so called safe, (i.e. no failure) region. Various suggestions may be found in the literature to describe this region. Most frequently the so called R6-curve as defined by

$$K_r = \cfrac{S_r}{\sqrt{\cfrac{8}{\pi^2} \ln \, \sec\left(\cfrac{\pi}{2}S_r\right)}}$$
(10)

is used where $K_r = K_I / K_{IC}$ and $S_r = \sigma / \sigma_{fail}$.

As a consequence of the fatigue crack growth, the failure behavior must be considered time variant. This in turn affects the estimates of the failure probability and the hazard function (which in fact expresses a conditional failure probability). If a structure is to be designed for a certain target failure probability, say 10^{-6}, periodic inspection (with and without repair) is to be applied to guarantee this value. From this a rational measure of inspection intervals maybe derived (see e.g. [25, 26]). As shown for example in Fig.3, this depends also on the fact whether or not the crack removal is perfect.

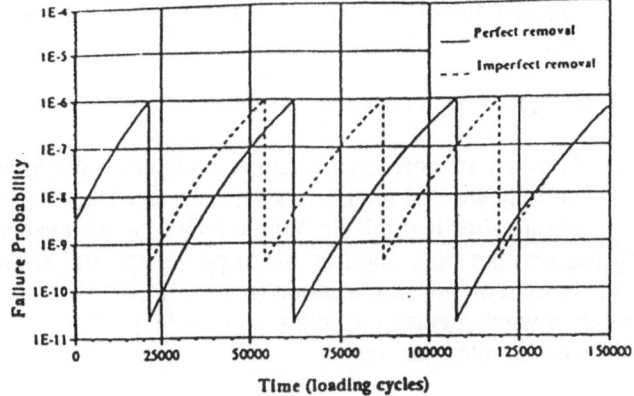

Figure 3: Failure probabilities accounting for high-cycle fatigue and periodic inspection, where an upper bound of 1 x 10^{-6} has been imposed [26].

Summarizing, some of the methods of stochastic fracture mechanics and fatigue have already successfully disseminated into engineering practice.

3.4 STOCHASTIC FINITE ELEMENTS (SFE)

In the course of the development of the procedures and methods in Stochastic Structural Mechanics, the effect of the random variability of material parameters, such as the Young's modulus, as well as the geometry, due to imperfections received increasing attention. Its practical implications, however, are still quite limited. For quantification of this variability, currently used deterministic Finite Element procedures are being expanded. This requires the modeling of the parameters as mentioned above by random fields which allows to take into account the special variability of the respective properties. The Young's modulus, for example is generally modeled by the following expression

$$E = E_o(1 + f(x))$$
(11)

where f(x) represents a zero mean random field with a particular autocorrelation function and E_0 is the mean value of the Young's modulus. The class of semi-analytical methods of solution to the problem requires generally series approximation. For the numerical solution, the MCS proves to be very useful. For a reliability analyses using Stochastic Finite Elements (SFE) the application of the response surface method (RSM) is most efficient. In this context the large number of random variables can be reduced to the most important ones by a sensitivity analysis (see e.g. [27]). This way only few time consuming SFE-computations are required. It allows, e.g. at the current stage, the dynamic analysis of structures of a size of about 400 DOF under stochastic loading assuming randomly distributed Young's modulus and yield stress respectively. It should be noted, however, that in some cases the effect on the reliability estimate as compared to utilizing deterministic properties appears to be not very significant, though.

In summarizing one can say that the procedures of SFE have developed now to such a stage that they may become of interest for the engineering practice in the near future. However, considerable efforts are still required in order to further reduce the computational efforts.

3.5. STOCHASTIC STRUCTURAL DYNAMICS

The current state of the area of stochastic structural dynamics is described in detail e.g. in [29 - 32]. It is well known that for linear systems under Gaussian excitation well developed procedures are already available. With reference to their capabilities to serve the engineering practice the early applications of power spectral analysis applied to linear systems in the field of wind engineering (see [33]) - which in fact form still the basis of most modern wind engineering codes - should be pointed out. The method is based on the following simple equation

$$S_y(\omega) = |H(\omega)|^2 S_x(\omega) \tag{12}$$

where $S_y(\omega)$ and $S_x(\omega)$ are the power spectral densities on the response and excitation respectively and $H(\omega)$ is the transfer function of the structure derived from the linear equation of motion; ω represents the circular frequency. As the integration over the power spectral density yields the variance of the process, the probability distribution for the response is completely determined for a given Gaussian input. For non-Gaussian excitation difficulties arise which are similar to those of linear systems as described below. For this purpose the utilization of higher order moments, higher order spectra, series approximations, etc. (see [34, 35]) are required and hence systems of higher dimension, as used in engineering practice, are quite difficult to analyze.

For practical application of nonlinear systems mainly approximate solutions are available (see e.g.[36]) such as the perturbation method, equivalent linearization and (advanced) MCS method. The direct MCS provides insufficient information on the tails of the distribution. The RSM is also applicable, however, not well suited for wide band excitation. The perturbation method, although applicable to structures of larger dimension, applies only to problems concerned with small nonlinearities.

Again, for most methods the high dimensionality of the problem and the requirement to treat strongly nonlinear systems (collapse failure condition) may cause major difficulties. Hence their utilization for most important practical problems of determining the survival probability of larger structures, for example under earthquake conditions, is still limited.

4. Applications

4.1 GENERAL REMARKS

Stochastic methods of structural mechanics - as already been pointed out above - are still not being used extensively within the engineering practice. Besides the fact that many decision makers, and also engineers, are still not too familiar with methods of probability and statistics, there is also some criticism articulated concerning stochastic methods. I.e. it is frequently claimed that:

(1) there is still an insufficient amount of data available to perform a credible, i.e. realistic analysis

(2) quite simplified structural models are being used, i.e. models which do not reveal the same sophistication as those used in deterministic design.

(3) there is a lack of procedures available to evaluate collapse probabilities

(4) by and large, the methods are still limited to small structures, i.e. to structures of small dimensions and hence not applicable to structures of larger size common to engineering practice.

(5) the additional computational efforts needed are considered to be still too high.

(6) there is a lack of easy-to-use software.

Despite of these reservations to use stochastic methods as analysis or design tools as expressed within the engineering community, the methods nevertheless disseminate, although slowly, steadily into various fields of application.

Due to space limitation only three representative areas of application are discussed in the following, i.e. aerospace, nuclear- and offshore structures.

4.2 AEROSPACE STRUCTURES

Current procedures for the design of aerospace structures still utilize the stress-strength method. By integrating design and testing aspects on the basis of the stress-strength theory and of statistical data on loads and materials, factors of safety (FoS) were defined in order to reach a specified structural reliability target for typical manned and unmanned space crafts (see e.g. [37, 38]). This approach was consolidated and expanded to include the effects of fatigue and fracture as well [39], however, the results of these efforts are still under discussion by the participating industry [40]. In [37, 38] the following approach, which in fact is now actually being used by the European Space Agency (ESA) and the European aerospace industry, was taken. For a set of spacecrafts, the mechanical dimensioning and justification files were screened to identify the various load types, the modes with lowest margins of safety and their relevant materials and construction types. Also identified was the number of such critical items of each type in the surveyed spacecrafts in relationship the spacecraft types and masses. Based on this information and assuming all related load and material distributions are known, the stress-strength approach can be used in combination with FoS's to compute component reliability. From this, assuming independence and the serial approach, the spacecraft system mechanical reliability was evaluated and compared to the reliability target (Fig.4).

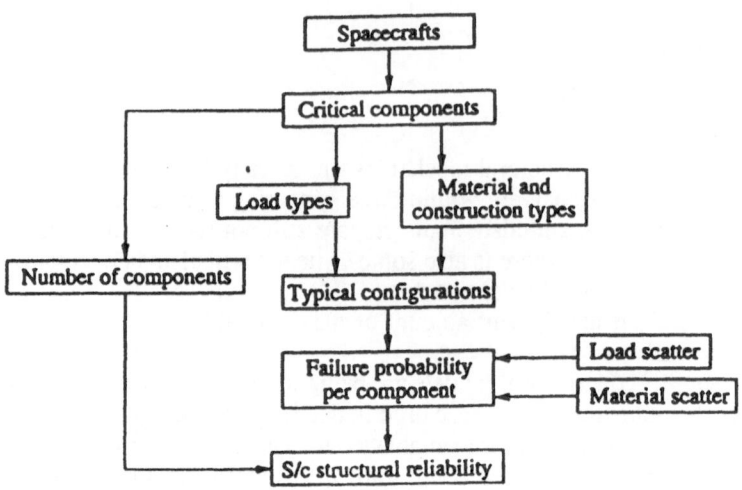

To apply this methodology an extensive survey and scatter analysis of materials has been performed using test data from the aerospace field. A similar survey has been performed regarding loads using flight data whenever possible but also by assessing scatter of analytically defined loadings. Load and material classes have been defined for each couple of which a central factor of safety had to be defined to comply with the target component reliability of $1.0 - 10^{-6}$. This target was derived from the fact that about 1000 critical items could be identified for typical 2200 kg mass spacecrafts and that, from the analysis of the successful program, it appeared that the level of spacecraft global mechanical reliability had to be put at 0.999.

Since some of the structural components are exposed to repeated loading, a quantitative relation between the expected failure rate and the load duration was established. In other words different designs have to be considered for achieving the same target failure probability for different design life times. Consequently this requires the consideration of time variant aspects within the concept of FoS. Hence a crack growth analysis based on linear elastic fracture mechanics as described in section 3.3. is applied. The crack length was taken as a measure of the deterioration stage of the structural component. In an expansion to the *Paris* law (see eq. (8)), the crack growth rate da/dN may be calculated by the so called extended *Forman* equation:

$$\frac{da}{dN} = \frac{C(1-R)^m (\Delta K)^n \left(\Delta K - \Delta K_{th}{}^P\right)}{\left[(1-R)K_C - \Delta K\right]^q} \tag{13}$$

where C is the growth rate coefficient, R is the stress ratio, ΔK is the stress intensity factor range, ΔK_{th} is the threshold of the stress intensity factor range, K_C is the critical stress intensity factor, and m, n, p and q are model parameters. The threshold of the stress intensity factor range ΔK_{th} is determined by:

$$\Delta K_{th} = (1 - C_o R)^d \Delta K_o \tag{14}$$

where C_0 and d are model parameters and ΔK_0 is the threshold of the stress intensity factor for R=0. The critical stress intensity factor K_C is calculated from the plane-strain fracture toughness K_{IC} taking into account the dependence on the specimen thickness. As failure criterion the *Two-Criteria Approach* by utilizing the *R6-Curve* is applied (see eq. (10)). Since by the Two-Criteria Approach the crack length can be related to the strength of the component, the stress-strength method as used in the definition of the FoS is feasible. Obviously, the FoS's had to be determined for different load durations. Since the initial (t=0) statistical description of stress and strength does not change, a separation of the safety factors in a time independent part (FoS at time t=0) and a time dependent part - the so called time function f(t), as a result of the crack propagation - is instrumental. This results in a matrix of time functions, which depend on the statistical description of the material as well as of the load.

In Fig. 5 the time function f(t) and the time dependent safety factors for a target failure probability of 10^{-6} are plotted for a certain set of input parameters.

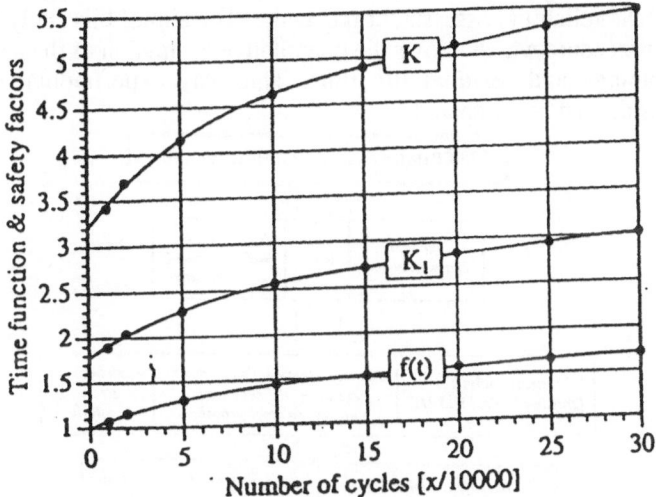

Figure 5: Time function f(t) and time-dependent safety factors K and K_1 [39]; K...Central Safety Factor; K_1...Nominal Safety Factor; f(t)...time function

Finally it should be mentioned that in the near future it is intended to introduce the - comparably more advanced - concept of partial safety factors for practical application in space craft design.

With respect to evaluating the dynamic response of space craft structures it is to be mentioned, that, although some interest in quantifying the effects of material and geometric uncertainties in terms of e.g. using the perturbation method for the dynamic response of space craft structures exist (see e.g. [41]), its application for practical design has still not materialized.

In summary, considering the wealth of stochastic methods available in structural analysis and design, aerospace industry uses still a very small part of it.

4.3 NUCLEAR STRUCTURES

In nuclear engineering the risks involved with this technology have been quantified at quite an early stage (see e.g.[42]). This, of course required the utilization of stochastic methods. However, these risk studies, by and large, did not utilize the respective state-of-the-art of procedures in stochastic structural mechanics in order to obtain credible estimates for the failure probabilities of such structures and components to be used in the fault and event tree analysis. But yet, it was a considerable progress when compared to other risk related industry, e.g. the chemical industry, which, despite of a number of spectacular failure events, is still reluctant to do so. It should be noted, however, that both analysis and design procedures for licensing are still performed on a deterministic basis. Just recently plant specific risk analyses are required both in the United States and in Germany. It should be noted that in these studies extremely simplified structural models are used. For a review of the state of development of stochastic methods of structural mechanics for the analysis and design of nuclear structures it is referred to e.g. [43]. In this context stochastic fracture analysis of pressure vessels, primary piping (PP) and containments play a major role. For this purpose probabilistic models and experimental evidence of fatigue crack propagation are suggested (see e.g.[44]). This combination is expressed in Fig.6 that shows the flow chart of the reliability assessment procedure. The aim of this study is to develope and verify a procedure which allows the quantitative estimation of the accumulation of structural damage and residual life time. This way experimental evidence and theoretical results can be compared.

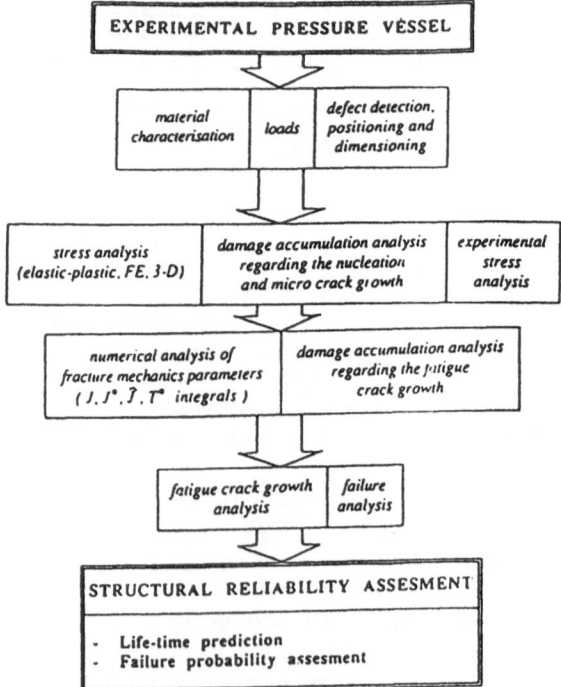

Figure 6. Applied structural reliability assessment procedure [44]

The Primary Piping (PP) (see Fig. 7) of nuclear power plants is also of great interest. Hence practical models to predict its failure probability have been developed (see e.g. [24, 43]).

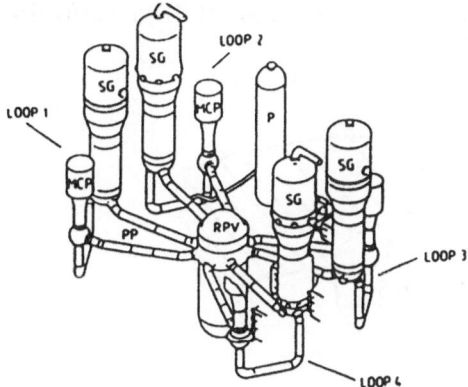

Figure 7: Schematic sketch of the primary piping of a nuclear power plant

During its design life the piping structure is expected to experience various internal and external loading conditions, a number of them associated with uncertainties in frequency of occurrence as well as in magnitude. Internal loading conditions are due to normal operation (e.g. start up of load ramp cases, etc.) as well as accidental loading (such as shut down, scram, etc.). External loading conditions may be due to accidental aircraft impact, earthquake, etc.. All these loading conditions have to be combined. Due to the generally large number of possible load cases, the generation of a "most likely" load history (by simulation) is recommended. A possible failure of the piping system under the loading conditions described above may be due to either brittle and/or ductile failure. Again, these two failure conditions may be combined in the so-called Two Criteria approach as described in section 3.3 (eq. (10)).

In-service conditions of structural piping systems may cause their deterioration. Hence, the failure behavior mentioned above must be considered time variant. Depending on the respective environmental conditions, this time variant effect may be due to fatigue, corrosion, aging, etc.. Naturally, the deterioration effects reduce the lifetime of a system - in some cases considerably. Since all piping systems in nuclear power plants - primary as well as secondary piping systems - contain weldings, their fabrication procedures, i.e. quality, inspection, etc., are of great interest for reliability estimation, both for fatigue and ultimate load failure modes. It is a well known fact, that despite most stringent quality control, it is not possible to avoid faults (e.g. flaws, gas inclusions, etc.) in welds during the fabrication process. Under static as well as cyclic loading of respective stress amplitudes, crack growth has to be expected. For the linear range the most frequently applied models are due to *Paris-Erdogan* (eq. (9)) as well as *Forman* (eq. (13)). For the nonlinear range the *Coffin-Manson* relation is frequently utilized. Due to the uncertainties in the material properties, reflected by the respective random parameters in the crack growth relations, the crack growth rate is also a random quantity. Based on this, the crack growth equations, and hence the crack growth behaviour, can be solved in various ways, i.e. by simulation and numerical integration or by solving the diffusion equation (see sect. 3.3). Numerical integration must be performed for every combination of simulated values of the random variables within the reliability analysis, whereas the solution of the diffusion equation leads directly to a probability density distribution of the crack length after a specified number of load applications.

Assuming a semi elliptical surface crack (see Fig.8), the limit state function (LSF) for the failure case "leak before break" (LBB) is easily defined i.e.

$$g(x) = t - X_4 = 0 \qquad\qquad (15)$$

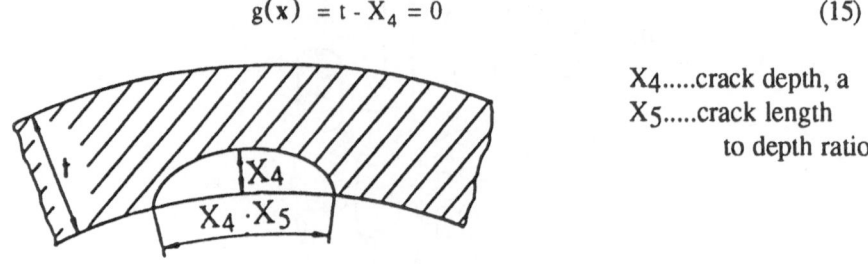

X₄.....crack depth, a
X₅.....crack length
        to depth ratio

Figure 8: Schematic sketch of a semi elliptical surface crack in a piping system

The LSF of the failure mode "break before leak" (BBL) may be derived accordingly. As a result the reliability estimates w.r.t. both failure conditions, i.e. LBB and BBL can be easily calculated as a function of the loading cycles.

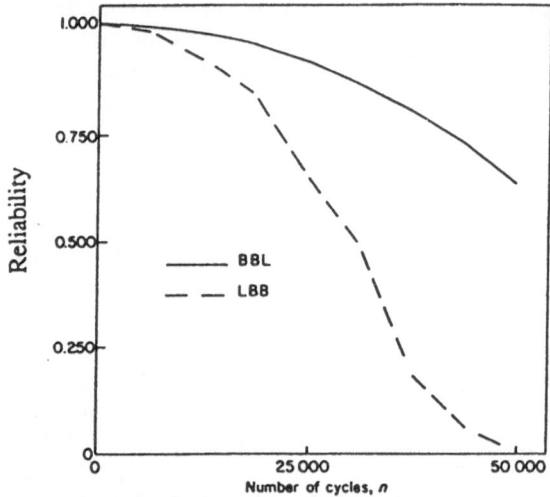

Figure 9: Reliability estimates (comparison of the BBL and LBB failure criteria) [24]

4.4 OFFSHORE STRUCTURES

Among the various technical fields, it is the area of offshore engineering where stochastic methods of structural mechanics are applied most frequently. This maybe for various reasons. Firstly it is a rather new field and hence not confronted with long traditions of analysis and design, secondly, total failures do not concern the public directly who is much less sensitive to the professional risks of third parties. For a recent detailed overview on risk analysis for design of offshore structures it is referred to e.g. [45]. The analysis and design of offshore structures requires the application of almost all tools and procedures as described in section 2. Offshore structures are designed either fixed, as steel-jacket (truss or space frame) or concrete gravity platforms, or floating, as tension leg platforms. (see Fig 10).

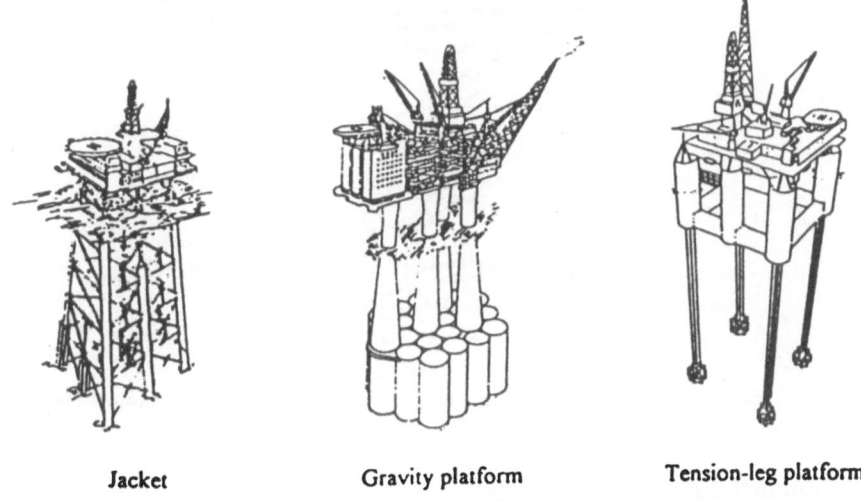

Jacket   Gravity platform   Tension-leg platform

Figure 10: Schematic sketches of various types of offshore structures (from [45])

The major type of loading of these platforms is due to wind generated gravity waves. The (Gaussian) ocean surface is characterized by certain types of wave (power) spectra (e.g. *Pierson-Moskowitz*, Jonswap Spectra, etc.). The extreme wave heights for static (collapse) analysis are modelled by extreme value distributions (Rayleigh-, Weibull distribution, etc). Depending on the geometrical characteristics of the platform the resulting wave forces on the piling or tethers is calculated by utilizing the so-called *Morison* equation which may be nonlinear and where a coefficient of variation of wave forces has to be taken into account. Again, even if the structural system is assumed to behave linearly, the response will be non Gaussian, a fact which is important for a subsequent reliability analysis.

Risk based analyses of offshore structures date already back to the late sixtieth and early seventieth (see e.g. [46-48]). These concepts, by and large, have been developed in detail over the last 20 years, although the limitations of the methods and tools as described in section 2 should be kept in mind. This applies particularly to the systems reliability aspect when collapse (ultimate strength) probabilities for large systems are to be sought. The reliability based design of such platforms is generally carried out by using the partial safety factor approach, which refers to the component level. But, particularly in the North Sea, fatigue and fracture are important failure modes to be considered in addition to the ultimate strength consideration. Again the procedures as discussed in section 3.3 are being used where initial crack distribution, quality assurance including inspection planning relating to crack damage and repair, play an important role. This leads also to time variant reliability consideration. Needless to say that, in order to achieve minimum costs at optimal safety levels, optimization procedures are being developed and applied. Reassessments of existing platforms and hence requalifications are also carried out by utilizing probabilistic methods.

In summary, within the field of offshore engineering, the status of application of stochastic methods appears to be encouraging. Last but not least, the offshore

industry is one of the very few technical fields in which decisions on acceptable risks
are being issued and discussed openly (Fig 11) [49].

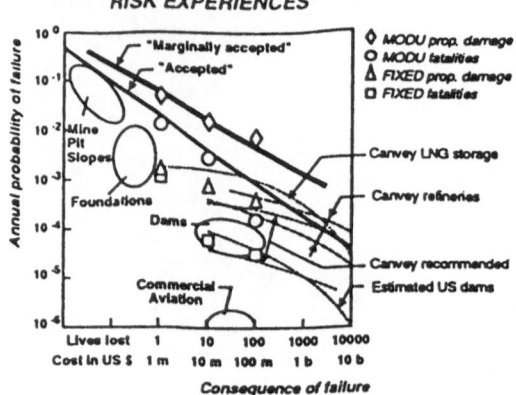

Figure 11: Comparison of experienced risk in the offshore and other industries [49]

5. Conclusions and Prospects

Compared to other well established classical fields of mechanics the utilization of
stochastic methods in mechanics can be still considered a quite recent branch of
science. Hence it is only natural that many of the procedures developed, so far, have
not yet surpassed the academic i.e. research stage. In fact, some of them never will
because of their inherent limitations, for example to systems of low dimension etc..
Their scientific merit is of course beyond question.
Based on the discussion of the various aspects of engineering application of stochastic
methods in mechanics the following can be said:
  - Sensitivity analyses enjoy an increasing popularity among practising
    engineers. They are utilized to identify the most important parameters for
    particular designs.
  - So far reliability methods are generally not yet used as a regular design tool.
  - Probabilistic methods, although not yet used for licensing procedures of
    technical systems, are utilized for codified design of civil engineering
    structures, offshore structures, etc.. Although these codes are still based on
    simplified procedures i.e. on the semi-probabilistic partial safety factor
    concept, it is expected that this may change the way of thinking of the
    engineering profession. Moreover this may develop into a sparkling point to
    use stochastic methods in structural analysis for the day-to-day design.
  - For this purpose, however, easy-to-use software development is required which
    provides the engineering practice with the necessary tools to work efficiently.
    Needless to say, that this type of software has to have the possibilities to
    directly communicate with existing commercial Finite Element software as
    currently used by the engineering practice.
Finally it should be stated, that the consideration of uncertainties, inherent in
the problems of the field of engineering mechanics, by stochastic methods certainly
has not revolutionized our profession, but there are very good reasons to expect that
the change of thinking will come about in a long-term evolution.

References

[1] MAYER M.:"Die Sicherheit der Bauwerke", J.Springer Verlag, Berlin, 1926 (in German).

[2] FREUDENTHAL A.M.:"Introductory Remarks", in Structural Safety and Reliability, A.M. Freudenthal (ed.), Pergamon Press, Oxford, 1972, pp 5-9.

[3] YANG J.N.:"Application of Reliability Methods to Fatigue, Quality Assurance and Maintenance" in Structural Safety and Reliability, G.I. Schuëller, M. Shinozuka and J.T.P. Yao (eds.), Balkema, Rotterdam, 1994, pp. 3 - 20.

[4] EUROCODE 1: "Basis of Design and Actions on Structures", 6th Draft, ENV 1991 - 1, CEN/TC250/N105, March 1993.

[5] RUBINSTEIN R.Y.: "Simulation and the Monte Carlo Method", J. Wiley, New York, 1981.

[6] KAHN H.: "Use of Different Monte Carlo Sampling Techniques", Symp. on Monte Carlo Methods, Ed. H.A. Mayer, Wiley, New York, USA, 1956, pp 146-190.

[7] SCHUËLLER G.I., STIX R.: "A Critical Appraisal of Methods to Determine Failure Probabilities", J. Structural Safety, 4, (1987), pp 293-309.

[8] PRADLWARTER H.J., SCHUËLLER G.I.: "On Advanced MCS Procedures in Stochastic Structural Dynamics", to appear: Journal Nonlinear Mechanics, 1996

[9] PRADLWARTER H.J., SCHUËLLER G.I., JEHLICKA P., STEINHILBER H.: "Structural Failure Probabilities of the HDR-Containment", J. Nuclear Engineering and Design, 128, 1991, pp 237-246.

[10] SCHUËLLER G.I. and BUCHER C.G.: "Computational Stochastic Structural Analysis - A Contribution to the Software Development for the Reliability Assessment of Structures under Dynamic Loading", Probabilistic Engineering Mechanics, Vol. 6, Nos. 3/4, 1991, pp 134-138.

[11] IWAN W.D. (1973): "A Generalization of the Concept of Equivalent Linearization", Int. J. Nonlinear Mechanics, 8, pp 279-298.

[12] PRADLWARTER H.J., SCHUËLLER G.I. and CHEN X.-W.: "Consideration of Non-Gaussian Response Properties by Use of Stochastic Equivalent Linearization", Proc., Third International Conference on Recent Advances in Structural Dynamics, 18 -22 July 1988, Southampton, M.Petyt, H.F. Wolfe and C. Mei (eds.), Vol. II, pp. 737-752.

[13] WEN Y.K. (1989): "Methods of Random Vibrations for Inelastic Structures", Appl. Mech. Review, 42(2), pp 39-52.

[14] BENAROYA H. and REHAK M. (1988): "Finite Element Methods in Probabilistic Structural Analysis: A Selective Review", Appl. Mech. Rev., 41(5), pp 201 - 213.

[15] PRADLWARTER H.J., SCHUËLLER G.I.: "A Practical Approach to Predict the Stochastic Response of Many-DOF-Systems Modelled by Finite Elements", in Nonlinear Stochastic Mechanics, N. Bellomo and F. Casciati (Eds.), Springer Verlag, Berlin, Heidelberg, 1992, pp. 427-437.

[16] NAESS A: "Prediction of Extreme Response of Nonlinear Structures by Extended Stochastic Linearization", Probabilistic Engineering Mechanics, 10 (1995) pp 153-160.

[17] SCHUËLLER G.I.: "Current Trends in Systems Reliability", Proc., 4th Int. Conference on Structural Safety and Reliability, (ICOSSAR'85), Kobe, Japan, I. Konishi, A.H.-S. Ang and M. Shinozuka (Ed.), IASSAR, New York, 1985, Volume I, pp. 139 - 148.

[18] MUROTSU Y., OKADA H., GRIMMELT M.J., YONEZAWA M., TAGUCHI K.: "Automatic Generation of Stochastically Dominant Failure Modes of Frame Structures" J. Structural Safety, Vol. 2, 1984, pp 17 -25.

[19] YAO J.T.P., KOZIN F., WEN Y.-K., YANG J.-N., SCHUËLLER G.I. and DITLEVSEN O.: "Stochastic Fatigue, Fracture and Damage Analysis", J. Structural Safety, Vol. 3+4, 1986, pp. 231-267.

[20] SOBCZYK K. and SPENCER B.F.: "Random Fatigue: From Data to Theory" Academic Press, Boston, 1992.

[21] BOGDANOFF J.L. and KOZIN F.: "Probabilistic Models of Cumulative Damage", John Wiley & Sons, New York, 1985.

[22] OSWALD G.F., SCHUËLLER G.I.: "Reliability of Deteriorating Structures", Int. Journ. Engineering Fracture Mechanics, Vol. 20, No. 3, pp. 479 - 488, 1984.

[23] MARSHALL W. (Ed.): "An Assessment of the integrity of PWR pressure vessels" United Kingdom Atomic Energy Authority UKAEA (Oct 1, 1976).

[24] SCHUËLLER G.I., TSURUI A., NIENSTEDT J.: "On the Failure Probability of Pipings", J. Nuclear Engineering and Design, 128, 1991, pp 201-206.

[25] DEODATIS G.: "Non-periodic Inspection of Fatigue- sensitive Structures by Bayesian Approach", in Struct. Safety and Reliability, (Ed.): Schuëller G.I., Shinozuka M. and Yao J.T.P., Balkema, Rotterdam, 1994, pp 997-1004.

[26] ROCHA M.M., SCHUËLLER G.I.: "Markov Chain Modelling of NDE-Techniques for Crack Inspection in Structural Components", in Struct. Safety and Reliability, (Ed.): Schuëller G.I., Shinozuka M. and Yao J.T.P., Balkema, Rotterdam, 1994, pp 1117 - 1124.

[27] BRENNER C.E.: "Ein Beitrag zur Zuverlässigkeitsanalyse von Strukturen unter Berücksichtigung von Systemunsicherheiten mit Hilfe der Methode der Stochastischen Finiten Elemente" (in German) Dissertation, University of Innsbruck, Innsbruck, Austria, 1995.

402   G.I. SCHUËLLER

[28]   PRADLWARTER H.J., SCHUËLLER G.I.: "Equivalent Linearization - A Suitable Tool to
        Analyse MDOF-Systems", *Probabilistic Engineering Mechanics*, 8, 1993, pp. 115 - 126.
[29]   LIN Y.K.: *"Probabilistic Theory of Structural Dynamics"*, McGraw-Hill, New York, 1967.
[30]   LIN Y.K. and CAI G.Q.: *"Probabilistic Structural Dynamics - Advanced Theory and
        Applications"*, McGraw Hill, New York, 1995.
[31]   SCHUËLLER G.I.(Ed.): *"Structural Dynamics - Recent Advances"*, Springer-Verlag, Berlin,
        Heidelberg, 1991.
[32]   SOONG T.T. and GRIGORIU M.: *"Random Vibration of Mechanical and Structural Systems"*,
        Prentice Hall, Englewood Cliffs, N.J., USA, 1993.
[33]   DAVENPORT A.G.: "The Application of Statistical Concepts to the Wind Loading of
        Structures", *Proc. Inst. Civ. Engrs.*, Vol 19, 1961, pp. 449 - 471.
[34]   SCHUËLLER G.I., BUCHER C.G.: "Non-Gaussian Response of Systems Under Dynamic
        Excitation", in *Stochastic Structural Dynamics - Progress in Theory and Applications*, S.T.
        Ariaratnam et al. (Ed.), Elsevier Appl. Science Publ., Barking, Essex, Engl., pp 219 - 239, 1988.
[35]   GRIGORIU M.: *"Applied Non-Gaussian Processes"*, Prentice Hall, Englewood Cliffs, N.J. USA,
        1995.
[36]   SCHUËLLER G.I., PRADLWARTER H.J., PANDEY M.D: "Method for reliability assessment
        of nonlinear systems under stochastic loading - a review", Proc. EURODYN'93, Vol. 2, T.
        Moan et al. (Eds.), A.A. Balkema Publ., Rotterdam, The Netherlands, 1993, pp. 751-760.
[37]   DE MOLLERAT T., VIDAL C.: "Evaluation of Design and Tests Safety Factors", Final Report
        of ESTEC Contract No. 6370/85/NL/PB, Cannes, 1986.
[38]   VIDAL C. DE MOLLERAT T., KLEIN M.: "Evaluation of Test and Design Factors", Proc.
        International Conference on Spacecraft Structures and Mechanical Testing, Noordwijk, The
        Netherlands, 19-21 October 1988, ESA SP-289, 1989.
[39]   KLEIN M., SCHUËLLER G.I., DEYMARIE P., MACKE M., COURRIAN P., CAPITANIO
        R.S.: "Probabilistic Approach to Structural Factors of Safety in Aerospace", Proc. International
        Conference on Spacecraft Structures and Mechanical Testing, Paris, France, Cépaduès-
        Editions, 1994, pp 679-693.
[40]   GUIDELINES OF RELIABILITY BASED FACTORS OF SAFETY FOR SPACE CRAFT
        STRUCTURAL DESIGN, Prelim. Draft, ESA/ESTEC Contr. No. 10197/92/NL/PP(SC),
        YMD/MK/95.228, Noordwijk, March 1995.
[41]   PRADLWARTER H.J., DIEZ R., KLEIN M., SCHUËLLER G.I.: "On Engineering Tools for
        Numerical Evaluation of Structural Scatter and Reliability", Proc. International Conference on
        Spacecraft Structures and Mechanical Testing, Paris, France, Cépaduès-Editions, 1994, pp 695 -
        708.
[42]   REACTOR SAFETY STUDY, WASH 1400, Washington D.C., US.NRC, 1975
[43]   SCHUËLLER G.I. and ANG A.H-S.: "Advances in Structural Reliability", *J. Nuclear
        Engineering and Design*, 134, 1992, pp. 121-140.
[44]   LUCIA A.C., ARMAN G. and JOVANOVIC A.: "Fatigue Crack Propagation: Probabilistic
        Models and Experimental Evidence, Vol. M, SMIRT 9, F.H. Wittmann (ed.), Balkema,
        Rotterdam 1987, pp 313 - 320.
[45]   MOAN T.: "Reliability and Risk Analysis for Design and Operations Planning of Offshore
        Structures", in: *Structural Safety and Reliability*, (Ed.): Schuëller G.I., Shinozuka and Yao J.T.P.,
        Balkema, Rotterdam 1994, Vol I, pp 21 - 43.
[46]   FREUDENTHAL A.M. and GAITHER W.S.: "Design Criteria for Fixed Offshore Structures"
        Prepr., Offshore Technology Conf., pp. 623 - 646, Houston, Texas, 1969.
[47]   MARSHALL P.W.: "Risk Evaluations for Offshore Structures", *J. Struct. Div., Proc. ASCE*,
        Dez. 1969.
[48]   BEA R.G.: "Selection of Environmental Criteria for Offshore Platform Design", Prepr. 5th Ann.
        Offshore Tech. Conf., Houston, 1973, Paper No. 1839, pp 186 - 193.
[49]   PAYER H.G., HUPPMANN H.H., JOCHUM C., MADSEN H.O., NITTINGER K., SHIBATA
        H., WILD W. and WINGENDER H.-J.: "Plenary Panel Discussion: How Safe is safe enough?",
        in *Structural Safety and Reliability*, G.I. Schuëller, M. Shinozuka and J.T.P. Yao, (Eds.),
        Balkema, Rotterdam, 1994, pp 57 - 74.

WIND INDUCED VIBRATION OF HIGH-RISE STRUCTURES

MASANUBO SHINOZUKA[1] and RUICHONG ZHANG[2]
[1] *Fred Champion Chair in Civil Engineering*
[2] *Research Assistant Professor*
Department of Civil Engineering
University of Southern California, Los Angeles, CA 90089.

Abstract

Stochastic dynamic responses of high-rise structures under turbulent wind is analyzed at two directions in this study. In particular, two types of structural configurations, namely, a mega-sub building and a TV tower, are considered. The wind speed consists of mean and turbulent parts. The turbulent part is idealized as a non-white stochastic process with spatial variation along the height of the structures. When the interaction between wind and structure is taken into consideration, the drag force represents the additive and parametric excitation with nonlinear colored noise, while the lift force the parametric excitation with nonlinear colored noise (updated Scanlan-Simiu model). Assuming that the wind direction is slightly off the direction normal to one side of the structures (the x-direction), it is shown that the x-direction vibration is generated primarily by the drag force due to the x-direction wind speed, while the y-direction vibration is generated primarily by the lift force exerted by vortex shedding resulting from the x-direction wind speed, and by the drag force due to the y-direction wind speed. With the aids of spectral representation method and numerical technique, the static and random vibration responses of the structures are then found for each simulated turbulent wind input. The response trajectories of the structures at specific heights are also derived. These results provide fundamentally important information for the vibration control analysis of the high-rise structures with passive and/or active mass dampers.

1. Introduction

With remarkable advances made over the past decades in engineering science including materials, geotechnical, and structural engineering, design and construction

403

A. Naess and S. Krenk (eds.), IUTAM Symposium on Advances in Nonlinear Stochastic Mechanics, 403–426.
© 1996 *Kluwer Academic Publishers.*

of high-rise structures become more common. To build super tall structures, e.g. buildings as high as 1,000 m, is now considered not impractical. One of the major issues in this endeavor is the problem ensuring the structural safety under seismic and wind loads. In addition to meeting the requirements for structural safety, the structural vibration also needs to be contained below the threshold of human comfort particularly during the sustained wind environment. To this end, many devices such as active and/or passive tuned mass damper systems are developed and used in the high-rise structures to suppress the vibration dynamically. However, as a structure gets taller and more massive, a heavier additional mass is required and a larger stroke of this mass is anticipated in case of passive systems, and reliable and powerful actuators together with advanced optimal control algorithm are needed in case of active systems. In either case, these requirements raise serious safety concerns as well as high cost of the implementation of the tuned mass damper system.

Recently, a mega-sub structural configuration for the super tall building is proposed by Feng and Mita (1994). The mega-structure not only supports the sub-structure but also resists external forces such as seismic and wind loads, whereas the sub-structure provides the usable space for office, commercial and residential purposes, depending on the specific usage of the building. The mass ratio between mega- and sub-structures can be designed to be much high compared with that in the conventional tuned mass damper so that part of kinetic energy of the mega-structure generated by the external loads such as earthquake and wind flows into the sub-structure. With the proper design of the dynamic characteristics of the mega-sub structure such as the mass ratio between the mega- and sub-structures, the peak displacement response of the mega-structure is significantly suppressed, while the peak acceleration response of the sub-structure is constrained within a prescribed level. This not only enhances the structural safety but also minimizes the acceleration-related risk for the objects supported by the sub-structures including human discomfort. To demonstrate the aforementioned advantages and provide the fundamental understanding for optimal design procedure, random vibration of the mega-sub structure under the turbulent wind is analyzed, which is one of the objectives in this study.

Also included in this study is the dynamic response analysis of another type of high-rise structure, namely a TV tower, under turbulent wind. In particular, the 310 m tall Nanjing TV tower, built at Nanjing, China in May, 1993, is under consideration. Its main structure consists of three prestressed concrete legs with hollow rectangular sections. There are large and small observation halls at the levels of 186 m and 240 m of the tower, respectively. The large hall is designed for common visitors, while the small hall is designed for the honored guests. Therefore, the wind-induced acceleration responses at both of the observation halls must satisfy the design requirement of human comfort (e.g., less than 0.15 m/sec^2). In case the prescribed requirement is not achieved, control devices have to be used to suppress the vibration within the tolerable range of human comfort.

Wind-induced vibration of the Nanjing TV tower with and without control devices has been analyzed by Qu and Chen (1993), Cheng, et al. (1993), Ding and Ren (1994), and Reinhorn et al. (1995). However, their analyses were focused on the along-wind vibration generated by drag force, which is computed on the basis of the turbulent wind speed without considering the spatial variation. As well known, for a large Reynolds number, a bluff body immersed in a wind flow sheds vortices in its wake. The vortex oscillation produces a new lift force on the vibrating structure in the direction normal to the flow, which is dependent on both shedding frequency and across-wind structural response (Simiu and Scanlan, 1986). Recently, the vortex-induced vibration mechanism is further explored by Billah (1989) and Goswami, et al (1993). Their studies indicate that the across-wind vibration is generated by the parametric excitation, which may result in an unexpectedly larger vibration than the along-wind vibration when the structural vibration in the across-wind direction gets synchronized with vortex shedding. For this reason, it is fundamentally important to analyze the across-wind vibration to the turbulent wind. In addition, the spatial vibration of turbulent wind speed along the structural height may have to be considered in the analysis in order for the problem at hand to be dealt with more physical consistency.

2. Theory

As shown in Figs. 1 and 2, the high-rise structures (mega-sub configuration building and TV tower) are subjected to the turbulent wind loads, in the direction slightly off the direction normal to one side of the structure (the x-direction). Since the y-direction wind speed component is very small, its vortex shedding effect in the x-direction is thus neglected in this study. Therefore, the x-direction vibration is primarily generated by the drag force due to the x-direction wind speed, while the y-direction vibration is generated primarily by the lift force exerted by the vortex shedding resulting from the x-direction wind speed, and by the drag force due to the y-direction wind speed. The torsional motion is not considered at the present study since it is believed to be not as important as the translational motion in this case. However, the refinement of the model, in which the coupling is taken into account among x-direction, y-direction and torsional vibration under the turbulent wind in an arbitrary direction, will be considered in the future.

2.1 GOVERNING EQUATIONS FOR MEGA-SUB STRUCTURE

As shown in Fig. 1, the relatively rigid mega-structure is modeled as a cantilever beam which is further discretized as a multi-degree freedom system resisting in bending, while the relatively soft sub-structure system is modeled as a series of shear-type structures, each of which is simplified as a mass connected to both upper and lower lumped mass of the mega-structure. Therefore, the governing equations can be written as

$$M_\alpha^m \ddot{W}_\alpha^m + \left(C_\alpha^m + C_\alpha^{s1}\right) \dot{W}_\alpha^m + \left(K_\alpha^m + K_\alpha^{s1}\right) W_\alpha^m - C_\alpha^{s2} \dot{W}_\alpha^s - K_\alpha^{s2} W_\alpha^s = F_\alpha \quad (1)$$

$$M_\alpha^s \ddot{W}_\alpha^s + C_\alpha^{s3} \dot{W}_\alpha^s + K_\alpha^{s3} W_\alpha^s - \left[C_\alpha^{s2}\right]^T \dot{W}_\alpha^m - \left[K_\alpha^{s2}\right]^T W_\alpha^m = 0 \quad (2)$$

In equations (1) and (2), superscript T denotes transpose of a matrix, subscript α denotes respectively the x- and y-directions, superscripts m and s (or β in general, representing either m or s) denote respectively the mega- and sub-structures, $W_\alpha^\beta = \left\{w_{\alpha,1}^\beta w_{\alpha,2}^\beta \cdots w_{\alpha,n}^\beta\right\}^T$ is the displacement vector, and $F_\alpha = \{f_{\alpha,1} f_{\alpha,2} \cdots f_{\alpha,n}\}^T$ is the external excitation vector. Furthermore, the i-th lumped mass of diagonal matrix $M_x^\beta \left(= M_y^\beta\right)$ is $m_i^m = m^m \left(h_i + h_{i+1}\right)/2$ and $m_i^s = m^s h_i$ with m^β being the mass per unit height, stiffness matrix K_α^m can be found as an inverse of compliance matrix whose ij-th (same as ji-th) component is found as $i^2 h_i^3 (3j - i)/(6E^m I_\alpha^m)$ for $j \geq i$ with EI being the bending rigidity of the idealized mega-structure, $C_\alpha^m = M_\alpha^m D_\alpha^m \varsigma_\alpha^m [D_\alpha^m]^T M_\alpha^m$ (Clough and Penzien, 1975) with the i-th damping ratio of diagonal matrix ς_α^m being $2\xi_{\alpha,i}^m \omega_i/m_i^m$ and D_α^m being the vibration mode matrix of mega-structure, components of K_α^{si} and C_α^{si} ($i = 1, 2, ..., n$) are zeros except that $k_{\alpha,ij}^{s2} = k_{\alpha,j}^s$ and $c_{\alpha,ij}^{s2} = c_{\alpha,j}^s$ for $j = i, i+1$ and $k_{\alpha,ii}^{s1} = k_{\alpha,i}^s + k_{\alpha,i+1}^s$, $c_{\alpha,ii}^{s1} = c_{\alpha,i}^s + c_{\alpha,i+1}^s$, $k_{\alpha,ii}^{s3} = 2k_{\alpha,i}^s$, $c_{\alpha,ii}^{s3} = 2c_{\alpha,i}^s$, in which $k_{\alpha,i}^s = 6E^s I_{\alpha,i}^s/h_i^3$ and $c_{\alpha,i}^s = 2\xi_{\alpha,i}^s (k_{\alpha,i} m_i^s)^{1/2}$.

2.2 GOVERNING EQUATION FOR TV TOWER

Based on previous investigations and measured data (Cheng, et al. 1993, and Ding and Ren, 1994), the Nanjing TV tower is modeled as 16 degrees of freedom system, as shown in Fig. 2. The governing equation can be written as

$$M_\alpha \ddot{W}_\alpha + C_\alpha \dot{W}_\alpha + K_\alpha W_\alpha = F_\alpha \quad (3)$$

with the similar definitions for M, W, K, F and α as in the last sub-section.

2.3 WIND LOADS

Assume the wind velocity profile in the z-direction as

$$U(z,t) = \overline{U}(z) + u(z,t) \tag{4}$$

where $\overline{U}$ denotes a steady mean wind speed at height z and can be found by either the power law or the logarithmic law, and $u(z,t)$ is the fluctuating random components with a specified spectrum (non-white noise) and its dominant along-wind wave length is much larger than the major dimensions of the structure in the x- and y-directions under consideration. Then, the x-direction force exerted on the i-th mass is the drag force, due to the x-direction wind speed, and can be found as (Simiu and Scanlan, 1986 and Vaicaitis, et al. 1973, 1975)

$$f_{x,i}(t) = \frac{1}{2}\rho C_D A_i \left[U(z_i,t) \cos\theta - \dot{w}_{x,i} \right]^2 \tag{5}$$

where θ is the angle away from direction normal to one side of the structure, ρ is the air density, A_i is the windward area of the structure, and C_D is the drag coefficient. The interaction between wind and structure is taken into account through the square of relative speed, which is due to the flexibility of the structure.

The y-direction force exerted on the i-th mass consists of two parts, as shown in equation (6) below. The first part is the lift force due to the x-direction wind speed involving the updated Scanlan-Simiu's model for vortex-induced vibration (Goswami, et al. 1993), and the second part is the darg force due to the y-direction wind speed.

$$
\begin{aligned}
f_{y,i}(t) = &\frac{1}{2}\rho A_i' (U(z_i,t)\cos\theta - \dot{w}_{x,i})^2 \left[Y_1(K) \frac{\dot{w}_{y,i}}{U(z_i,t)\cos\theta - \dot{w}_{x,i}} \right. \\
&+ Y_2(K) \frac{(w_{y,i})^2 \dot{w}_{y,i}}{b_i^2 (U(z_i,t)\cos\theta - \dot{w}_{x,i})} \\
&\left. + J_1(K) \frac{w_{y,i}}{b_i} + J_2(K) \frac{w_{y,i}}{b_i} \sin(2\omega_{s,i}t) \right] \\
&+ \frac{1}{2}\rho C_D A_i' [U(z_i,t)\sin\theta - \dot{w}_{y,i}]^2
\end{aligned} \tag{6}
$$

Also, in equation (6), A'_i is the across-wind area, b_i is the width of the structure, $K = \omega_1 b_i/U(z_i, t)$ is the reduced frequency with ω_1 denoting the fundamental frequency of the structure under consideration, the terms associated with coefficients Y_1 and Y_2 are linear and nonlinear aeroelastic damping terms, which lend the commonly observed self-excited, self-limiting character to the model, the term associated with J_1 is an aeroelastic stiffness term which provides for a shift in the structural response frequency, and the term associated with J_2 is a parametric stiffness coupling between the near wake and the structure, which excites the structure at twice the shedding frequency $\omega_{s,i}(= 2\pi S_i U(z_i, t)/b_i$, with S_i denoting Strouhal number at the i-th lumped mass)as suggested first by Billah (1989). In reality, Y_1 and J_1 terms do not usually produce significant effects, while J_2 term becomes very important only when the structural vibration in the across-wind direction gets synchronized with vortex shedding. It is noted that the lift force model in equation (6) was established by Goswami, et al. (1993) on the basis of the cylinder under the turbulent wind loads, which is modeled as single degree of freedom system. We assume that the use of the model in the current study for multiple degrees of freedom system will not significantly distort the fundamental physical behavior of the structural responses. Since the structural responses to the turbulent wind loads are primarily contributed by its fundamental vibration mode (see e.g. Simiu and Scanlan, 1986), the reduced frequency K is calculated on the basis of fundamental frequency of the structure ω_1. For the sake of convenience in practical use of coefficients Y_i and $J_i(i = 1, 2)$ provided in Goswami et al. (1993), the reduced frequency K may also simplified as $K = \omega_1 b/\overline{U}(z_i)$.

2.4 CHARACTERISTICS OF WIND LOADS AND STRUCTURAL RESPONSE

As shown in equation (5), the drag force is a function of wind speed (composed of mean and turbulent random parts) and structural responses, which accounts for the interaction between wind and structures. Therefore, the x-direction vibration of the TV tower is generated by additive and parametric excitations with non-white colored noise. Since the turbulent wind speed is in most cases much less than its corresponding mean part, the consideration of interaction between wind and structure in the present model does not affect the structural responses significantly.

Similarly, the interaction between wind and structure through the relative wind speed $(U(z_i, t)cos\theta - \dot{w}_{x,i})$ in the lift force expression of equation (6) does not affect the across-wind vibration responses significantly either and thus can be neglected in the practical numerical analysis (Vaicaitis, et al. 1973, 1975). Consequently, the y-direction vibration is independent on the x-direction vibration. It is noted that he interaction between the wind (in the near wake) and the structure in the y-direction vibration is still taken into account through the lift force model in equation (6), even if relative speed $(U(z_i,t)\cos\theta - \dot{w}_{x,i})$ in equation (6) is replaced by absolute wind speed $U(z_i,t)\cos\theta$.

As can be seen from equation (6), the lift force itself is the parametric excitation with nonlinear colored noise. Theoretically, as long as the structure is motionless in the y-direction at the beginning of the turbulent wind at $\theta = 0°$, the parametric excitation of the lift force disappears, leaving the structure motionless in that direction forever. However, as long as the y-direction vibration starts, even if it is very small in magnitude due to the drag force with small $\theta(\neq 0)$, it will grow larger and larger due to the parametric excitation, especially when synchronization takes place, which may possibly lead to the instability of the system. It is noted that the J_2 term in the left force of equation (6) represents the fundamental instability mechanism under the synchronization condition. On the other hand, the nonlinear aeroelastic damping term (the Y_2 term in equation (6)) provides the instability saturation effort to counteract the instability effects of J_2 term to a certain extent.

3. Digital Simulation and Numerical Technique

Although many methods have been developed to solve various stochastic differential equations with additive and/or parametric (multiplicative)excitations, they are mostly restricted to simple cases such as the white noise excitation and/or with weak nonlinearity. The extension of methods is also limited due, primarily, to the assumptions that must be introduced for analytical expediences which often distort the physics involved, thus resulting in quite misleading, if not totally wrong, answers to the physical problems at hand. Instead, Monte Carlo approach is always an worthwhile alternative to pursue the solution in a hybrid mode as much as possible so that the problem solving process proceeds analytically as well as in Monte Carlo simulation, whichever is more appropriate, in each phase of the solution process. For the problem at hand, digital simulation of the wind loads and numerical solution for the structural responses are combined to obtain necessary response statistics. Specifically, with the aid of spectral representation method (e.g. Shinozuka and Deodatis, 1988), the turbulent component of wind velocity is generated by

$$u(z,t) = \sqrt{2}\sum_{i=1}^{N_1}\sum_{j=1}^{N_2}\sqrt{2S(\omega_i,k_j)\Delta\omega\Delta k}$$

$$[\cos(\omega_i t + k_j z + \phi_{ij}) + \cos(\omega_i t - k_j z + \phi'_{ij})]$$

(7)

where $S(\omega_i, k_j)$ is the spectral density function of $u(z, t)$ in the frequency-wave number domain, ϕ_{ij} and ϕ'_{ij} are random phase angles, independent each other and uniformly distributed between 0 and 2π, and the intervals, $\Delta\omega$ and Δk, are taken respectively as ω_u/N_1 and k_u/N_2 in which ω_u and k_u are respectively upper cut-off frequency and wave number. Substituting equation (7) into the equations (5) and (6) and using Runge-Kutta integration method, the structural responses of both time histories and trajectories are found.

4. Numerical Examples

4.1 MEGA-SUB AND CONVENTIONAL BUILDINGS

Firstly, dynamic responses of two high-rise buildings with the same amount of total mass subjected to the turbulent wind are analyzed. One is the conventional building, modeled as a cantilever beam and further discretized as a multi-degree freedom system, while the other is the mega-sub configuration building. The mean wind speed is assumed to follow the logarithmic law, i.e. $\overline{U}(z) = \overline{U}_{10}\sqrt{K'}\ln(z/z_0)/k$. The spectral density function for the turbulent component of wind speed, suggested first by Devanport and used by Vaicaitis, et al. (1975) for the use in the spectral representation method, is

$$S(k, \omega) = \frac{K'\Phi^2}{2\pi^2} \frac{|\omega|}{[1 + \Phi^2\omega^2/(2\pi\overline{U}_{10})^2]^{4/3}} \cdot \frac{\epsilon|\omega|}{\pi(\epsilon\omega^2 + k^2)} \qquad (8)$$

The parameters for the Mega-sub configuration building and wind loads are selected as follows as well as in Table 1.
Buildings
Height: $H = 200\ m$; Length: $b = 80\ m$; Width: $d = 40\ m$; Mass of conventional building: $m^c = 4 \times 10^5 kg/m$; Mass of mega-sub buildings: $m^m = m^s = 4 \times 10^5\ kg/m$; Bending rigidity of buildings: $EI = 2.4 \times 10^{14} N \cdot m$.
Turbulent Wind
Drag coefficient: $C_D = 1.2$; Density of air: $\rho = 1.225\ kg/m^3$; Surface drag coefficient: $K' = 0.03$; von Karman's constant: $k = 0.4$; Roughness length: $z_0 = 1\ m$; Coefficients in Power Spectrum: $\Phi = 1200\ m$; $\epsilon = 0.37\ m/sec$; Strouhal number: $S = 0.142$; Shedding Frequency: $\omega_s = 2\pi S U(z_i, t)/b$; Wind direction: $\theta = 5°$; Wind speed at 10 m height: $\overline{U}_{10} = 30\ m/sec(67\ mile/hr)$.
As an example, one realization of the simulated turbulent wind speed, specifically the spatial variation at frozen time and time history at frozen height location for the turbulent wind, is shown in Fig. 3.

The x- and y-directions (the along-wind and the across-wind) responses at the top buildings are shown in Figs. 4 and 5, respectively. It can be seen clearly from these figures that the displacement response of the mega-structure is efficiently suppressed compared with that of conventional structure, implying that the mega-sub structure has more reliable than the conventional one. At the meantime, the acceleration response of the sub-structure is significantly reduced compared with that of the conventional structure. Therefore, the acceleration-related human discomfort for the present mega-sub structure is better than that of conventional one. The trajectories of corresponding responses at top of buildings are shown in Fig. 6.

The overall performance in statistics for mega-sub structure in suppressing the wind-induced vibration is also examined. Specifically, as a function of the ratio of natural frequencies between sub and mega structures, the mean and standard deviation for the ratio of maximum along-wind displacement responses between mega and conventional structures and for the ratio of maximum along-wind acceleration responses between sub and conventional structures are plotted in Fig. 7. As seen in Fig. 7, the optimal design for the mega-sub structure is for the ratio of natural frequencies between sub and mega structures being around 0.5. At this natural frequency ratio, the ratio of maximum displacement responses keeps lowest, while the ratio of maximum acceleration responses is at a relatively small value, which implies that both the structural safety and the human comfort are improved for the present mega-sub structure in comparison with the conventional structure.

4.2 NANJING TV TOWER

Due to the specific cross-section of the TV tower, the vortex shedding effect is considered only at the locations of the small and the large observation halls. This means the coefficients Y_1, Y_2, J_1 and J_2 are selected to be zero except those at two halls, as seen in Table 2. The stiffness matrix $[K]$ is given by Reinhorn et al. (1995), by modeling the TV tower as a set of segments with approximately uniform properties. The parameters of power spectrum for turbulent wind are selected as follows: $\epsilon = 0.046 \ m/sec$ and $\Phi = 1200 \ m$. The mean wind speed is assumed to follow the power law, i.e. $\overline{U}(z) = \overline{U}_{10}(z/10)^\alpha$ with $\alpha = 0.16$ and $\overline{U}_{10} = 20.7 m/sec$ (46.3 mile/hr).

The wind-induced displacement and acceleration responses of the TV tower were found, as shown in Fig. 8-11. Specifically, as shown in Fig. 11, the peak acceleration of small observation hall is larger than $0.15 \ m/sec^2$, a requirement of human comfort. Therefore, the control devices may have to be activated so that the corresponding acceleration responses can be reduced below $0.15 \ m/sec^2$.

As seen from Figs. 10 and 11 and compared with Figs. 8 and 9, the y-direction displacement response are much smaller, while the y-direction peak acceleration is comparable with the x-direction acceleration, indicating that the vortex-induced y-direction vibration is important, as far as the acceleration-based human comfort is concerned. This can be seen clearly from Fig. 12, in which the trajectories of the displacement, velocity and acceleration of the TV tower at three height locations are depicted.

5. Concluding Remarks

The wind-induced random vibration of the high-rise structures is simulated under the turbulent wind. The effectiveness of suppressing the wind-induced vibration of mega-sub structure is demonstrated in this study, which may not be carried out using other available theoretical approaches. Optimal design for the mega-sub structure subjected to the turbulent wind is also examined. The simulated results of Nanjing TV tower provide an important input information for testing the effectiveness of control devices installed. The most important issue investigated in this study is the vortex-induced vibration with the proper use of the updated Scanlan-Simiu model. This is an attempt to have better understanding of the vortex-induced vibration response for the multi degrees of structural systems. The spatial variation of turbulent wind is also considered in the present analysis. Further improvement on the mathematical model will be carried out later.

6. Acknowledgements

This work was partially supported by the National Science Foundation under Grant CBS-9223234. The authors would like to express thanks to Professor K.Y.R. Billah of Stevens Institute of Technology and Professor M.Q. Feng of University of California at Irvine for their constructive suggestions to the problem under consideration.

7. References

1. Billah, K.Y.R.: A study of vortex-induced vibration, *PhD dissertation*, Princeton University, Princeton, New Jersey, (1989).
2. Cheng, W.R. et al.:The analysis of dynamic characteristics of Nanjing TV tower, *Southeast University*, (1992).
3. Cheng, W.R. et al.: Structural control of Nanjing TV tower under wind excitation, (1993).

4. Clough, R.W. and Penzien, J.: *Dynamics of Structures*, McGraw-Hill Inc., (1975).

5. Ding, Dajun and Ren, Zhenhua: Design testing and analyses of the Nanjing TV tower, *Concrete International*, pp. 42-44, (1994).

6. Feng, M.Q. and Mita, A.: Vibration control of tall building using mega-sub structure configuration, accepted for publication in *Journal of Engineering Mechanics*, (1994).

7. Goswami, I., Scanlan, R.H. and Jones, N.P.: Vortex-induced vibration of circular cylinders. I: Experimental data, *Journal of Engineering Mechanics*, Vol. 119, No. 11, pp. 2270-2287, (1993).

8. Goswami, I., Scanlan, R.H. and Jones, N.P.: Vortex-induced vibration of circular cylinders. II; New model, *Journal of Engineering Mechanics*, Vol. 119, No. 11, pp. 2288-2302, (1993).

9. Qu, W.L. and Cheng, W.R.: Control for turbulent wind vibration response of Nanjing TV tower by semi-active TMD, *Third Asia-Pacific Symposium on Wind Engineering*, (1993).

10. Shinozuka, M., and Deodatis, G.: Stochastic process models for earthquake ground motion, *Probabilistic Engineering Mechanics*, Vol. 3, No. 3, pp. 114-123, (1988).

11. Simiu, E. and Scanlan, R.H., *Wind Effects on Structures*, Second Edition, John Wiley & Sons, New York, (1986).

12. Vaicaitis, R., Shinozuka, M. and Takeno, M.: Parametric study of wind loading on structures, *Journal of the Structural Division*, Vol. 99, No. ST3, pp. 453-468, (1973).

13. Vaicaitis, R., Shinozuka, M. and Takeno, M.: Response analysis of tall buildings to wind loading, *Journal of the Structural Division*, Vol. 101, No. ST3, pp. 585-600, (1975).

TABLE 1: Coefficients of lift force for high-rise buildings

No. of Masses	Height (m)	Y_1	Y_2	J_1	J_2
1	50	10	-100000	0	2200
2	100	10	-100000	0	1500
3	150	10	-100	0	400
4	200	10	-1000	0	500

TABLE 2: Parameters of Nanjing TV tower and turbulent wind

No.	Height	Mass	Width	Area	Drag Coef.	Y_1	Y_2	J_1	J_2	Strouhal
i	$h_i(m)$	$m_i(t)$	$b_i(m)$	$A_i(m^2)$	C_D					No. S_i
1	10.1	3992.9	28.56	4.13	1.39	0	0	0	0	0
2	32.2	3186.7	24.345	11.28	1.39	0	0	0	0	0
3	58.6	2820.1	20.228	21.55	1.39	0	0	0	0	0
4	80.2	2319.8	18.75	66.38	1.39	0	0	0	0	0
5	101.8	1917.9	17.1	175.58	1.39	0	0	0	0	0
6	119.8	1624.5	15.45	213.51	1.39	0	0	0	0	0
7	137.8	1628.1	14.7	183.85	1.39	0	0	0	0	0
8	158.6	1322.3	13.95	182.51	1.39	0	0	0	0	0
9	171.8	3395.3	13.42	233.65	1.20	0	0	0	0	0
10	185.8	5678.6	13.42	277.38	1.20	10	-10^6	0	1700	0.2
11	199.2	1512.4	13.42	271.35	1.20	0	0	0	0	0
12	240.4	1254.0	6.0	323.73	1.20	10	-10^6	0	1700	0.2
13	270.1	165.1	3.5	387.18	1.30	0	0	0	0	0
14	286.1	18.7	1.8	469.5	1.30	0	0	0	0	0
15	299.1	12.0	1.1	536.02	1.30	0	0	0	0	0
16	310.1	4.0	0.75	413.24	1.30	0	0	0	0	0

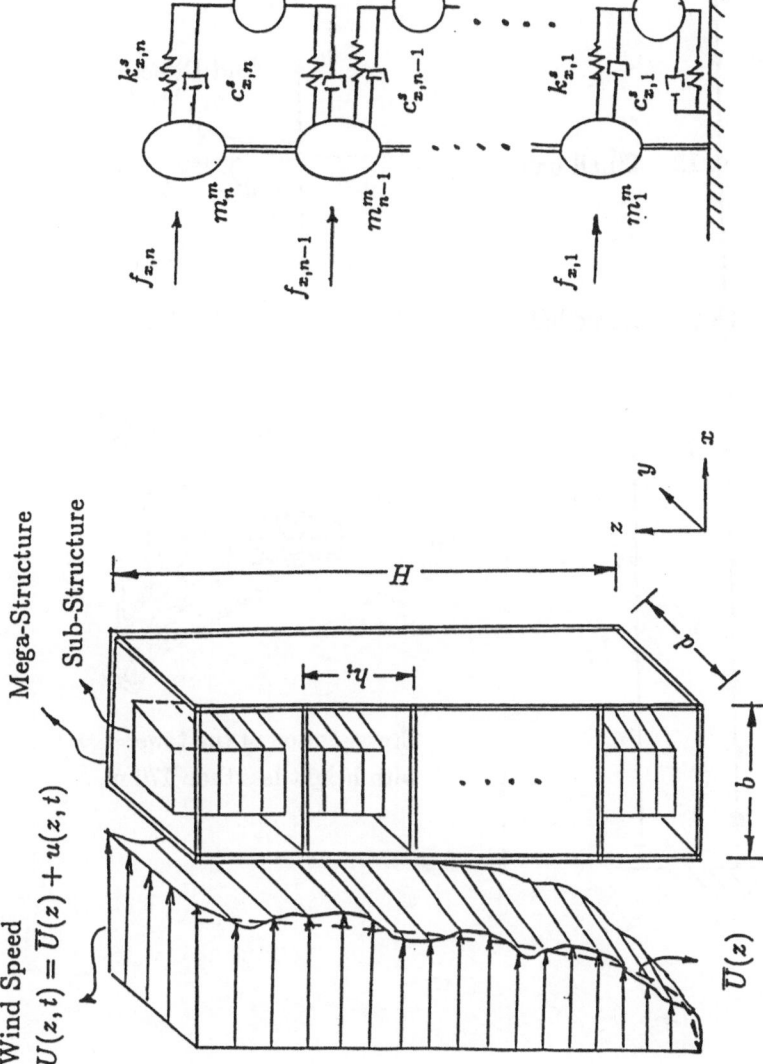

Figure 1. Schematic of mega-sub structure subjected to turbulent wind

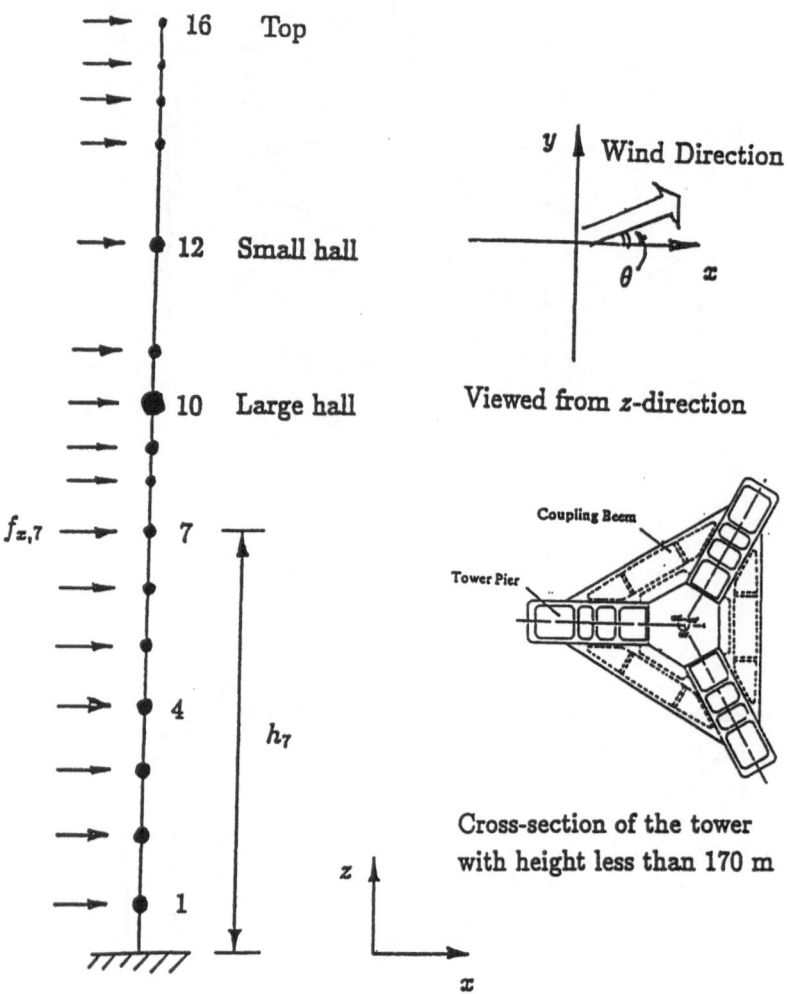

Figure 2. Schematic of Nanjing TV tower subjected to turbulent wind

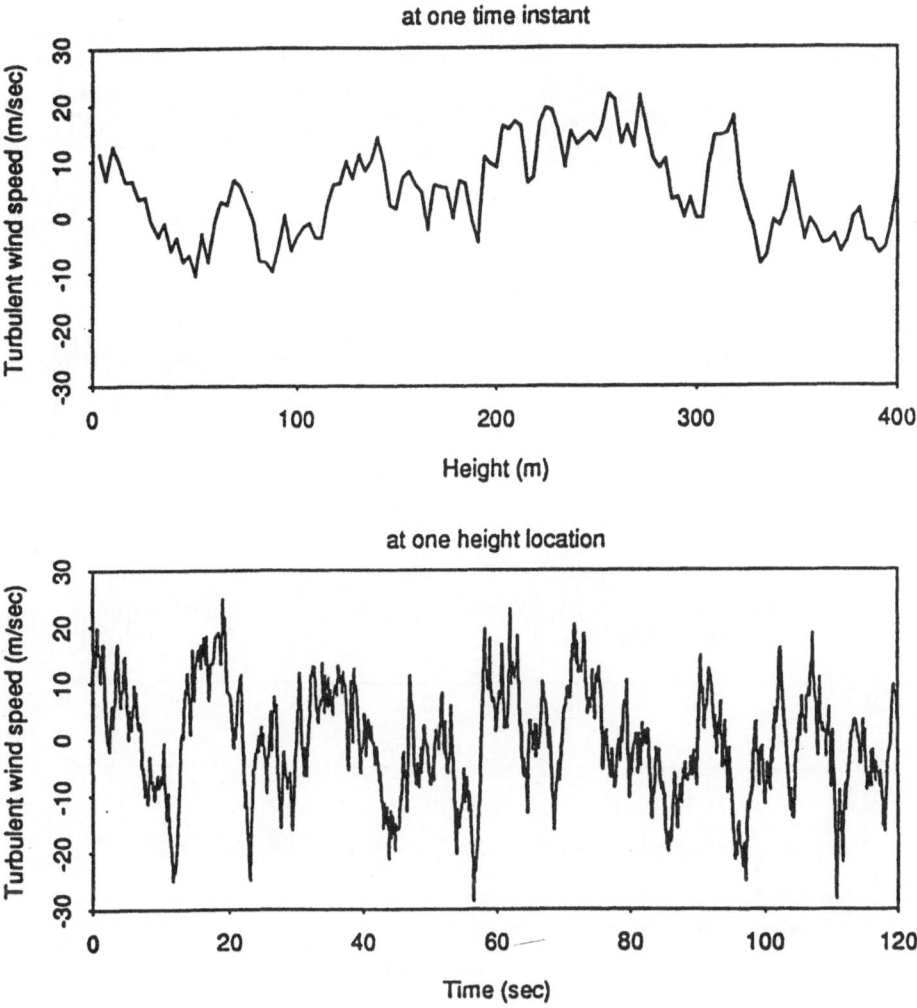

Figure 3. One realization of turbulent wind speed

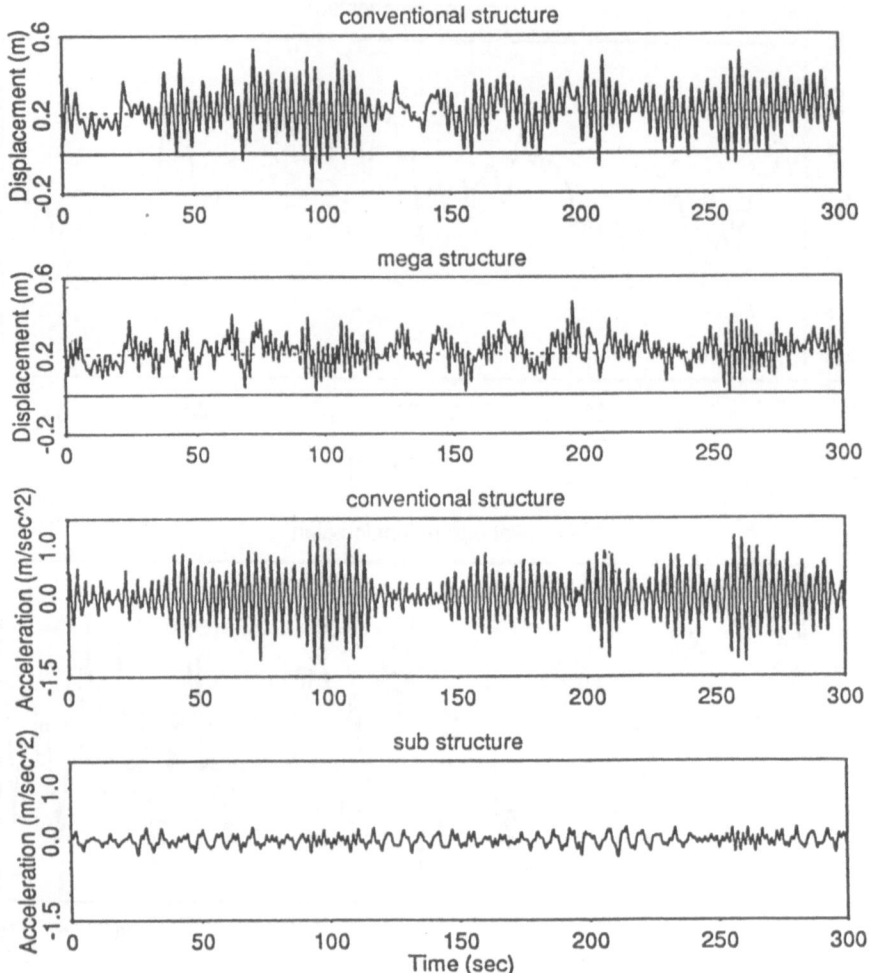

Figure 4. The x-direction responses at top of buildings

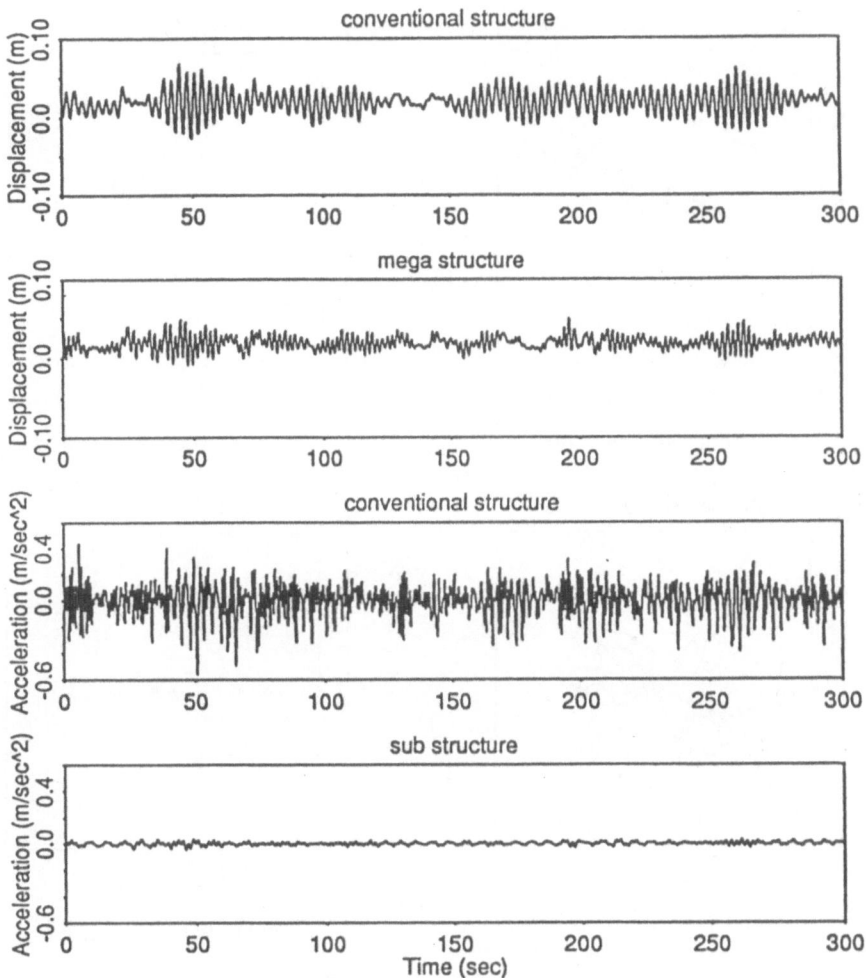

Figure 5. The *y*-direction responses at top of buildings

420 M. SHINOZUKA AND R. ZHANG

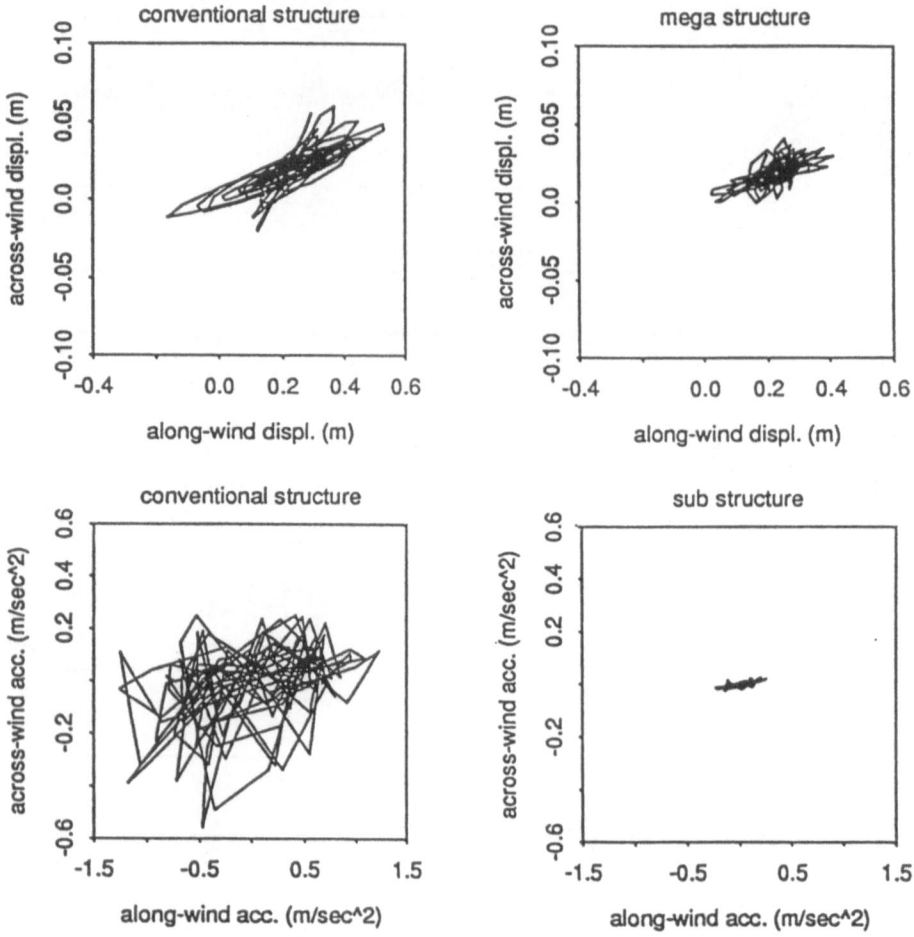

Figure 6. Trajectories of displacement and acceleration responses at top of buildings

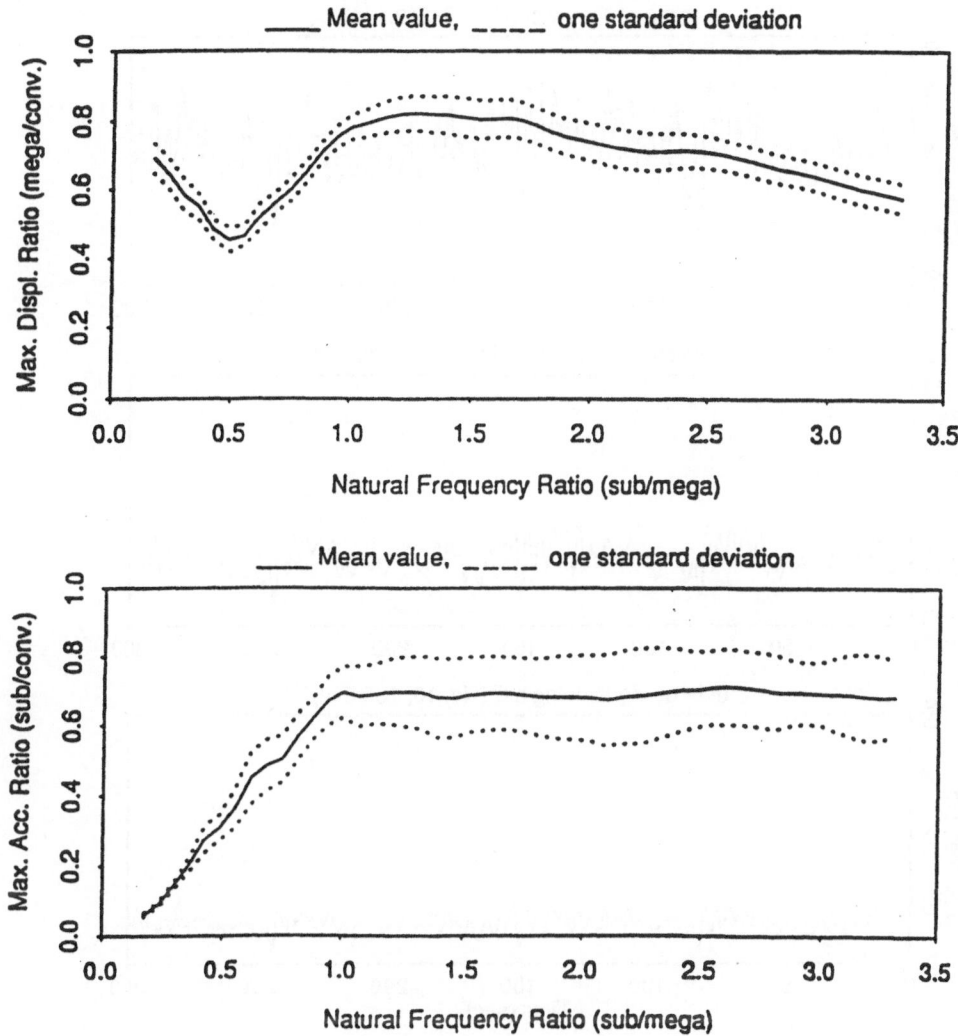

Figure 7. Statistical maximum *x*-direction response ratios at top of buildings vs natural frequency ratio (average over a sample of size 35)

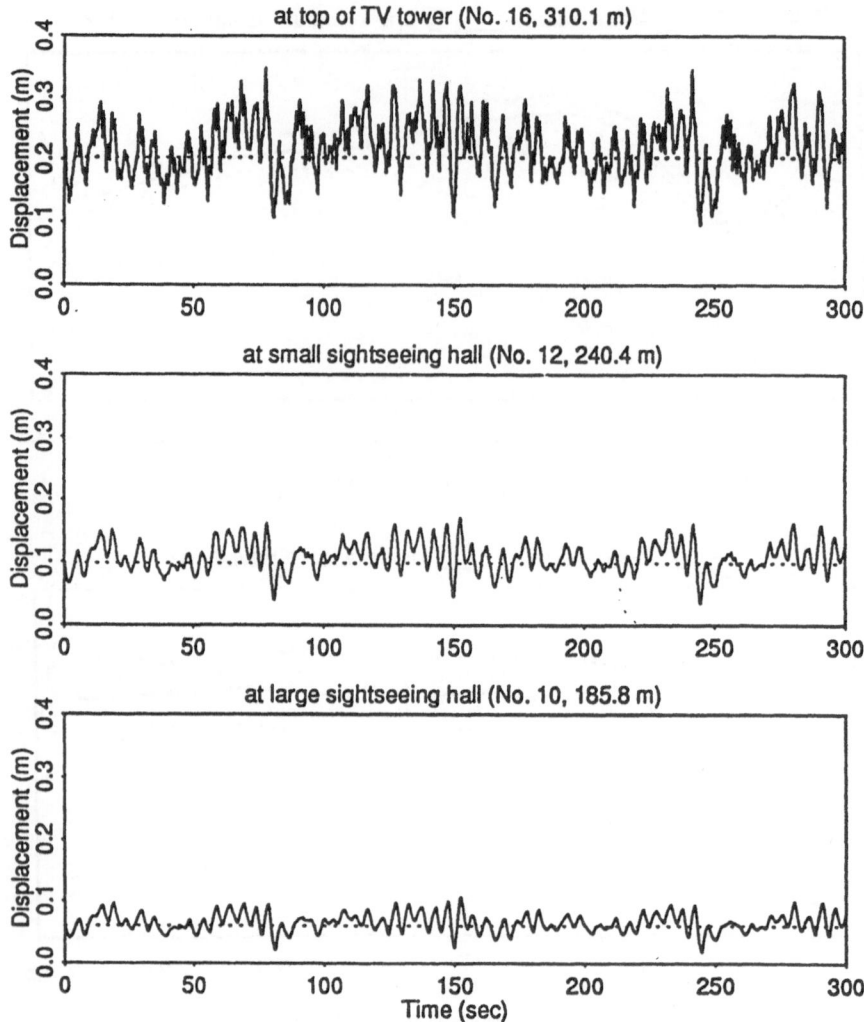

Figure 8. The *x*-direction displacement responses of TV tower at three height locations (solid line: total response; dotted line: static response)

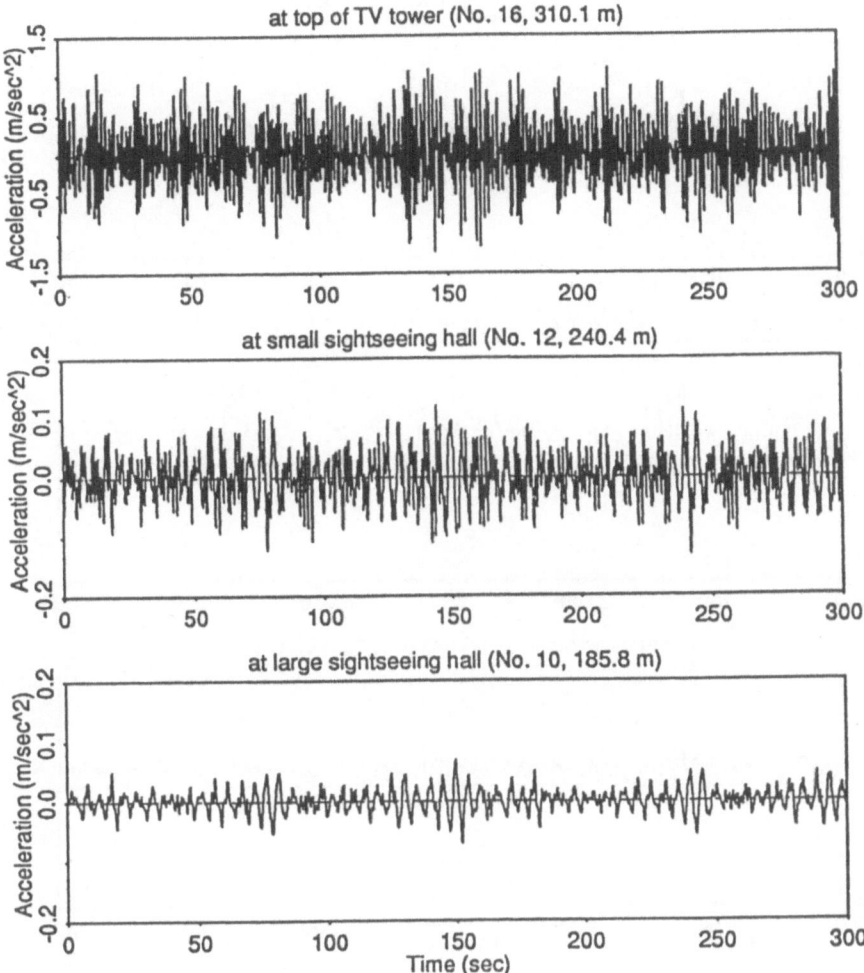

Figure 9. The x-direction acceleration responses of TV tower at three height locations

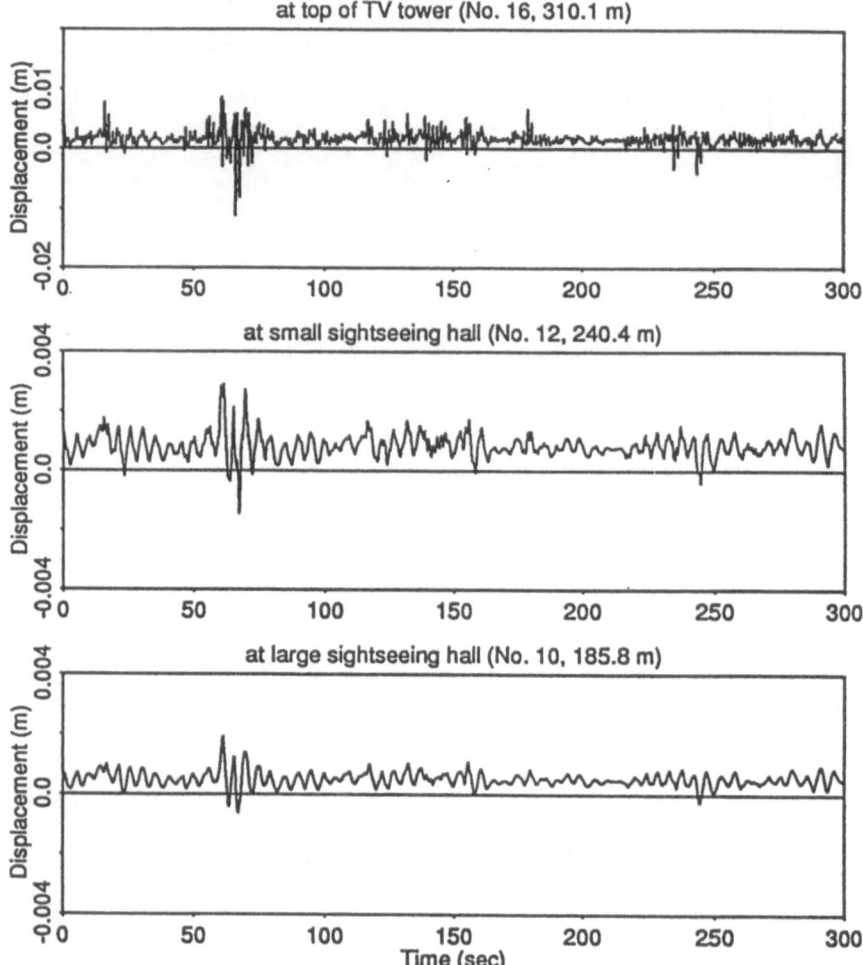

Figure 10. The *y*-direction displacement responses of TV tower at three height locations

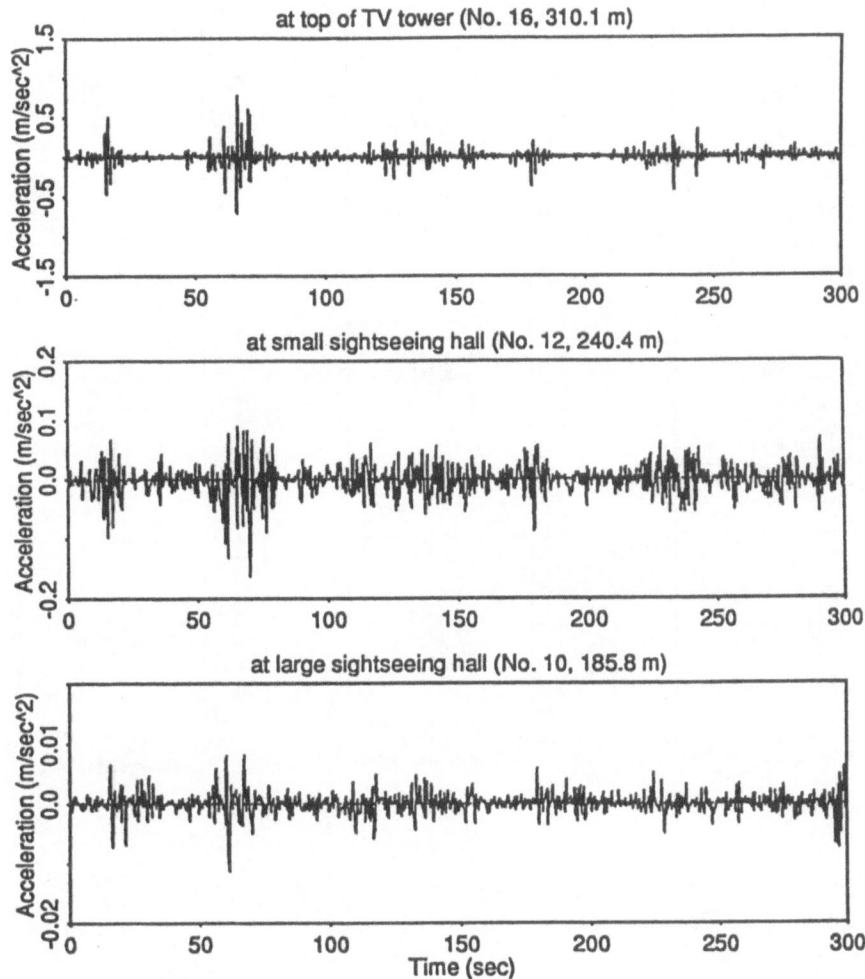

Figure 11. The *y*-direction acceleration responses of TV tower at three height locations

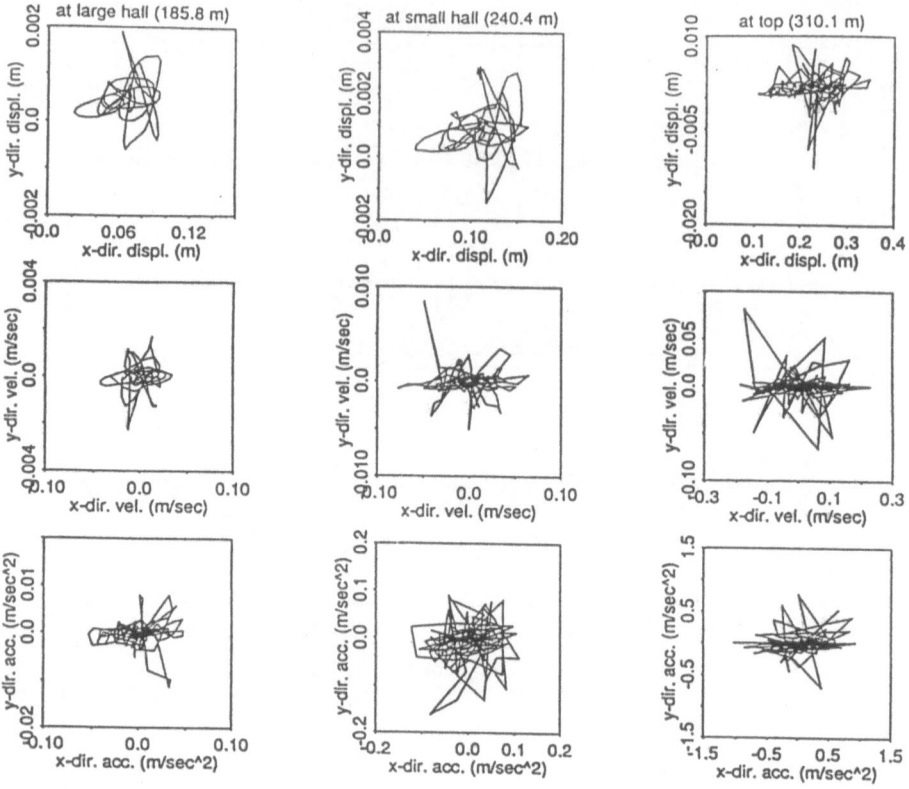

Figure 12. Trajectories of the displacement, velocity and acceleration responses of TV tower at three height locations

A NEW TOOL FOR THE INVESTIGATION OF A CLASS OF NONLINEAR STOCHASTIC DIFFERENTIAL EQUATIONS: THE MELNIKOV PROCESS

E. SIMIU and M. FRANASZEK
National Institute of Standards and Technology
Gaithersburg, MD 20899, USA

1. Introduction

The Melnikov process, a construct rooted in chaotic dynamics theory, was recently developed as a tool for the investigation of a broad class of nonlinear stochastic differential equations [1-6]. This paper briefly reviews the stochastic Melnikov-based approach and applications to (i) oceanography, (ii) open-loop control of stochastic nonlinear systems, and (iii) snap-through of buckled beams with distributed mass and distributed random loading.

2. Melnikov Processes

2.1 DYNAMICAL SYSTEMS

The stochastic Melnikov-based approach reviewed in this paper is applicable, among others, to systems described by the equation

$$\ddot{z} = f(z) + \epsilon[\beta \dot{z} + \gamma G(t)] \tag{1}$$

where $|\epsilon| \ll 1$, and the function f and the random process $G(t)$ are assumed to be sufficiently smooth and bounded. (This restriction on $G(t)$ may be in practice be removed: if $G(t)$ is not smooth and/or bounded, it may be replaced by a bounded and sufficiently smooth process $G_a(t)$ that approximates $G(t)$ as closely as desired over any any arbitrarily large, though finite, time interval. For example, one may approximate broadband Gaussian noise by a sum of a large number of bounded harmonic terms with random frequencies and phases [7]; white noise by broadband noise with constant spectrum and very large cut-off frequency [4]; square wave coin-toss dichotomous noise by a series of sums with a random (coin-toss) parameter and a non-random parameter defining the position of the square wave on the time axis, each sum in the series representing an arbitrarily close Fourier approximation of a square wave [5].) In addition, it is assumed that the unperturbed system has two hyperbolic critical points z_+ and z_-, not necessarily distinct, and that there is an

427

A. Naess and S. Krenk (eds.), IUTAM Symposium on Advances in Nonlinear Stochastic Mechanics, 427–436.
© 1996 *Kluwer Academic Publishers. Printed in the Netherlands.*

orbit $\gamma(t)$ which connects them. (The orbit is called homoclinic if the two points coincide, and heteroclinic otherwise.) In particular this assumption holds for systems with $f(z)=dV(z)/dz$, where $V(z)$ is a two- or multi-well potential. Examples, in addition to others given later in the paper, are the Duffing-Holmes equation, the Josephson junction, and models of vessel capsizing in beam seas [2].

Equation 1 with $\epsilon=0$ is referred to as the unperturbed system.

2.2 MELNIKOV-BASED NECESSARY CONDITONS FOR ESCAPES

2.2.1 *Persistence Theorem*

For the system approximating Eq. 1 [i.e., Eq. 1 in which $G_a(t)$ is substituted for $G(t)$], the persistence theorem can be used to show that, for sufficiently small ϵ, the hyperbolic critical points persist in a phase space slice through the approximating system's extended phase space [8]. This means that the perturbed system will possess separated stable and unstable manifolds.

2.2.2 *Application of Smale-Birkhoff Theorem*

The Smale-Birkhoff theorem states that a necessary condition for the occurrence of chaos (i.e., for sensitivity to initial conditions) is that the stable and unstable manifolds of system (1) intersect transversely [8]. Crossing of a potential barrier (i.e., escapes from regions of phase space associated with the interior of a potential well) can occur only via lobes resulting from the transverse intersections of the system's stable and unstable manifolds [8]. The existence of such intersections is the a necessary condition for chaotic transport. In particular, for systems with stochastic excitation, it is a necessary condition for the occurrence of escapes. Therein lies the connection between chaotic dynamics and the study of escapes in nonlinear multistable stochastic differential equations [1,4].

To first order, the Melnikov process is a measure of the distance beween the stable and unstable manifolds of the stochastic system approximating Eq. 1. If a realization of the Melnikov process has simple zeros, the corresponding stable and unstable manifolds intersect transversely.

It can be shown that: (a) the process $G_a(t)$ induces a Melnikov process arbitrarily close to the process

$$M(t) = -\beta \int_{-\infty}^{\infty} \dot{z}_s^{\,2}(\tau)d\tau + \gamma \int_{-\infty}^{\infty} h(\tau)G(t-\tau)d\tau \qquad (2)$$

where $\dot{z}_s$ is the ordinate in the $z,\dot{z}$ phase plane of the unperturbed system's heteroclinic or homoclinic orbit; (b) the filter in the convolution integral of Eq. 2 is $h(t) = \dot{z}_s(-t)$; and (c) the mean zero upcrossing time of the Melnikov process, τ_u, is a lower bound for the system's mean escape time, τ_e [1,3,4]. To increase τ_e it is therefore necessary to increase τ_u. If the marginal distribution of the noise $G(t)$ is bounded, for values of β, γ such that $M(t) \leq 0$ for $-\infty < t < \infty$ no escapes can ever occur. In some applications the mean value and spectral density of $M(t)$ are useful. They are, respectively, $-\beta K$, where K denotes the value of the first integral of Eq. 2, and

$$2\pi \Psi_M(\omega) = 2\pi \Psi(\omega) \gamma^2 S^2(\omega). \tag{3}$$

$S(\omega)$, the Fourier transform of $h(t)$, is known as the Melnikov relative scale factor.

Finally, we note that although in theory the results just reviewed are valid for small ϵ, in practice they were found to hold even for relatively large ϵ [9].

2.3 EXTENSIONS TO OTHER TYPES OF DYNAMICAL SYSTEMS

Results similar to those just reviewed have been extended to: systems with multiplicative noise [4]; a class of systems of two first-order stochastic differential equations with a slowly-varying parameter; and a stochastic partial differential equation whose deterministic counterpart was first studied by Marsden and Holmes [10] -- see Sections 3 to 5. Similar results can be obtained for multi-degree of freedom systems whose unperturbed counterparts have homo/heteroclinic orbits.

3. Model of Along-shore Currents Due to Randomly Fluctuating Winds Over Continental Shelf With Periodic Corrugations Normal to Coastline

This problem was studied by Allen et al. (1991) [11] for the case of harmonically fluctuating wind forcing. In the absence of friction and forcing this model exhibits homoclinic orbits due to the presence of ocean bottom corrugations normal to the coastline, which correspond in the mathematical model to potential wells separated by a barrier. Under excitation by wind with low frequency *harmonic* fluctuations, and in the presence of friction, for low wind speeds the steady-state motion of a fluid particle will occur within a well for all time. However, if the wind speeds are sufficiently strong, the system's Melnikov function will have simple zeros, and the particle can behave chaotically, that is, move back and forth across the potential barrier in an apparently random fashion [11].

The equations of motion governing the current motion belong to a class of deterministic second order systems with a slowly varying parameter studied by Wiggins and Holmes (1987) [12]. An extension to the stochastic case was reported in [3], and allowed consideration of the more realistic case where the alongshore currents are excited by random wind fluctuations. The fluctuations induce a Melnikov process (that is, an ensemble of Melnikov functions). Assume for example that the excitation is Gaussian. Then, with probability one, the Melnikov process will have simple zeros, and escapes across the potential barrier are possible -- provided that one waits a sufficiently long time. However, the probability that escapes will occur within a specified finite time interval is less than one. Using estimates of the mean zero upcrossing time of the Melnikov process, a weak upper bound for this probability was estimated in [3,4].

4. Melnikov-based Open-loop Control

The performance of certain nonlinear stochastic systems is deemed acceptable if, during a specified time interval, the systems have sufficiently low probabilities of escape from a region of phase space associated with a potential well. An open-loop control method for reducing these probabilities was proposed in [13]. The method is applicable to stochastic systems defined in Sections 2.1 and 2.3. The Melnikov relative scale factors, defined in Section 2.2, are system properties containing information on the frequencies of the random forcing spectral components that are most effective in inducing escapes. An ideal open-loop control force applied to the system would be equal to the negative of a fraction of the exciting force from which the ineffective components have been filtered out. Limitations inherent in any practical control system make it impossible to achieve such an ideal control. Nevertheless, numerical simulations summarized in this section show that, substantial advantages can be achieved in some cases by designing control systems that take into account the information contained in the Melnikov scale factors.

4.1 NUMERICAL SIMULATIONS

4.1.1 *Dynamical System and Excitation*
To illustrate the application of Melnikov-based open-loop control we assume that our system is described by Eq. 1 in which $f(z)=x-x^3$ (i.e., Eq. 1 is the Duffing-Holmes equation), and that the spectral density of the forcing $G(t)$ is

$$2\pi\Psi(\omega)=\begin{cases} 0.03990\,\ell n(\omega)+0.12829 & 0.04\leq\omega\leq 0.4 \\ 0.05755\,\ell n(\omega)+0.14493 & 0.4\leq\omega\leq 1.2 \\ -0.38301[\,\ell n(\omega)]^2+1.06192\,\ell n(\omega)-0.02941 & 1.2\leq\omega\leq15.4 \end{cases}$$

(Fig. 1). A rescaled version of this spectrum approximates low-frequency fluctuations of the horizontal wind speed [3]. For our system the Melnikov relative scale factor is $S(\omega) = (2)^{1/2}\pi\omega sech(\pi\omega/2)$ (Fig. 2a), and $K=4/3$ [1]. The spectral density of the Melnikov process for $\gamma=1$ (see Eq. 3), is shown in Fig. 2b.

4.1.2 *Types of Open-loop Control*
One possible type of open-loop control force has the expression $-\epsilon\gamma_a G(t-t_o)$ and seeks to counteract the excitation by applying a control force proportional and of opposite sign to $\gamma G(t)$. We refer to this as type *(a)* control. The smaller the lag t_o, the more effective the control.

A more efficient open-loop control force is one in which the information provided by the Melnikov relative scale factor $S(\omega)$ is utilized as follows. Figures 1 and 2 show that, owing to their suppression by $S^2(\omega)$, spectral components with frequencies $0 \leq \omega \leq \omega_1$, where $\omega_1 < 0.3$, say, and frequencies $\omega > \omega_2$ where $\omega_2 = 2.5$, say, contribute little to the spectral density of the Melnikov process. We refer to these components as *ineffective*. Our objective is to increase the system's mean exit time τ_e, and as indicated earlier to accomplish this we must increase the mean zero upcrossing time of the system's Melnikov process, τ_u, that is, we must reduce the spectral density of the controlled system's Melnikov process. We can do so -- more effectively than by applying a control force of type *(a)* -- by passing the signal $-\gamma_b G(t-\tau_o)$ through an ideal filter that suppresses the ineffective components. We refer to this control force as type *(b)*.

We now consider a third type of control force, obtained by passing the signal $-\epsilon \gamma_c G(t-t_o)$ through a realistic, practical filter, the impulse response of which is shown in Fig. 3 (a=0.1, b=2.25). This control force is referred to as type *(c)*.

The type *(c)* control force can be improved upon by passing the signal $-\epsilon \gamma_d G(t-t_o)$ through the filter of case *(c)*, and then suppressing from the output all ineffective Fourier components while leaving the other components unchanged. We refer to this force as type *(d)*.

4.1.3 Simulation Results

In all the simulations we assumed $t_o = 0.1$, and $\gamma_b = 0.5$, $\gamma_d = 0.5$, $\epsilon = 0.1$, $\beta = 0.45$. Control force *(a)* was chosen so that it has the same average power as force *(b)*; this yielded $\gamma_a = 0.195$. A similar criterion applied to the forces *(c)*, *(d)* resulted in $\gamma_c = 0.167$. (To within a constant the average power is the variance of the control force.) Simulation results are summarized in Fig. 4. For example, Fig. 4 shows that, given the external excitation $\sigma = \epsilon \gamma = 0.15$, the escape rate reduction due to the use of a control force type *(b)* is about 20 times larger than that due to a control force type *(a)* having the same average power; and control force type *(d)* is almost five times more effective than control force type *(c)* with the same power. Note that the effectiveness of the control force increases as σ decreases.

The simulation results show that the information inherent in the Melnikov scale factors can be utilized to obtain relatively effective open-loop control systems aimed at reducing escape rates. The extent to which this is the case depends upon the spectral density of the excitation, the system characteristics as reflected by the Melnikov relative scale factor, the lag time t_o, and the properties of the filter being used.

5. Snap-through of Buckled Column with Continuous Mass Distribution, Excited by Distributed Stochastic Load

We now illustrate the application of the Melnikov approach to a spatially-extended stochastic dynamical system. We obtain a stochastic counterpart of the Melnikov necessary condition for chaos --and snap-through-- derived by Holmes and Marsden

(1981) [10] for the harmonic loading case. As in Section 3, our approach can yield a lower bound for the probability that snap-through cannot occur during a specified time interval. For excitations with finite-tailed marginal distribution, a simple criterion can be obtained that guarantees the non-occurrence of snap-through.

Assume that: (a) the mechanical properties of the column are uniform over its length, (b) the material is linearly elastic, (c) following the initial, static deformation of the column due to buckling the distance between the column supports is fixed, and (d) the column deformations are sufficiently small that, in the Taylor expansion of the projection of the elemental deformed column length on the line joining the column supports, terms of power higher than two can be neglected. The equation of motion of the column is then [10, 14]

$$z_{tt} + z_{yyyy} + \{\Gamma - \xi \int_0^1 z_y^2(\varsigma,t)d\varsigma\}z_{yy} = \epsilon\{R(y,t) - \beta z_t\} \qquad (4a)$$

$$R(y,t) = \gamma(y)cos(\omega_o t) + \rho(y)G(t) \qquad (4b)$$

where $z(y,t)=Z(Y,\tau)/\Delta$, Z=deflection at time τ, Y=coordinate along column length ℓ, $y=Y/\ell$, $\Delta=Z_o(\ell/2)$ is the static deflection of the column $Z_o(Y)$ at midlength, t and τ=dimensionless and dimensional time, respectively, $\Gamma=P_o\ell^2/EI$, E=Young's modulus, I=moment of inertia of column cross-section with respect to weak axis,

$$P_o=P_{cr}+[EA/2\ell]\int_0^\ell (dZ_o/dY)^2 dY, \qquad (4c)$$

$P_{cr}=k\pi^2 EI/\ell^2$ is Euler's critical buckling load, k=coefficient dependent upon the boundary conditions (for columns hinged at both ends $k=1$), A=cross-sectional area, $\xi=\frac{1}{2}\Delta^2 A/I$, $\epsilon\beta=c\ell^2/[mEI]^{1/2}$, c=viscous damping coefficient, m=column mass per unit length, $t=\omega_1\tau$ is the nondimensional time, $\omega_1^2=(EI/\ell^4 m)$, $\epsilon\gamma(y)=f(Y)\ell^4 m/(EI\Delta)$, $f(Y)$=amplitude of harmonic force per unit length, $G(t)$=nondimensional nonperiodic function, $\epsilon\rho(y)=s(Y)\ell^4 m/(EI\Delta)$, $s(Y)$=measure of nonperiodic force per unit length. Both ends of the column are assumed to be hinged, i.e., the boundary conditions are $z(0,t) = z(1,t) = z_{yy}(0,t) = z_{yy}(1,t) = 0$. The initial deflection $Z(Y,0)=Z_o(Y)$. For our boundary conditions $Z_o(Y)=\Delta sin(\pi Y/\ell)$. It can be easily verified that $\Gamma = \pi^2+\pi^2\xi/2$.

The eigenvalues of the linearized, unforced equation are [10]

$$\lambda_j=\pm\pi j(\Gamma-\pi^2 j^2)^{1/2}, \quad j=1,2,... \qquad (5a)$$

From the expression of Γ given earlier it follows that $\Gamma\geq\pi^2$. Since we assume the deflections are small, $\pi^2<\Gamma<4\pi^2$. Therefore the solution $z=0$ has one positive and one negative eigenvalue and the system with $\epsilon=0$ and $\xi>0$ has two nontrivial buckled equilibrium states. The system also has pure imaginary eigenvalues

$$\lambda_n = \pm \pi n (\Gamma - \pi^2 n^2)^{1/2}, \quad n = 2,3,... \tag{5b}$$

The expansion of $z(y,t)$ in the eigenfunctions of the linearized problem

$$z(y,t) = \sum_{j=1}^{\infty} a_j(t) \sin(j\pi y),$$

used with the Galerkin method, yields

$$\ddot{a}_j + \epsilon \beta \dot{a}_j + (j\pi)^2 \{(j\pi)^2 - [\Gamma - (\xi \pi^2/2) \sum_{k=1,2,..} k^2 a_k^2]\} a_j = \epsilon [2\phi_j \cos(\omega_o t) + \psi_j G(t)] \tag{6}$$

where $\phi_j = \int_0^1 \gamma(y) \sin(j\pi y) dy$, $\psi_j = \int_0^1 \rho(y) \sin(j\pi y) dy$.

The unperturbed counterpart of Eq. 4a has homoclinic orbit coordinates [11]

$$z_o(y,t) = (2)^{1/2} \sin(\pi y) sech[t\pi(\Gamma - \pi^2)^{1/2}]$$

$$\dot{z}_o(y,t) = -(2)^{1/2} \pi (\Gamma - \pi^2)^{1/2} \sin(\pi y) sech[t\pi(\Gamma - \pi^2)^{1/2}] tanh[t\pi(\Gamma - \pi^2)^{1/2}].$$

The Melnikov function for the harmonically excited system can be written as

$$M(t) = \int_{-\infty}^{\infty} \int_0^1 [R(y,\theta) \dot{z}_o(y,\theta - t) - \beta \dot{z}_o^2(y,\theta - t)] dy d\theta \tag{7}$$

where $R(y,t)$ is given by Eq. 4b.

We now consider Eq. 6 and let $\rho(y) = 0$. If the non-resonance condition $\omega_o^2 \neq \lambda_j^2$ holds, Eq. 6 has unique solutions of $O(\epsilon)$; otherwise the linearized counterparts of Eqs. 6 would have solutions of $O(1)$. This would violate a basic assumption of Melnikov theory [10]. If $\rho(y) \neq 0$, for excitations $G(t)$ with continuous spectral density it can be shown that the solutions of the linearized counterparts of the Galerkin equations are of $O(\epsilon)^{1/2}$ [15]. For sufficiently small ϵ those solutions will be as small as desired, and non-resonance conditions associated with $G(t)$ are not required for the assumptions of Melnikov theory to be satisfied.

For the particular case of dichotomous coin-toss square wave noise, following steps similar to those of [5], it can be shown that Eq. 7 yields the following criterion guaranteeing the non-occurrence of snap-through

$$\rho_o \leq 2.584 \xi^{1/2} \beta \tag{8}$$

The validity of this criterion was verified by numerical simulations via the system's Galerkin equations. For additional details, see [6,16].

6. Acknowledgments

Partial support for E.S. by the Office of Naval Research, Ocean Engineering Division (Contract No. N00014-94-0028) is acknowledged with thanks. Dr. T. Swean served as project monitor. M.F. served as NIST Guest Researcher on leave from the Institute of Physics, Cracow Pedagogical University, Cracow, Poland.

7. References

1. Frey, M. and Simiu E., "Noise-induced chaos and phase space flux," *Physica D* **63**, 321-340, 1993.
2. Hsieh,, S.R., Troesch, A.W., and Shaw, S.W., "A nonlinear probabilistic method for predicting vessel capsizing in random beam seas," *Proc. Royal Soc. London A*, **446**, pp. 195-211, 1994.
3. Simiu, E., "Melnikov Process for Stochastically Perturbed Slowly Varying Oscillator: Application to a Model of Wind-driven Coastal Currents," *J. Appl. Mech.* (in press).
4. Simiu, E. and Frey, M., "Melnikov Processes and Noise-induced Exits from a Well," *J. Eng.Mech.*, Feb. 1996 (in press).
5. Simiu, E., and Hagwood, C., "Exits in Second-Order Nonlinear Systems Driven by Dichotomous Noise," *Proc., 2nd Int. Conf. Comp. Stoch. Mech.*, (P. Spanos, ed.), pp. 395-401, Balkema, 1995.
6. Franaszek, M., and Simiu, E., "Noise-induced Snap-through of Buckled Column With Continuously Distributed Mass: A Chaotic Dynamics Approach," submitted to *Int. J. Non-linear Mech.*
7. Shinozuka, M., "Simulation of Multivariate and Multidimensional Random Processes, *J. Acoust. Soc. Amer.*, **49**, 347-357, 1971.
8. Wiggins, S. (1990). *Introduction to Applied Nonlinear Dynamical Systems and Chaos*, New York: Springer-Verlag.
9. Moon, F.C., *Chaotic Vibrations*, New York: John Wiley and Sons, 1987.
10. Holmes, P. and Marsden, J., "A Partial Differential Equation with Infinitely Many Periodic Orbits: Chaotic Oscillations of a Forced Beam," *Arch. Rat. Mech. Analys.*, **76** 135-166, 1985.
11. Allen, J.S., Samelson, R.M. and Newberger, P.A., "Chaos in a Model of Forced Quai-geostrophic Flow Over Topography: an Application of Melnikov's Method," *J. Fluid Mech.*, **226**, 511-547, 1991.
12. Wiggins, S. and Holmes, P., "Homoclinic Orbits in Slowly Varying Oscillators," *SIAM J. Math. Anal.* **18** 612-629; Errata: **19** 1254-1255, 1987.
13. E. Simiu and M. Franaszek, "Melnikov-based Open-loop Control of Escape for a Class of Nonlinear Systems," Proc., Symp. on Vibr. Contr. Stoch. Dynam. Syst., ASME, (L. Bergman, ed.), Sept. 1995.
14. Tseng, W.Y. and Dugundji, J., "Nonlinear Vibrations of a Buckled Beam Under Harmonic Excitation," *J. Appl. Mech.*, 467-476, 1971.
15. Meirovich, L., *Analytical Methods in Vibration*, Elsevier, New York, 1964.
16. Sivathanu, Y., Hagwood, C., and Simiu, E., "Exits in multistable systems excited by coin-toss square wave dichotomous noise: a chaotic dynamics approach," *Phys. Rev. E* (to be published).

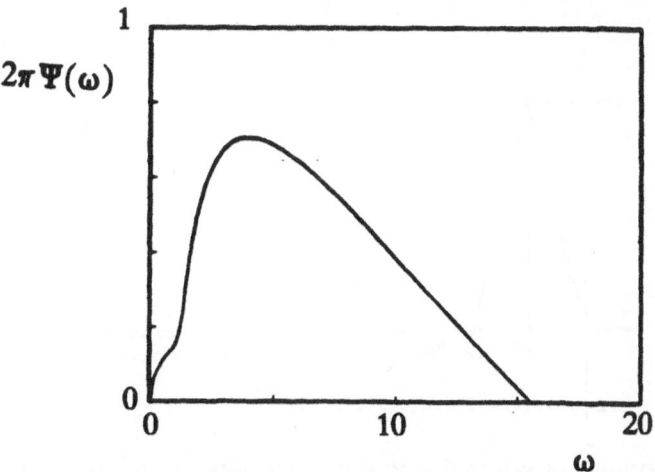

Fig. 1. Spectral density of uncontrolled system's excitation G(t)

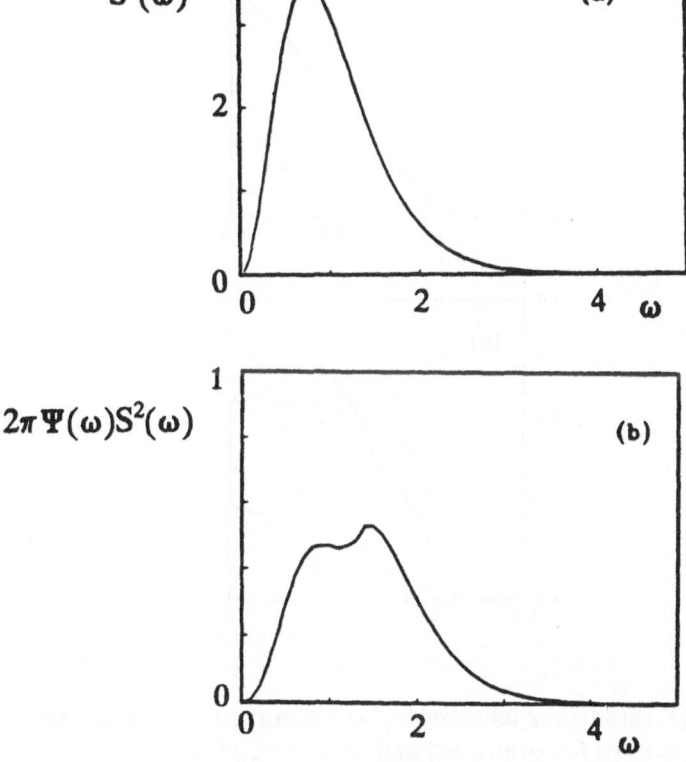

Fig. 2. (a) Melnikov relative scale factor;
(b) spectral density of Melnikov process.

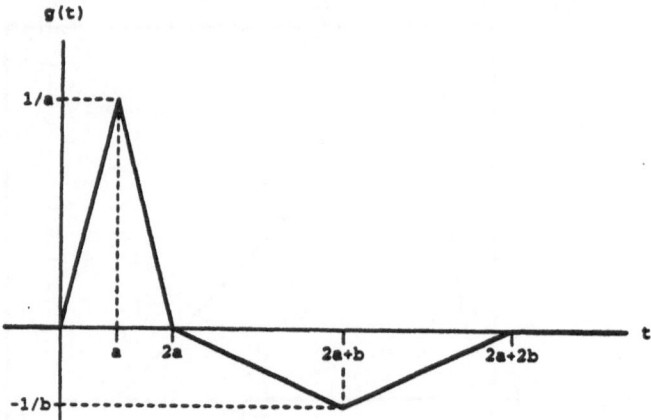

Fig. 3. Impulse response function of filter with initial response and recoil

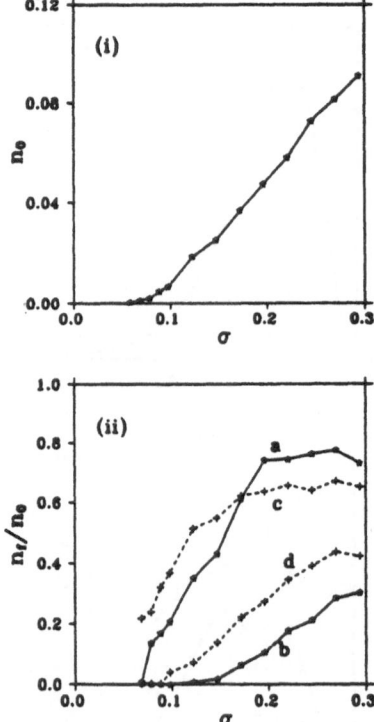

Fig. 4 (a) Escape rate n_o for uncontrolled oscillator; (b) ratio n_f/n_o between escape rates for controlled and uncontrolled oscillator.

Non-Stationary Response of Stochastic Systems via Maximum Entropy Principle

K. SOBCZYK
J. TRĘBICKI

Center of Mechanics, Institute of Fundamental
Technological Research, Polish Academy of Sciences
ul. Świętokrzyska 21, 00-049 Warsaw.

1. Introduction

The principle of maximum entropy states that of all distributions (probability densities) that satisfy the appropriate moment constraints one should choose the distribution having the largest informational (Shannon) entropy. Since the entropy characterizes a global randomness of a random quantity in question, the principle of maximum entropy means that the maximum entropy distribution is maximally noncommittal with regard to the missing information. Due to this reason in statistics the maximum entropy distributions have been proposed to serve as the most unbiased prior distributions in Bayesian inferences. The principle has also been successfully applied in many other fields including reliability estimation of randomly vibrating systems. However, in all these studies the prior information was presented in the form of given (constant) moments.

The principle of maximum entropy has recently been developed to the problems of stochastic dynamics of physical systems when the prior information is given in the form of moment equations (Sobczyk (1991), Sobczyk and Trębicki (1990),(1993)). More specially, we provided a maximum entropy approach to a general class of stochastic nonlinear systems (with constraints in the form of the moment equations) and we showed the effectiveness of the method in determination of the stationary probability distribution of nonlinear vibratory systems.

In this paper we report how the maximum entropy idea can be extended to characterize non-stationary probability distributions of stochastic nonlinear systems. In this case, the variational problem for the entropy functional includes time-dependent constraints in the form of differential equations for moments. We present a scheme for treatment such a problem via time-dependent Lagrange multipliers. The efficiency and goodness of the method is shown by considering some specific stochastic systems of first and second order.

A. Naess and S. Krenk (eds.), IUTAM Symposium on Advances in Nonlinear Stochastic Mechanics, 437–448.
© *1996 Kluwer Academic Publishers.*

2. Idea of Classical Maximum Entropy Principle

It has been a natural desire for the long time to recognize how a partial information contained in statistical moments (of unknown probability distribution) can be used - in the best rational way - to determine the corresponding probability distribution. This is, in essence, the classical moment problem of mathematical analysis (Ahiezer and Krein (1962)) consisting in finding a positive function $f(x)$ from the knowledge of its power moments

$$\int_a^b x^k f(x)\mathrm{d}x = m_k , \quad k = 0,1,2,....$$ (1)

The conditions under which the function $f(x)$ can be recovered from its moments (1) has been extensively studied in the mathematical literature (e.g. Frontini and Tagliani (1994)); they depend on the interval on which $f(x)$ is defined (if $x \in (-\infty, +\infty)$ - Hamburger moment problem; if $x \in [0, +\infty)$ - Stiltjes moment problem; if $x \in [a,b]$, a,b - finite, we have Hausdorff problem).

In practice, only a finite number of moments (say, $M+1$) is usually available. Clearly, there exists then an infinite variety of functions whose $M+1$ moments coincide, and a *unique* reconstruction of $f(x)$ is impossible. There are some approximations of $f(x)$ in the form of sequences of functions $f_n(x)$ such that (1) holds for each $f_n(x)$, $n=0,1,2,...M$ and the sequence $f_n(x)$ converges (in some defined sense) to $f(x)$ as n tends to infinity (e.g. Padé approximations).

The classical maximum entropy method treats the functions $f(x)$ as a probability density and its approximations $f_n(x)$, $n=0,1,2,...M$ are determined from the condition of maximization of entropy

$$\mathrm{H}[f(x)] = -\int f(x)\ln f(x)\,\mathrm{d}x$$ (2)

under the condition that the first $M+1$ moments of $f(x)$ be equal to the true moments m_n, $n=0,1,2,...M$, $(m_0 = 1)$. Using the method of Lagrange multipliers, one looks for the maximum of the extended functional

$$\tilde{\mathrm{H}}[f(x)] = -\int f(x)\ln f(x)\mathrm{d}x + \sum_{n=0}^M \lambda_n (\int x^n f(x)\mathrm{d}x - m_n)$$ (3)

where λ_n are the Lagrange multipliers. Functional differentiation of $\tilde{\mathrm{H}}$ with respect to $f(x)$ leads to the following maximum entropy distribution when $M+1$ moments are given:

$$f_M \equiv p(x) = C\exp\{-\sum_{n=1}^M \lambda_n x^n\}$$ (4)

Unknown multipliers λ_n are determined from the moment constraints $(C = \exp[-\lambda_0 - 1])$

$$\int x^n f_M(x) \, dx = m_n, \quad n = 0, 1, ..., M, \quad m_0 = 1 \tag{5}$$

Equations (5) constitute a nonlinear system of $M+1$ unknown equations for $M+1$ unknown multipliers $\lambda_0, \lambda_1, ..., \lambda_M$. When the multipliers obtained from system (5) are substituted into (4) we get explicitly the approximation $f_M(x)$ of the true probability density $f(x)$. The idea briefly sketched above (originating in statistical physics - Ingarden (1963), Jaynes (1957)) has constituted a methodical base for variety of applications, including the first-passage distribution of vibrating systems (Lin (1970), Spencer and Bergman (1986)).

3. Maximum Entropy Principle for Stochastic Systems

3.1 GENERAL SCHEME

There is no need to restrict the idea of maximum entropy to its classical pattern indicated above. This idea should be applicable to any problem with incomplete data, including analysis of stochastic differential systems. Indeed, when we deel with stochastic systems the probability distributions of the solution process are achievable only in the simplest cases. Most often, all that we are able to obtain from given stochastic (generally, nonlinear) equations are the moments of the solution, or the equations for moments. This partial information can be used (via maximum entropy principle) to determine the approximate probability distribution of the solution process. The extension of the idea of maximum entropy to stochastic systems has recently been presented by the authors in (Sobczyk and Trębicki (1990), (1993)) where emphasis was concentrated on stationary probability distributions. It turns out that the method of maximum entropy can be effectively constructed to non-stationary distributions as well.

Let the system of interest be governed by the following stochastic Itô equation for the vector process $\mathbf{Y}(t) = [Y_1(t), ..., Y_n(t)]$

$$d\,\mathbf{Y}(t) = \mathbf{F}[\mathbf{Y}(t)] dt + \sigma[\mathbf{Y}(t)] d\,\mathbf{W}(t, \gamma) \tag{6}$$

where $\mathbf{W}(t, \gamma) = [W_1(t, \gamma), ..., W_m(t, \gamma)]$ is the m-dimensional Wiener process. Under known conditions (e.g. Sobczyk (1991)) the solution of (1) is a diffusion Markov process with the following drift vector $\mathbf{A}(\mathbf{y})$ and diffusion matrix $\mathbf{B}(\mathbf{y})$

$$\mathbf{A}(\mathbf{y}) = \mathbf{F}(\mathbf{y}) \quad, \quad \mathbf{B}(\mathbf{y}) = \sigma(\mathbf{y})\sigma^{\mathrm{T}}(\mathbf{y}) \tag{7}$$

or, within the Stratonovich interpretation of the equation with white noise

$$\tilde{\mathbf{A}}(\mathbf{y}) = \mathbf{F}(\mathbf{y}) + \frac{1}{2}\sigma(\mathbf{y})\frac{\partial \sigma(\mathbf{y})}{\partial \mathbf{y}} \quad, \quad \tilde{\mathbf{B}}(\mathbf{y}) = \mathbf{B}(\mathbf{y}) \tag{8}$$

The equations for moments are derived easily by use of Itô formula to the function $h_{\mathbf{k}} = Y_1^{k_1}...Y_n^{k_n}$ of the solution and taking the average. The symbol $\mathbf{k}$ denotes here the multi-index i.e. $\mathbf{k}=(k_1,...,k_n)$; we will denote: $|\mathbf{k}|=k_1+...+k_n$ and $|\mathbf{k}|=1,2,..,K$.

The moments of process $\mathbf{Y}(t)$ at time t are defined as usual

$$m_{\mathbf{k}} = \left\langle Y_1^{k_1} Y_2^{k_2}...Y_n^{k_n} \right\rangle_f = \left\langle h_{\mathbf{k}}(\mathbf{Y}(t)) \right\rangle_f \tag{9}$$

where $<\cdot>_f$ denotes the mean value of the quantity indicated, i.e. $<\cdot>_f$ is the integral of $y_1^{k_1}...y_n^{k_n}$ with respect to the true probability density $f(\mathbf{y};t)=f(y_1,...,y_n; t)$ of the solution process. The general form of the moment equations is

$$\frac{d\,m_{\mathbf{k}}(t)}{dt} = \sum_i \left\langle F_i(\mathbf{y}) \frac{\partial h_{\mathbf{k}}(\mathbf{y})}{\partial Y_i} \right\rangle_f + \frac{1}{2}\sum_l \sum_{i,j} \left\langle \sigma_{il}(\mathbf{y})\sigma_{jl}(\mathbf{y}) \frac{\partial^2 h_{\mathbf{k}}(\mathbf{y})}{\partial Y_i \partial Y_j} \right\rangle_f \tag{10}$$

The initial conditions $m_{\mathbf{k}}(t_0)$ are specified from the given probability density $f(\mathbf{y},t_0)$ of the initial condition $\mathbf{Y}_0(\gamma)$.

If $F_i(\mathbf{Y})$ and $\sigma_{ij}(\mathbf{Y})$ are polynomials with respect to $_1,...,Y_n$, equations (10) can be represented symbolically as the following infinite hierarchy of equations

$$\frac{d\,m_{\mathbf{k}}(t)}{dt} = g_{\mathbf{k}}(m_1,...,m_{\mathbf{k}},.....) \quad , \quad |\mathbf{k}| = 1,2,... \tag{11}$$

where $g_{\mathbf{k}}$ are functions of moments specified on the basis of given stochastic system. The finite set of moment equations usually considered is

$$\frac{d\,m_{\mathbf{k}}(t)}{dt} = g_{\mathbf{k}}(m_1,...,m_r) \quad , \quad |r| \geq |\mathbf{k}| \tag{12}$$

where $|\mathbf{k}|=1,2,...,K$, and K is a specified number - the highest order of the moment (9) on the left-hand side of eq. (10) or, in particularly, (11), (12). Since $|r| \geq |\mathbf{k}|$, in general the system (12) is not closed. The same property has a finite collection of equations from hierarchy (10).

According to the spirit of the maximum entropy principle the approximate probability density $p(\mathbf{y};t)$ of the stochastic process $\mathbf{Y}(t)$ governed by general system (6) is determined as a result of maximization of the entropy functional

$$H[p] = -\int p(\mathbf{y};t)\ln p(\mathbf{y};t)d\mathbf{y} \tag{13}$$

under constraints (10) and normalization condition

$$\int p(\mathbf{y};t)d\mathbf{y} = 1 \tag{14}$$

The integration in (13) and (14) is extended over the range of the possible values $Y(t)$ for each t.

Let us notice that constraints (10) and (14) in the maximum entropy scheme can be represented as

$$\frac{d\,m_k(t)}{dt} = \langle G_k(y) \rangle_p \tag{15}$$

$$\int p(y;t) - 1 = 0 \tag{16}$$

where

$$G_k(y) = \sum_i F_i(y)\frac{\partial h_k(y)}{\partial y_i} + \frac{1}{2}\sum_l\sum_{i,j}\sigma_{il}(y)\sigma_{jl}(y)\frac{\partial^2 h_k(y)}{\partial y_i \partial y_j} \tag{17}$$

It has been shown (Trębicki and Sobczyk (1995)) that the probability density which maximizes entropy functional (13) under constraints (15), (16) has the form

$$p(y;t) = C(t)\exp\{-\sum_{|k|=1}^{K}\lambda_k(t)G_k(y)\} \tag{18}$$

where $C(t)$ is the normalizing factor equal to $e^{-\lambda_0(t)-1}$. Functions $\lambda_k(t)$ being unknown Lagrange multipliers are determined by substituting density (18) into constraints (15) and (16). This means that all moments (including moments m_r, when $|r| > |k|$) occurring in set of moment equations (15), which is not closed, are calculated with the use of probability density (18); this is just the maximum entropy closure.

The system of equations for unknown Lagrange multipliers $\lambda_k(t)$ can be generally represented as

$$\int G_k(y)\,p(y;\lambda_k(t))\,dy = \frac{d\,m_k(t)}{dt} \tag{19}$$

where $|k| = 1,2,...,K$, and by the definition (9) of $m_k(t)$:

$$\frac{d\,m_k(t)}{dt} = \int h_k(y)\frac{\partial}{\partial t}p(y;t)\,dy \tag{20}$$

Equations (19), (20) due to (18) include differentiation of $\lambda_k(t)$ and, therefore, constitute a system of nonlinear differential equations for $\lambda_k(t)$. To obtain a standard representation of this system some transformation are required (Trębicki and Sobczyk (1995)).

The initial conditions $\lambda(t_0) = [\lambda_k(t_0)]_{|k|=1,2,...,K}$ associated with system of differential equations for $\lambda_k(t)$ are determined from equations

$$\int [h_{\mathbf{k}}(y) - m_{\mathbf{k}}(t_0)]p(y;t_0)d\,y = 0 \qquad (21)$$

where $m_{\mathbf{k}}(t_0)$ are given values of moments of initial condition $\mathbf{Y}(t_0) = \mathbf{Y}_0$.

The maximum entropy method for stochastic systems presented above for general non-stationary situation, reduces the problem of approximation of nonstationary probability distribution of the response to solving a system of nonlinear differential (deterministic) equations for the Lagrange multiplier functions $\lambda_{\mathbf{k}}(t)$. This system of equations can be solved by available numerical methods. However, to make the calculations easier the method can be modified.

3.2 DISCRETIZATION OF MOMENT EQUATIONS

The system of moment equations associated with the stochastic system in question has the form

$$\frac{d\,m_{\mathbf{k}}(t)}{d\,t} = \frac{d}{d\,t}\langle \mathbf{Y}^{\mathbf{k}}(t)\rangle = \int G_{\mathbf{k}}(y)\,f(y;t)d\,y \qquad (22)$$

where, as above, $\mathbf{k} = (k_1,...,k_n)$ and $\mathbf{Y}^{\mathbf{k}} = Y_1^{k_1}Y_2^{k_2}...Y_n^{k_n}$.
Let us discretize system (22) using (for instance) the Euler scheme

$$\frac{d}{d\,t}\langle \mathbf{Y}^{\mathbf{k}}(t)\rangle = \frac{1}{\Delta t}[\langle \mathbf{Y}^{\mathbf{k}}(t+\Delta t)\rangle - \langle \mathbf{Y}^{\mathbf{k}}(t)\rangle] \qquad (23)$$

System of moment equations (12) takes the form

$$\langle \mathbf{Y}^{\mathbf{k}}(t+\Delta t)\rangle = \langle \mathbf{Y}^{\mathbf{k}}(t)\rangle + \Delta t\int G_{\mathbf{k}}(y)f(y;t)d\,y$$
$$\langle \mathbf{Y}^{\mathbf{k}}(t_0)\rangle = m_{\mathbf{k}}(t_0) \qquad (24)$$

Assuming that the considered time interval is T and denoting by I the number of iterations we get $\Delta t = T/I$, and $\Delta t = t_i - t_{i-1}$, $i=1,...,I$. System (24) can be written as

$$m_{\mathbf{k}}(t_i) = m_{\mathbf{k}}(t_{i-1}) + \Delta t\int G_{\mathbf{k}}(y)f(y;t_{i-1})d\,y \qquad (25)$$

with $m_{\mathbf{k}}(t_0)$ assumed to be given initial condition. Therefore, the variational - maximum entropy problem can be formulated as: in order to determine the approximation $p(y;t)$ of the true density $f(y;t)$ in the discretization points t_i, $i=1,2,...,I$ we maximize the entropy

$$H_i[p] = -\int p(y;t_i)\ln p(y;t_i)d\,y \qquad (26)$$

under the discretized moment constraints (25). In this way the original maximum entropy problem (13)-(17) for stochastic systems is reduced to the iterative sequence of classical maximum entropy problems (26), (25). Of course, to start the numerical procedure we determine first $p(y; t_0)$ via maximum entropy algorithm with constraint $m_k(t_0)$.

If a stochastic system under consideration includes only polynomial nonlinearities, the functions $G_k(y)$ have the form

$$G_k(y) = \sum_{|j|=1}^{L} \alpha_j y^j \tag{27}$$

where $L \geq K$ and α_j are coefficients, then constraints (25) can be represented as

$$m_k(t_i) = m_k(t_{i-1}) + \Delta t \sum_{|j|=1}^{L} \alpha_j m_j(t_{i-1}) \tag{28}$$

$|k| = 1, 2, ..., K$; $|j| = 1, 2, ..., L$. When maximum entropy method is used in calculations, the moments of order $|j|$ higher than K occurring in moment constraints (25) are approximated in the following way

$$m_j(t) = \int y^j p(y; t) dy \quad , \quad |j| \geq K \tag{29}$$

where $p(y; t)$ is the maximum entropy density corresponding to moment constraints with the highest moment order being K. This can be regarded as the maximum entropy closure.

Often the initial conditions for stochastic equations under consideration (6) are assigned in terms of the probability density $f(y; t)$. If so, then this density is used at the start of calculations of (29). To calculate moments (29) one has to perform a multi - fold integration of the integrand. This can be simplified by establishing relationships between these integrals for moments of order higher then K. For example, for $(n=1)$ we have (Sobczyk and Trębicki (1990))

$$m_j = \frac{1}{K\lambda_K}[(j-K+1)m_{j-K} - \sum_{i=1}^{K} i\lambda_i(t) m_{j-K+i}] \quad , \quad j \geq K \tag{30}$$

4. Illustrative examples

4.1 FIRST ORDER SYSTEM

In order to verify the effectiveness of the maximum entropy approach we consider first-order stochastic equation

$$\frac{dY(t)}{dt} + c_1 Y(t) \cos Y(t) - c_2 Y^2(t) \sin Y(t) + c_3 Y^3(t) = D\xi(t) \quad , \quad Y(t_0) = Y_0 \tag{31}$$

where D is a positive constant and $\xi(t)$ is a standard white noise process; c_1, c_2, c_3 are positive constants. The Itô stochastic equation corresponding to (31) is

$$\mathrm{d}Y(t) = \left(c_2 Y^2(t)\sin Y(t) - c_1 Y(t)\cos Y(t) - c_3 Y^3(t)\right)\mathrm{d}t + D\,\mathrm{d}W(t), \quad Y(t_0) = Y_0 \qquad (32)$$

According to (10) the equations for moments are $(k=1,2,...,K)$

$$\frac{\mathrm{d}m_k(t)}{\mathrm{d}t} = k\left\langle Y^k[c_2 Y(t)\sin Y(t) + c_1\cos Y(t)]\right\rangle - c_3 k m_{k+2}(t) + \frac{1}{2}k(k+1)D^2 m_{k-2} \qquad (33)$$

with initial condition $m_k(t_0) = m_k^{(0)}$. The results of application of the maximum entropy method with four first moment equations $(K=4)$ from (33) as the constraints have been compared with Monte-Carlo simulation results for various instant of time. Figure 1 shows, for $t_1 = 1.0, t_2 = 3.0, t_3 = 5.0, t_4 = 7.0$, the comparison of empirical distribution function obtained from simulated trajectories of (31) and distribution function follows maximum entropy method. It is seen that the agreement is very satisfactory. The constants c_1, c_2, c_3, D in the equation (31) are: $c_1 = c_2 = c_3 = D = 1.0$. The initial condition was assumed to be Gaussian distribution with following two first moments: $m_1^{(0)} = 2.0$, $m_2^{(0)} = 2.05$. Numerical calculations indicate that around time $t_4 = 7.0$ the behaviour of the system is near to the stationary state. Figure 2 illustrates the maximum entropy density function for t_1, t_2, t_3, t_4 given above. Figure 3 shows the evolution in time of the entropy $H_t[p]$, mean value $m_Y(t) = \langle Y(t)\rangle$ and standard deviation $\sigma_Y(t)$ of $Y(t)$ calculated with the use of the maximum entropy distribution. This figure presents also the empirical mean value and standard deviation obtained from simulated trajectories.

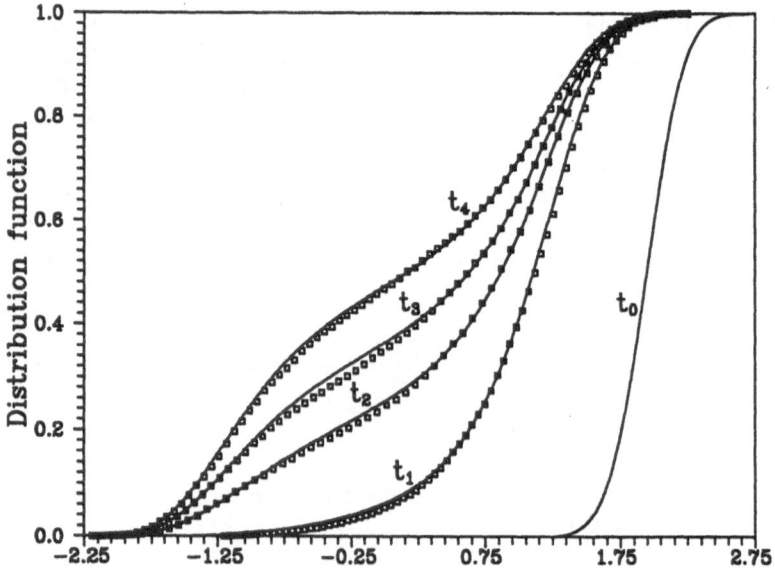

Figure 1. Maximum entropy density function of $Y(t)$ for various instants of time. Comparison between distribution functions based on maximum entropy method (———) and on simulation results (□□□□□□); .

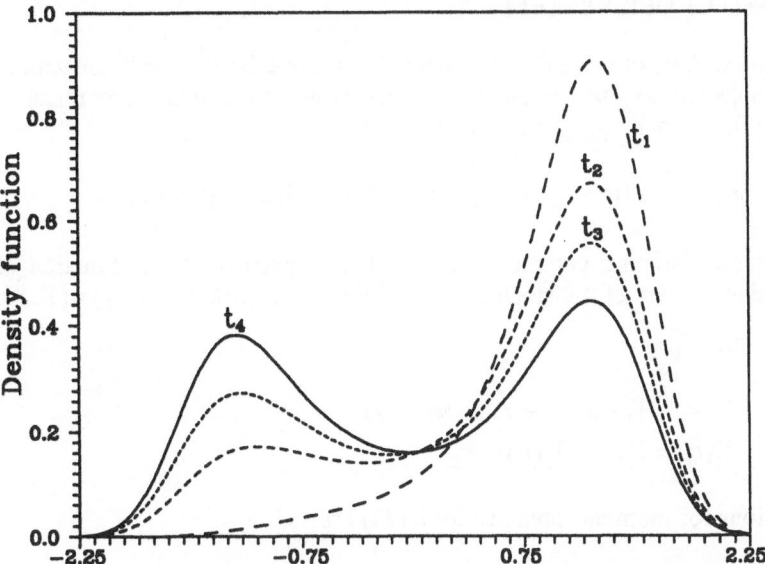

Figure 2. Maximum entropy density function of $Y(t)$ for various instants of time; t_1, t_2, t_3, t_4 as in Fig.1.

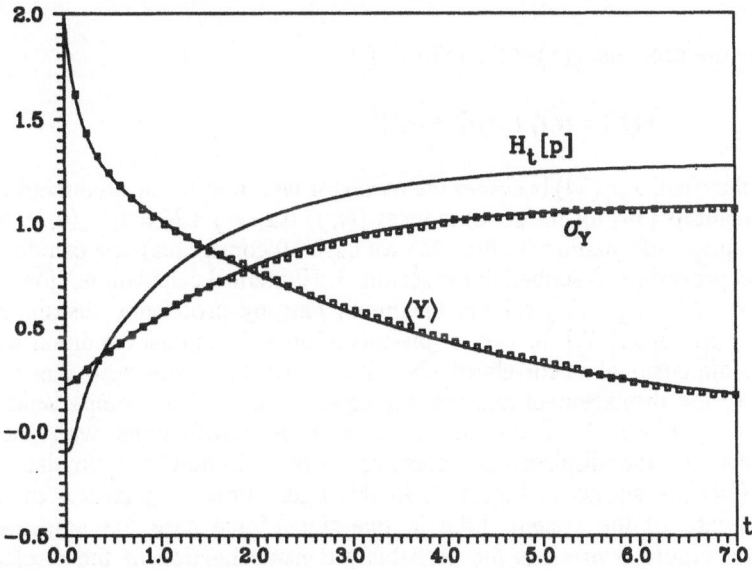

Figure 3. Mean value, standard deviation and entropy of $Y(t)$ versus time predicted from maximum entropy method (———) and from simulation results (□□□□□□).

4.2 SECOND ORDER SYSTEM

As the example of second order system let us consider the oscillator with non-linear stiffness subjected to the stationary Gaussian white noise with zero mean value and correlation function $K_\xi(t_1, t_2) = 2D\delta(t_2 - t_1)$, i.e.

$$\ddot{Y}(t) + \beta\dot{Y}(t) + g(Y) = \xi(t, \gamma), \qquad Y(t_0) = Y_{10}, \quad \dot{Y}(t_0) = Y_{20} \tag{34}$$

where β is the damping parameter and $g(y)$ is a given nonlinear function of y. The corresponding system of Itô stochastic equations for the process $[Y_1, Y_2] = [Y, \dot{Y}]$ is

$$\dot{Y}_1 = Y_2$$
$$\dot{Y}_2 = -\beta Y_2 - g(Y_1) + \sqrt{2D}\, dW(t, \gamma) \tag{35}$$
$$Y_1(t_0) = Y_{10}, \qquad Y_2(t_0) = Y_{20}$$

The equations for moments obtained from (33) are

$$\frac{d}{dt}\langle Y_1^i Y_2^j \rangle = \langle i Y_1^{i-1} Y_2^{j+1} - j[\beta Y_2 + g(Y_1)]Y_1^i Y_2^{j-1} + Dj(j-1)Y_1^i Y_2^{j-2} \rangle$$
$$\equiv \langle G_{ij}(Y_1, Y_2) \rangle, \qquad \langle Y_1^i(t_0)Y_2^j(t_0) \rangle = m_{ij}(t_0) \tag{36}$$

Let us take the function $g(y)$ in the following form

$$g(Y_1) = \alpha_1 Y_1 + \alpha_2 Y_1^2 + \alpha_3 Y_1^3 \tag{37}$$

If $\alpha_2 \neq 0$ the oscillator (34) becomes the oscillator with non-linear asymmetric stiffens. Taking equations (34) for the set of indexes $\{(i, j): 0 \leq i + j \leq K = 4\} \setminus \{(0,3), (0,4), (1,3), (1,2)\}$ as a prior information (in this case we have 10 constraints) one can directly follow the general procedure described in the section 3. Numerical calculations (for $\beta = 1$, $D = 1$, $\alpha_1 = 1, \alpha_2 = -2.85, \alpha_3 = 1$) yield the maximum entropy probability distributions of the process in question $[Y_1, Y_2]$ for various instants of time. The initial condition was assumed to be two-dimensional uncorrelated Gaussian distribution with zero mean values and variances of the displacement and velocity equal to 0.05. The comparisons (for $t_1 = 1$, $t_2 = 2$, $t_3 = 3$, $t_4 = 6.5$) of the maximum entropy distributions with the empirical distributions of the displacement obtained from Monte-Carlo simulations of the equation (34) are shown in Figure 4. In this figure time t_4 is closely to time of the stationary state of the system. Like in one-dimensional case the agreement is very satisfactory. Figure 5 presents the probability density function of the displacement for t_1, t_2, t_3, t_4 given above. Figure 6 shows the mean value and standard deviation of the displacement - curves (1), (2), respectively, and mean value and the standard deviation of the velocity - curves (3), (4) - respectively. In numerical calculations the distribution of the velocity (from initial to stationary state) was assumed to be Gaussian.

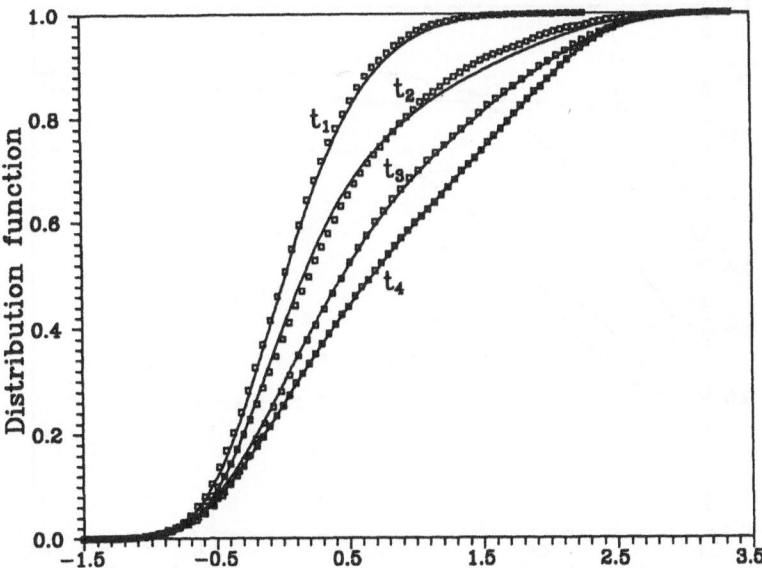

Figure 4. Maximum entropy distribution function of displacement for various instants of time. Comparison between maximum entropy method (——) and empirical distribution function from simulation results (□□□□□□).

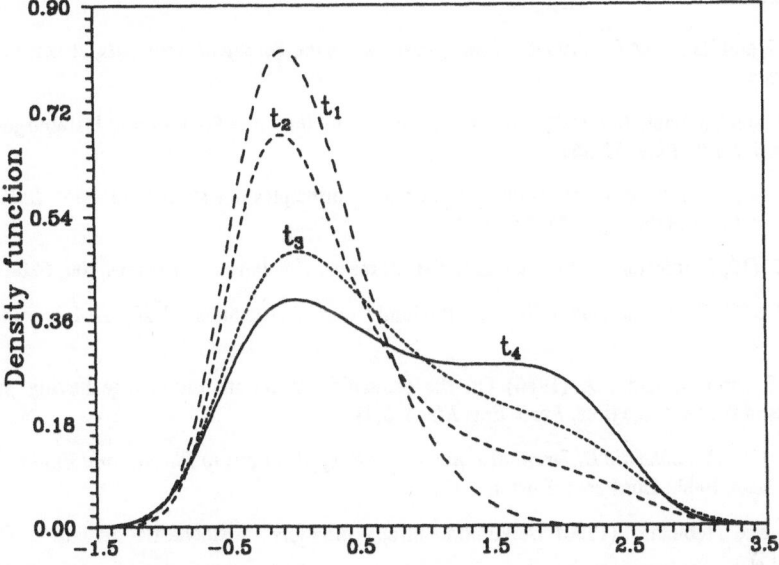

Figure 5. Maximum entropy density function of displacement for various instants of time; t_1, t_2, t_3, t_4 as in Figure 4.

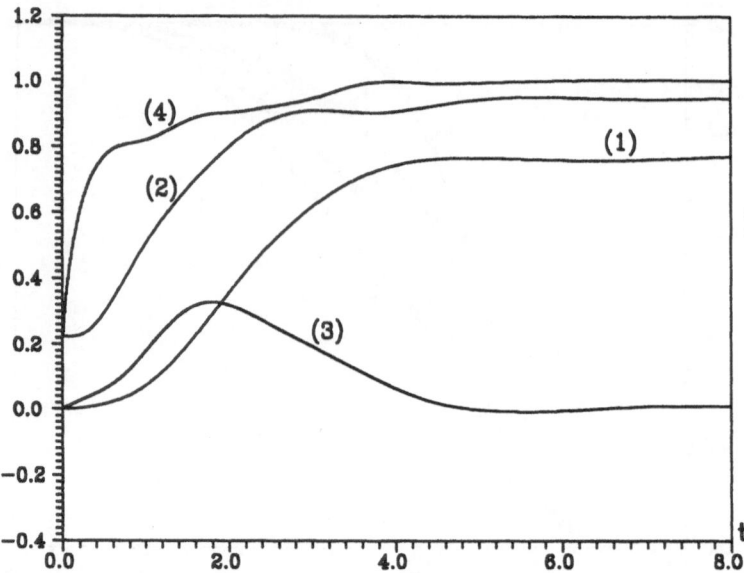

Figure 6. Mean value - (1), standard deviation - (2) of the displacement and mean value - (3) and standard deviation - (4) of the velocity versus time calculated from maximum entropy density function.

References

Ahiezer, N.I. and Krein, M.G. (1962) *Some Questions in the Theory of Moments*, Amer. Math. Soc., Providence.

Frontini, M. and Tagliani, A. (1994) Maximum entropy in the finite Stielties and Hamburger moment problem, *J. Math. Phys.* 12(35).

Ingarden, R.S. (1963) Information theory and variational principles in statistical theories, *Bull. Acad. Polo.., Ser. Math. Astr. Phys.*, 11, 541-547.

Jaynes, E. T. (1957) Information theory and statistical mechanics, *Physical Review*, 106, 620-630

Lin, Y. K. (1970) First-excursion failure of randomly excited structures, *P. II, AIAA Jouranl*, 10(8), 1888-1890

Spencer, B.F. and Bergman, L.A. (1986) On the estimation of failure probability having prescribed moments of first passage time, *Prob. Eng. Mech.* 3(1)

Sobczyk, K. (1991) *Stochastic Differential Equations with Applications to Physics and Engineering*, Kluwer Acad. Publ., Dordrecht, Boston

Sobczyk, K. and Trębicki, J. (1990) Maximum entropy principle in stochastic dynamics, *Prob. Eng. Mech.* 3(5)

Sobczyk, K. and Trębicki, J. (1993) Maximum entropy principle and non-linear stochastic oscillators, *Physica A*, 193, 448-468

Trębicki, J. and Sobczyk, K. (1995) Maximum entropy principle and non-stationary distributions of stochastic systems. *Prob. Eng. Mech*, (submitted for publication)

LYAPUNOV EXPONENTS AND INFORMATION DIMENSIONS OF MULTI-DEGREE-OF-FREEDOM SYSTEMS UNDER DETERMINISTIC AND STATIONARY RANDOM EXCITATIONS

C.W.S. TO[†] and M.L. LIU[§]
Department of Mechanical Engineering

[†]*University of Western Ontario, London, Ontario, Canada N6A 5B9*
[§]*Lakehead University, Thunder Bay, Ontario, Canada P7B 5E1*

1. Introduction

The important concept of Lyapunov exponent has emerged in many fields in the last decade. It plays a crucial role in the determination of bifurcations and chaotic motions in nonlinear systems. Strategies for its numerical computation of multi-degree-of-freedom (MDOF) nonlinear systems under deterministic excitations are available in the literature [1-2]. For nonlinear systems under stochastic excitations, techniques available for the determination of Lyapunov exponents are very limited. They are confined to single degree-of-freedom (DOF) systems under stationary random excitations. For two DOF systems with small nonlinearities and under stationary random excitations of small intensities it is restricted to non-resonant cases. Essentially, these techniques have their basis on the work due to Khasminskii [3].

While nonlinear systems under small and large deterministic forces, and nonlinear systems under small stochastic excitations can be found in many engineering applications, there are nonlinear dynamic engineering systems excited by very large deterministic and stochastic forces. The thrust from the propulsion system of a rocket or a jet engine of an aircraft and the force of atmospheric turbulence exerted on the wings of an aeroplane are typical examples. To be able to design these systems more reliably and economically it is imperative that the question of stability and the nature of motion, be it ordered or chaotic, be addressed qualitatively and quantitatively. Furthermore, it is also of interest to know the consequence of a nonlinear system disturbed by a large deterministic and stationary random force. Thus, there is a need to develop analytical techniques and numerical strategies for the determination of Lyapunov exponents of MDOF nonlinear systems under large deterministic and stationary random excitations.

A numerical strategy for the computation of averages of Lyapunov exponents and information dimensions of MDOF nonlinear systems under small or large deterministic and random excitations is presented. It makes use of the numerical algorithm of Wolf *et al.* [2] and a digital simulation technique of To and Zhang [4]. The proposed strategy

449

A. Naess and S. Krenk (eds.), IUTAM Symposium on Advances in Nonlinear Stochastic Mechanics, 449–458.
© 1996 *Kluwer Academic Publishers.*

is applied to study the characteristic motion of a Duffing oscillator and a two DOF nonlinear system under deterministic and stationary random excitations.

In the next section, equations for Lyapunov exponents and information dimensions are introduced. Section 3 deals with the proposed strategy. Section 4 is concerned with the application of the strategy to a Duffing oscillator, while Section 5 considers the two DOF nonlinear system. Concluding remarks are included in Section 6.

2. Lyapunov Exponents and Information Dimensions

Consider a MDOF dynamic system under deterministic excitations. Its governing equations of motion may be expressed in matrix form as

$$M \ddot{X} + C \dot{X} + K X - P \qquad (1)$$

where X is the displacement vector, while the overdot and double overdot denote the first and second time derivatives, respectively. The mass, damping and stiffness matrices of the system are represented by M, C and K, respectively. These matrices are, in general, nonlinear functions of displacements and velocities. In addition, P is the deterministic excitation vector. Assuming the number of DOF for the system defined by Equation (1) is n_D, then a state vector may be written as

$$Y - [\, X, \dot{X}, t \,]^T \qquad (2)$$

where the superscript T denotes "transpose of", and the dimension of Y is $n = 2n_D + 1$. Applying the state vector above, Equation (1) reduces to

$$\dot{Y} - f(Y, b) \qquad (3)$$

with f(.,.) being a vector function of Y and b a vector of parameters.

Denoting the solution of Equation (3) by $Y^*(t)$, which is obtained by employing a numerical integration scheme, such as the fourth order Runge-Kutta (RK4) algorithm, and a vector representing the variation from $Y^*(t)$ by $\Gamma(t)$, it can be shown that $\Gamma(t)$ or simply Γ satisfies [2]

$$\dot{\Gamma} - A(Y^*, b)\Gamma \quad , \quad A_{ij} - \left. \frac{\partial f_i}{\partial Y_j} \right|_{Y - Y^*} . \qquad (4a,b)$$

Measuring the variation vector Γ at time instants t_0, t_1, t_2, ..., t_M with M being a large integer, the Lyapunov exponent λ_1 can be determined by [2]

$$\lambda_1 = \frac{1}{t_M - t_0} \sum_{k=0}^{M-1} \log_2 \left| \frac{\Gamma(t_{k+1})}{\Gamma(t_k)} \right| \quad , \quad \lambda_1 \geq \lambda_2 \geq \ldots \lambda_n . \qquad (5a,b)$$

The Lyapunov spectrum in Equation (5b) is intimately related to the conjectured information dimension d_I, which is a *static* property of an attracting set, by the following equation

$$d_I = j + \left(\sum_{i=1}^{j} \lambda_i \right) |\lambda_{j+1}|^{-1} \qquad (6)$$

where the integer j is defined by the condition that

$$\sum_{i=1}^{j} \lambda_i > 0, \quad \sum_{i=1}^{j+1} \lambda_i < 0 . \qquad (7)$$

3. Strategy for Numerical Studies

In Equation (1) the excitation vector P is deterministic. Therefore, the Lyapunov exponents and information dimensions of MDOF systems defined in the last section can be computed with the numerical algorithm and digital computer program written in Fortran language [2]. When a zero mean Gaussian white noise vector process $w(t)$ is added to the right hand side (RHS) of Equation (1), application of the Monte Carlo simulation (MCS) technique for its solution becomes necessary. The MCS technique requires generation of series of pseudo-random numbers. In the presently proposed strategy such technique is employed. The approaches [4] of implementing the generated pseudo-random numbers are adopted. These are: (i) by treating the pseudo-random numbers as impulses that are applied to the system at t_0, t_1, ...; and (ii) by adding the pseudo-random numbers to the RHS of Equation (3), which is now assumed to be discretized, such that they are numerically integrated together with the discretized P. Both approaches require correction factors due to the fact that discrete white noise processes are employed instead of continuous ones.

The Lyapunov exponents for every realization, λ_{ir}, are computed by the digital computer program given in reference [2]. Once all realizations are computed, averages of the Lyapunov exponents are determined by

$$<\lambda_i> = \frac{1}{N} \sum_{r=1}^{N} \lambda_{ir} \qquad (8)$$

where the angular brackets denote average, the second subscript r is the number of realizations and the remaining symbols have already been defined above. The information dimension of the MDOF system under deterministic and stochastic excitations is evaluated by applying Equation (6), in which λ_i is replaced by $<\lambda_i>$ defined in Equation (8).

4. Forced Duffing Oscillator

The governing equation of motion for the Duffing oscillator under deterministic and stochastic excitations is given by

$$\ddot{x} + \beta\dot{x} + \alpha x + \varepsilon x^3 = F\cos(\omega t) + w(t) \tag{9}$$

where β, α and ε are system parameters; $w(t)$ is the zero mean Gaussian white noise process; F and ω are the amplitude and angular frequency of the deterministic excitation, respectively.

By writing the state vector $Y = [Y_1, Y_2, Y_3]^T = [x, dx/dt, t]^T$ and considering deterministic excitation only for the time being, equations similar to Equations (3) and (4) may be obtained. In this way, Equations (5) and (6) may be applied to compute the Lyapunov exponents and information dimension of the oscillator.

In the investigation, the system parameters are $\beta = 0.1$, $\alpha = \varepsilon = 1.0$. The angular frequency of the deterministic excitation is $\omega = 1.0$. Initial displacement and velocity are 3.5 and 0.0, respectively. Three different loading conditions are considered. These are: (a) deterministic excitation only, in which the amplitude F is varied from 1.0 to 256.0; (b) stochastic excitation only, in which the spectral intensity of the zero mean Gaussian white noise process S_0 is varied from 1.0 to 9.8×10^4; and (c) deterministic and stochastic excitations, in which $F^2 = 1250.0$ while S_0 varies from 1.0 to 9.8×10^4. Larger values of S_0 were found to cause numerical overflow. In applying the RK4 scheme, a time step size $\Delta t = T/256$ is employed throughout the investigation, where T is the period of the excitation. In the present oscillator, with $\omega = 1.0$ the time step size is $\Delta t = 2\pi/256 = 0.02454$. With such a time step size a time duration of 85 forcing periods ($85T = 85 \times 2\pi = 534.0708$) is considered. Of the 85 periods the first 10 are disregarded during the calculation of the Lyapunov exponents λ_1, λ_2 and λ_3 by means of Equation (5a).

For the cases that include the stationary white noise process two considerations should be noted. Firstly, in the present study at least 22016 pseudo-random numbers are generated to form a realization and the number of realizations used is $N = 75$. This amounts to at least 1.65×10^6 random numbers in total for every information dimension calculation. To verify the results, $N = 100$ and 150 have also been considered. However, no significant discrepancy was observed. The second consideration is the treatment of the pseudo-random numbers. As noted in Section 3, two different approaches may be followed. For the Duffing oscillator studied here, the procedure of incorporating the pseudo-random numbers as a series of impulses is employed. This procedure is efficient [4] because the series is translated into "initial conditions":

$$Y_1(t_{k+0}) - Y_1(t_{k-0})$$
$$Y_2(t_{k+0}) - Y_2(t_{k-0}) + B_1 \bar{w}(t_k) \qquad (10)$$
$$Y_3(t_{k+0}) - Y_3(t_{k-0})$$

where $\bar{w}(t_k)$ is the k-th pseudo-random number in the series. Note that the notation $\bar{w}(t_k)$ is used to differentiate discrete white noise from the continuous white noise process $w(t)$. The correction factor B_1 may be determined through numerical experiments. In the present study, the steps to determine the correction factor B_1 are as follows.

For the Duffing oscillator under a zero mean Gaussian white noise excitation, the exact variance of displacement $\sigma_x{}^2$ expressed in terms of the parabolic cylinder function D is computed by the following equation [5]

$$\sigma_x{}^2 - \sqrt{\frac{\pi S_0}{2\varepsilon\beta} \; \frac{D_{-3/2}(\sqrt{\rho})}{D_{-1/2}(\sqrt{\rho})}} \qquad (11)$$

where the parameter ρ is $\rho = \alpha^2\beta/2\pi\varepsilon S_0$, while a Fortran code has been written to generate a large number of pseudo-random numbers (150 realizations $\times$ 25600 numbers in each set). A number of different time step sizes has also been studied. For a given time step size, computational experiments have been performed to determine the associated correction factor B_1 that gives accurate estimate compared with the exact variance $\sigma_x{}^2$. In this manner, with the selected time step size $\Delta t = 0.02454$, the correction factor $B_1 = 0.136$ has been applied in the computation. Typically, calculated averages of Lyapunov exponents and information dimensions are presented in Figures 1 through 3.

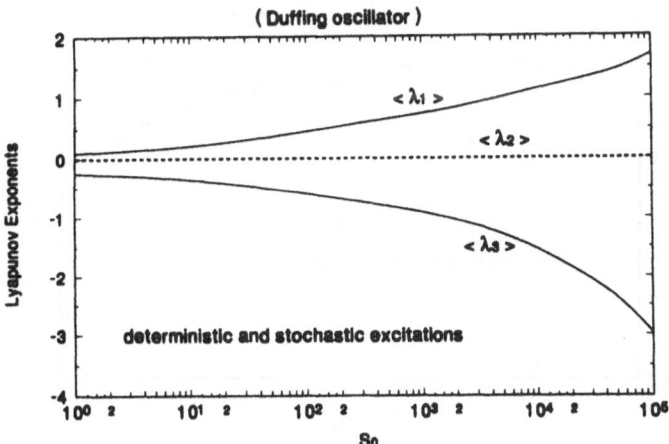

Figure 1. Influence of spectral intensity on averages of Lyapunov exponents.

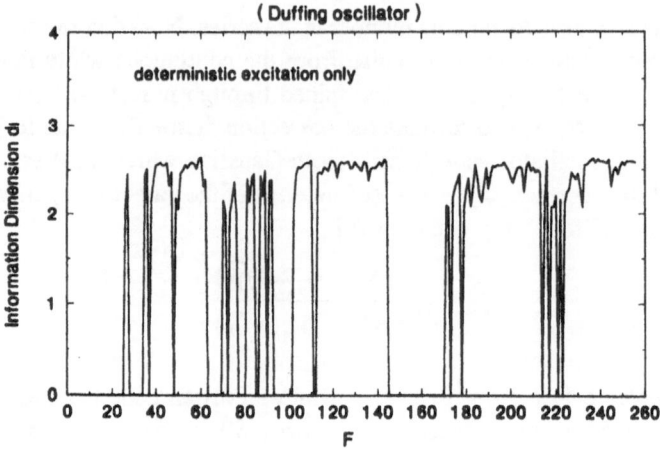

Figure 2. Influence of amplitude of excitation on information dimension.

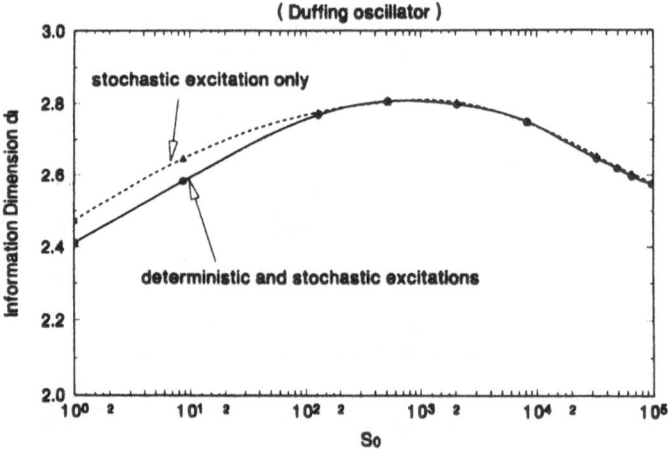

Figure 3. Influence of spectral intensity on information dimension.

5. System with Nonsymmetric Nonlinear Stiffnesses

The two DOF nonsymmetric nonlinear system under a deterministic excitation has the following governing equations [6] in terms of relative displacements $X_1 = x_1 - x_0$ and $X_2 = x_2 - x_1$

$$
\begin{bmatrix} 1 & 0 \\ 0 & 1 \end{bmatrix} \begin{Bmatrix} \ddot{X}_1 \\ \ddot{X}_2 \end{Bmatrix} + \begin{bmatrix} 2\zeta_1 W & -2\mu\zeta_2 \\ -2\zeta_1 W & 2(1+\mu)\zeta_2 \end{bmatrix} \begin{Bmatrix} \dot{X}_1 \\ \dot{X}_2 \end{Bmatrix} + \begin{bmatrix} W^2 & -\mu \\ -W^2 & 1+\mu \end{bmatrix} \begin{Bmatrix} X_1 \\ X_2 \end{Bmatrix}
$$

$$
+ \begin{Bmatrix} \mu\eta X_2^2 - \mu e X_2^3 \\ (1-\mu)\eta X_2^2 + (1+\mu)e X_2^3 \end{Bmatrix} - \begin{Bmatrix} -\ddot{x}_0 \\ 0 \end{Bmatrix} - \begin{Bmatrix} F\cos(\omega t) \\ 0 \end{Bmatrix} \qquad (12)
$$

or in state vector form similar to Equation (2), such that $Y = [Y_1, Y_2, Y_3, Y_4, Y_5]^T = [X_1, X_2, dX_1/dt, dX_2/dt, t]^T$. This matrix equation may be applied to investigate the large motions experienced by the equipment on board of an aeroplane that is approximated as a primary system. Note that here t is the dimensionless time and, henceforth, all quantities are nondimensional. The remaining symbols are defined as

$$
\omega_1^2 - k_1/m_1 \, , \ \omega_2^2 - k_2/m_2 \, , \ 2\zeta_1\omega_1 - c_1/m_1 \, , \ 2\zeta_2\omega_2 - c_2/m_2 \, ,
$$
$$
\eta - \eta'/m_2\omega_2^2 \, , \ \varepsilon - \varepsilon'/m_2\omega_2^2 \, , \ \mu - m_2/m_1 \, , \ W - \omega_1/\omega_2 \, . \qquad (13)
$$

In the investigation, the following parameters have been chosen: $\zeta_1 = \zeta_2 = 0.1$, and $\mu = W = 1.0$. These values have been kept constant throughout all the cases studied. The corresponding linear system, which is obtained by setting $\eta = \varepsilon = 0.0$ in Equation (12), has two undamped natural frequencies: 0.618 and 1.618. The angular frequency of the forcing function $\omega = 1.618$, and initial relative displacements and velocities are zero.

Similar to the Duffing oscillator, three loading conditions are studied: (a) for the deterministic excitation case F is varied from 1.0 to 256.0; (b) for the stochastic excitation case S_0 is varied from 1.0 to 7.5×10^7; and (c) for the case with both deterministic and stochastic excitations, $F = 215.0$ and S_0 is varied from 1.0 to 7.5×10^7. Under every loading condition, two different sets of nonlinearity parameters have been considered. They are: $\eta = -2$ and $\varepsilon = 3$; and $\eta = -4$ and $\varepsilon = 5$. In applying the RK4 numerical integration scheme, time step size equal to T/256 has been selected. That is, $\Delta t = 0.01517$. This time step size was kept constant throughout all computations for the two DOF system. The time duration considered in the computation was 85 periods of the forcing function. That is, 330.0808. The first 10 periods were discarded during the computation of Lyapunov exponents λ_1 through λ_5 and information dimension. For loading conditions (b) and (c), pseudo-random numbers were generated. In every set at least 22016 pseudo-random numbers were generated for every realization, and 75 realizations were included in the computation of every set of Lyapunov exponents and information dimension. Note that in the computation, the pseudo-random numbers were added to the RHS of the discrete state vector equation instead of treating them as impulses. That is, the third of the discrete state vector equation for this system becomes

$$\dot{Y}_3(t_k) = -W^2 Y_1(t_k) + \mu Y_2(t_k) - \mu\eta\, Y_2(t_k)^2 + \mu\varepsilon\, Y_2(t_k)^3 - 2W\zeta_1 Y_3(t_k)$$
$$+ 2\mu\zeta_2 Y_4(t_k) + F\cos(\omega\, Y_5(t_k)) + B_2\overline{w}(t_k) \tag{14}$$

where B_2 is a correction factor that is different from B_1. In the present case, $\Delta t = 0.01517$ and $B_2 = 8.1191$ have been selected and determined, respectively. Typically, computed results are given in Figures 4 through 6.

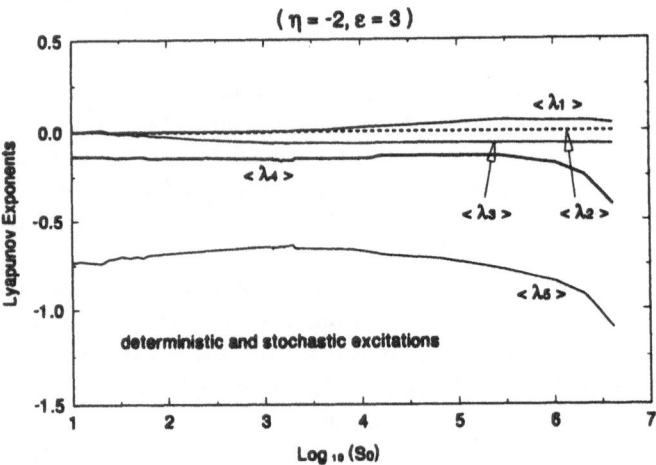

Figure 4. Influence of spectral intensity on averages of Lyapunov exponents.

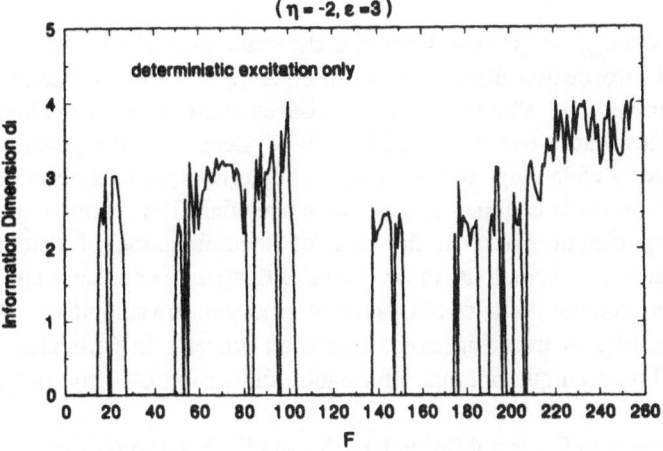

Figure 5. Influence of amplitude of excitation on information dimension.

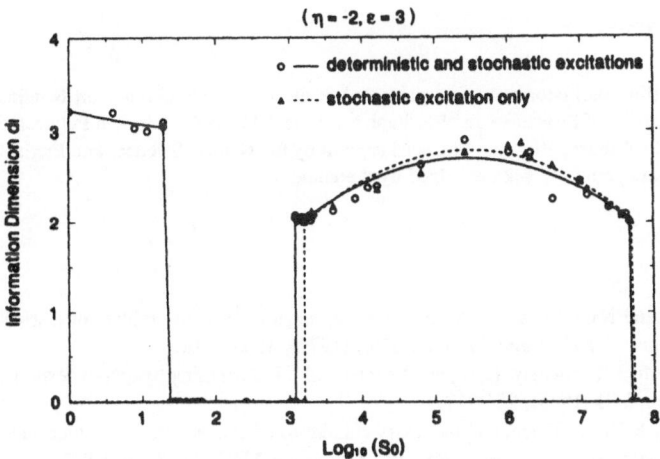

Figure 6. Influence of spectral intensity on information dimension.

6. Concluding Remarks

A numerical strategy is proposed for the determination of averages of Lyapunov exponents and information dimensions of multi-degree-of-freedom nonlinear systems under deterministic and stationary white noise excitations. It is applied to study a Duffing oscillator and a two degree-of-freedom system with nonsymmetric nonlinear stiffnesses. Over a wide range of magnitudes of deterministic excitation the Lyapunov exponents and information dimensions are very irregular. They show ranges of ordered motions among chaotic ones. In the case of a combination of deterministic and stationary white noise excitations, the averages of Lyapunov exponents and information dimensions based on the averages of Lyapunov exponents are smooth over a wide range of spectral densities of the white noise excitation process. In particular, the Duffing oscillator exhibits a unique optimal information dimension over the range of random force studied.

With reference to Figures 4 through 6, the two DOF system exhibits trends similar to those of the Duffing oscillator, except that within the range of $\log_{10}(S_0) \sim 1.4$ and 3.1 in Figure 6 the system seems to have regular motions since in this range the information dimensions are zero.

In passing, it is believed that the proposed strategy is a useful alternative for characterizing motions of MDOF nonlinear systems under a combination of deterministic and stationary white noise excitations.

Acknowledgements

Results of the Duffing oscillator were first presented at the Fourth Conference on Nonlinear Vibrations, Stability, and Dynamics of Structures and Mechanisms, June 7-11, 1992, Virginia Polytechnic Institute and State University, Blacksburg, Virginia. Financial support by the Natural Sciences and Engineering Research Council of Canada is gratefully acknowledged by the authors.

References

1.   Shimada, I., and Nagashima, T.: A numerical approach to ergodic problem of dissipative dynamical systems, *Progress of Theoretical Physics* **61**(6) (1979), 1605-1616.
2.   Wolf, A., Swift, J.B., Swiney, H.L., and Vasano, J.A.: Determining Lyapunov exponents from a time series, *Physica* **16D** (1985), 285-317.
3.   Khasminiskii, R.Z.: Sufficient and necessary conditions of almost sure asymptotic stability of a linear stochastic system, *Theory of Probability and Application* **12**(1) (1967), 144-147.
4.   To, C.W.S., and Zhang, S.W.: On the techniques for digital simulation of random response of nonlinear oscillators, *Journal of Sound and Vibration* **131**(1) (1989), 168-173.
5.   Nigam, N.C.: *Introduction to Random Vibrations*, The MIT Press, Cambridge, Massachusetts, 1983.
6.   Kimura, K., and Sakata, M.: Nonstationary response analysis of a nonsymmetric nonlinear multi-degree-of-freedom system to nonwhite random excitation, *Japan Society of Mechanical Engineers, International Journal* **31**(4) (1988), 690-697.

Randomly excited vibratory systems with variable structure

J. TRĘBICKI

*Center of Mechanics, Institute of Fundamental
Technological Research, Polish Academy of Sciences
ul. Świętokrzyska 21, 00-049 Warsaw, Poland*

1. Introduction

The object of the paper is the analysis of a vibratory system that consists of many randomly excited different vibratory sub-systems. Each of such a vibratory sub-system and its analytical description can be identified as a certain *structure* (or *mode*) of the main system. Randomness of the environment, in which the system works, or randomness hidden in its elements involves the random switching between the structures. The appropriate stochastic differential equations give the global probability description of the system with random switching between different sub-systems (*structures*).

The systems with variable structure contain wide and various classes of engineering systems. Typical examples of these systems are complicated multibody robots and multifunction machinery where stages of work can change in random way because of random disturbances. Another example can be control systems where the connections between working elements change according to control signals or values of some state coordinates (or combination of them). Characteristic examples are also: systems of searching for signals and observing their trajectories, systems where the vanishing of the signal (break of the control) and searching for the lost signal (return to control) is random sequence of different structures. They are also systems containing internal damages resulting in the inability to work of the systems at random moment of time. Specific classes of these systems are systems with degradation where during the exploitation the parameters of the system (e.g. stiffness) change their values (e.g. decreases) in the way dependent on the state coordinates.

The main problem is: how we can assign the probabilistic characterization of the system that is constrained with many subsystems. Obviously, this problem can be solved using known methods consist in solving the system from one structure to another with appropriate continuos conditions. In this way, we have sequence of boundary value's problems. The method here presented based on the solution of the Fokker-Planck-Kolmogorov equation which includes additional functions (cf. [1,2]) containing information about boundary conditions. In words, the classical problem of the processes considered in the given boundaries will be replaced by the process considered in infinity space. Wide class of practical problems connected with the statistical analysis of engineering systems can be led to the consideration systems with variable structure. In the paper [3] the application of the theory of stochastic systems with variable structure was used to identification of the parameters of complicated mechanical systems.

A. Naess and S. Krenk (eds.), IUTAM Symposium on Advances in Nonlinear Stochastic Mechanics, 459–468.
© *1996 Kluwer Academic Publishers.*

2. Stochastic differential equations system.
Functions of annihilation and creation

Let N be a number of the structures (vibrators) of considered system. In general, the evolution of such a system can be described by the extended state vector $[\mathbf{Y}(t), L(t)]$ where $\mathbf{Y}(t) \in R^n$ is a picewice smooth stochastic vector process describing the main system, and $L(t)$ is a stochastic point process describing the succession of the change of the structure. Let $q=1,2,...,N$ denotes the index of the fixed structure and n_q means dimension of the state vector $\mathbf{Y}(t)$ in this structure. If the process $\mathbf{Y}(t)$ is in the q-th structure we will note it as $\mathbf{Y}^{(q)}(t)$. The different structures can have different dimensions n_q, therefore we assume that maximal dimension of the process $\mathbf{Y}(t)$ is equal to $n = \max\{n_q : q = 1, 2, .., N\}$.

Let each of the structure be represented in the form of multidimensional Itô stochastic differential equation

$$d\mathbf{Y}^{(q)}(t) = \mathbf{A}^{(q)}(\mathbf{Y}^{(q)}, t)dt + \mathbf{B}^{(q)}(\mathbf{Y}^{(q)}, t)d\mathbf{W}^{(q)}(t), \quad \mathbf{Y}^{(q)}(t_0) = \mathbf{Y}_0^{(q)} \tag{1}$$

where $\mathbf{A}^{(q)}(\mathbf{Y}^{(q)}, t) = [A_k^{(q)}(\mathbf{Y}^{(q)}, t)]_{k=1}^n$ is the vector and $\mathbf{B}^{(q)}(\mathbf{Y}^{(q)}, t) = [B_{ik}^{(q)}(\mathbf{Y}^{(q)}, t)]_{j,k=1}^n$ is the matrix of nonlinear (or linear) deterministic functions of the process $\mathbf{Y}^{(q)}(t) \in R^n$, respectively; $\mathbf{W}^{(q)}(t) = [W_k^{(q)}(t)]_{k=1}^n$ is the vector of mutually independent standard Wiener processes.

If functions $A_k^{(q)}(\mathbf{y}, t)$ and $B_k^{(q)}(\mathbf{y}, t)$ satisfy appropriate conditions (cf. [5]) then the process $\mathbf{Y}^{(q)}(t)$ is the diffusion Markov processes which full characteristic is given by the probability density function $f_q(\mathbf{y}, t)$ and transition probability density function $f_q(\mathbf{y}, t | \mathbf{y}', t')$, where $\mathbf{y}, \mathbf{y}'$ belong to admissible region $D_q(\mathbf{y})$ for the process $\mathbf{Y}^{(q)}(t)$.

Transition from the q-th structure to the r-th structure will be treated as an annihilation (vanishing) of the realizations of the process $\mathbf{Y}^{(q)}(t)$ and creation (appearing) of the realizations of the process $\mathbf{Y}^{(r)}(t)$. In general, this transition can take place on the boundary $\partial D_{qr}(\mathbf{y})$ between structure q-th and r-th (i.e. on the part or on the whole n_q – hypersurface $\partial D_q(\mathbf{y})$ bounded the admissible region $D_q(\mathbf{y})$) or in the given sub-region $\omega_{qr}(\mathbf{y}) \subseteq D_q(\mathbf{y})$ when coordinates of the process $\mathbf{Y}^{(q)}(t)$ reach required values. In these cases we will say that we have localized (with respect to state coordinates) change of the structure. In other cases this transition can take place in independent way of the state coordinates with any intensity in whole region $D_q(\mathbf{y})$. Then we will say that we have unlocalized change of the structure.

The functions characterizing the annihilation of the trajectories in the q-th structure and functions characterizing creation of the trajectories in considered structure will be called annihilation and creation functions, respectively. The main process $\mathbf{Y}(t)$ is fully described by the set of all equations (1) together with appropriate functions of annihilation and creation. The forms of these functions depend on the way of annihilation or creation (localized or unlocalized) and will be discussed in further analysis.

Potentially the trajectories from the q-th structure can go to other arbitrary structure from the set $I = \{1, 2, ..., N\} \setminus \{q\}$. Structures available from the q-th structure are treated as the active structures for the q-th structure. Total number of these structures implies the number of possible ways of leaving the q-th structure. Symbolically the space of annihilation of realizations in the q-th structure can be presented as $S_q(\mathbf{y}) = \bigcup S_{qr}(\mathbf{y})$ where $S_{qr}(\mathbf{y})$ symbolically means $\partial D_{qr}(\mathbf{y})$ or $\omega_{qr}(\mathbf{y})$. The similar formulae can be given for spaces of creation of realizations in the q-th structure.

The annihilation of realizations of the process $\mathbf{Y}^{(q)}(t)$ and transition to the r-th structure be characterized by local function of annihilation $\tilde{c}_{qr}(\mathbf{y}, t)$, $\tilde{c}_{qq}(\mathbf{y}, t) = 0$ The creation in the q-th structure of the realizations from arbitrary r-structure be characterized by local function of creation $\tilde{d}_{rq}(\mathbf{y},t)$, $\tilde{d}_{rr}(\mathbf{y},t)=0$. These functions are closely connected with coming in and coming out the probability mass (trajectories) in the point $\mathbf{y}$ of the state space in the q-th structure. Integrals from these functions

$$\tilde{z}_{qr}(\mathbf{y},t) = \int \tilde{c}_{qr}(\mathbf{y},t)\,\mathrm{d}\mathbf{y} \quad , \qquad \tilde{u}_{rq}(\mathbf{y},t) = \int \tilde{d}_{rq}(\mathbf{y},t)\,\mathrm{d}\mathbf{y} \tag{2}$$

characterize intensities of annihilation and creation for unit time and they are interpreted as local flux of annihilation $\tilde{z}_{qr}(\mathbf{y},t)$ and local flux of creation $\tilde{u}_{rq}(\mathbf{y},t)$. Taking into account all active structures for the structure q we can introduce more general quantitative characteristics of annihilation and creation phenomena in the form of global function of annihilation $\tilde{c}_q(\mathbf{y},t)$ and global function of creation $\tilde{d}_q(\mathbf{y},t)$, where

$$\tilde{c}_q(\mathbf{y},t) = \sum_{r=1}^{N} \tilde{c}_{qr}(\mathbf{y},t) \quad , \qquad \tilde{d}_q(\mathbf{y},t) = \sum_{r=1}^{N} \tilde{d}_{rq}(\mathbf{y},t) \tag{3}$$

In analogous way global flux of annihilation $\tilde{z}_q(\mathbf{y},t)$ and global flux of creation $\tilde{u}_q(\mathbf{y},t)$ can be described

$$\tilde{z}_q(t) = \sum_{r=1}^{N} \tilde{z}_{qr}(t) = \int \tilde{c}_q(\mathbf{y},t)\,\mathrm{d}\mathbf{y}, \quad \tilde{u}_q(t) = \sum_{r=1}^{N} \tilde{u}_{rq}(t) = \int \tilde{c}_q(\mathbf{y},t)\,\mathrm{d}\mathbf{y} \tag{4}$$

Presented above global functions and fluxes of annihilation and creation described total possibility annihilation and creation which takes place in the q-th structure.

3. Extended Fokker-Planck-Kolmogorov equation

Let us take the point $\mathbf{y}$ of the state space in the q-th structure and an environment $\Delta V(\mathbf{y})$ of this point. Because of annihilation and creation of realizations within $\Delta V(\mathbf{y})$, the change of the probability inside $\Delta V(\mathbf{y})$ during the time Δt is equal to

$$\Delta P_1 = -\Delta t \oint_{\Delta S} [\mathbf{n}_q^0]^{\mathrm{T}} \tilde{\Pi}^{(q)}(\mathbf{y},t)\,\mathrm{d}\mathbf{y} + \Delta t \Delta V(\mathbf{y})[-\tilde{c}_q(\mathbf{y},t) + \tilde{d}_q(\mathbf{y},t)] \tag{5}$$

where $\tilde{\Pi}^{(q)}(\mathbf{y},t)$ denotes the vector of the probability flux with components

$$\tilde{\pi}_k^{(q)}(\mathbf{y},t) = a_k^{(q)}(\mathbf{y},t)\tilde{f}_q(\mathbf{y},t) - \frac{1}{2}\mathrm{div}[b_{ki}^{(q)}(\mathbf{y},t)\tilde{f}_q(\mathbf{y},t)] \quad , \quad k,i = 1,2,\ldots,n_q \tag{6}$$

$a_k^{(q)}(\mathbf{y},t)$ are elements of the drift vector $\mathbf{a}^{(q)}(\mathbf{y},t) = \mathbf{A}^{(q)}(\mathbf{y},t)$ and $b_k^{(q)}(\mathbf{y},t)$ are elements of the diffusion matrix $\mathbf{b}^{(q)}(\mathbf{y},t) = [\mathbf{B}^{(q)}]^{\mathrm{T}} \mathbf{B}^{(q)}$; $\mathrm{div}[\cdot]$ denotes divergence of the given function. The first term of formulae (5) describes quantity of the probability coming out from the environment $\Delta V(\mathbf{y})$ trough its surface $\Delta S(\mathbf{y})$. The second term of this sum takes into account change of the probability described by global functions of annihilation $\tilde{c}_q(\mathbf{y},t)$ and creation $\tilde{d}_q(\mathbf{y},t)$, respectively.
On the other hand the change of probability within $\Delta V(\mathbf{y})$ can be written as

$$\Delta P_2 = \Delta V(\mathbf{y})[\tilde{f}_q(\mathbf{y},t+\Delta t) - \tilde{f}_q(\mathbf{y},t)] \tag{7}$$

On the strength of conservation probability law in the environment $\Delta Q(\mathbf{y})$

$$\Delta P_1 = \Delta P_2 \tag{8}$$

Dividing both sides of equation follows formulae (7), (5), (7) by $\Delta V(\mathbf{y})\Delta t$ and limiting $\Delta V(\mathbf{y})$ and Δt in this equation to zero we obtain the extended F-P-K equation (cf. [1,2]) in the form

$$\frac{\partial \tilde{f}_q(\mathbf{y},t)}{\partial t} = -\operatorname{div}\tilde{\Pi}^{(q)}(\mathbf{y},t) - \tilde{c}_q(\mathbf{y},t) + \tilde{d}_q(\mathbf{y},t) \tag{9}$$

under boundary $\tilde{f}_q(\pm\infty,t), \partial \tilde{f}_q(\mathbf{y},t)/\partial\mathbf{y}|_{y=\pm\infty}=0$ and initial $\tilde{f}_q(\mathbf{y},t_0) = \tilde{f}_0^{(q)}(\mathbf{y}_0)$ conditions, respectively. To write finally form of the equation (8) well known connections were used

$$\frac{\partial \tilde{f}_q(\mathbf{y},t)}{\partial t} = \lim_{\Delta t \to 0} \frac{1}{\Delta t}[\tilde{f}_q(\mathbf{y},t+\Delta t) - \tilde{f}_q(\mathbf{y},t)]$$

$$\operatorname{div}\tilde{\Pi}^{(q)}(\mathbf{y},t) = \lim_{\Delta V \to 0} \frac{1}{\Delta V} \oint_{\Delta S} [\mathbf{n}_q^0]^{\mathrm{T}} \tilde{\Pi}^{(q)}(\mathbf{y},t)\,\mathrm{d}\mathbf{y} \tag{10}$$

It should be underlined that form of drawing and way of obtaining of the equation (9) is analogous to the case of general diffusion equation (cf. [4]).
From physical point of view (because of annihilation) $f_q(\mathbf{y},t)$ is the probability density function of non annihilated realizations. Therefore this functions do not satisfy normalization condition, which means that $\int \tilde{f}_q(\mathbf{y},t)\mathrm{d}\mathbf{y}<1$. In this way, the probability $P_q(t)=\int \tilde{f}_q(\mathbf{y},t)\mathrm{d}\mathbf{y}$ of the duration of the system in the q-th structure becomes very important characteristic. Ordinary differential equation

$$\dot{P}_q(t) = -\int_{-\infty}^{+\infty} \tilde{c}_q(\mathbf{y},t)\mathrm{d}\mathbf{y} + \int_{-\infty}^{+\infty} \tilde{d}_q(\mathbf{y},t)\mathrm{d}\mathbf{y} \quad , \quad \sum_{q=1}^{N} P_q(t_0) = 1 \tag{11}$$

describing an evolution of the probability $P_q(t)$ is obtained by integration of the equation (9) in the whole space taking into account that $\int_{-\infty}^{+\infty} \operatorname{div}\tilde{\Pi}^{(q)}(\mathbf{y},t)\mathrm{d}\mathbf{y}=0$ because of boundary condition.
For normalized probability density function $f_q(\mathbf{y},t) = \tilde{f}_q(\mathbf{y},t)/P_q(t)$ the extended F-P-K equation (9) after simple transformations takes the following form

$$\frac{\partial f_q(\mathbf{y},t)}{\partial t} = -\operatorname{div}\Pi^{(q)}(\mathbf{y},t) - c_q(\mathbf{y},t) + d_q(\mathbf{y},t) - \frac{\dot{P}_q(t)}{P_q(t)} f_q(\mathbf{y},t) \tag{12}$$

where $\Pi^{(q)}(\mathbf{y},t) = \tilde{\Pi}^{(q)}(\mathbf{y},t)/P_q(t)$ is the normalized probability flux and functions

$$c_q(\mathbf{y},t) = \sum_{r=1}^{N} c_{qr}(\mathbf{y},t) \quad , \quad d_q(\mathbf{y},t) = \sum_{r=1}^{N} \frac{P_r(t)}{P_q(t)} d_{rq}(\mathbf{y},t) \tag{13}$$

are respectively global normalized functions of annihilation and creation based on local normalized function of annihilation $c_{qr}(\mathbf{y},t) = \tilde{c}_{qr}(\mathbf{y},t)/P_q(t)$ and local normalized function of creation $d_{rq}(\mathbf{y},t) = \tilde{d}_{rq}(\mathbf{y},t)/P_r(t)$. Simultaneously the equation (11) can be transformed to the form

$$\dot{P}_q(t) = -\sum_{r=1}^{N}\left[P_q(t)z_{qr}(t) - P_r(t)u_{rq}(t)\right] \quad , \quad \sum_{q=1}^{N}P_q(t_0) = 1 \tag{14}$$

where functions $z_{qr}(t) = \int_{-\infty}^{+\infty}c_{qr}(\mathbf{y},t)\,\mathrm{d}\mathbf{y}$ and $u_{qr}(t) = \int_{-\infty}^{+\infty}d_{rq}(\mathbf{y},t)\,\mathrm{d}\mathbf{y}$ are local normalized fluxes of annihilation and creation, respectively.

Taking into consideration (14) in equation (12) the extended F-P-K equation for normalized density function $f^{(q)}(\mathbf{y},t)$ can be rewritten as follows

$$\frac{\partial f_q(\mathbf{y},t)}{\partial t} = -\operatorname{div}\Pi^{(q)}(\mathbf{y},t) - c_q(\mathbf{y},t) + d_q(\mathbf{y},t) - [z_q(t) - u_q(t)]f_q(\mathbf{y},t) \tag{15}$$

where functions $z_q(t)$ and $u_q(t)$ are appropriate sum of normalized fluxes of annihilation and creation, respectively. These functions express global normalized annihilation and creation fluxes of the probability mass in the q-th structure.

4. Forms of annihilation and creation functions.

It was noticed that analytical representation of the functions of annihilation (or creation) depends on its localized or unlocalized character. To express the form of annihilation function connected with the boundary conditions for the process $\mathbf{Y}^{(q)}(t)$ let us at first analyze the situation when the annihilation and creation of the trajectories is localized and occurs on the surface $\partial D_{qr}(\mathbf{y}) \subseteq \partial D_q(\mathbf{y})$ described by the equation $\partial D_{qr}(\mathbf{y}) - \zeta_{qr}(t) = 0$ where $\partial D_{qr}(\mathbf{y})$ is a scalar function of components of the vector $\mathbf{y}$ and $\zeta_{qr}(t)$ is given real function. In practical problems the surface $\partial D_{qr}(\mathbf{y})$ can be taken as a sphere (part of the sphere) or surface (part of the surface) of n_q-dimensional polyhedron with walls perpendicular to the state coordinates of the process.

When annihilation occurs, the normal component of the vector $\tilde{\Pi}^{(q)}(\mathbf{y},t)$ to the surface $\partial D_{qr}(\mathbf{y})$ is positive. It means that annihilation of trajectories on this surface takes place when part of the vector $\tilde{\Pi}^{(q)}(\mathbf{y},t)$ is directed outside to the space $D_q(\mathbf{y})$. Thus the local function of annihilation $\tilde{c}_{qr}(\mathbf{y},t)$ is proportional to the probability flux and takes the following form

$$\tilde{c}_{qr}(\mathbf{y},t) = \delta[\partial D_{qr}(\mathbf{y}) - \zeta_{qr}(t)]\mathbf{n}^0_{qr}[\tilde{\Pi}^{(q)}(\mathbf{y},t)]^{\mathrm{T}} \tag{16}$$

where $\delta\,(.)$ means delta function, $\mathbf{n}^0_{qr}$ is the outward to the surface $\partial D_{qr}(\mathbf{y})$ unit normal vector in horizontal notation. Formally the form of this function can be assign using unit function $\mathbf{1}[.]$ to presentation density function and flux of the probability in the q-th structure as

$$f_q(\mathbf{y},t) = \tilde{f}_q(\mathbf{y},t)\mathbf{1}[\partial D_{qr}(\mathbf{y}) - \zeta_{qr}(t)] \quad , \quad \Pi^{(q)}(\mathbf{y},t) = \tilde{\Pi}^{(q)}(\mathbf{y},t)\mathbf{1}[\partial D_{qr}(\mathbf{y}) - \zeta_{qr}(t)] \tag{17}$$

Then applied above formulae in the equation $\partial f_q(\mathbf{y},t)/\partial t = -\operatorname{div}\Pi^{(q)}(\mathbf{y},t)$ we obtain for the function $\tilde{f}_q(\mathbf{y},t)$ the F-P-K equation with additional term given in (16). The local function of creation $\tilde{d}_{rq}(\mathbf{y},t)$ in the q-th structure from the r-th structure has the similar form as $\tilde{c}_{qr}(\mathbf{y},t)$, but based on the flux of the probability $\bar{\Pi}^{(r)}(\mathbf{y},t)$ in the r-th structure.

In the case of annihilation (or creation) of realizations in n_q-dimensional space $\omega_{qr}(\mathbf{y}) \subseteq D_q(\mathbf{y})$ the functions $\tilde{c}_{qr}(\mathbf{y},t)$ and $\tilde{d}_{rq}(\mathbf{y},t)$ can be modelled for $\mathbf{y} \in \omega_{qr}(\mathbf{y})$ as

$$\tilde{c}_{qr}(\mathbf{y},t) = \varphi_{qr}(\mathbf{y},t)\tilde{f}_q(\mathbf{y},t) \quad , \quad \tilde{d}_{rq}(\mathbf{y},t) = \varphi_{rq}(\mathbf{y},t)\tilde{f}_r(\mathbf{y},t) \tag{18}$$

where functions $\varphi_{qr}(\mathbf{y},t)$, $\varphi_{rq}(\mathbf{y},t)$ take values from interval $(0,1)$ and characterize respectively the intensity of annihilation and creation in the space $\omega_{qr}(\mathbf{y})$. They should be obtained experimentally or assumed in the forms which are adequate to given engineering problem. For example, if process $L(t)$ describing the transition from q-th structure to the r-th structure is the Markov process with intensity of transition $\mu_{qr}(\mathbf{y},t)$ then $\varphi_{qr}(\mathbf{y},t) = \mu_{qr}(\mathbf{y},t)$. If the process of transition between structures makes simple Markov chain with intensity of transition $\mu_{qr}(t)$ independent on state coordinates then change of the structure has unlocalized character and $\varphi_{qr}(\mathbf{y},t) = \mu_{qr}(t)$.

In general functions $\tilde{c}_{qr}(\mathbf{y},t)$ and $\tilde{d}_{rq}(\mathbf{y},t)$ should be functionally connected with probability densities $\tilde{f}_q(\mathbf{y},t)$ and $\tilde{f}_r(\mathbf{y},t)$. This dependence can be symbolically written as

$$\tilde{c}_{qr}(\mathbf{y},t) = V_{qr}[\tilde{f}_q(\mathbf{y},t)] \quad , \quad \tilde{d}_{qr}(\mathbf{y},t) = U_{rq}[\tilde{f}_r(\mathbf{y},t)] \tag{19}$$

where $V_{qr}[.]$ and $U_{rq}[.]$ are given operators shapes of which depend on the kind of the annihilation and creation spaces and on the physical behavior of given problem.

5. Moments and equations for moments

The moments of the vector $\mathbf{Y}^{(q)}(t) = [Y_1 Y_2 ... Y_n]$ for normalized density function $f_q(\mathbf{y},t)$ have the classical definition

$$m_{\mathbf{k}}^{(q)}(t) =: E[\mathbf{M}_{\mathbf{k}}(\mathbf{y})] = \int_{-\infty}^{+\infty} \mathbf{M}_{\mathbf{k}}(\mathbf{y}) f_q(\mathbf{y},t) d\mathbf{y} \quad , \quad |\mathbf{k}| = 1,2,...K \tag{20}$$

where $E[.]$ denotes the mean value; $\mathbf{k}$ denotes the multi-index i.e. $\mathbf{k} = (k_1 k_2 ... k_n)$ and $|\mathbf{k}| := k_1 + k_2 + ... + k_n$; $k_j = 0,1,...,n$ $\mathbf{M}_{\mathbf{k}}(\mathbf{y}) = y_1^{k_1} y_2^{k_2} ... y_n^{k_n}$. Formula (15) defined in the q-th structure moments of the order $|\mathbf{k}| = 1,2,...K$ where K is the given number.

The moments $m_{\mathbf{k}}(t)$ of order $|\mathbf{k}|$ of the main process $\mathbf{Y}(t)$ are given as the following sum

$$m_{\mathbf{k}}(t) = \sum_{q=1}^{N} P_q(t) m_{\mathbf{k}}^{(q)}(t) \quad , \quad |\mathbf{k}| = 1,2,...,K \tag{21}$$

where $P_q(t)$ is the probability of the duration of the process $\mathbf{Y}(t)$ in the q-th structure. To obtain the system of ordinary differential equations for normalized moments $m_{\mathbf{k}}^{(q)}(t)$ the both sides of the extended F-P-K equation will be multiplied by $\mathbf{M}_{\mathbf{k}}^{(q)}(\mathbf{y})$ and then integrated over whole space. Hence

$$\frac{d m_{\mathbf{k}}^{(q)}}{dt} = -\int_{-\infty}^{+\infty} \mathbf{M}_{\mathbf{k}}^{(q)}(\mathbf{y})\{\operatorname{div}\Pi^{(q)}(\mathbf{y},t) + c_q(\mathbf{y},t) - d_q(\mathbf{y},t)\} d\mathbf{y} - m_{\mathbf{k}}^{(q)}(t)[z_q(t) - u_q(t)] \tag{22}$$

with initial condition $m_{\mathbf{k}}^{(q)}(t_0) = m_{\mathbf{k}_0}^{(q)}$ where $m_{\mathbf{k}_0}^{(q)}$ are moments of the distribution $f_q(\mathbf{y}, t_0)$ of the vector $\mathbf{Y}^{(q)}(t_0)$. To simplify the appearance of the further formulae and equations let us rewrite eq. (22) in the form

$$\frac{d m_{\mathbf{k}}^{(q)}(t)}{dt} = \Psi_{\mathbf{k}}^{(q)}[f_q(\mathbf{y}, t); f_r(\mathbf{y}, t); t] \quad , \quad |\mathbf{k}| = 1, 2, \ldots, K \tag{23}$$

where term $\Psi_{\mathbf{k}}^{(q)}[f_q(\mathbf{y}, t); f_r(\mathbf{y}, t); t]$ symbolically denotes the right hand side of the equation (22) and indicates that this equation based on the exact probability density function $f_q(\mathbf{y}, t)$ of the q-th structure and includes exact density functions $f_r(\mathbf{y}, t)$ of all r-structures from which the trajectories in the q-th structure are created.

Because finding analytical nonstationary solution of nonlinear F-P-K equation is very difficult task, therefore we will use an approximation method to express probability density function $f_q(\mathbf{y}, t)$ in question. Let us discretize the eq. (23) using one step approximation of the moment derivative (i.e. Euler scheme)

$$m_{\mathbf{k}}^{(q)}(t_i) = m_{\mathbf{k}}^{(q)}(t_{i-1}) + \Delta t \, \Psi_{\mathbf{k}}^{(q)}[p_q(\mathbf{y}, t_{i-1}); p_r(\mathbf{y}, t_{i-1}); t_{i-1}] \quad , \quad i = 1, 2, \ldots \tag{24}$$

where Δt is the discretization step, $t_i = t_{i-1} + \Delta t$. Functions $p_q(\mathbf{y}, t)$ and $p_r(\mathbf{y}, t)$ are approximations of the true densities $f_q(\mathbf{y}, t)$, $f_r(\mathbf{y}, t)$ respectively. Taking moments $m_{\mathbf{k}}^{(q)}(t_i)$ as a constraints in the maximum entropy method we obtain instead of true density $f_q(\mathbf{y}, t)$ its approximation $p_q(\mathbf{y}, t)$ in the well known form

$$p_q(\mathbf{y}, t_i) = C^{-1}(t_i) \exp\{-\sum_{|\mathbf{k}|=1}^{K} \lambda_{\mathbf{k}}^{(q)}(t_i) y_1^{k_1} y_2^{k_2} \ldots y_{n_q}^{k_{n_q}}\} = C^{-1}(t_i) p_q^*(\mathbf{y}, t_i) \tag{25}$$

where $C(t_i) = \int_{-\infty}^{+\infty} p_q^*(\mathbf{y}, t_i) d\mathbf{y}$ is the normalization constant and $\lambda_{\mathbf{k}}^{(q)}(t_i)$ are unknown Lagrange multipliers calculated numerically from the system of nonlinear equations

$$\int_{-\infty}^{+\infty} [y_1^{k_1} y_2^{k_2} \ldots y_{n_q}^{k_{n_q}} - m_{\mathbf{k}}^{(q)}(t_i)] p_q^*(\mathbf{y}, t_i) d\mathbf{y} = 0 \quad , \quad |\mathbf{k}| = 1, 2, \ldots, K \tag{26}$$

It should be underlined that in (24) all moments of order higher than K are calculated using maximum entropy density $p_q(\mathbf{y}, t_i)$ obtained for moments up to order K. It is *maximum entropy closure method*, (non-Gaussian closure).

When change of the structure occurs on the given hypersurface the appropriate conditions of the boundary regularity should be considered (cf. [1,2]).

Having prescribed moments $m_{\mathbf{k}}^{(q)}(t_i)$ and probability $P_q(t)$ of duration in the q-th structure we calculate from (21) moments of the main process $\mathbf{Y}(t)$ and we adopt for these moments maximum entropy method to obtain approximation $p(\mathbf{y}, t_i)$ of the true density $f(\mathbf{y}, t_i)$.

6. Illustrative examples

To illustrate the application of the theory of the systems with variable structure let us consider the system formed from two oscillators, i.e. $N=2$. In each of these structures the behavior of the system is described by Itô equation $(Y_1^{(q)} = Y^{(q)}, Y_2^{(q)} = \dot{Y}^{(q)})$

$$\dot{Y}_1^{(q)}(t) = Y_2^{(q)}(t) \quad , \quad \dot{Y}_2^{(q)}(t) = -[\beta_q Y_2^{(q)} + g_q(Y_1^{(q)})] + \sqrt{2D_q} \, dW^{(q)}(t) \quad , \quad q = 1, 2 \tag{27}$$

where β_q constant damping coefficients, $g_q(.)$ given function of displacement.
Let change of the structure has unlocalized character independent on state coordinates. It means that annihilation and creation of trajectories of the processes $Y^{(q)}(t)$ can occur within whole spaces admissible for these processes. Transition from structure I-st to structure II-ed occurs with intensity $\mu_{12}(t)$, from structure II-ed to I-st with intensity $\mu_{12}(t)$. Functions $\mu_{12}(t)$ and $\mu_{21}(t)$ should be assumed or estimated from experiment. According to formula (18) the functions of annihilation and creation for normalized density $f_q(y,t)$ are following: $c_{qr}(y,t) = \mu_{qr}(t)f_q(y,t)$, $d_{rq}(y,t) = \mu_{rq}(t)f_r(y,t)$, q, $r=1,2$. According to formula (14) the probability of the duration in fixed structure is governed by equation

$$\dot{P}_q(t) = -P_q(t)\mu_{qr}(t) + P_r(t)\mu_{rq}(t) \quad , \quad q,r = 1,2 \quad , \quad P_1(t_0) + P_2(t_0) = 1 \qquad (28)$$

Because $P_2(t) = 1 - P_1(t)$, then the equation for $P_1(t)$ takes form $\dot{P}_1(t) = -\mu(t)P_1(t) + \mu_{21}(t)$ (where $\mu(t) = \mu_{12}(t) + \mu_{12}(t)$) and its solution in the simplest case (if $\mu_{12}(t) = \mu_{12} = \text{const}$ and $\mu_{12}(t) = \mu_{12} = \text{const}$) is $P_1(t) = P_1(t_0)\exp(-\mu t) + \mu_{12}[1 - e^{\mu t}]/\mu$. In the stationary case (if $t \to +\infty$) for constants μ_{12}, μ_{21} probabilities P_1 and P_2 take values $P_1 = \mu_{21}/\mu$ and $P_2 = \mu_{12}/\mu$, respectively.
According to (15) the F-P-K equation for normalized density function $f_q(y,t)$ is

$$\frac{\partial f_q(y,t)}{\partial t} = -\operatorname{div}\Pi^{(q)}(y,t) - \frac{P_r(t)}{P_q(t)}\mu_{rq}[f_q(y,t) - f_r(y,t)] \quad , \quad q,r = 1,2 \qquad (29)$$

Equations for moments of density $f_q(y,t)$ follow equation (22) and take form

$$\dot{m}_k^{(q)}(t) = E[G_k(y)] - \frac{P_r(t)}{P_q(t)}\mu_{rq}[m_k^{(q)}(t) - m_k^{(r)}(t)] \quad , \quad q,r = 1,2 \qquad (30)$$

where $E[G_k(y)] = -\int_{-\infty}^{+\infty} y_1^{k_1} y_2^{k_2} \operatorname{div}\Pi^{(q)}(y,t)\,dy$. After calculations

$$E[G_k(y)] = E[k_1 y_1^{k_1-1} y_2^{k_2+1} - k_2(\beta_q + g_q(y_1))y_1^{k_1} y_2^{k_2-1} + D_q k_2(k_2-1)y_1^{k_1} y_2^{k_2-1}] \qquad (31)$$

To find approximations $p_q(y,t)$ of densities $f_q(y,t)$ in each structure and approximation $p(y,t)$ of density $f(y,t)$ of main process $Y(t)$ we use method described in section 5 based on discretization of equation (30) and maximum entropy closure.

6.1 TWO LINEAR OSCILLATORS

Let us consider the behavior of the described above system when the oscillators are linear with following damping, stiffness and intensities of Wiener processes:
$\beta_1 = 1.5$, $g_1(y_1) = y_1 - 1.25$, $D_1 = 0.25$, $\beta_2 = 1$, $g_2(y_1) = 0.25y_1 + 1.25$, $D_2 = 0.2$
The oscillator in the first structure has harder stiffness characteristic as oscillator in the second structure. Change between oscillators occurs in the way independent of state coordinates (pure Markov chain) with the constant intensities $\mu_{12} = \mu_{21} = 0.1$.
Figure 1 presents for different moments of time $t_1 = 3.0, t_2 = 5.0, t_3 = 25.0$ the maximum entropy densities of the displacement obtained for four moments (21) of the main process $Y(t)$. For time t_3 the local Gaussian densities (dashed line) of the first (I-st) and second (II-ed) structure are also given. The behavior of the system around time t_3 is close to the

stationary case. Initial conditions in both structures were assumed to be Gaussian distributions with zero mean values and variances of displacement and velocity equal to 0.1. Initial probabilities of duration in each of the structures were following: $P_1(t_0) = 0.25$ $P_2(t_0) = 0.75$, $t_0 = 0$.

Figure 2 illustrates the evolution versus time of the probabilities $P_1(t), P_2(t)$ and coefficient of the skewness $\vartheta_1(t) = m_3 / \sigma^3$ and kurtosis $\vartheta_2(t) = m_4 / \sigma^4 - 3$ (σ - standard deviation) for displacement distribution of the main process $\mathbf{Y}(t)$.

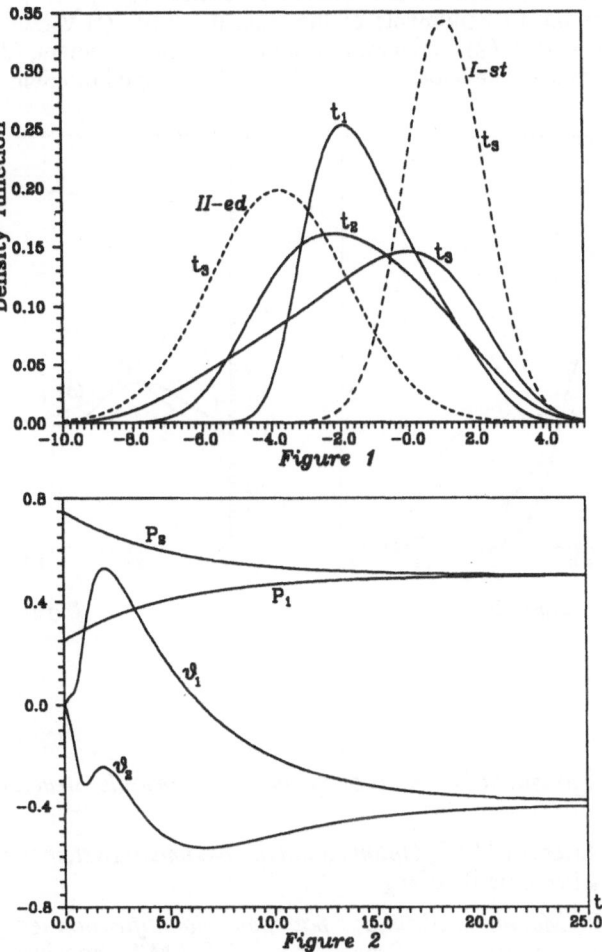

Figure 1

Figure 2

6.2 LINEAR AND NONLINEAR OSCILLATOR

As next example will be considered system formed with linear oscillator (*I*-st structure) and non-linear oscillator (*II*-ed structure). The characteristic of these oscillators are following: $\beta_1 = 2.0$, $g_1(y_1) = 2y_1$, $\beta_2 = 1.0$, $g_2(y_1) = 2y_1 + y_1^3 - 2.85y_1^2$ and $D_1 = D_2 = 1$. As before, change of the structure has character of the pure Markov chain but with constant intensities $\mu_{12} = \mu_{21} = 0.01$. Initial probability of duration in each structure is: $P_1(t_0) = 0.5$, $P_2(t_0) = 0.5$. To investigate the behavior of such a system ten discretized moments equation from hierarchy (30) were taken as constraints in the maximum entropy method resulting in density (25). In detail, the nonstationary distribution of nonlinear oscillator in the *II*-ed structure was considered using maximum entropy method in the paper [6].

Figure 3 shows at time $t_1 = 8.0$ the maximum entropy displacement distribution (solid line) obtained for four moments (21) of the main process $\mathbf{Y}(t)$. The dashed curves present the displacement distribution in each of the structure.

Figure 4 illustrates the first moments of the main process $\mathbf{Y}(t)$ versus time: (1) -mean value of the displacement, (2) -standard deviation of the displacement, (3) -mean value of the velocity, (4) -standard deviation of the velocity, (5) - mixed moment $E[Y_1 Y_2]$.

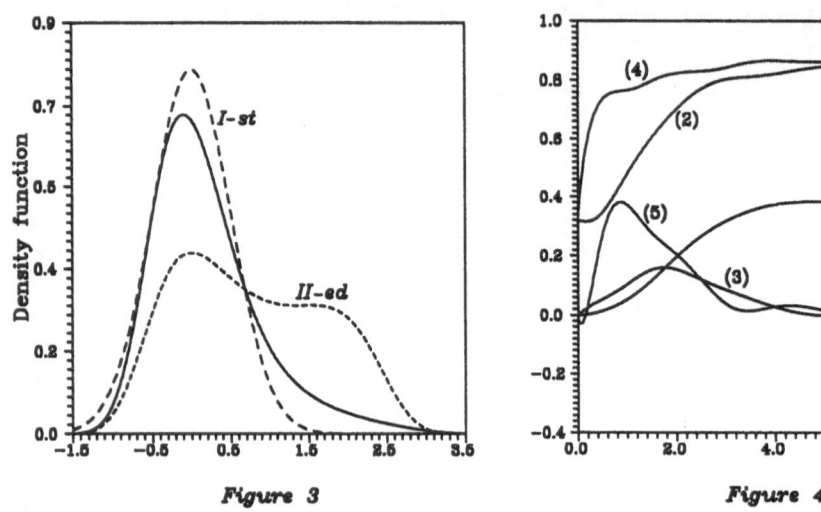

Figure 3   Figure 4

References

1. Kazakov I.E., *Statistical Dynamic of System with Variable Structure*, Nauka, 1997, (in Russian)

2. Kazakov I.E., Artem'ev V.M., *Optimization of Dynamical System with Variable Structure*, Mir, 1980, (in Russian)

3. Kazimierczyk P., *Maximum-likelihood parametric identification technique for objects of randomly varying structure*, Proceedings of IUTAM Symposium, pp. 321-332, Turin 1992

4. Vladimirov V.S., *Equations of Mathematical Physics*, Moscow, Nauka 1988, (in Russian)

5. Gihman I., Skorochod A.V., *Stochastic Differential Equations*, New York 1972.

6. K.Sobczyk, J. Trębicki *Non-stationary response of stochastic systems via maximum entropy principle*, Proceedings of IUTAM Symposium, Trondheim 1995, (submitted to publication)

STABILITY AND INVARIANT MEASURES
OF PERTURBED DYNAMICAL SYSTEMS

Walter V. Wedig
University of Karlsruhe
D-76128 Karlsruhe, FRG

Abstract:

For stability investigations of perturbed dynamical systems it is suitable to utilize the associated invariant measures in order to determine Lyapunov exponents of
higher numerical accuracy. The densities of the invariant measures can be calculated
by the stationary solutions of Fokker-Planck equations. The paper presents iterative
schemes to solve these parabolic equations numerically and check the obtained results by means of Monte-Carlo simulations. Both methods are applied to oscillators
perturbed by non-normal processes with bounded realizations.

1. Stability problems in mechanics

To introduce stability problems of interest [1,2], let us consider a dynamical system
described by the following ordinary differential equations:

$$\ddot{X}_{k,t} + 2D_k\omega_k\dot{X}_{k,t} + \omega_k^2(X_{k,t} + \Gamma_t\sum_{\ell=1}^{n}\varrho_{k\ell}X_{\ell,t}) = 0, \tag{1}$$

$$\Gamma_t = \arctan Z_t, \quad \dot{Z}_t + \omega_g Z_t = \sigma\dot{W}_t, \quad k = 1, \ldots, n. \tag{2}$$

Herein, the parameters ω_k denote the natural frequencies, D_k are damping measures
and $X_{k,t}$ are n scalar state processes of the system. Subscript t represents the time
dependency. Time derivatives are abbreviated by dots. The equations (1) are coupled
by the amplification coefficients $\varrho_{k\ell}$ $(k, \ell = 1, 2, \ldots, n)$ of the parametric perturbation
Γ_t, which is defined in (2). Accordingly, a low-pass process Z_t is generated from
normalized white noise $\dot{W}_t$ by a scalar filter equation with the limiting frequency ω_g
and the intensity parameter σ. Applying the arctan-transformation to Z_t gives the
parametric perturbation Γ_t within $|\Gamma_t| < \pi/2$.

For a typical application in structural mechanics (see Figure 1), let us consider
the lateral vibrations of a uniform beam under axial loading $P(t)$. Assuming that

A. Naess and S. Krenk (eds.), IUTAM Symposium on Advances in Nonlinear Stochastic Mechanics, 469–478.
© 1996 *Kluwer Academic Publishers.*

longitudinal waves $u(x,t)$ can be approximated by $u(t) = -P(t)\ell/EA$, the lateral motion $w(x,t)$ is described by the following boundary value problem [3]:

$$EIw_{xxxx} + \beta w_t + \mu w_{tt} - \tfrac{1}{l}EA[u(t) + \tfrac{1}{2}\int_0^l w_{xx}^2 dx]w_{xx} = 0, \qquad (3)$$

$$w(0,t) = w(l,t) = w_{xx}(0,t) = w_{xx}(l,t) = 0, \qquad 0 \le x \le l. \qquad (4)$$

Herein, EI is the bending stiffness of the beam, β is a damping coefficient, μ denotes the mass per length ℓ and EA is the axial stiffness of the beam. The first natural mode $\sin \pi x/\ell$ satisfies all boundary conditions in (4) and approximates the beam vibrations by $w(x,t) = T(t)\sin \pi x/\ell$. Insertion into the partial integro-differential equation (3) reduces to the nonlinear ordinary differential equation

$$\ddot{T}(t) + 2D\omega_1\dot{T}(t) + \omega_1^2[1 + \frac{Al}{I\pi^2}u(t) + \gamma T^2(t)]T(t) = 0, \qquad (5)$$

where $T(t)$ determines the time behaviour of the first mode, $\omega_1^2 = \pi^4 EI/(\mu\ell^4)$ is the first natural frequency of the beam and γ denotes the coefficient of its cubic restoring. Linearized by $\gamma = 0$, the equation (5) is decoupled from the higher modes describing a linear oscillator, parametrically excited by the axial end displacement $u(t)$ or by the axial end loading $P(t)$, respectively.

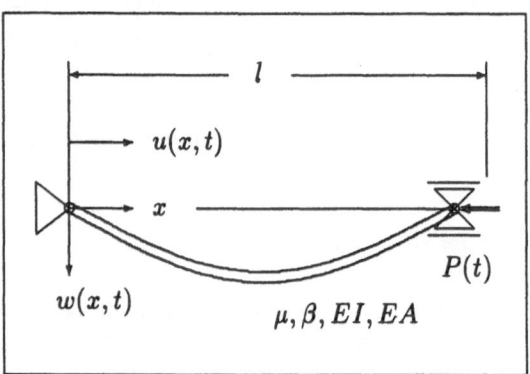

Figure 1: Beam under axial loading

2. Density of the parametric excitation process

To eliminate the low-pass process Z_t we apply Itô's calculus to (2) in order to derive the following transformed stochastic differential equation:

$$d\Gamma_t = -(\omega_g + \sigma^2\cos^2\Gamma_t)\sin\Gamma_t\cos\Gamma_t dt + \sigma^2\cos^2\Gamma_t\, dW_t. \qquad (6)$$

The Fokker-Planck equation, associated to (6), has the stationary form

$$\frac{\partial}{\partial\gamma}[(\omega_g + \sigma^2 \cos^2\gamma)\sin\gamma\cos\gamma\, p(\gamma)] + \frac{1}{2}\sigma^2\frac{\partial^2}{\partial\gamma^2}[\cos^4\gamma\, p(\gamma)] = 0. \qquad (7)$$

It is valid in the angle range $|\gamma| < \pi/2$ and solved by the density function

$$p(\gamma) = \frac{1}{\sqrt{2\pi}\,\sigma_z\cos^2\gamma}\exp(-\frac{\tan^2\gamma}{2\sigma_z^2}), \qquad \sigma_z = \frac{\sigma}{\sqrt{2\omega_g}}. \qquad (8)$$

The density (8) can also be derived by means of the gaussian distribution

$$p(z) = \frac{1}{\sqrt{2\pi}\,\sigma_z}\exp(-\frac{z^2}{2\sigma_z^2}), \qquad\qquad |z| < \infty, \qquad (9)$$

which can easily be transformed to (8) by applying the variable substitution $\gamma = \arctan z$. Figure 2 shows some numerical evaluations of the densities (8) and (9) inside the range $|\gamma| < \pi/2$ and $|z| < \pi/2$, respectively. For small variances σ_z of the low-pass process Z_t, there is a good coincidence between both densities, $p(z)$ and $p(\gamma)$. However, deviations between both become stronger for increasing intensities σ_z, and the non-normal density $p(\gamma)$ degenerates finally to a bi-modal distribution with two peaks near the singularities at $\gamma = \pm\pi/2$.

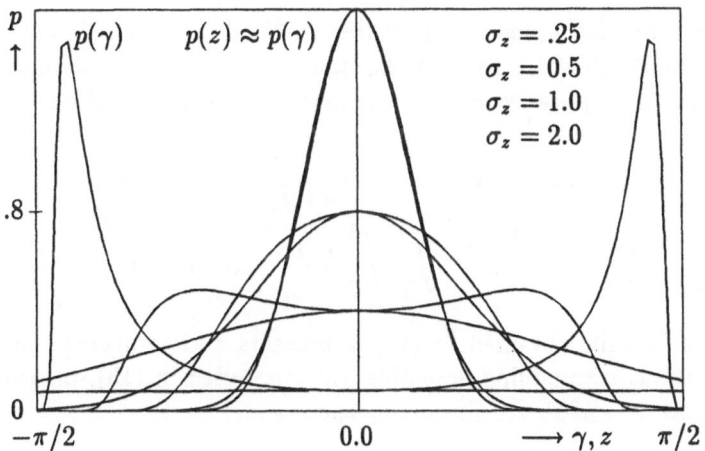

Figure 2: Densities with and without limitation

3. Stability analysis by invariant measures

To investigate the destabilizing effect of the parametric perturbation Γ_t we apply

Khasminskii's projection [4] on hyperspheres introducing corresponding spherical coordinates into the stability equations (1). They split up one transient solution part, which describes the time behaviour of the radius of the hypersphere deciding the stability of the dynamic system. The remaining solution parts rotate permanently on the surface of the hyperspheres. Applying suitable modulo-operators they are bounded and become stationary inside one angle range, in which they can be evaluated in order to obtain the time-invariant characteristics of the angle processes determining the invariant measure of the dynamical system.

In the two-dimensional case of the decoupled oscillator equation

$$\ddot{X}_t + 2D_k\omega_k\dot{X}_t + \omega_k^2(1 + \varrho\Gamma_t)X_t = 0, \tag{10}$$

the projection [5] is performed by means of polar coordinates replacing the displacement X_t and the dimensionless velocity $\dot{X}_t/\omega_1$ of the oscillator by the phase angle Ψ_t and the amplitude process A_t, as follows:

$$\Psi_t = \arctan\frac{\dot{X}_t}{\omega_1 X_t}, \qquad A_t = \sqrt{X_t^2 + (\dot{X}_t/\omega_1)^2}. \tag{11}$$

The transformed equations, obtained for the circle processes (11), are

$$\dot{\Psi}_t = -g(\Psi_t, \Phi_t), \qquad g(\Psi_t, \Gamma_t) = \omega_1(1 + D\sin 2\Psi_t + \varrho\Gamma_t\cos^2\Psi_t), \tag{12}$$
$$\dot{A}_t = -h(\Psi_t, \Phi_t)A_t, \qquad h(\Psi_t, \Gamma_t) = \omega_1(2D\sin^2\Psi_t + \tfrac{1}{2}\varrho\Gamma_t\sin 2\Psi_t). \tag{13}$$

Both equations are decoupled such that the amplitude equation (13) can be integrated for any initial value A_0, given at the time $t_0 = 0$. Inserting the integrated form of (13) into the multiplicative ergodic theorem of Osceledec [6,7] leads to

$$\lambda = \lim_{t\to\infty}\frac{1}{t}\log\frac{A_t}{A_0} = -\lim_{t\to\infty}\frac{1}{t}\int_0^t h(\Psi_\tau, \Gamma_\tau)d\tau, \tag{14}$$

$$\lim_{t\to\infty}\frac{1}{t}\log\frac{A_t}{A_0} = -\int_{\frac{-\pi}{2}}^{\frac{\pi}{2}}\int_{\frac{-\pi}{2}}^{\frac{\pi}{2}} h(\psi, \gamma)p(\psi, \gamma)d\psi d\gamma. \tag{15}$$

Obviously, the result, obtained in (14), represents a time average, which can be replaced by the corresponding ensemble average, noted in (15), provided that the angle processes Ψ_t and Γ_t are stationary and ergodic.

4. Fokker-Planck equation of the invariant measure

It is important to note that the Lyapunov exponent λ of the form (15) is determined by a finite integral defined by the two-dimensional density $p(\psi, \gamma)$ of the invariant measure of the dynamical system. This density can be calculated in solving the stationary Fokker-Planck equation

$$\frac{\sigma^2}{2}\frac{\partial^2}{\partial\gamma^2}[\cos^4\gamma\,p(\psi,\gamma)] + \frac{\partial}{\partial\gamma}[(\omega_g + \sigma^2\cos^2\gamma)\frac{1}{2}\sin 2\gamma\,p(\psi,\gamma)]$$

$$+\omega_1\frac{\partial}{\partial\psi}[(1 + D\sin 2\psi + \varrho\gamma\cos^2\psi)p(\psi,\gamma)] = 0, \tag{16}$$

which is associated to the angle equations (6) and (12). The solution of (16) has to satisfy the normalization condition and the periodicity conditions on all boundary lines of the two-dimensional angle range $|\psi| \leq \pi/2$ and $|\gamma| \leq \pi/2$, as follows:

$$p(\psi,\gamma = +\frac{\pi}{2}) = p(\psi,\gamma = -\frac{\pi}{2}), \qquad p(\psi = +\frac{\pi}{2},\gamma) = p(\psi = -\frac{\pi}{2},\gamma), \tag{17}$$

$$\frac{\partial}{\partial\gamma}[p(\psi,\gamma)]|_{\gamma=+\frac{\pi}{2}} = \frac{\partial}{\partial\gamma}[p(\psi,\gamma)]|_{\gamma=-\frac{\pi}{2}}, \qquad \int_{\pi/2}^{\pi/2}\int_{-\pi/2}^{\pi/2} p(\psi,\gamma)d\psi d\gamma = 1. \tag{18}$$

Equivalent to the periodicity at $\gamma = \pm\pi/2$ are the boundary conditions

$$p(\psi,\gamma = +\pi/2) = 0, \qquad p(\psi,\gamma = -\pi/2) = 0, \tag{19}$$

which follow from the fact that the marginal density $p(\gamma)$ is vanishing on the boundary lines $\gamma = \pm\pi/2$ in correspondence to (8).

Finally, it is worth to mention that the Fokker-Planck equation (16) can easily be integrated with respect to γ. Because of the periodicity at $\gamma = \pm\pi/2$, this integration leads to the following marginal forms:

$$\omega_1\int_{\pi/2}^{\pi/2}\frac{\partial}{\partial\psi}[(1 + D\sin 2\psi + \varrho\gamma\cos^2\psi)p(\psi,\gamma)]d\gamma = 0, \tag{20}$$

$$(1 + D\sin 2\psi)p(\psi) + \varrho\cos^2\psi\int_{-\pi/2}^{\pi/2}\gamma p(\psi,\gamma)d\gamma = C. \tag{21}$$

Interchanging differentiation and integration in (20) one can perform the integration with respect to ψ in order to obtain the result (21). Herein, C is a constant of integration, determined by the normalization condition in (18). Obviously, (21) represents a stationarity condition, which has to be fulfilled for each ψ-line in $|\psi| \leq \pi/2$.

5. Monte-Carlo simulations of the invariant measures

To obtain first results for the two-dimensional density $p(\psi,\gamma)$, we apply Monte-Carlo simulations to the angle processes Ψ_t and Γ_t. For this purpose, we discretize the angle equations (6) and (12) by a forward Euler scheme, which gives the following recurrence equations for the numerical simulation:

$$\Gamma_{n+1} = \Gamma_n - (\omega_g + \sigma^2 \cos^2 \Gamma_n) \sin \Gamma_n \cos \Gamma_n \, \Delta t + \sigma \cos^2 \Gamma_n \sqrt{\Delta t} \, R_n, \qquad (22)$$

$$\Psi_{n+1} = \Psi_n - \omega_1 [1 + D \sin 2\Psi_n + \varrho \Gamma_n \cos^2 \Psi_n] \Delta t, \qquad n = 1, 2, \ldots. \qquad (23)$$

Herein, Δt is the selected time step and R_n is a sequence of pseudo-random numbers, which are normally distributed with zero mean and the normalized mean square $E(R_n^2) = 1$. The numbers R_n are generated by the Box-Muller transformation

$$R_n = \sqrt{-2 \log U_n} \, \cos 2\pi V_n, \qquad n = 0, 1, 2, \ldots, \qquad (24)$$

where U_n and V_n are independent sequences. Both, U_n and V_n, are uniformly distributed in $0 \leq U_n, V_n < 1$ and produced by linear congruence generators.

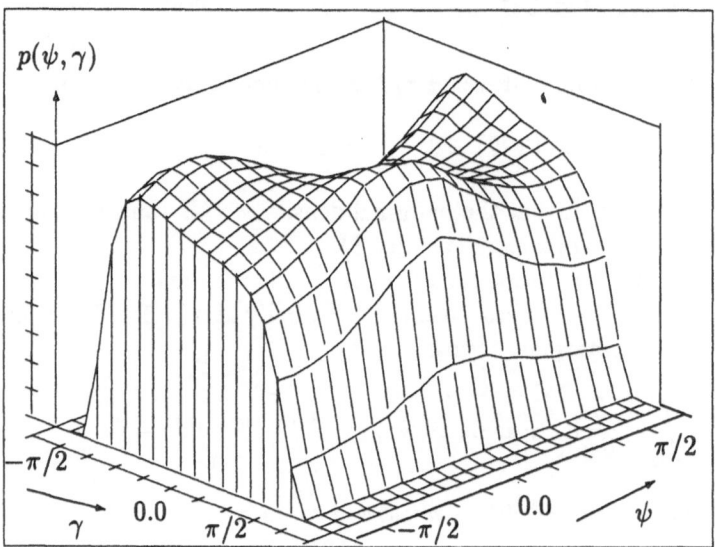

Figure 3: Three-dimensional picture of the angle density

Figure 3 shows a typical result of Monte-Carlo simulations. It is obtained for $\Delta t = 0.01s$, $\omega_1 = 1s^{-1}$, $\omega_g = 1s^{-1}$, $D = .1$, $\sigma_z = .8$, $\varrho = .5$ and for $N_s = 100,000,000$ sample points, which are evaluated in a grid of 40×40 elements of area in the two-dimensional range $|\psi|, |\gamma| \leq \pi/2$. In this angle domain, both angle processes are limited by applying the modulo operator $\mathrm{mod}(\pi)$ to Ψ_n and Γ_n. This is justified because of the periodicity of the time-invariant characteristics of the system angle Ψ_t, which rotates clockwise, and the fact, that $\gamma = \pm\pi/2$ represents a natural boundary of the excitation angle Γ_t in the sense of Feller [8].

6. Iterative solution of the Fokker-Planck equation

The numerical results, obtained in Figure 3 by Monte-Carlo simulations, can considerably be improved by means of the stationary solutions of the Fokker-Planck equation, which has been derived in (16), previously. For iterative solutions, this parabolic differential equation is discretized with respect to the system angle ψ applying the step size $\Delta\psi$ and the Euler scheme

$$\frac{\partial}{\partial\psi}p(\psi,\gamma) = \frac{1}{\Delta\psi}[\bar{p}(\gamma) - p(\gamma)], \qquad \text{with} \quad \Delta\psi = -\frac{\pi}{K}. \tag{25}$$

Insertion of (25) into the Fokker-Planck equation (16) results into the explicit scheme

$$\bar{p}(\gamma) = p(\gamma) - \frac{\Delta\psi}{g(\psi,\gamma)}[f(\psi,\gamma) + a(\gamma)]p(\gamma)$$

$$- \frac{\Delta\psi}{g(\psi,\gamma)}[b(\gamma)\frac{\partial p(\gamma)}{\partial\gamma} + c(\gamma)\frac{\partial^2 p(\gamma)}{\partial\gamma^2}]. \tag{26}$$

Herein, $g(\psi,\gamma)$ and $f(\psi,\gamma)$ are terms of the system angle equation (12), given by

$$g(\psi,\gamma) = \omega_1(1 + D\sin 2\psi + \varrho\gamma\cos^2\psi), \tag{27}$$
$$f(\psi,\gamma) = \omega_1(2D\cos 2\psi - \varrho\gamma\sin 2\psi), \tag{28}$$

meanwhile $a(\gamma)$, $b(\gamma)$ and $c(\gamma)$ belong to the parametric excitation equation (6).

$$a(\gamma) = \omega_g\cos 2\gamma - \sigma^2(4\cos^2\gamma - 3)\cos^2\gamma, \tag{29}$$
$$b(\gamma) = \tfrac{1}{2}(\omega_g - 3\sigma^2\cos^2\gamma)\sin 2\gamma, \tag{30}$$
$$c(\gamma) = \tfrac{1}{2}\sigma^2\cos^4\gamma, \qquad |\gamma| \leq \pi/2. \tag{31}$$

The iteration process in (26) is started e.g. at the line $\psi = \pi/2$ with any arbitrary density distribution $p(\gamma)$ within the range $-\pi/2 \leq \gamma < \pi/2$, given e.g. by a uniformly distributed density, which possesses the integration constant C, calculated by the stationarity condition (21).

Subsequently, one determines the derivatives of $p(\gamma)$, which are needed in the right-hand side of the recurrence formula (26). This is performed e.g. by means of the fourth order central differences

$$\frac{\partial p(\gamma)}{\partial\gamma}|_0 = \frac{1}{12\Delta\gamma}(p_{-2} - 8p_{-1} + 8p_{+1} - p_{+2}), \quad \Delta\gamma = \frac{\pi}{L}, \tag{32}$$

$$\frac{\partial^2 p}{\partial\gamma^2}|_0 = \frac{1}{12\Delta\gamma^2}(-p_{-2} + 16p_{-1} - 30p_0 + 16p_{+1} - p_{+2}), \tag{33}$$

applying the angle step size $\Delta\gamma = \pi/L$ of a second discretization with respect to the excitation angle γ. Note, that both central differences are to be evaluated taking into account the periodicity condition $p(\gamma) = p(\gamma + \pi)$.

Knowing the starting density $p(\gamma)$ and its first two derivatives at the line $\psi = \pi/2$ one calculates the next distribution $\bar{p}(\gamma)$ at the line $\psi = \pi/2 - \pi/K$ by means of the explicit iteration scheme (26). Finally, the calculated $\bar{p}(\gamma)$ is moved a bit up or down in order to satisfy the same stationarity constant C, as before. This eliminates the drift behaviour, which is inherent in parabolic differential equations. In the same manner, the iteration scheme is continued into the negative ψ-direction for all following ψ-lines until a stationary solution is obtained, which satisfies the periodicity condition at the lines $\psi = \pi/2$ and $\psi = -\pi/2$. The explicit iteration is regular for positive definite coefficients $g(\psi, \gamma) > 0$. Naturally, the iteration will be destabilized, when $g(\psi, \gamma)$ changes its sign and becomes negative inside the angle range $|\psi|, |\gamma| \leq \pi/2$. In this singular case, the iteration has to be modified by an implicit scheme applying corresponding backward differences to (26).

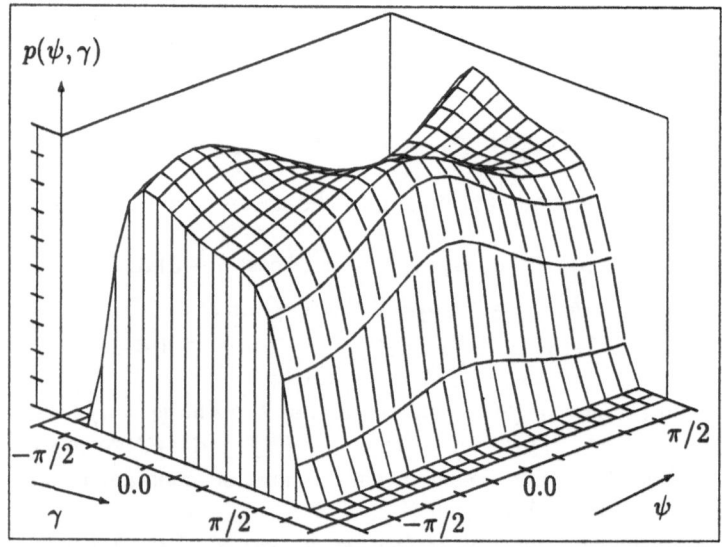

Figure 4: Stationary solution of the Fokker-Planck equation

Figure 4 presents a three-dimensional picture of the density $p(\psi, \gamma)$, calculated for the parameters $\omega_1 = 1$, $\omega_g = 1$, $D = .1$, $\sigma_z = .8$ and $\varrho = .5$. The discretization of the angle range $|\psi|, |\gamma| \leq \pi/2$ was performed by $L \times K$-elements of area with $L = 40$ and $K = 1600$. The comparison of the above calculated density $p(\psi, \gamma)$ with the simulated one in Figure 3, shows a good coincidence between both two-dimensional distribution densities.

7. Numerical tests by separable density solutions

For the special case of vanishing amplifications $\varrho = 0$, both angle processes Ψ_t and Γ_t are completely decoupled and statistically independent. Correspondingly, the two-dimensional density is separable and possesses the stationary form

$$p(\psi, \gamma) = \frac{\sqrt{1 - D^2}}{\pi(1 + D \sin 2\psi)} \frac{1}{\sqrt{2\pi}\sigma_z} \frac{1}{\cos^2 \gamma} \exp(-\frac{\tan^2 \gamma}{2\sigma_z^2}). \qquad (34)$$

This special closed-form solution of the Fokker-Planck equation (16) is utilized to check the numerical results obtained by the iteration scheme. Figure 5 shows a comparison of both, the analytical solution (34) and the iterative solution by (26) for $D = .1$, $\sigma_z = .8$, $\varrho = 0$ and the three fixed angle values $\psi = -\pi/2, 0.0, -\pi/2$. The associated curves coincide graphically. Deviations between both, the analytical evaluations and the iterative solutions, are not visible.

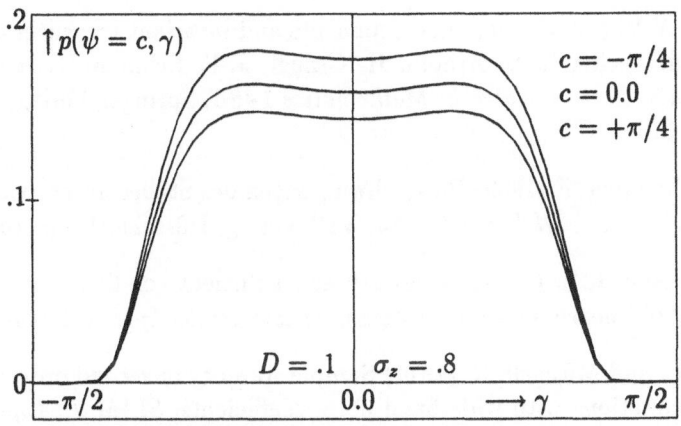

Figure 5: Density intersections for the special case $\varrho = 0$

8. Conclusions

For stability investigations, it is important to calculate the density of the invariant measures of perturbed dynamical systems. For this purpose, the paper presents an iterative scheme, which solves the associated stationary Fokker-Planck equations by means of explicit Euler schemes applied into the negative direction of the system angle. The drift behaviour, inherent in parabolic equations, is eliminated by means of stationarity conditions, which are additionally introduced into the iteration scheme.

The iteration method is applied to oscillators, which are perturbed by non-normal processes with bounded realizations. The obtained results are checked by means of Monte-carlo simulations and by special closed-form density solutions. The iterative scheme, proposed in the paper, is restricted to regular Fokker-Planck equations; i.e. the coefficient of the highest derivative with respect to the system angle is positive definite. Extensions to singular cases and to higher-dimensional problems are of actual interest.

9. References

1. Wedig, W.V (1991) Dynamic stability of beams under axial forces - Lyapunov exponents for general fluctuating loads, in W.B. Krätzib et al. (eds), *Proceedings of Eurodyn '90, Conference on Structural Dynamics*, A.A. Balkema, Rotterdam, **1**, pp. 141-148.

2. Wedig, W.V. (1991) Lyapunov exponents and invariant measures of equilibria and limit cycles, in L. Arnold, H. Crauel, J.-P. Eckmann (eds.), *Lyapunov Exponents, Lecture Notes in Mathematics* **1486**, Springer-Verlag, Heidelberg, pp. 309-321.

3. Weidenhammer, F.(1969) Biegeschwingungen des Stabes unter axial pulsierender Zufallslast, *VDI-Berichte* **135**, VDI-Verlag, Düsseldorf, pp. 101-107.

4. Khasminskii, R.Z. (1974) Necessary and sufficient conditions for asymptotic stability of linear stochastic systems. *Theory Prob. Appl.* **12**, 144-147.

5. Kozin, F. and Mitchell, R. (1974) Sample stability of second order linear differential equations with wide band noise coefficients. *SIAM J. Appl. Math.* **27**, 571-605.

6. Oseldec, V.I. (1968) A multiplicative ergodic theorem, Lyapunov characteristic numbers for dynamical systems, *Trans. Moscow Math. Soc.* **19**, 197-231.

7. Arnold, L. and Wihstutz, V. (1985) Lyapunov exponents: a survey, in L. Arnold, V. Wihstutz (eds), *Lyapunov Exponents, Lecture Notes in Mathematics* **1186**, Springer-Verlag, Heidelberg, pp. 1-26.

8. Feller, W. (1952) The parabolic differential equations and the associated semigroups of transformations, *Annals of Mathematics* **55**(3), 468-519.

The paper is dedicated to Prof. Dr. rer. nat. F. Weidenhammer,
University of Karlsruhe, on the occasion of his 75th birthday.

NEW INSIGHTS ON THE APPLICATION OF MOMENT CLOSURE METHODS TO NONLINEAR STOCHASTIC SYSTEMS

S. F. WOJTKIEWICZ

Department of Aeronautical and Astronautical Engineering,
University of Illinois at Urbana-Champaign, Urbana, IL 61801, U.S.A.

B. F. SPENCER, JR.

Department of Civil Engineering and Geological Sciences,
University of Notre Dame, Notre Dame, IN 46556, U.S.A.

L. A. BERGMAN

Department of Aeronautical and Astronautical Engineering,
University of Illinois at Urbana-Champaign, Urbana, IL 61801, U.S.A.

1. Abstract

The cumulant-neglect closure method is briefly outlined and subsequently applied to two Duffing systems, one exhibiting a unimodal and the other a bimodal response probability density function. The closure results are compared at stationarity to the exact solution over a broad range of parameters, and some connections are drawn between the accuracy of cumulant-neglect closure and the choice of system parameters. Finally, characteristic equations governing all stationary solutions of the closed system of moments equations are obtained, and the stability of the resulting solutions is ascertained. From this analysis, it is determined whether the closure results are physically consistent; that is, if the stationary closure results can be reached by letting the system evolve from arbitrary initial conditions.

2. Introduction

In a previous paper [1], cumulant-neglect closure was employed to examine the nonstationary moment response of the Duffing oscillator subject to additive and multiplicative white noise excitations. Cumulant-neglect closure schemes through eighth order were applied to systems having (1) a positive linear component of stiffness, known to exhibit a unimodal stationary response probability density function; and (2) a negative linear component of stiffness, known to exhibit a bimodal stationary response probability density function, both symmetric about the origin. One of the more striking results reported in that paper was the apparent failure of cumulant-neglect closure of any order to accurately predict the response of the bimodal system, either through its evolution or at stationarity. Comparisons were made with post-processed results from finite element solutions of the Fokker-Planck equation [2-3] and with the exact stationary moments

A. Naess and S. Krenk (eds.), IUTAM Symposium on Advances in Nonlinear Stochastic Mechanics, 479–488.
© *1996 Kluwer Academic Publishers.*

obtained from the first order probability density function [4]. At the time the work was presented, the authors noted that more questions had been raised than answered during the course of their research. Indeed, there was concern over several issues, among which were the apparent failure of cumulant-neglect closure results to converge to the "exact" solution with increasing order of the closure scheme and the inability in some cases to find a stable fixed point of the closed system of moment equations despite *a priori* knowledge that the system possesses finite, stationary moments.

This paper examines some, but not all, of those issues. Attention is focused here on the stability and convergence issues raised in the previous work. A thorough stability analysis is performed in which all fixed points produced by a given closure scheme are determined, and their stability is subsequently ascertained. In addition, the accuracy of cumulant neglect closure schemes up to order 8 were analyzed over a large parameter domain. Some physical arguments are made to explain why cumulant neglect closure schemes work better in certain parameter regimes.

3. Problem Formulation

Consider the time invariant dynamical system

$$\dot{\underline{X}} = \underline{m}(\underline{X}) + \mathbf{G}(\underline{X})\underline{W}(t) \tag{1}$$

where $\underline{X}$ is an R^n-valued stochastic process, $\underline{m}(\underline{X}) \in R^n$, $\underline{W}(t)$ is a m-dimensional vector Gaussian white noise process, and $\mathbf{G}(\underline{X})$ is a $n \times m$ matrix where $G_{ij} \in R$. The white noise excitation vector is fully defined by its first and second moments,

$$E[\underline{W}(t)] = 0, \qquad E[\underline{W}(t_1)\underline{W}(t_2)] = 2\mathbf{D}\delta(\tau), \tag{2}$$

where $\tau = t_1 - t_2$, and $\mathbf{D}$ is a square diagonal matrix of dimension m.

It is well established that the response of the aforementioned system forms a sample continuous Markov process of dimension n or, equivalently, an *n-state* vector diffusion process. Ito's differential rule can be applied to the vector diffusion process to derive the evolution equations for the moments of the system,

$$dg = \left[\frac{\partial g}{\partial t} + \sum_{i=1}^{n} m_i \frac{\partial g}{\partial x_i} + \frac{1}{2}\sum_{i=1}^{n}\sum_{j=1}^{n} H_{ij}\frac{\partial^2 g}{\partial x_i \partial x_j}\right]dt + \sum_{i=1}^{n}\sum_{k=1}^{m} G_{ik}\frac{\partial g}{\partial x_i}dB_k \tag{3}$$

where $g(\underline{X}, t) = \prod_{i=1}^{n} X_i^{s_i}$ and $\mathbf{H} = 2\mathbf{G}\mathbf{D}\mathbf{G}^T$. The resulting moment equations can be shown to be of the form

$$\frac{d}{dt}\alpha(s_1, s_2, ..., s_n) = E\left[\sum_{i=1}^{n} m_i \frac{\partial g}{\partial x_i}\right] + \frac{1}{2}\sum_{i=1}^{n}\sum_{j=1}^{n} E\left[H_{ij}\frac{\partial^2 g}{\partial x_i \partial x_j}\right] \tag{4}$$

where $\alpha(s_1, s_2, ..., s_n) = E[g(\underline{X}, t)]$.

The moment evolution equations possess several important characteristics that can cause their solution to be a formidable task. First of all, they comprise an infinite dimensional linear system of first order differential equations in which higher order moments consistently appear in the equations for lower order moments, forming a so-called infinite hierarchy of equations. Consequently, a heuristic approach must be adopted to close the system at some finite dimension, rendering the system nonlinear. The next section of the paper discusses one approach to close the infinite hierarchy of moment equations.

4. Cumulant-Neglect Closure Method

The cumulant-neglect closure method was developed independently by Wu and Lin [5] and Ibrahim [6] as a means of closing the system of moment equations. Illustrative applications of the method can also be found in [7-9], for example.

In this approach, the moment evolution equations are closed by using the fact that response cumulants of increasing order possess a diminishing amount of information, [10], and can thus be neglected. This allows for the determination of relationships between response moments above the closure level and moments below it. These relationships are determined through a multistep process. First, the characteristic function is expressed in terms of those response cumulants at or below the threshold level,

$$\phi(\underset{\sim}{u}) = \exp\left\{ j \sum_{s_1 = 1}^{n} u_{s_1} \lambda(X_{s_1}) + \frac{j^2}{2} \sum_{s_1 = 1}^{n} \sum_{s_2 = 1}^{n} u_{s_1} u_{s_2} \lambda(X_{s_1}, X_{s_2}) + \right.$$

$$\left. \cdots + \frac{j^p}{p!} \sum_{s_1 = 1}^{n} \cdots \sum_{s_p = 1}^{n} u_1^{s_1} u_2^{s_2} \cdots u_n^{s_p} \lambda(X_{s_1}, X_{s_2}, \ldots, X_{s_p}) \right\} \tag{5}$$

where $j = \sqrt{-1}$, n is the number of states in the response process, p is the threshold or closure level, $\phi(\underset{\sim}{u})$ is the characteristic function, and λ represents the cumulant operator (12). The relationship between the moments and the characteristic function of the response process is given by

$$\alpha(s_1, s_2, \ldots, s_n) = \frac{1}{j^{\Sigma}} \left[\frac{\partial^{\Sigma}}{\partial u_1^{s_1} \partial u_2^{s_2} \cdots \partial u_n^{s_n}} \phi(\underset{\sim}{u}) \right] \Bigg|_{\underset{\sim}{u} = 0} \tag{6}$$

where $\Sigma = \sum_{i=1}^{n} s_i$ and α is as defined earlier. Cumulants of order less than or equal to the threshold level can then be written in terms of the moments of order less than or equal to the threshold level by inverting the relationships obtained by the application of (5).

Once the lower order cumulants are known in terms of the lower order moments, they are substituted into the expressions for the response moments of order greater than the closure level which were obtained by application of (6). The system of moment equations can now be closed, and the right hand side of (4) can be calculated. For a more thorough review of closure methods, the interested reader is referred to the texts by Bolotin [11], Soong and Grigoriu [12], Lin and Cai [13], and the references therein.

The finite system of equations resulting from the application of the cumulant-neglect closure method is nonlinear and involves expressions which become extremely large and complex as the number of cumulants retained or the number of states of the dynamical system increases. Derivation by hand quickly becomes intractable. To facilitate generation and solution of the closed set of moment equations obtained by the cumulant-neglect closure approach, a computational algorithm has been been developed by the authors to automate the closure procedure using $MAPLE^{\circledR}$ [14]. For a more complete description of the algorithm, the interested reader is referred to [3].

Rather than performing time-integration on the closed set of moment equations, one can set the derivative of the moments equal to zero and solve the resulting set of nonlinear algebraic equations directly for the stationary moments. To do so, the fixed points of the closed system of moment equations must be found. For the two state Duffing systems to be discussed here, the fixed points can be expressed in terms of the stationary mean square displacement, A, and characteristic polynomial equations involving A, assuming that zero mean displacement solutions are germane, which is precisely correct. The development of the characteristic equations follows that found in [5,15]; however, as a consequence of the automation procedure, the results shown herein are more general than those of either earlier work and are extended, for the first time, to include eighth order closure. The characteristic equations for each closure scheme, as well as the formulas relating the remaining stationary moments to the mean square displacement, are contained in [3].

Although the Duffing system is guaranteed to have a unique stationary solution, the approximate system represented by the closed set of moment equations has no such property. This non-uniqueness of solutions for the stationary moments is obvious from the polynomial characteristic equations and has been previously reported in [8,9]. Extraneous solutions, therein, were eliminated based on arguments such as positive definiteness of mean squares and satisfaction of a Cauchy-Scharwz inequality (e.g., $\alpha(1, 1)^2 \leq \alpha(2, 0)\alpha(0, 2)$). Those arguments, however, are of limited use since they provide no information concerning the stability of the fixed point. Langley [16] distinguished the multiple fixed points obtained by equivalent linearization by using the dynamic stability of each solution.

Here, a similar methodology will be applied. First, the possible values of the mean square displacement, A, were found by solving the appropriate characteristic equation. Next, the stationary moments were found by substituting a candidate value of the mean square displacement into the relationships for the remaining moments. The nonlinear system of closed moment equations was then linearized about the resulting fixed point, and the eigenspectrum of the Jacobian was analyzed to determine the stability of that fixed point. This procedure was repeated for each fixed point computed including those which were physically unrealizable (e.g., $\alpha(2, 0) < 0$ or $Im(\alpha(2, 0)) \neq 0$).

In order to gain insight into the transient instability shown by the eighth order cumulant neglect closure scheme, two systems taken from [1] will be reanalyzed using the aforementioned stability analysis. In addition, the stationary closure results will be compared to the exact solution over a wide range of system parameters. This parameter study will show that the instability phenomenon displayed in the systems of [1] is a persistent feature of the system of moment equations closed by cumulant-neglect closure and not a by-product of poor parameter choice.

5. Results

The Duffing oscillator can be effectively used to model the hardening stiffness effect found in many mechanical systems. The systems to be analyzed in this study are described by the stochastic differential equations

$$
\begin{bmatrix} \dot{X}_1 \\ \dot{X}_2 \end{bmatrix} = \begin{bmatrix} X_2 \\ -\gamma\omega_o^2 X_1 - \epsilon\omega_o^2 X_1^3 - 2\zeta\omega_o X_2 \end{bmatrix} + \begin{bmatrix} 0 \\ 1 \end{bmatrix} W(t) \tag{7}
$$

with $W(t)$ a Gaussian white noise stochastic process such that

$$
E[W(t)] = 0, \qquad E[W(t_1)W(t_2)] = 2D\delta(\tau), \tag{8}
$$

$\tau = t_2 - t_1$, and where ω_o would be the undamped natural frequency of the system if there were no nonlinearity, γ determines the sign of the linear restoring term, ζ is a damping parameter, and ϵ controls the magnitude of the nonlinearity. For the systems herein, D and ω_o were chosen to be one. An analytical solution [4] exists for the stationary probability density function governing the response of the system defined by (7)-(8) and is given by

$$
p(x_1, x_2) = C_o \exp\left(-\frac{2\zeta\omega_o}{D}\left(\frac{x_2^2}{2} + \frac{\gamma\omega_o^2 x_1^2}{2} + \frac{\epsilon\omega_o^2 x_1^4}{4}\right)\right) \tag{9}
$$

where C_o is a normalization constant. The exact response moments of arbitrary order can be calculated from this density function to assess the accuracy of the closure results.

Alternatively, Ito's rule can be applied to obtain the set of moment evolution equations,

$$
\begin{aligned}
\dot{\alpha}(p, q) = {} & p\alpha(p-1, q+1) - \gamma\omega_o^2 q\alpha(p+1, q-1) - 2\zeta\omega_o q\alpha(p, q) \\
& - \epsilon\omega_o^2 q\alpha(p+3, q-1) + Dq(q-1)\alpha(p, q-2)
\end{aligned} \tag{10}
$$

where $E[X_1^p X_2^q] = \alpha(p, q)$. The algorithm described in the previous section has been used to perform the closure of the moment evolution equations (10).

A variant of this problem has been previously studied in [5,15] using cumulant-neglect closure schemes of order up through six. It is desired to assess the effects of closure schemes of higher order upon the accuracy of the approximation to the solution of this problem.

Cumulant-neglect closure schemes of order two (Gaussian closure), four, six, and eight have been implemented, as outlined previously. It was found for both Duffing systems that the second order cumulant neglect closure (Gaussian closure) produces a single stable fixed point that also satisfies the positive semi-definiteness requirement, $E[X_1^2] = \alpha(2, 0) \geq 0$ for each pair (ζ, ε) in the domain of interest. The accuracy of the second order closure results was assessed by computing the relative error between the closure result and the exact solution for the mean square displacement,

$$\% \text{ error} = \frac{\left| (\alpha(2, 0)_{closure} - \alpha(2, 0)_{exact}) \right|}{\alpha(2, 0)_{exact}} \times 100 \% . \tag{11}$$

The error for both the unimodal and bimodal systems is shown in Figure 1. It is interesting to note the opposite trends for the unimodal and bimodal cases. In the unimodal case, the error decreases with decreasing ε as one would expect since the unimodal system degenerates into a simple linear oscillator as $\varepsilon \to 0$. However, the opposite is true for the bimodal case as the error decreases with larger nonlinearities. This decrease is due to the instability of the system which occurs as $\varepsilon \to 0$.

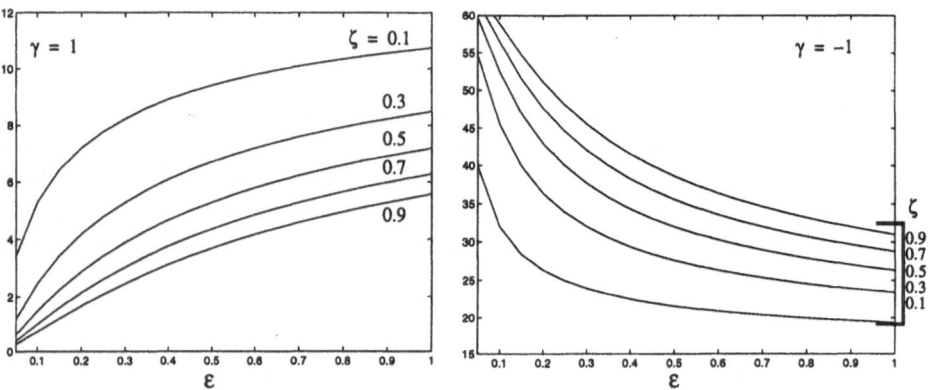

Fig. 1 Percent Error in Mean Square Displacement Closure Result, $\alpha(2, 0) = E[X_1^2]$, for Second Order Cumulant Neglect Closure.

Unlike for the case of second order closure, the fourth, sixth and eighth order closure schemes all produced multiple, possible, physically realizable solutions. This multiplicity of roots has been reported before; however, extraneous solutions were eliminated based on arguments such as positive definiteness of mean squares. Here, the extraneous fixed points were eliminated based on their transient instability. For each level of cumulant-neglect closure level used, all possible fixed points of both the unimodal and bimodal Duffing systems were obtained from the characteristic equations, and their stability was subsequently ascertained. Several of the more interesting and informative cases will be discussed here to shed some light on the stability questions raised in earlier work.

The four branches of possible solutions produced by sixth order closure for the bimodal case are shown along with the exact solution in Figure 2 over the parameter space. It should be noted that the two physically unrealizable branches, $\alpha(2, 0) < 0$, as well as one of the physically realizable branches are unstable throughout while the branch of closure solutions best mirroring the exact solution is stable everywhere in the domain of interest. The relative error between the closure solutions and the exact solution was

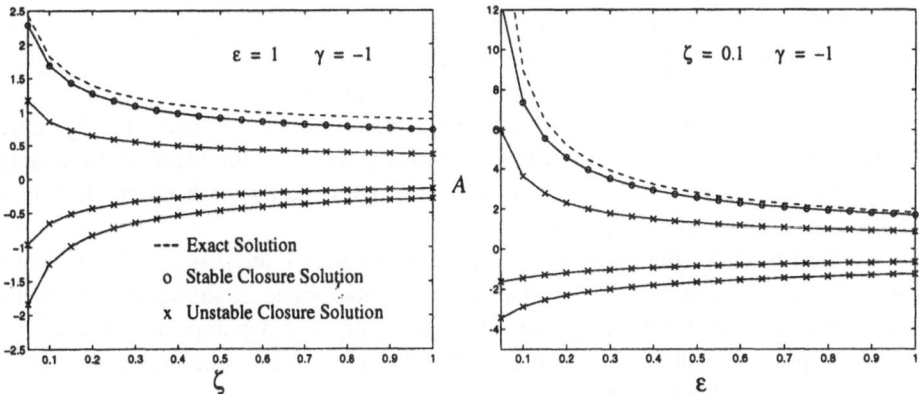

Fig. 2 Multiple Solutions for Mean Square Displacement, $\alpha(2, 0) = E[X_1^2]$, obtained by Sixth Order Cumulant Neglect Closure.

computed and is shown in Figure 3. The parameter trends in accuracy discussed earlier persist for the sixth order closure case. In addition, the sixth order closure results show improved accuracy over the second order closure results for both the unimodal and bimodal cases as one would hope. Although not shown, the fourth order cumulant neglect closure scheme did produce a single stable, fixed point for both the unimodal and bimodal systems over the domain of interest with the scheme's accuracy lying between that of Gaussian and sixth order closure schemes.

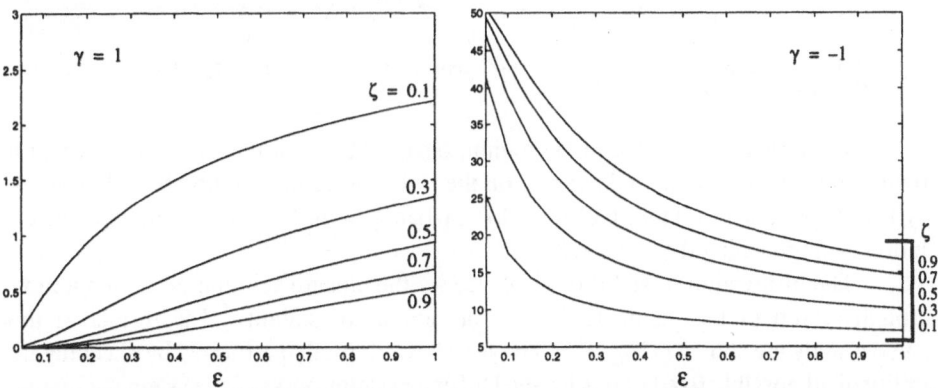

Fig. 3 Percent Error in Mean Square Displacement Closure Result, $A = E[X_1^2]$, for Sixth Order Cumulant Neglect Closure.

The question of the stability of cumulant neglect closure results arose when eighth order closure was applied to the bimodal system in [1], and the transient response of the moment equations failed to converge to a steady state value. The time integration scheme was immediately held suspect; however, this possibility was eliminated by shrinking the integration time step to a prohibitively small value, 10^{-7} seconds, and noting that the response became unstable at the same point in time as with the larger time step.

To more definitively answer the stability question, the analysis previously described was applied to both Duffing systems closed by eighth order cumulant neglect closure. The unimodal Duffing system closed by eighth order closure displayed a single stable fixed point throughout. The eighth order closure result was only slightly more accurate than the closure schemes of level two, four, and six. This observation of only slight gains in accuracy obtained by employing eighth rather sixth order closure is consistent with the findings of [5,15].

The results for the bimodal system are much more interesting. In Figure 4, the eighth order closure solutions for the mean square displacement are shown along with the exact solution for the bimodal system. From the plot, it is apparent that four of the branches of possible solutions remain unstable throughout while one branch possesses both stable and unstable portions.

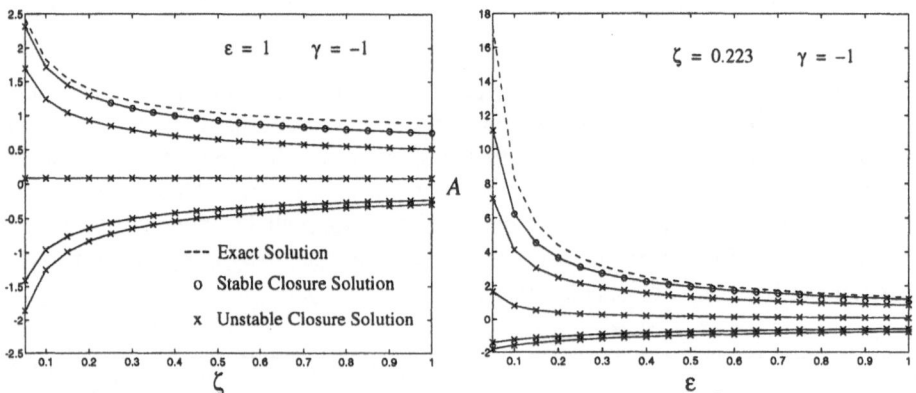

Fig. 4 Multiple Solutions for Mean Square Displacement, $\alpha(2, 0) = E[X_1^2]$, for Eighth Order Cumulant Neglect Closure.

The branch is unstable in the region, $0 \le \zeta \le 0.223$, for the case $\varepsilon = 1$ which is shown in the left hand plot, and the plot on the right shows the solution branch is stable only for the parameter region, $0.05 \le \varepsilon \le 0.75$, when $\zeta = 0.223$, even though the closure solution is approaching the exact solution.

The bifurcation of stability displayed by the closure solution prompted a more thorough search of the parameter space for regions of stability. The results of this intensive study are shown in Figure 5. The search was initially performed by checking the stability of all possible fixed points for the Duffing systems possessing system parameters in the domain, $0 \le \varepsilon \le 1$, $0 \le \zeta \le 1$. This domain was reduced to that shown in the plot by focusing on the region where the bifurcation appeared to be occurring. Since there is no stability boundary of the actual physical system coincident with that of Figure 5, it is believed that the boundary arises from an unwanted and unpredictable feature of the approximation of the physical system obtained from cumulant neglect closure.

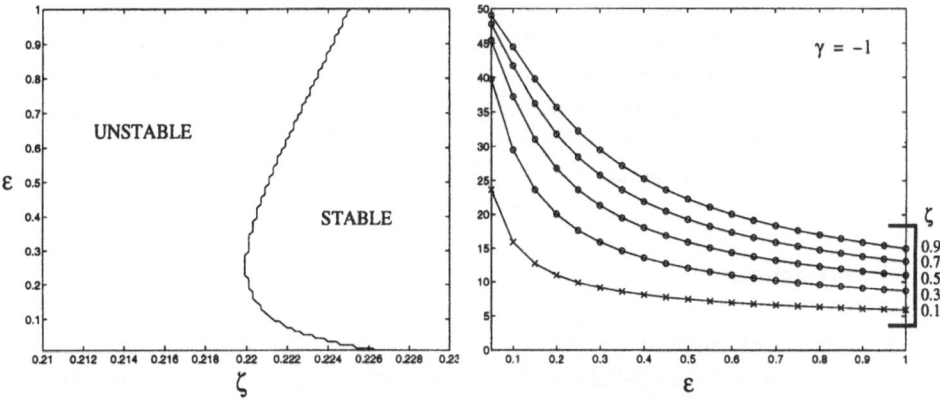

Fig. 5 Stability Boundary for Eighth Order Cumulant Neglect Closure Scheme

Fig. 6 Percent Error in Mean Square Displacement Closure Result, $A = E[X_1^2]$, for Eighth Order Cumulant Neglect Closure.

The relative error of the eighth order cumulant neglect closure was computed and is shown in Figure 6. The qualitative trends in accuracy are consistent with the lower levels of closure; however, it is important to note that, for $\zeta < 0.22$, all closure solutions are dynamically unstable. This instability makes the closure solutions physically inconsistent since they cannot be reached by transient evolution even when starting in an infinitesimal neighborhood of the fixed point. Further, in the region $0.22 \leq \zeta \leq 0.23$, both stable and unstable closure solutions exist which are dependent on the magnitude of the nonlinearity.

6. Conclusions

The limits of the range of applicability of the cumulant-neglect closure method were examined through the study two nonlinear systems. First, a unimodal Duffing oscillator subjected to a white noise excitation was considered primarily to verify our computational ageratum. Second and more interestingly, a bimodal Duffing oscillator was analyzed. This system possesses a highly non-Gaussian stationary response. It was found that physically consistent solutions for the system's stationary moments did not always exist when the system's moment equations were closed by employing cumulant-neglect closure. Further, it was found that when a physically consistent solution did exist the results obtained by the cumulant-neglect closure were poor approximations for the system response moments in large portions of the parameter space.

To gain a better understanding of the absence of a stable solution of the closed set of evolutionary moment equations, the characteristic equations governing these solutions was derived, and the set of possible fixed points were determined for each level of closure employed. Each of the resulting solutions was analyzed and classified according to its transient stability. This analysis confirmed the divergent solutions found from numerical integration of the equations as reported in [1].

Several important observations about the behavior of the cumulant-neglect closure schemes can be gleaned from these relatively innocuous-looking systems. First, there is no guarantee that cumulant-neglect closure will give a unique stationary solution of the closed system of moment equations. Second, cumulant-neglect closure can lead to dynamically unstable results thus making the approximation physically inconsistent; and finally, cumulant-neglect closure appears to perform significantly better when applied to systems exhibiting unimodal rather than multimodal response distributions.

7. Acknowledgment

This work has been supported in part by National Science Foundation contracts ECS-9224828, ECS-9411589, CEE-92000-4N, and MSS-94000-1N, the latter two through the National Center for Supercomputing Applications at the University of Illinois at Urbana-Champaign.

8. References

1. Bergman, L.A., Wojtkiewicz, S.F., Johnson, E.A., and Spencer, Jr., B.F. (1995) Some Reflections on the Efficacy of Moment Closure Methods, in P.D. Spanos (ed.), *Computational Stochastic Dynamics*, A.A. Balkema, Rotterdam, 87-95.
2. Spencer, B.F., Jr. and Bergman, L.A. (1993) On the Numerical Solution of the Fokker–Planck Equation for Nonlinear Stochastic Systems, *Nonlinear Dynamics* 4, 357-372.
3. Wojtkiewicz, S.F., Bergman, L.A., and Spencer, Jr., B.F. (1995) On the Cumulant-Neglect Closure Method in Stochastic Dynamics, *International Journal of Nonlinear Mechanics*, submitted for publication.
4. Caughey, T.K., (1971) Nonlinear Theory of Random Vibrations, in Chia-Shun Yih, (ed.), *Advances in Applied Mechanics*, Academic Press, New York, 11, 209–253.
5. Wu, W.F. and Lin, Y.K. (1984) Cumulant-Neglect Closure for Nonlinear Oscillators Under Parametric and External Excitations. *International Journal of Nonlinear Mechanics*, 19, 349-362.
6. Ibrahim, R.A. (1985) *Parametric Random Vibration* Research Studies Press, Great Britain.
7. Pawleta, M. and Socha, L. (1990) Cumulant-Neglect Closure of Nonstationary Solutions of Stochastic System, *Journal of Applied Mechanics*, 57, 776-779.
8. Sun, J.-Q. and Hsu, C.S. (1987) Cumulant-Neglect Closure Method for Nonlinear Systems Under Random Excitations, *Journal of Applied Mechanics*, 54, 649-655.
9. Fan, F.G. and Ahmadi, G. (1990) On Loss of Accuracy and Non-Uniqueness of Solutions Generated by the Equivalent Linearization and Cumulant-Neglect Methods, *Journal of Sound and Vibration*, 137:3, 385-401.
10. Gardiner, C.W. (1983) *Handbook of Stochastic Methods*, Springer Verlag, Heidelberg.
11. Bolotin, V.V. (1984) *Random Vibrations of Elastic Systems*, Martinus Nijhoff, The Hague.
12. Soong, T. and Grigoriu, M. (1993) *Random Vibration of Mechanical and Structural Systems*, Prentice Hall, Englewood Cliffs, New Jersey.
13. Lin, Y.K. and Cai, G.Q. (1995) *Probabilistic Structural Dynamics: Advanced Theory and Applications*, McGraw Hill, New York.
14. Char, B.W., Geddes, K.O., Gonnet, G.H., Monagan, M.B., and Watt, S.M., (1991) *MAPLE V Language Reference Manual*, Springer-Verlag, New York.
15. Roberts, J.B. and Spanos, P.D. (1990) *Random Vibration and Statistical Linearization.*, Wiley, New York.
16. Langley, R.S. (1988) An Investigation of Multiple Solutions Yielded by the Equivalent Linearization Technique, *Journal of Sound and Vibration*, 127:2, 271-281.

UNIFIED ANALYSIS OF COMPLEX NONLINEAR MOTIONS VIA DENSITIES

S.C.S. YIM and H. LIN
Department of Civil Engineering
Oregon State University
Corvallis, OR 97331, U.S.A.

Abstract

In this study analyses of deterministic and randomly perturbed complex nonlinear responses of nonlinear systems using densities are illustrated from an engineering perspective. Motivations to examine deterministic nonlinear dynamical systems via densities are first discussed. Pertinent mathematical backgrounds and techniques common to the analyses of both deterministic chaos and random chaotic processes are reviewed. Probability densities of nonlinear responses are computed by solving the Fokker-Planck equation (FPE) numerically to examine stochastic properties of random chaotic responses. It is shown that, by introducing random perturbations in the otherwise deterministic (periodic) excitation, the boundaries separating co-existing attractors associated with the corresponding deterministic system become blurred, and the existence of the attractors can be depicted by the evolution of a unique density in (physical) phase space. Transient and asymptotic behaviors of the densities reveal the relative stability of the various attractors. Mathematical theories useful for stability analysis of the attractors are discussed. Finally, asymptotically stable co-existing "periodic" and "chaotic" motions of a nonlinear dynamical system subjected to periodic excitation with random perturbations are demonstrated via an engineering example.

1. Introduction

Two recent trends in structure design make the use of (probability) density indispensable. One is the conversion of existing design codes to a reliability-based format. Another is the increasing demand for safe performance of a new generation of structures in the (possibly highly) nonlinear range.

Prior to the 1980's, most design codes for structures were deterministic with safety factors selected based on experience. With the maturing of reliability theory and the availability of an increasing amount of field data, many on-going efforts have been devoted to converting deterministic-based codes to probabilistic-based ones (e.g. partial safety factors in API LRFD code for offshore structures). Probability densities of response processes are needed in computing the reliability indices of the structures.

A. Naess and S. Krenk (eds.), IUTAM Symposium on Advances in Nonlinear Stochastic Mechanics, 489–498.
© 1996 *Kluwer Academic Publishers.*

Concurrently, the need for economical design and construction in often increasingly hostile environments requires structures to be more compliant and to perform in highly nonlinear modes. Under certain conditions, nonlinear system responses may become chaotic. Although chaotic motions under regular (periodic) excitations are deterministic, they possess many stochastic properties that can be efficiently characterized via densities. Also, excitations in many natural environments (e.g. aerospace and ocean) contain intrinsic random components. Thus, in addition to the dominant periodic component, random perturbations need to be included in modelling of the excitations and the response analyses.

2. Background

Chaotic behavior in a periodically driven deterministic nonlinear system has been of great interest to researchers in engineering [1,2]. Criteria for occurrence of chaotic response in purely deterministic nonlinear systems have been developed [3,4,5]. Applications of global analysis techniques on stability analyses of ship roll motion have been carried out by Thompson et al. [6] using basin boundaries erosion, and Falzarano et al. [7] using lobe dynamics to explain "unexpected capsizing".

It is well known that chaotic responses are sensitive to small variations in initial conditions and system parameters [1-3]. This "unstable" characteristic makes it difficult to take into account chaotic responses in the design of nonlinear structures using conventional deterministic methods. On the other hand, chaotic responses possess many "stable" stochastic properties. In fact, under properly selected measure spaces, chaotic attractors possess invariant densities (and measures) and ergodic (and mixing) properties. However, even these stable properties are difficult to employ in practical engineering applications because the invariant sets of chaotic attractors have fractal dimensions and thus have Lebesgue measure zero in physical phase space commonly used in classical stochastic analyses and reliability calculations.

Nevertheless, under realistic field environments, purely deterministic (periodic) excitations rarely exist and random perturbations are often inevitable. Thus structures have to be modelled as randomly perturbed systems. Meunier and Verga [8] studied these so called noisy dynamical systems and concluded that deterministic analysis techniques via topological concepts may not be useful and global behaviors including bifurcation phenomenon should be studied from a stochastic perspective.

Stochastic extensions of a few analysis techniques originally developed for deterministic chaotic systems have been derived. In particular, the effects of weak noise perturbations on chaotic behavior have been investigated via Melnikov process and phase-space flux approaches [9-12].

This study focuses on extensions of analysis techniques based on ergodic theory developed for deterministic systems using densities to extract and analyze the stable characteristics of randomly perturbed complex nonlinear (chaotic) motions. The unifying nature of these analysis techniques is emphasized.

3. Deterministic Chaotic Motions

Deterministic chaotic motions of a nonlinear system are often studied via modern nonlinear dynamic approach. Stability and global bifurcation behaviors of such a system can be illustrated by examining a periodically driven nonlinear oscillator.

3.1 GOVERNING EQUATION

The system is first cast in Hamiltonian form, with dissipative and external forcing terms (cx_2 and $A\cos\omega t$) included in a separated term as "small" external excitations:

$$\begin{Bmatrix} \dot{x}_1 \\ \dot{x}_2 \end{Bmatrix} = \begin{Bmatrix} x_2 \\ x_1 - x_1^3 \end{Bmatrix} + \begin{Bmatrix} 0 \\ -cx_2 + A\cos\omega t \end{Bmatrix} \tag{1}$$

3.2 MELNIKOV CRITERION

A Melnikov function measures the distance between perturbed stable and unstable manifolds of the Hamiltonian. It is equal to zero when there exist transverse intersections of these two manifolds, indicating the existence of chaos. A necessary criterion for chaos in terms of system parameters (A, ω, c) has been derived [13]:

$$c \leq \left[\frac{3\pi\omega\,\mathrm{sech}\dfrac{\pi\omega}{2}}{2\sqrt{2}} \right] A \tag{2}$$

4. Random Chaotic Dynamics

Random perturbations in the external excitation can be represented by a Gaussian white noise $\eta(t)$ with intensity "modulation parameter" κ.

4.1 GOVERNING EQUATION

The governing stochastic equation for the corresponding system with random perturbations is

$$\begin{Bmatrix} \dot{x}_1 \\ \dot{x}_2 \end{Bmatrix} = \begin{Bmatrix} x_2 \\ x_1 - x_1^3 \end{Bmatrix} + \begin{Bmatrix} 0 \\ -cx_2 + A\cos\omega t \end{Bmatrix} + \begin{Bmatrix} 0 \\ \eta(t) \end{Bmatrix} \tag{3}$$

4.2 STOCHASTIC MELNIKOV CRITERION

The deterministic Melnikov criterion (Eq.(2)) can be extended to take into account presence of random perturbations. Specifically, with the presence of a zero-mean white noise $\eta(t)$ (Eq.(3)), a generalized (Gaussian distributed) stochastic Melnikov process and its variance can be obtained through the transfer function associated with the convolution integrals. In the presence of random noise, the generalized stochastic Melnikov criterion can be performed in a mean-square sense

$$c^2 \leq \frac{9}{8}\pi^2\omega^2 \, sech^2\frac{\pi\omega}{2} \, A^2 + \frac{9}{16}\sigma_M^2, \tag{4}$$

Equation (4) indicates that (in the mean-square sense) the presence of zero-mean random noise enlarges the possible chaotic domain in parameter space. Its influence on the Melnikov criterion is shown in Figure 1. Note that, for the system considered, the deterministic Melnikov criterion can be recovered in the limit as the intensity of the random perturbation approaches zero. Thus the deterministic criterion represents a degenerated case of the stochastic version.

5. Analysis of Nonlinear Behaviors Via Densities

Results of computer experiments on deterministic chaotic systems have suggested a wealth of stochastic phenomena of complex nonlinear system behaviors. Common stochastic properties among deterministic and randomly perturbed systems including existence of invariant densities (and measures), stochastic stability and asymptotic periodicity have been observed. Thus, stochastic concepts and ergodic theory can be applied to chaotic systems. To gain an understanding of the relationships between deterministic systems exhibiting complex nonlinear behaviors and stochastic systems, knowledge of basic stochastic calculus, ergodic and operator theories is helpful. A few key elements pertinent to engineering interests are briefly described below.

5.1 STOCHASTIC CALCULUS AND FOKKER-PLANCK EQUATION

In classical nonlinear random vibration analysis, a Markov process approach is often used to examine the response of nonlinear systems to white noise or filtered-white noise excitations. Equation (3) can be cast into a stochastic differential equation based on Ito calculus [14].

The response probability density of a stochastic differential equation is governed by a statistically equivalent deterministic partial differential equation called the Fokker-Planck equation (FPE). An advantage to this approach is that there is no limit to the degree of nonlinearity in the system considered. However, closed-form solutions are rarely available.

The associated FPE to Eq.(3) may be written as [14]

$$\frac{\partial P(X,t)}{\partial t} = -\frac{\partial}{\partial x_1}\{x_2 P(X,t)\} - \frac{\partial}{\partial x_2}\{(-cx_2 + x_1 - x_1^3$$

$$+ A\cos\omega t)P(X,t)\} + \frac{\kappa}{2}\frac{\partial^2}{\partial x_2^2}P(X,t) \tag{5}$$

where $P(X,t)$ denotes the probability density, x_2 and $(-cx_2 + x_1 - x_1^3 + A\cos\omega t)$ are the two entries in the corresponding drift vector, and κ is the only non-zero coefficient in the 2×2 diffusion matrix [15]. The periodic excitation in the drift vector (Eq.(5)) implies that the steady-state probability density is periodic with period $2\pi/\omega$ in time [16,17]. Equation (5) can be solved for the probability density using a path integral solution numerical procedure [18].

5.2 MARKOV AND FROBENIUS-PERRON OPERATORS

The Markov operator is a primary tool in studying the flow of densities of stochastic systems. The Frobenius-Perron operator, which may be considered a deterministic restriction of the Markov operator, is useful for examining the flow of densities of corresponding deterministic systems. Ergodic theories concerning asymptotic behaviors of densities often apply equally well under both deterministic and stochastic settings. In particular, many analytical results applicable to deterministic flows described by Frobenius-Perron operators concerning chaotic behaviors evolving under the influence of periodic excitations have direct extensions to their corresponding stochastic counterparts under periodic excitations with random perturbations. The theoretical foundation for analyses of Markov and Frobenius-Perron operators are described in details in [19]. Interested readers should consult the text directly.

5.3 ASYMPTOTIC PROPERTIES OF DENSITIES

5.3.1 *Stochastic Properties of Deterministic Transformations*
There are three levels of irregular response (chaotic) behaviors a deterministic (measure-preserving) transformation of nonlinear systems can induce -- ergodicity, mixing, and exactness. The Frobenius-Perron operator is an efficient means to discern these behaviors. Preservation of an initial measure by a nonlinear transformation corresponds to the fact that the constant density is a stationary density of the Frobenius-Perron operator. Ergodicity requires that the constant density be the unique stationary density of the Frobenius-Perron operator, while mixing and exactness correspond to two different types of stability of the stationary density. However, the measure spaces suitable for analyzing these deterministic attractors and their stochastic properties are usually different for each (possibly co-existing) attractor and are often of measure zero

with respect to Lebesgue measure associated with the (physical) state space, which is of practical engineering interest. This undesirable (measure zero) property renders direct application of the stochastic analyses of the deterministic attractors impractical for conventional reliability analyses.

5.3.2 *Relationships between Deterministic and Stochastically Perturbed Systems*

As mentioned earlier, in the presence of random perturbations, the response of a nonlinear system can be described by a stochastic differential equation. The corresponding response behaviors can be examined via the evolution of the density governed by the Fokker-Planck equation. Solutions of the FPE are equivalent to the flows of densities governed by a semigroup of Markov operators.

When the relative intensity of the stochastic perturbations in the excitation reduces to zero, the FPE reduces to the Liouville equation and the semigroup of Markov operators reduces to a semigroup of Frobenius-Perron operators. Under certain smoothness conditions (topological structural stability), properties associated with the deterministic system can be recovered as degenerated cases of stochastic results. The limiting behavior when the intensity of the random perturbation approaches zero (i.e. structural stability of dynamical systems), which is called the "Kolmogorov Problem" in the mathematical community, is currently an active area of research.

Major advantages of having random perturbations as a component of the excitation include that: (1) it represents a better model of the physical system, and (2) the resulting densities governing the responses belong to the physical state space of engineering interest. In fact the resulting densities are the familiar ones studied extensively in classical random vibrations, hence can be used directly in reliability calculations. Thus it is important to examine nonlinear complex (chaotic) response behaviors from a stochastic setting.

5.3.3 *Asymptotic Stability of Densities*

While solutions to the FPE can be computed efficiently via numerical (finite-element or path-integral) procedures, stability characteristics of the responses may be proficiently extracted by examining properties of their associated semi-group of Markov operators.

A particular useful mathematical result is the Foguel Alternative Theorem, which states that: (1) a continuous stochastic semigroup of Markov operators possesses either an asymptotic stationary density or sweeping properties for the FPE; and (2) if there exists a time shift t_o such that the densities corresponding to t and $t + t_o$ are identical, then the semigroup of Markov operators is periodic and a stationary density which is equal to the averaged value of the densities over the period t_o exists. The conditions under which the theorem holds are described in [19].

Although, as many mathematical theorems, validation procedures of particular engineering systems satisfying the conditions of this theorem may not be trivial to develop and implement, the availability of such a mathematical framework and theorems are often helpful in interpreting numerical results. To demonstrate the usefulness of the Foguel Alternative Theorem, a numerical example illustrating periodicity and asymptotic stability of an engineering system is presented next.

6. Numerical Example

By including effects of water-on-deck, the roll motion of a ship under beam seas can be characterized by the following equation [7,12]

$$\ddot{x} + \alpha\dot{x} + \alpha_q\dot{x}|\dot{x}| - x + \beta^2 x^3 = A\cos(\Omega\tau + \Psi) + \eta(\tau) \qquad (6)$$

where $\eta(t)$ is a zero-mean, delta-correlated white noise with intensity κ. In the homoclinic region [12], the presence of weak noise expedites the occurrence of chaos (Fig. 2), and the steady-state probability density can depict the existing response attractors (Fig.3).

Periodic samples of the evolution of the density in state space governing the roll motion response are shown in Figure 4. Starting from quiescent condition (Fig.4a), the density gradually evolved into a steady state which contains both (randomly perturbed) periodic and chaotic motions (Fig.4b-f). By the Foguel Alternative Theorem, the presence of periodicity in the evolution of the density implies the existence of an invariant density, which can be obtained by averaging the density over one period of the steady-state response. Thus the system is asymptotically stable.

It is important to observe that the stochastic analysis via the FPE formulation can rapidly capture the essential behaviors and the stable attractors (both periodic and chaotic) with relatively little efforts. In deterministic analysis, the amount of analytical and computational efforts involved in identifying the co-existing attractors and examining their stability could be substantial.

7. Summary and Concluding Remarks

Analyses of deterministic and randomly perturbed chaotic responses of nonlinear systems using densities have been illustrated from an engineering perspective. Motivations and pertinent mathematical backgrounds and techniques common to the analyses of both deterministic chaos and random chaotic processes using densities have been discusses and reviewed.

A stochastic version of the Melnikov criterion has been presented. It is shown that the deterministic Melnikov criterion can be recovered from the stochastic criterion in the limit as the intensity of the random perturbation approaches zero.

Complex nonlinear (chaotic) behaviors of a deterministic (measure-preserving) transformation of nonlinear systems possess many stochastic properties including ergodicity, mixing, and exactness. However, the measure spaces suitable for analyzing these deterministic attractors and their stochastic properties are usually different for each (possibly co-existing) attractor and are often not of practical engineering interest.

Probability densities of nonlinear responses of an engineering system have been computed by solving the Fokker-Planck equation (FPE) numerically to examine stochastic properties of random chaotic responses. The resulting densities are the familiar ones studied extensively in classical random vibrations, hence can be used directly in reliability calculations. It is observed that the stochastic analysis via the FPE formulation can rapidly capture the essential behaviors and the stable attractors (both periodic and chaotic) with relatively little efforts.

Transient and asymptotic behaviors of these densities have been employed to reveal the relative stability of the various attractors. The Foguel Alternative Theorem, which is particularly useful for stability analysis of the attractors, has been presented. Periodicity and asymptotic stability of the unique density containing both perturbed "periodic" and "chaotic" motions of the ship roll system have been utilized to demonstrate applications of the theorem.

Finally, the stochastic analyses described here, can be considered (when applicable) generalized versions of their corresponding deterministic analyses applied to chaotic motions, thus providing a unifying approach to the analysis of complex nonlinear motions.

8. Acknowledgement

The authors gratefully acknowledge the financial support from the United States Office of Naval Research Grant No.N00014-92-1221.

9. References

1. Moon, F. C., *Chaotic Vibrations*, John Wiley & Sons (1987).
2. Thompson, J.M.T. and Stewart, H.B., *Nonlinear Dynamics and Chaos*, John Wiley & Sons (1986).
3. Guckenheimer, J. and Holmes P., *Nonlinear Oscillations, Dynamical Systems, and Bifurcations of Vector Fields*, Springer-Verlag (1983).
4. Wiggins, S., *Global Bifurcations and Chaos: Analytical Methods*, Springer-Verlag (1988).
5. Wiggins, S., *Introduction to Applied Nonlinear Dynamical Systems and Chaos*, Springer-Verlag (1990).
6. Thompson, J.M.T., Rainey, R.C.T. and Soliman, M.S., *Phil Trans Roy Soc London*, **332**, 149 (1990).
7. Falzarano, J., Shaw S.W. and Troesch, A.W., *Int J of Bifurcation and Chaos*, **2**, 101, (1992).
8. Meunier, C., and Verga, A.D., *J Stats Phys*, **50**, 345 (1988).
9. Frey, M. and Simiu, E., *Proc 9th Engrg Mech Conf, ASCE*, 660 (1992).
10. Frey, M. and Simiu, E., *Proc 2nd Comp Stoch Mech Conf*, 113 (1995).
11. Hsieh, S-R., Troesch, A.W. and Shaw S.W., *Proc R Soc Lond. A.*, **446**, 195, (1994).
12. Lin, H. and Yim, S.C.S., *Applied Ocean Research*, (to appear).
13. Yim, S.C.S. and Lin, H., *Applied Chaos*, John Wiley & Sons (1992).
14. Risken, H., *The Fokker-Planck Equation*, Springer-Verlag (1984).
15. Gardiner, C.W., *Handbook of Stochastic Methods*, Springer-Verlag (1985).
16. Lin, Y.K., *Probabilistic Theory of Structural Dynamics*, McGraw-Hill (1967).
17. Stratonovich, R.L., *Topics in the Theory of Random Noise* (1967).
18. Wissel, C., *Z Physik B*, **35**, 185 (1979).
19. Lasota, A., and Mackey, M.C., *Chaos, Fractals, and Noise*, 2nd Ed., Springer-Verlag (1994).

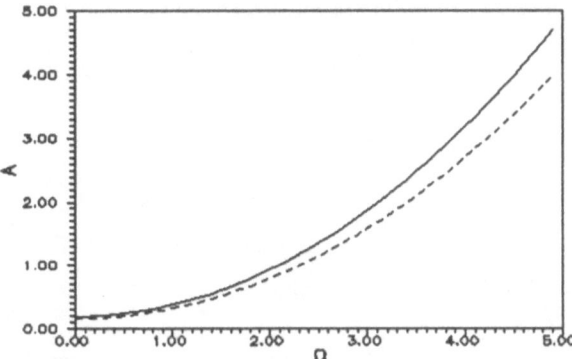

Fig.1: Stochastic Melnikov criterion: zero-mean white noise expands the possible chaotic domain (dashed line)

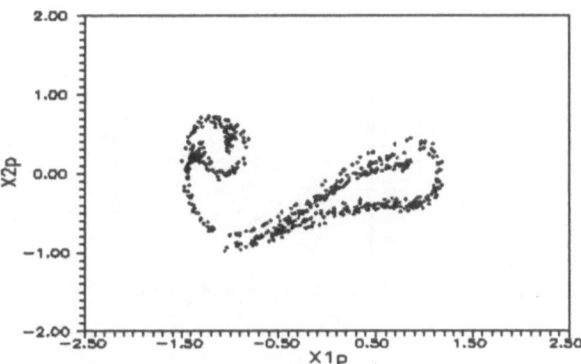

Fig. 2: Noise-induced chaotic response

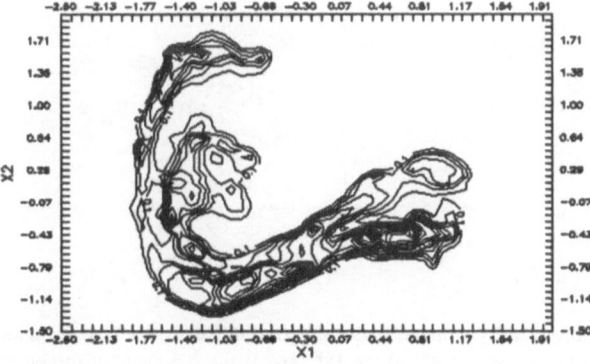

Fig. 3: Coexisting chaotic and periodic attractors

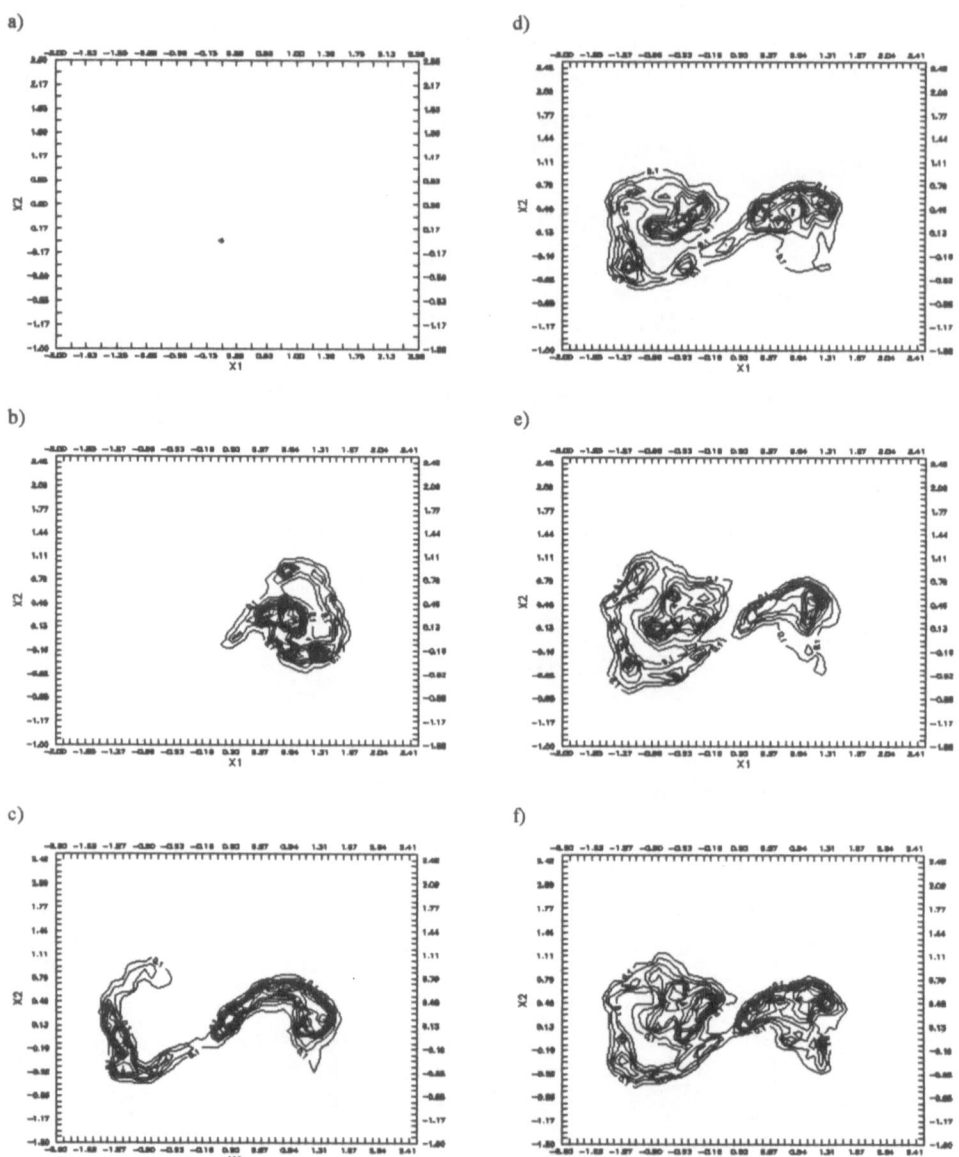

Fig. 4   Evolution of joint probability density: contour maps of probability densities at a) initiation with
           (0.0,0.0); at b) the 1ˢᵗ, c) the 2ⁿᵈ, d) the 4ᵗʰ, e) the 18ᵗʰ, and f) the 20ᵗʰ cycle of forcing period,
           $(A,\omega,\alpha_e,\kappa)=(0.27,1.0,0.185,0.003)$

SOME RECENT ADVANCES IN THEORY OF STOCHASTICALLY EXCITED DISSIPATIVE HAMILTONIAN SYSTEMS

W. Q. ZHU
Department of Mechanics, Zhejiang University, Hangzhou 310027, P. R. China

Abstract

Some recent advances in the theory of stochastically excited dissipative Hamiltonian systems made by the author and his co-workers are summarized. It is shown that the structure of the solution and the energy partition among various degrees of freedom of a stochastically excited dissipative Hamiltonian system depend upon the integrability and resonance of the Hamiltonian system modified by the Wong-Zakai correction terms. Three procedures, i. e. , one for obtaining exact stationary solution, equivalent nonlinear system method and stochastic averaging method, for predicting the response of stochastically excited dissipative Hamiltonian systems are presented. It is pointed out that all presently available exact stationary solutions of nonlinear stochastic systems can be obtained by the present procedure as special cases and that the Stratonovich stochastic averaging and the stochastic averaging of energy envelope are included in the present stochastic averaging of quasi-Hamiltonian systems as two special cases.

1. Introduction

In dealing with the response of finite DOF nonlinear systems to stochastic excitatios, a system is usually described by a set of second order ordinary differential equations, which are derived from Newton's second law or Lagrangian equations. However, the Hamiltonian formulation of nonlinear stochastic systems, i. e. , stochastically excited dissipative Hamiltonian systems, offers some advantages, especially in dealing with exact stationary solutions of nonlinear stochastic systems (Soize, 1988; Zhu, Cai and Lin, 1990).

In the present paper, some recent progresses in the theory of stochastically excited dissipative Hamiltonian systems made by the present author and his co-workers are reviewed. After a brief introduction to some concepts in Hamiltonian dynamics, the paper presents the procedures for obtaining the exact stationary solutions and the equivalent nonlinear systems for stochastically excited dissipative Hamiltonian systems and the stochastic averaging equations for quasi-Hamiltonian systems. One common feature of these procedures is that they are applicable to MDOF strongly nonlinear stochastic systems as well as quasi-linear stochastic systems.

499

A. Naess and S. Krenk (eds.), IUTAM Symposium on Advances in Nonlinear Stochastic Mechanics, 499–510.
© 1996 Kluwer Academic Publishers.

2. Integrable and Nonintegrable Hamiltonian Systems

Consider an n DOF Hamiltonian system governed by the following n pairs of Hamiltonian equations of motion

$$\dot{q}_i = \frac{\partial H}{\partial p_i} \tag{1a}$$

$$\dot{p}_i = -\frac{\partial H}{\partial q_i} \tag{1b}$$

$$i = 1, 2, \cdots, n$$

where q_i and p_i are generalized displacements and momenta, respectively; $H = H(\mathbf{q}, \mathbf{p})$ is a Hamiltonian with continuous first order derivatives. A Hamiltonian system of n DOF is said to be integrable or completely integrable if there exist n independent integrals of motion, $\mathbf{II} = [H_1, H_2, \cdots H_n]^T$, which are in involution. This last term means that H_i all commute with each other, i.e.,

$$[H_i, H_j] = 0 \tag{2}$$

where

$$[H_i, H_j] = \frac{\partial H_i}{\partial p_k}\frac{\partial H_j}{\partial q_k} - \frac{\partial H_i}{\partial q_k}\frac{\partial H_j}{\partial p_k} \tag{3}$$

$$k = 1, 2, \cdots, n$$

is the Poisson bracket of H_i and H_j.

The integrability of a Hamiltonian system can also be stated as follows. A Hamiltonian system of n DOF governed by Eqs. (1a) and (1b) is said to be integrable or completely integrable if there exist a set of canonical transformations

$$I_i = I_i(\mathbf{q}, \mathbf{p}) \tag{4a}$$
$$\theta_i = \theta_i(\mathbf{q}, \mathbf{p}) \tag{4b}$$
$$i = 1, 2, \cdots, n$$

such that the new Hamiltonian equations of motion are of the following canonical form

$$\dot{I}_i = -\frac{\partial}{\partial \theta_i}H(\mathbf{I}) = 0 \tag{5a}$$

$$\dot{\theta}_i = \frac{\partial}{\partial I_i}H(\mathbf{I}) = \omega_i(\mathbf{I}) \tag{5b}$$

$$i = 1, 2, \cdots, n$$

where I_i and ω_i are action variables and frequencies associated with each degree of freedom, respectively; θ_i are angle variables conjugate to I_i; $H(\mathbf{I})$ is the transformed Hamiltonian independent of θ_i. Eqs. (5a) and (5b) are then trivially integrated to yield

$$I_i = const. \tag{6a}$$
$$\theta_i = \omega_i(\mathbf{I})t + \delta_i \tag{6b}$$
$$i = 1, 2, \cdots, n$$

where δ_i are some arbitrary constants. Note that I_i and δ_i constitute the set of 2n constants of integration. I_i are a set of such special constants that once they are known, the other set of constants, δ_i, can be obtained immediately. Thus, I_i can be regarded as n special independent integrals of motion which satisfy Eq. (2).

The phase space orbits of an n DOF integrable Hamiltonian system will lie on an n-D torus. If the ω_i are not rationally related, the orbits on the torus are almost periodic and the system is termed nonresonant. In this case, a sin-

gle orbit will eventually cover the torus uniformly, that is, the flow is ergodic
on the torus (Tabor, 1989). The orbits on the torus will be periodic or closed
and the system is called resonant or completely resonant if the ω_i are related by
$n-1$ equations of the following form

$$k_i^u \, \omega_i = 0 \tag{7}$$

$$i=1,2,\cdots,n; \qquad u=1,2,\cdots,n-1$$

where k_i^u are integers and not all zero for a fixed u. Obviously, n frequencies of
an integrable Hamiltonian system may satisfy α $(1\leqslant\alpha<n-1)$ resonant relations of form (7). In this case, the orbits are periodic on some subtorus and
almost periodic on the complementary subtorus, and the system is termed partially resonant.

A Hamiltonian system of n DOF is said to be nonintegrable or completely
nonintegrable if there is only one independent integral of motion, i. e. , Hamiltonian. The evolution of a nonintegrable system takes it, with equal probabli-
ty, through all states which are accessible from the starting point subject to
the constant of energy conservation, that is, the flow is ergodic on the energy
shell (Binney, et al, 1992).

Using Poisson bracket, the following two conclusions can be drawn (Zhu
and Yang, 1995a).

I . For an integrable Hamiltonian system of n DOF, if a function F of state
variables is in involution with the Hamiltonian, i. e. ,

$$[F,H]=0 \tag{8}$$

then F can be any functional of n independent integrals H_i of motion, $F=F$
(H), or of n action variables, $F=F(I)$, in nonresonant case; or any functional of n action variables and α combinations of angles variables, $F=F(I,\psi)$, ψ
$=[\psi_1,\psi_2,\cdots\psi_a]^T$, $\psi_u=k_i^u\theta_i$, in resonant case with α resonant relations of form
(7).

II . For a nonintegrable Hamiltonian system of n DOF, if a function F of state
variables satisfies Eq. (8), then it can be any functional of the Hamiltonian, F
$=F(H)$.

3. Stochastically Excited Dissipative Hamiltonian Systems

Consider a stochastically excited dissipative Hamiltonian system of n DOF governed by the following equations of motion

$$\dot{Q}_i = \frac{\partial H'}{\partial P_i} \tag{9a}$$

$$\dot{P}_i = -\frac{\partial H'}{\partial Q_i} - c_{ij}\frac{\partial H'}{\partial P_j} + f_{ik}W_k(t) \tag{9b}$$

$$i,j=1,2,\cdots,n; \quad k=1,2,\cdots,m$$

where Q_i and P_i are generalized displacements and momenta, respectively; H'
$=H'(Q,P)$ is a twice differentiable Hamiltonian; $c_{ij}=c_{ij}(Q,P)$ are differentiable functions; $f_{ik}=f_{ik}(Q,P)$ are twice differentiable functions; $W_k(t)$ are
Gaussian white noises in the sense of Stratonovich with correlation functions

$$E[W_k(t)W_l(t+\tau)]=2D_{kl}\delta(\tau) \tag{10}$$

$$k,l=q,2,\cdots,n$$

The system governed by Eqs. (9a) and (9b) is generally nonlinear. The
first summation terms on the right hand side of Eq. (9b) may represent a set
of linear and (or) nonlinear damping mechanisms, while the second summa-

tion terms may include external (additive) and (or) parametric (multiplicative) excitations of Gaussian white noises.

Eqs. (9a) and (9b) are equivalent to the following Itô equations

$$dQ_i = \frac{\partial H'}{\partial P_i} dt \tag{11a}$$

$$dP_i = (-\frac{\partial H'}{\partial Q_i} - c_{ij}\frac{\partial H'}{\partial P_j} + D_{kl}f_{jl}\frac{\partial f_{ik}}{\partial P_j})dt + f_{ik}dB_k(t) \tag{11b}$$

$$i,j=1,2,\cdots,n; \quad k,l=1,2,\cdots,m$$

where $B_k(t)$ are the Wiener processes. The double summation terms on the right hand side of Eq. (11b) are known as the Wong-Zakai correction terms. These terms usually can be split into two parts: one has the effect of modifying the conservative forces and another modifying the damping forces (Cai and Lin, 1988b). The first part can be combined with $- \partial H'/\partial Q_i$ to form an overall effective conservative force $- \partial H/\partial Q_i$ with a modified Hamiltonian $H = H(Q,P)$ and with $\partial H/\partial P_i = \partial H'/\partial P_i$. The second part may be combined with $-c_{ij} \partial H'/\partial P_j$ to constitute an effective damping force $-m_{ij}\partial H/\partial P_j$ with $m_{ij} = m_{ij}(Q, P)$. With these accomplished, Eqs. (11a) and (11b) can be rewritten as

$$dQ_i = \frac{\partial H}{\partial P_i} dt \tag{12a}$$

$$dP_i = -(\frac{\partial H}{\partial Q_i} + m_{ij}\frac{\partial H}{\partial P_j})dt + f_{ik}dB_k(t) \tag{12b}$$

$$i,j=1,2,\cdots,n; \quad k=1,2,\cdots,m$$

4. Exact Stationary Solutions

The reduced FPK equation governing the stationary probability density $p = p(q,p)$ associated with Itô equations (12a) and (12b) can be written as

$$[p,H] + \frac{\partial}{\partial p_i}(m_{ij}\frac{\partial H}{\partial p_j}p) + \frac{\partial^2}{\partial p_i \partial p_j}(b_{ij}^{(i)}p) = 0 \tag{13}$$

where $b_{ij} = 2(fDf^T)_{ij}$ and $b_{ij} = b_{ij}^{(i)} + b_{ji}^{(i)}$. The Poisson bracket of p and H in Eqs. (13) is associated with circulatory probability flow while the remaining two terms with potential probability flow. Eq. (13) is solved subject to the following boundary conditions

$$\frac{\partial H}{\partial p_i}p = 0 \tag{14a}$$

$$(\frac{\partial H}{\partial q_i} + m_{ij}\frac{\partial H}{\partial p_j})p + \frac{1}{2}\frac{\partial}{\partial p_j}(b_{ij}p) = 0 \tag{14b}$$

$$(q,p) \in S$$

which imply vanishing probability flow at the boundaries. The solution of Eq. (13) depends upon the integrability and resonance of the modified Hamiltonian system governed by Eqs. (12a) and (12b) with $m_{ij} = f_{ik} = 0$ (Zhu and Yang, 1995a)

4. NONINTEGRABLE CASE

If the modified Hamiltonian system is nonintegrable, then, based on conclusion I in section 2, the solution of Eq. (13) with $m_{ij} = b_{ij} = 0$ can be any functional of H. Therefore, if we can find a particular functional p (H) which makes the remaining two terms in Eq. (13) vanish, we shall have found a solution of Eq. (13) with $m_{ij} \neq 0$ and $b_{ij} \neq 0$. Taking account of nonnegativeness of probability density and boundary conditions (14a) and (14b), the solution of Eq. (13) can be assumed of the form

$$p(\mathbf{q},\mathbf{p}) = C \, exp[-\lambda(H)]|_{H=H(\mathbf{q},\mathbf{p})} \qquad (15)$$

where C is a normalization constant and $\lambda(H)$ is a functional to be determined. Substituting Eq. (15) into Eq. (13) and taking account of boundary conditions (14a) and (14b) lead to

$$m_{ij}\frac{\partial H}{\partial p_j} + \frac{\partial b_{ij}^{(i)}}{\partial p_j} - b_{ij}^{(i)}\frac{\partial H}{\partial p_j}\frac{d\lambda}{dH} = 0 \qquad (16)$$

$$i,j = 1,2,\cdots,n$$

These are a set of n first order linear ordinary differential equations for λ and they represent vanishing potential probability flows in n directions. To find a consistent functional $\lambda(H)$ satisfying all these n equations, certain restrictions must be imposed on the parameters of the dampings and stochastic excitions. Generally, if

$$\frac{d\lambda}{dH} = (m_{ij}\frac{\partial H}{\partial p_j} + \frac{\partial b_{ij}^{(i)}}{\partial p_j})/(b_{ij}^{(i)}\frac{\partial H}{\partial p_j}) = h(H) \qquad (17)$$

are the same for all $i = 1,2,\cdots,n$, then the exact solution of Eq. (13) is

$$p(\mathbf{q},\mathbf{p}) = C \, exp[-\int_0^H h(u)du]|_{H=H(\mathbf{q},\mathbf{p})} \qquad (18)$$

4.2 INTEGRABLE AND NONRESONANT CASE

If the modified Hamiltonian system is integrable and nonresonant, then, following a similar reasoning, the exact solution of Eq. (13) is of the form

$$p(\mathbf{q},\mathbf{p}) = C \, exp[-\lambda(\mathbf{H})]|_{H=H(\mathbf{q},\mathbf{p})} \qquad (19)$$

or

$$p(\mathbf{q},\mathbf{p}) = C \, exp[-\lambda(\mathbf{I})]|_{I=I(\mathbf{q},\mathbf{p})} \qquad (20)$$

where $\lambda(\mathbf{H})$ is the solution of the following set of n first order linear partial differential equations

$$m_{ij}\frac{\partial H}{\partial p_j} + \frac{\partial}{\partial p_j}b^{(i)}ij - b_{ij}^{(i)}\frac{\partial H_l}{\partial p_j}\frac{\partial \lambda}{\partial H_l} = 0 \qquad (21)$$

$$i,j,l = 1,2,\cdots,n$$

If $\partial\lambda/\partial H_l$ can be obtained as functionals of H_k and satisfy the following compatibility conditions

$$\frac{\partial^2 \lambda}{\partial H_l \partial H_k} = \frac{\partial^2 \lambda}{\partial H_k \partial H_l} \qquad (22)$$

$$k,l = 1,2,\cdots,n$$

then the solution of (21) is

$$\lambda = \lambda_0 + \int_0^{(H_1,H_2,\cdots,H_n)} \frac{\partial \lambda}{\partial H_l}dH_l \qquad (23)$$

where $\lambda_0 = \lambda(0)$, the second term on the right hand side is a line integral and the integrand is a summation over $l = 1,2,\cdots,n$. $\lambda(\mathbf{I})$ in Eq. (20) can be obtained in a similar way.

4.3 INTEGRABLE AND RESONANT CASE

If the modified Hamiltonian system is integrable and resonant with α resonant relations of the form (7), then, following a similar reasoning, the exact solution of Eq. (13) is of the form

$$p(\mathbf{q},\mathbf{p})=\rho(\mathbf{I},\mathbf{\psi}) \tag{24}$$

where $\mathbf{I}=\mathbf{I}(\mathbf{q},\mathbf{p})$, $\mathbf{\psi}=\mathbf{\psi}(\mathbf{q},\mathbf{p})$ and ρ is the solution of the following set of n first order linear partial differential equations

$$(m_{ij}\frac{\partial H}{\partial p_j}+\frac{\partial b_{ij}^{(i)}}{\partial p_j})\rho+b_{ij}^{(i)}(\frac{\partial I_i}{\partial p_j}\frac{\partial \rho}{\partial I_i}+\frac{\partial \psi_u}{\partial p_j}\frac{\partial \rho}{\partial \psi_u})=0 \tag{25}$$

$$i,j,l=1,2,\cdots,n; \quad u=1,2,\cdots,\alpha$$

It should be noted that solution (15) and solutions (19), (20) and (24) have different physical implications. Solution (15) has the property of energy equipartition among various degrees of freedom and only the total energy of the system is controlled by the damping forces and stochastic excitations. Solution (19), (20) or (24), on the other hand, implies that both the total energy of the system and its partition among various degrees of freedom can be adjusted by the magnitudes and distributions of the damping forces and stochastic excitaitons. It may be emphasized that the restrictions imposed on the system parameters for the exact stationary solution to exist for a stochastically excited dissipative integrable Hamiltonian system are less severe than those for a nonintegrable Hamiltonian system. Note also that all the exact stationary solutions obtained todate for nonlinear stochastic systems are of the form (15) and the exact stationary solutions of linear systems subject to external stochastic excitations are of the form (20) or (24). The exact stationary solutions (19), (20) and (24) for nonlinear stochastic systems are new and break the limitation of energy equipartition in nonlinear stochastic systems.

The procedure for obtaining the exact stationary solutions of the system governed by Eqs. (9a) and (9b) can be further extended to a more general class of systems (Zhu and Yang, 1995a).

5. Equivalent Nonlinear System Method

The conditions for the exact stationary solutions to exist are usually severely restrictive and may not be satisfied by many engineering systems. For a given system of the form (9a) and (9b), however, it may be possible to find an equivalent nonlinear system which has an exact stationary solution and whose behavior is close to that of the original one in some statistical sense (Lutes, 1970). The equivalent nonlinear system and the procedure to find it also depend upon the integrability and resonance of the modified Hamiltonian system of the given system (Zhu and Lei, 1995; Zhu, Soong and Lei, 1994).

5.1 NONINTEGRABLE CASE

Suppose we are given an n DOF stochastically excited dissipative Hamiltonian system whose Itô equations are of the form

$$dQ_i=\frac{\partial H}{\partial P_i}dt \tag{26a}$$

$$dP_i=-(\frac{\partial H}{\partial Q_i}+m'_{ij}\frac{\partial H}{\partial P_j})dt+f_{ik}dB_k(t) \tag{26b}$$

$$i,j=1,2,\cdots,n; \quad k=1,2,\cdots,m$$

which are the same as Eqs. (12a) and (12b) except for the term m'_{ij}. Assume that the Hamiltonian system governed by Eqs. (26a) and (26b) with $m'_{ij}=f_{ik}$

$= 0$ is nonintegrable. The goal is to find an equivalent system of the form
(12a) and (12b) which has an exact stationary solution of the form (18). The
differene between the given system and the equivalent one is in damping forces
and can be expressed as

$$\varepsilon_i = (m'_{ij} - hb^{(i)}_{ij})\frac{\partial H}{\partial P_j} + \frac{\partial b^{(i)}_{ij}}{\partial P_j} \tag{27}$$

$$i,j = 1,2,\cdots,n$$

Once $h = h(H)$ is found based on some criterion, Eq. (18) can be regarded as
the approximate stationary probability density for the given system.

Three criteria have been proposed for this purpose (Zhu, Soong and Lei,
1994). The first one is to minimize the mean square of ε with respect to h
(H). It means that if $h(H)$ is replaced by $h(H) + \mu\eta(H)$ where μ is a constant, then $E[\varepsilon_i\varepsilon_i]$, as function of μ, is minimized by letting $\mu = 0$, i.e.,

$$\frac{\partial}{\partial\mu}E[\varepsilon_i\varepsilon_i]\big|_{\mu=0} = 0 \tag{28}$$

Eq. (28) is equivalent to $\delta E[\varepsilon_i\varepsilon_i]/\delta h = 0$, i.e.,

$$\int[(m'_{ij} - hb^{(i)}_{ij})\frac{\partial H}{\partial p_j} + \frac{\partial b^{(i)}_{ij}}{\partial p_j}](-b^{(i)}_{ij}\frac{\partial H}{\partial p_j})p(\mathbf{q},\mathbf{p})d\mathbf{q}d\mathbf{p} = 0 \tag{29}$$

The 2n-fold integral in Eq. (29) can be simplified by a transformation from $\mathbf{q}$,
$\mathbf{p}$ to $\mathbf{q}$, H, $p_2,\cdots,p_n$, i.e.,

$$\int_0^H p(H)dH\int_\Omega\{[(m'_{ij} - hb^{(i)}_{ij})\frac{\partial H}{\partial p_j} + \frac{\partial b^{(i)}_{ij}}{\partial p_j}](-b^{(i)}_{ij}\frac{\partial H}{\partial p_j})/\frac{\partial H}{\partial p_1}\}d\mathbf{q}dp_2\cdots dp_n = 0 \tag{30}$$

where the domain Ω of integration is
$$\Omega = \{(q_1,\cdots,q_n,p_2,\cdots,p_n) \mid H(q_1,\cdots,q_n,0,p_2,\cdots,p_n) \leqslant H\} \tag{31}$$
Since $p(H)$ is still unknown, to proceed further, Eq. (30) is replaced by a
more restrictive sufficient condition requiring that the integration on $\mathbf{q}, p_2,\cdots,$
p_n vanishes for every H (Cai and Lin, 1988b). This condition leads to an expression for $h(H)$, i.e.,

$$h(H) = \frac{\int_\Omega[(m'_{ij}\frac{\partial H}{\partial p_j} + \frac{\partial b^{(i)}_{ij}}{\partial p_j})(b^{(i)}_{ij}\frac{\partial H}{\partial p_j})/\frac{\partial H}{\partial p_1}]d\mathbf{q}dp_2\cdots dp_n}{\int_\Omega[(b^{(i)}_{ij}\frac{\partial H}{\partial p_j})(b^{(i)}_{ij}\frac{\partial H}{\partial p_j})/\frac{\partial H}{\partial p_1}]d\mathbf{q}dp_2\cdots dp_n} \tag{32}$$

where p_1 in the integrand should be replaced by $\mathbf{q}$, H, $p_2,\cdots,p_n$.

Based on the other two criteria, i.e., dissipation energy balancing and
minimum mean squared defficiency of dissipated energies, $h(H)$ can be obtained similarly (Zhu, Soong and Lei, 1994).

5.2 INTEGRABLE CASE

Now assume that the Hamiltonian system governed by Eqs. (26a) and (26b)
with $m'_{ij} = f_{ik} = 0$ is integrable and nonresonant and has n independent integrals
H_i of motion. In this case the equivalent system we are looking for is of the
form (12a) and (12b) with unknown m_{ij}. In order that the equivalent system
has the exact stationary solution of the form (19) where λ is defined by Eq.
(23), usuall $(n^2 + n)/2$ restrictions must be imposed on the m_{ij}. So the number of independent unknowns is $(n^2 - n)/2$. Let these independent unknowns
be $\varsigma = [\varsigma_1, \varsigma_2\cdots\varsigma_{(n^2-n)/2}]^T$. The difference between the given and equivalent systems can be expressed as

$$\varepsilon_i = [m'_{ij} - m_{ij}(\zeta)]\frac{\partial H}{\partial P_i} \tag{33}$$

$$i, j = 1, 2, \cdots, n$$

Our task is to determine ζ such that the behavior of the equivalent system is as close as possible to that of the given system in some statistical sense. For this purpose, the three criteria employed in section 5.1 can also be used here. For example, the first criterion is to minimize $E[\varepsilon_i \varepsilon_i]$ with respect to ζ. It means that if $\zeta_r(\mathbf{q}, \mathbf{p})$ is replaced by $\zeta_r(\mathbf{q}, \mathbf{p}) + \mu_r \eta_r(\mathbf{q}, \mathbf{p})$, where μ_r are constants, then $E[\varepsilon_i \varepsilon_i]$, as function of μ_r, is minimized by letting $\mu_r = 0$, i.e.,

$$\frac{\partial}{\partial \mu_r} E[\varepsilon_i \varepsilon_i]|_{\mu_r = 0} = 0 \tag{34}$$

$$i = 1, 2, \cdots, n; \quad r = 1, 2, \cdots, (n^2 - n)/2$$

Eq. (34) is equivalent to $\delta E[\varepsilon_i \varepsilon_i]/\delta \zeta_r = 0$, i.e.,

$$\int [(m'_{ij} - m_{ij}(\zeta))\frac{\partial H}{\partial p_j}][-\frac{\delta}{\delta \zeta_r} m_{ij}(\zeta)\frac{\partial H}{\partial p_j}]p(\mathbf{q}, \mathbf{p})d\mathbf{q}d\mathbf{q} = 0 \tag{35}$$

$$i, j = 1, 2, \cdots, n; \quad r = 1, 2, \cdots, (n^2 - n)/2$$

The 2n-fold integral in Eq. (35) can be simplified by a transformation from $\mathbf{q}$, $\mathbf{p}$ to $\mathbf{q}$, $\mathbf{H}$, i.e.,

$$\int p(\mathbf{H})d\mathbf{H}\int_{\Omega'}\{[(m'_{ij} - m_{ij}(\zeta))\frac{\partial H}{\partial p_j}][-\frac{\delta}{\delta \zeta_r} m_{ij}(\zeta)\frac{\partial H}{\partial p_j}]/|\frac{\partial \mathbf{H}}{\partial \mathbf{p}}|\}d\mathbf{q} = 0 \tag{36}$$

$$i, j = 1, 2, \cdots, n; \quad r = 1, 2, \cdots, (n^2 - n)/2$$

where $|\partial \mathbf{H}/\partial \mathbf{p}|$ is the absolute value of the Jacobian determinant of the transformation from $\mathbf{p}$ to $\mathbf{H}$, and the domain Ω' of integration is

$$\Omega' = \{\mathbf{q} \mid \bigcap_{i=1}^{n} H_i(\mathbf{q}, 0) \leqslant H_i\} \tag{37}$$

Since $p(\mathbf{H})$ is still unknown, to proceed further, Eq. (36) is replaced by more restrictive sufficient conditions requiring that the integrations on $\mathbf{q}$ vanish for every $\mathbf{H}$. These conditions lead to

$$\int_{\Omega'}\{[(m'_{ij} - m_{ij}(\zeta))\frac{\partial H}{\partial p_j}][-\frac{\delta}{\delta \zeta_r} m_{ij}(\zeta)\frac{\partial H}{\partial p_j}]/|\frac{\partial \mathbf{H}}{\partial \mathbf{p}}|\}d\mathbf{q} = 0 \tag{38}$$

$$i, j = 1, 2, \cdots, n; \quad r = 1, 2, \cdots, (n^2 - n)/2$$

where $\mathbf{p}$ in the integrand should be replaced by $\mathbf{q}, \mathbf{H}$. Once these $(n^2 - n)/2$ equations are solved for ζ_r, m_{ij} can be determined. Then $\partial \lambda/\partial H_i$, λ and $p(\mathbf{q}, \mathbf{p})$ can be obtained successively from Eqs. (21), (23) and (19).

Note that if the equivalent system is a linear one subject to external stochastic excitations, then λ is a linear function of H_k and conditions (22) are satisfied automatically. In this case, no restriction is necessary for the equivalent system having exact stationary solution, the number of independent unknown m_{ij} are n^2 and the present method is reduced to the method of equivalent linearization.

6. Stochastic Averaging Method

Now suppose that the dampings are light and stochastic excitations are weak so that in Eqs. (12a) and (12b)

$$m_{ij} = \varepsilon \overline{m_{ij}}, \quad f_{ik} = \varepsilon^{1/2}\overline{f_{ik}} \tag{39}$$

where ε is a positive small parameter. Then the system governed by Eqs. (12a) and (12b) is termed a quasi-Hamiltonian one. Three versions of stochastically averaged equations can be developed depending on the integrability and resonance of the modified Hamiltonian system (Zhu and Yang, 1995b,c).

6.1 NONINTEGRABLE CASE

If the modified Hamiltonian system governed by Eqs. (12a) and (12b) with ε $=0$ is nonintegrable, then the Hamiltonian H is the only independent integral of motion of the modified Hamiltonian system. For the quasi-Hamiltonian system governed by Eqs. (12a) and (12b), H is the only slowly varying process. The Itô equation for H is obtained from Eqs. (12a) and (12b) by using the Itô differential rule as follows

$$dH = (-m_{ij}\frac{\partial H}{\partial P_j}\frac{\partial H}{\partial P_i} + D_{kl}f_{ik}f_{jl}\frac{\partial^2 H}{\partial P_i \partial P_j})dt + \frac{\partial H}{\partial P_i}f_{ik}dB_k(t) \qquad (40)$$

According to a theorem due to Khasminskii (1968), the H process converges in probability to a 1-D diffusion process as $\varepsilon \to 0$, in a time interval $0 \leqslant t \leqslant T$, where $T \sim 0(\varepsilon^{-1})$. The Itô equation for this diffusion process is obtained by applying time averaging to Eq. (40). Making use the first equation in Eq. (12a) and taking account of the ergodic property of the flow of nonintegrable Hamiltonian systems on the energy shell, the time averaging can be replaced by phase space averaging. Thus, an averaged FPK equation for the transition probability density $p(H, t|H_0)$ of the diffusion process H can be obtained as follows

$$\frac{\partial p}{\partial t} = -\frac{\partial}{\partial H}[U(H)p] + \frac{1}{2}\frac{\partial^2}{\partial H^2}[V^2(H)p] \qquad (41)$$

where

$$U(H) = \frac{1}{T(H)V_{\Omega_1}}\int_\Omega [(-m_{ij}\frac{\partial H}{\partial p_j}\frac{\partial H}{\partial p_i} + D_{kl}f_{ik}f_{jl}\frac{\partial^2 H}{\partial p_i \partial p_j})/\frac{\partial H}{\partial p_1}]dq_1 \cdots dq_n dp_2 \cdots dp_n$$
$$(42a)$$

$$V^2(H) = \frac{1}{T(H)V_{\Omega_1}}\int_\Omega [2D_{kl}f_{ik}f_{jl}\frac{\partial H}{\partial p_i}\frac{\partial H}{\partial p_j}/\frac{\partial H}{\partial p_1}]dq_1 \cdots dq_n dp_2 \cdots dp_n \qquad (42b)$$

$$T(H) = \frac{1}{V_{\Omega_1}}\int_\Omega (1/\frac{\partial H}{\partial p_1})dq_1 \cdots dq_n dp_2 \cdots dp_n \qquad (42c)$$

$$V_{\Omega_1} = \int_{\Omega_1} dq_2 \cdots dq_n dp_2 \cdots dp_n \qquad (42d)$$

$$\Omega = \{(q_1, \cdots q_n, p_2, \cdots p_n)|H(q_1, \cdots q_n, 0, p_2, \cdots, p_n) \leqslant H\} \qquad (43a)$$

$$\Omega_1 = \{(q_2, \cdots q_n, p_2, \cdots, p_n)|H(0, q_2, \cdots, q_n, 0, p_2, \cdots, p_n) \leqslant H\} \qquad (43b)$$

The stationary solution of FPK equation (41) subject to the conditions of vanishing probability flow at boundaries can be readily obtained as follows

$$p(H) = C \exp[-\lambda(H)] \qquad (44)$$

where

$$\lambda(H) = \int_0^H \{[\frac{dV^2(x)}{dx} - 2U(x)]/V^2(x)\}dx \qquad (45)$$

$$C^{-1} = \int_0^\infty \exp[-\lambda(H)]dH \qquad (46)$$

The stationary probability density of displacements and velocities is then

$$p(\mathbf{q},\mathbf{p}) = \frac{p(H)}{T(H)}|_{H=H(\mathbf{q},\mathbf{p})} \qquad (47)$$

6.2 INTEGRABLE AND NONRESONANT CASE

If the modified Hamiltonian system governed by Eqs. (12a) and (12b) with ε $=0$ is integrable and nonresonant, then, there are n slowly varying processes I_r or H_r in the quasi-Hamiltonian system governed by Eqs. (12a) and (12b). The Itô equations for I_r or H_r are obtained from Eqs. (12a) and (12b) by using the Itô differential rule, e.g.,

$$dI_r = (-m_{ij}\frac{\partial H}{\partial P_j}\frac{\partial I_r}{\partial P_i} + D_{kl}f_{ik}f_{jl}\frac{\partial^2 I_r}{\partial P_i\partial P_j})dt + \frac{\partial I_r}{\partial P_i}f_{ik}dB_k(t) \tag{48}$$

$$r,i,j=1,2,\cdots,n; \quad k,l=1,2,\cdots,m$$

Based on the theorem due to Khasminskii (1968), the I_r or H_r converge in probability to an n-D diffusion process as $\varepsilon \to 0$, in a time interval $0 \leqslant t \leqslant T$, where $T \sim 0(\varepsilon^{-1})$. The Itô equations for this n-D diffusion process is obtained by applying time averaging to Eq. (48). Since the phase flow of a nonresonant integrable Hamiltonian system is ergodic on the n-D torus, the time averaging can be replaced by phase space averaging. Thus, the averaged FPK equation governing the transition probability density $p(I,t|I_0)$ of the n-D diffusion process I is

$$\frac{\partial p}{\partial t} = -\frac{\partial}{\partial I_r}[a_r(I)p] + \frac{1}{2}\frac{\partial^2}{\partial I_r\partial I_s}[b_{rs}(I)p] \tag{49}$$

where

$$a_r(I) = \frac{1}{(2\pi)^n}\int_0^{2\pi}(-m_{ij}\frac{\partial H}{\partial p_j}\frac{\partial I_r}{\partial p_i} + D_{kl}f_{ik}f_{jl}\frac{\partial^2 I_r}{\partial p_i\partial p_j})d\theta \tag{50a}$$

$$b_{rs}(I) = \frac{1}{(2\pi)^n}\int_0^{2\pi}2D_{kl}f_{ik}f_{jl}\frac{\partial I_r}{\partial p_i}\frac{\partial I_s}{\partial p_j}d\theta \tag{50b}$$

One feature of the FPK equation (49) is that there is only potential probability flow and no circulatory probability flow. Under the conditions of vanishing probability flow at the boundaries, the averaged systems belong to the class of stationary potential (Stratonovich, 1963). The stationary solution of averaged FPK equation (49) is of the form

$$p(I) = C\,exp[-\lambda(I)] \tag{51}$$

where $\lambda(I)$ satisfies the following set of n first order linear partial differential equations

$$b_{rs}\frac{\partial \lambda}{\partial I_s} = \frac{\partial b_{rs}}{\partial I_s} - 2a_r \tag{52}$$

$$r,s=1,2,\cdots,n$$

If $\partial\lambda/\partial I_s$ satisfy the compatibility equations of form (22), then the solution is

$$\lambda = \lambda_0 + \int_0^{(I_1,I_2,\cdots,I_n)}\frac{\partial \lambda}{\partial I_s}dI_s \tag{53}$$

where $\lambda_0 = \lambda(0)$, the second term on the right hand side is a line integral and the integrand is a summation over $s=1,2,\cdots,n$.

The averaged FPK equation governing the transition probability density p $(II,t|II_0)$ of n-D diffusion process II is of the same form of Eq. (49) and thus the stationary solution can be obtained similarly (Zhu and Yang, 1995c).

6.3 INTEGRABLE AND RESONANT CASE

If the modified Hamiltonian system governed by Eqs. (12a) and (12b) with ε $=0$ is integrable and has α resonant relations of the following form

$$k_i^u \omega_i = 0_u(\varepsilon) \tag{54}$$

$$u = 1, 2, \cdots, \alpha$$

then there are $n+\alpha$ slowly varying processes in the quasi-Hamiltonian system governed by Eqs. (12a) and (12b), i. e. , $\mathbf{I}$ and ψ. The Itô equations for $\mathbf{I}$ are the same as those in Eq. (48) and the Itô equations for ψ are obtained through appropriate combinations of those for Θ_i, which are obtained from Eqs. (12a) and (12b) by using the Itô differential rule.

$$d\psi_u = 0_u(\varepsilon) + (-m_{ij}\frac{\partial H}{\partial P_j}\frac{\partial \psi_u}{\partial P_i} + D_{kl}f_{ik}f_{jl}\frac{\partial^2 \psi_u}{\partial P_i \partial P_j})dt + \frac{\partial \psi_u}{\partial P_i}f_{ik}dB_k(t) \tag{55}$$

$$u = 1, 2, \cdots, \alpha; \quad i, j = 1, 2, \cdots, n; \quad k, l = 1, 2, \cdots, m$$

Based on the Khasminskii's theorem (Khasminskii, 1968), the $\mathbf{I}, \psi$ converge in probability to an $(n+\alpha)$-D diffusion precess as $\varepsilon \to 0$, in a time interval $0 \leqslant t \leqslant$ T, where $T \sim 0(\varepsilon^{-1})$. The Itô equations for this $(n+\alpha)$-D diffusion process are obtained by applying time averaging to Eqs. (48) and (55). Since the phase flow of the Hamiltonian system in this case is ergodic on an $(n-\alpha)$-D subtorus, the time averaging can be replaced by phase space averaging with respect to some $n-\alpha$ angle variables, say $\theta_1 = [\theta_1, \theta_2, \cdots, \theta_{n-\alpha}]^T$. Thus, the averaged FPK equation governing the transition probability density p $(\mathbf{I}, \psi, t | \mathbf{I}_0, \psi_0)$ is

$$\frac{\partial p}{\partial t} = -\frac{\partial}{\partial I_r}(a_r p) - \frac{\partial}{\partial \psi_u}(a_u p) + \frac{1}{2}\frac{\partial^2}{\partial I_r \partial I_s}(b_{rs} p)$$
$$+ \frac{\partial^2}{\partial I_r \partial \psi_u}(b_{ru} p) + \frac{1}{2}\frac{\partial^2}{\partial \psi_u \partial \psi_v}(b_{uv} p) \tag{56}$$

where

$$a_r = a_r(\mathbf{I}, \psi) = \frac{1}{(2\pi)^{n-\alpha}}\int_0^{2\pi}(-m_{ij}\frac{\partial H}{\partial p_j}\frac{\partial I_r}{\partial p_i} + D_{kl}f_{ik}f_{jl}\frac{\partial^2 I_r}{\partial p_i \partial p_j})d\theta_1 \tag{57a}$$

$$a_u = a_u(\mathbf{I}, \psi) = \frac{1}{(2\pi)^{n-\alpha}}\int_0^{2\pi}(0_u(\varepsilon) - m_{ij}\frac{\partial H}{\partial p_j}\frac{\partial \psi_u}{\partial p_i} + D_{kl}f_{ik}f_{jl}\frac{\partial^2 \psi_u}{\partial p_i \partial p_j})d\theta_1 \tag{57b}$$

$$b_{rs} = b_{rs}(\mathbf{I}, \psi) = \frac{1}{(2\pi)^{n-\alpha}}\int_0^{2\pi}2D_{kl}f_{ik}f_{jl}\frac{\partial I_r}{\partial p_i}\frac{\partial I_s}{\partial p_j}d\theta_1 \tag{57c}$$

$$b_{ru} = b_{ru}(\mathbf{I}, \psi) = \frac{1}{(2\pi)^{n-\alpha}}\int_0^{2\pi}2D_{kl}f_{ik}f_{jl}\frac{\partial I_r}{\partial p_i}\frac{\partial \psi_u}{\partial p_j}d\theta_1 \tag{57d}$$

$$b_{uv} = b_{uv}(\mathbf{I}, \psi) = \frac{1}{(2\pi)^{n-\alpha}}\int_0^{2\pi}2D_{kl}f_{ik}f_{jl}\frac{\partial \psi_u}{\partial p_i}\frac{\partial \psi_v}{\partial p_j}d\theta_1 \tag{57e}$$

$$r, s, i, j = 1, 2, \cdots, n; \quad k, l = 1, 2, \cdots, m; \quad u, v = 1, 2, \cdots, \alpha$$

The exact stationary solution of the averaged FPK equation (56) is difficult to obtain. An approach for obtaining the approximate stationary solution of Eq. (56) for very small ε was proposed by Zhu and Yang (1995c).

The stochastic averaging method developed for quasi-Hamiltonian systems (9a) and (9b) can be further extended to a more general class of systems (Zhu and Yang, 1995c). Obviously, for single DOF, the stochastic averaging of quasi-Hamiltonian systems is reduced to the stochastic averaging of energy envelope(Khasminskii, 1964). For MDOF quasi-linear systems, the action variables can be replaced by amplitudes and the stochastic averaging of quasi-integrable-Hamiltonian systems is reduced to the Stratonovich stochastic averaging(Stratonovich, 1963).

7. Acknowledgement

The support from National Natural Science Foundation of China is acknowledged.

8. References

Binney, J. J. et al. (1992) *The Theory of Critical Phenomena, An Introduction to the Renormalization Group*, Clarendon Press, Oxford.

Cai, G. Q. and Lin, Y. K. (1988a) On exact stationary solutions of equivalent non-linear stochastic systems, *Int. J. Non-Linear Mech.* **23**, 315-325.

Cai, G. Q. and Lin, Y. K. (1988b) A new approximate solution technique for randomly excited non-linear oscillators, *Int. J. Non-Linear Mech.* **23**, 409-420.

Khasminskii, R. Z. (1964) On the behavior of a conservative system with friction and small random noise (in Russian). *Prikladnaya Mate matika i Mechanica* (Appl. Math. Mech.), **28**, 1126-1130.

Khasminskii, R. Z. (1968) On the averaging principle for stochastic differential Itô equation (in Russian), *Kibernetica* **4**, 260-279.

Lutes, L. D. (1970) Approximate technique for treating random vibration of hysteretic systems, *J. Acoust. Soc. Am.* **48**, 299-306.

Soize, C. (1988) Steady state solution of Fokker-Planck equation in higher dimension, *Probabilistic Engineering Mechanics* **3**, 196-206.

Stratonovich, R. L. (1963) *Topics in the Theory of Random Noise*, Vol. 1, Gordon and Breach, New York.

Tabor, M. (1989) *Chaos and Integrability in Nonlinear Dynamics*, John Wiley & Sons, New York.

Zhu, W. Q., Cai, G. Q. and Lin, Y. K. (1990) On exact stationary solutions of stochasticlly perturbed Hamiltonian systems, *Probabilistic Engineering Mechanics* **5**, 84-89.

Zhu, W. Q. and Lei, Y. (1995) Equivalent nonlinear system method for stochastically excited dissipative integrable Hamiltonian systems, submitted to *ASME J. Appl. Mech.*

Zhu, W. Q., Soong, T. T. and Lei, Y. (1994) Equivalent nonlinear system method for stochastically excited Hamiltonian systems, ASME, *J. Appl. Mech.* **61**, 618-623.

Zhu, W. Q. and Yang, Y. Q. (1995a) Exact stationary solutions of stochastically excited and dissipated integrable Hamiltonian systems, to appear in ASME *J. Appl. Mech.*

Zhu, W. Q. and Yang, Y. Q. (1995b) Stochastic averaging of quasi-nonintegrable-Hamiltonian systems, submitted to ASME *J. Appl. Mech.*

Zhu, W. Q. and Yang, Y. Q. (1995c) Stochastic averaging of quasi-integrable Hamil-tonian systems, submitted to ASME *J. Appl. Mech.*

Mechanics

SOLID MECHANICS AND ITS APPLICATIONS

Series Editor: G.M.L. Gladwell

Aims and Scope of the Series

The fundamental questions arising in mechanics are: *Why?*, *How?*, and *How much?* The aim of this series is to provide lucid accounts written by authoritative researchers giving vision and insight in answering these questions on the subject of mechanics as it relates to solids. The scope of the series covers the entire spectrum of solid mechanics. Thus it includes the foundation of mechanics; variational formulations; computational mechanics; statics, kinematics and dynamics of rigid and elastic bodies; vibrations of solids and structures; dynamical systems and chaos; the theories of elasticity, plasticity and viscoelasticity; composite materials; rods, beams, shells and membranes; structural control and stability; soils, rocks and geomechanics; fracture; tribology; experimental mechanics; biomechanics and machine design.

Kluwer Academic Publishers – Dordrecht / Boston / London

Mechanics

SOLID MECHANICS AND ITS APPLICATIONS
Series Editor: G.M.L. Gladwell

Kluwer Academic Publishers – Dordrecht / Boston / London

Mechanics

Kluwer Academic Publishers – Dordrecht / Boston / London

Mechanics

FLUID MECHANICS AND ITS APPLICATIONS
Series Editor: R. Moreau

Kluwer Academic Publishers – Dordrecht / Boston / London